Environmental Footprints and Eco-design of Products and Processes

Series editor

Subramanian Senthilkannan Muthu, SGS Hong Kong Limited,
Hong Kong, Hong Kong SAR

More information about this series at http://www.springer.com/series/13340

Obulisamy Parthiba Karthikeyan
Kirsten Heimann
Subramanian Senthilkannan Muthu
Editors

Recycling of Solid Waste for Biofuels and Bio-chemicals

Springer

Editors
Obulisamy Parthiba Karthikeyan
James Cook University
Townsville, QLD
Australia

Kirsten Heimann
James Cook University
Townsville, QLD
Australia

Subramanian Senthilkannan Muthu
Environmental Services Manager-Asia
SGS Hong Kong Limited
Hong Kong
Hong Kong

ISSN 2345-7651 ISSN 2345-766X (electronic)
Environmental Footprints and Eco-design of Products and Processes
ISBN 978-981-10-0148-2 ISBN 978-981-10-0150-5 (eBook)
DOI 10.1007/978-981-10-0150-5

Library of Congress Control Number: 2016942490

Printed on acid-free paper

This Springer imprint is published by Springer Nature
The registered company is Springer Science+Business Media Singapore Pte Ltd.

Preface

Global per capita solid waste generation has more than doubled over the first 15 years of this millennium and is expected to reach 2.2 billion tons per year by 2025, produced by 4.3 billion urban residents. As the human population is also forecasted to experience an unprecedented growth from currently ~7 to ~9 billion by 2050, the management of solid waste is becoming an increasingly pressing problem now and in near future. It is estimated that growing population will encounter food (~70 %), fuel (50 %), and freshwater (30 %) demands, resources that are already limiting in many regions of the world. Furthermore, vast ranges of chemical feedstock are also required to maintain current-day products and to meet their rates of production. The non-renewable and finite nature of the energy required for their production and destruction leads to an escalation of this spiral. Although waste generation is typically inversely correlated with level of income, developing nations forecast to produce significantly high amount of solid waste generation, as most of the expected population growth is to occur there. This places developing countries at an ever-increasing disadvantage concerning adequate infrastructure (including waste collection), accommodating growing costs of municipal solid waste management, and experiencing health issues and environmental pollution associated with current solid waste disposal practices, i.e., mainly open dumping. Moreover, globally, landfills are the third largest emitter of methane, a highly potent greenhouse gas, which is intolerable in view of climate instability (droughts and flooding) and warming global temperatures further placing immense pressure on providing food security, worldwide.

Yet, in spite of the looming threat to energy, food, and unpolluted freshwater security, globally mankind continues to dispose of on average ~1.3 billion tonnes of food waste, ~130 million tonnes of non-degradable fossil fuel-derived plastics, and 52 million tonnes of metals in landfills annually, despite the rising costs for solid waste management. Although the actual composition of municipal solid wastes are region- and income-influenced, with low-income nations having higher proportions of organic waste, their management costs for urban centers have been forecast to increase by ~80 % over the next decade to ~375 USD per annum, an increase expected to be four and five times higher in middle-income and developing nations, respectively. Significant municipal solid waste management cost increases together with the realization of wasted energy, and limits of non-renewable resources are, however, likely to sway the balance in favor of recycling approaches, which should become more cost-competitive and offer co-income development through the generation of renewable products produced in the various processes.

As municipal solid waste is complex in its composition, which varies with region and level of income, it is wise to investigate a variety of methods for recycling of solids wastes into renewable products, particularly fuel and energy. Given the above, the general undersupply of energy/fuel in developing nations, especially in rural and remote areas, offers another attractive incentive for vigorously facilitating appropriate waste recycling options in these areas. Economic (at least long term) and environmental sustainability are, however, key criteria for method implementation, irrespective of nation. In recognition of the growing problem of municipal solid waste management and finite resources of energy, this book presents strategies and case studies on waste to energy conversion options for a variety of wastes, as well as odor management arising from colocation in close proximity to residents.

Chapter "Prospects of Biomethanation in Indian Urban Solid Waste: Stepping Towards a Sustainable Future" provides a succinct overview of urbanization and waste statistics globally and for developing nations such as India. As India is predicted to be the nation with the highest population growth, emphasis is placed on hurdles that impede sustainable solid waste management and regionalized examples are provided for common undesired practices such as illegal dump sites, failed implementations including reasons why, and successful ones. It briefly introduces process technologies and their advantages and disadvantages for managing solid wastes. The focus of this chapter is biomethanation (anaerobic digestion) of solid waste. Therefore, this chapter describes the different stages of microbial degradation of organic wastes and parameters that influence the digestion process, including operational parameters and anaerobic digester types. This chapter closes in detailing efforts undertaken by the Indian government to enable an organic waste to energy program and describes anaerobic digester types designed for volumes of organic waste generated in rural and suburban locations around India aiming to service the immediate need for energy of these communities and larger projects for energy generation between 5 and 7 MW funded by the Indian government. This chapter concludes that biomethanation is a sustainable solution for reducing organic wastes, which requires assistance through government funding, policies, private and public sectors, and education/participation by the public for implementation.

Chapter "Status and Prospects of Municipal Solid Waste to Energy Technologies in China" reviews municipal solid waste (MSW) management to energy strategies applied in China, the world's largest developing country. It strategically compares global advances with the current status in China. China's MSW management has to cope with rapid urbanization accompanied by a large increase in MSW generation due to improved living standards, which also result in sharply rising energy demands. MSW management has only recently been adopted in China (over the last 30–40 years) and is characterized additionally by large scales and rapid increase in waste generation. This chapter provides a concise treatment of factors influencing waste classification in different regions of China and the implementation of a zero-waste hierarchy aiming to reduce MSW through separating waste for recycling, reuse, and waste to energy types at the source. With landfilling representing around 70 % of the countries MSW management strategy, technical differences between the four main types are summarized. As energy generation from waste is the driver of China's MSW strategy, this chapter discusses the combined heat to power generation approaches at landfills and regional deployment of particular incinerator types. It discusses government responsibility for managing MSW, policies, and regulations, as well as economics and incentives for waste to energy approaches.

Chapter "GHGs Emissions and Sustainable Solid Waste Management" focuses on greenhouse gas (GHG) emissions emanating from waste management practices and impacts on climate. It provides an introduction on greenhouse gas emissions and climate consequences and highlights the flow on effects of increased urbanization and waste generation of the global population. It reviews zero-order and first-order models for waste generated GHG emissions and applies these to case studies of Panama and Thailand and Chennai, India, respectively. Based on the case studies, it is concluded that zero-order models are inaccurate compared to first-order models; however, zero-order models can provide estimates in situations were solid waste management data sets are not available. It reviews life cycle analyses to evaluate effects of solid waste management options on GHGs emissions. It is demonstrated that source reduction and recycling have significant GHG emission savings. For solid waste management of OECD countries, it is shown that mechanical biological treatments and incineration coupled with waste to energy recovery are the most promising approaches for reducing GHG emissions. Pyrolysis of waste materials and the production of biochar and renewable energy are identified as another promising approach for lowering GHG emissions from the agricultural sector. It evaluates landfill gas collection for case studies in Thailand, mechanical biological treatment in China and Nigeria, and gasification, landfill gas, and anaerobic digestion technologies in Indonesia. It is shown that all test methods yield significant reductions in GHG emissions. This chapter concludes that skilled human resources, technical capacities, and enforcement of national policies are the main obstacles for the implementation of GHG emission-conscious solid waste management practices, particularly in developing nations.

Chapters "Recent Advances of Anaerobic Digestion for Energy Recovery" and "Anaerobic Digestion of Solid Waste: A Focus on Microbial Community Structures" review treatment of organic waste by anaerobic digestion in detail. The process, i.e., the degradation phases occurring stepwise in the degradation process, is introduced in detail, as are system and system efficiencies and improvements, incorporation of efficiency monitoring through process analytical technologies, and details molecular tools to evaluate the microbial community, the latter being one of the drivers for system performance and management. As the efficiency of the anaerobic digestion process is strongly influenced by temperature, type of feedstock, pH level, retention time, carbon to nitrogen (C/N ratio), and volatile fatty acid (VFA) concentration, these factors are discussed. With regard to temperature, the review identifies that a phase-separated system (e.g., thermophilic followed by mesophilic conditions) would be beneficial for achieving higher biogas production. Anaerobic digestion systems are introduced with regard to feedstock applicability, advantages, and draw backs and are divided into wet/low residual solid reactors and dry/high residual solid systems. Regardless of the system, hydrolysis efficiency is identified as the rate limiting step. These chapters discuss mechanical, ultrasonic, thermochemical, and microbial/enzymatic pretreatments and their efficiencies in increasing the bioavailability of complex organic matters to microbes and phase-separated system approaches to increase biogas yields. Chapter "Recent Advances of Anaerobic Digestion for Energy Recovery" introduces anaerobic digestion systems and identifies phase-separated systems to improve the efficiency of the process. Retention time directly impacts on efficiency, the pros and cons of upflow anaerobic sludge blanket, expanded granular sludge bed reactors, attached growth reactors, membrane reactors, and hybrid models. Ultimately, however, the microbial community composition governs the efficiency of the entire complex process, which is the focus of Chapter "Anaerobic Digestion of Solid Waste: A Focus on Microbial Community Structures" for the different phases. As such, monitoring of changes to these would benefit models that can then be used to calculate efficiencies. Both chapters detail several molecular techniques suitable for obtaining these detailed inputs for improving process efficiencies. In addition, Chapter "Anaerobic Digestion of Solid Waste: A Focus on Microbial Community Structures" presents techno-economic outcomes of a pilot- and industrial-scale case study and for the anaerobic digestion of food waste, which in both cases proved profitable.

Chapter "Recycling of Livestock Manure into Bioenergy" reviews the outcomes of anaerobic digestion of pig, cattle, and poultry manure and details changes in the microbial communities at different stages of the process. The details of animal head production per country are detailed based on 2013 data, with manure production being calculated. The literature is reviewed with regard to meso- and thermophilic cattle manure biogas production under single feed with cofeeding conditions and a two-phase system operating a pretreatment step at 68 °C and a formal methane yielding phase at 55 °C. For single mesophilic pig manure digestion, the review suggests applying grass silage to pig manure ratio of 1:1, resulting in reduced lag times and increased biogas yields. For chicken manure due to the high ammonia levels, ammonia-stripping is recommended, as the study shows that continuous stirred tank reactors can then be operated under both meso- and thermophilic conditions. Under high ammonia, the mesothermic anaerobic digester showed a much more resilient microflora, as the system recovered after inhibition in contrast to the thermophilic one. This chapter closes in providing a proposed chicken manure anaerobic treatment path, providing potential energy yields for sustainable electricity and thermal heat generation.

Indonesia and Malaysia are the biggest palm oil producers, globally, and the negative environmental and ecological impacts have been discussed for a long time, in particular when considering production of biodiesel from palm oil. This approach has now been suspended, as palm oil has a myriad of applications for products fetching a much higher market value. In this light, Chapter "Anaerobic Digestion of Palm Oil Mill Residues for Energy Generation" presents a review of energy generation of palm oil mill effluent using anaerobic digestion. To describe the palm oil mill effluent characteristics, this chapter introduces the reader first to palm oil production pathway, before introducing the palm oil mill effluent treatments. It presents an evolutionary perspective from treatment of effluents in open lagoon pond systems with a primary focus on treatment of the wastewaters to meet requirements for discharge to more advanced and controlled anaerobic systems aiming to provide renewable energy as an additional benefit to treatment of wastewaters for discharge purposes. This chapter details benefits of biogas recovery and puts in perspective the importance of emission trading schemes, i.e., provided through the Kyoto Protocol until 2012, for modernization of anaerobic digestion approaches for bioenergy recovery from oil palm mill effluents in developing nations.

A case study on the benefits of leachate recirculation in BioReactor Landfill (BRL) is presented in Chapter "Landfill Bioreactor Technology for Waste Management". It is demonstrated that leachate recirculation positively affects waste stabilization reducing required times by >90 % and thereby reducing the risk of environmental impacts through accidental leachate leakage, which typically increase with the time required for the stabilization process to complete. This chapter details leachate recirculation also increase volumes of biogas produced, making biogas recovery economically attractive. Primary areas of consideration are that no landfill will equal another, and, as such, leachate recirculation specifics must be designed with the waste composition in mind. The case study presents a leachate recirculation waste stabilization laboratory assessment using a trickle bed reactor. Predicted outcomes are based on modeling on biogas yields and production times for landfills characterized by low biodegradable wastes.

Ammonia–nitrogen contents and removal are key issues in bioreactor landfill operations, significantly influencing monitoring requirements and affecting reclamation. Chapter "Biotransformation of Nitrogen in Landfills" focuses on laboratory- and field-scale application case studies for using novel mechanisms such as SHARON (Single reactor system for high activity ammonia removal over nitrite), ANAMMOX (Anaerobic Ammonium Oxidation), CANON (Complete Autotrophic Nitrogen removal Over Nitrite), and OLAND (Oxygen Limited Autotrophic Nitrification and Denitrification) for the removal of ammonia-nitrogen. With a total nitrogen removal of 84 % and ammonia–nitrogen removal efficacy of 71 % at nitrogen loading rate of 1.2 kg $N/m^3/day$ over 147 days, a combined process of SHARON–ANAMMOX yields promising results for waste nitrogen management of leachates in bioreactor landfills, although full-scale in situ operation still needs to be demonstrated and the effect of environmental parameter fluctuation on the combined processes still require further research.

In addition to waste management and renewable energy issues, both being pressing problems for mankind, the economical production of renewable hydrocarbons for conventional combustion engines also needs urgent attention. Chapter "Biofuel Production Technology and Engineering" discusses butanol production from organic waste using fermentation by Clostridia bacteria and its use as a potential hydrocarbon additive/replacement of conventional fossil fuel-derived petroleum. While some *Clostridium* wild-type species can hydrolyze lignocellulose directly, these are not the strains capable of the acetone–butanol–ethanol (ABE) producing pathway, making them unsuitable for the renewable butanol biofuel market. This chapter introduces mathematical models that can simulate potential outcomes of different processes. For example, it is shown that in a three-stage continuous stirred tank reactor, decreasing dilution rates led to a 120 % improvement on butanol yields but also resulted in lower productivity. It remains to be seen which factor, butanol yield and ease of purification or productivity, governs production cost. This chapter highlights the need for a system-integrated approach for evaluating the suitability and economics of the ABE process under different configurations for the production of butanol from organic municipal solid waste.

Chapter "Fast Pyrolysis of Agricultural Wastes for Bio-fuel and Bio-char" deals with another approach of managing solid waste, fast pyrolysis of waste, which yields liquid biofuel, which can also be used for the production of chemicals, and an additional product, biochar. This chapter discusses reactor types for fast pyrolysis the properties, challenges, and opportunities for the bio-oil and biochar, the current status of fast pyrolysis applications, and the energy and economics of the process. Configuration, differences, and application advantages/disadvantages are discussed for bubbling fluidized bed, circulating fluidized bed, rotating cone, augur, and ablative reactors, and novel configurations, such as vacuum pyrolysis, fixed bed, entrained flow reactors, microwave pyrolyzer, plasma, and solar reactors. The most marketable reactor designs are bubbling fluidized bed, circulating fluidized bed, and auger reactors, with fluidized bed reactors producing higher bio-oil yields compared to auger reactors. Bio-oil properties are summarized against engine performance and fuel standard criteria for combustion engines. Although bio-oil is more suitable industrial burners, due to the high acidity, viscosity, oxygen and water contents, and heterogeneity of contributing compounds, it can still be used in pilot-ignited medium-speed diesel engines, provided, due to the corrosiveness of bio-oil, injector and fuel pumps are made of stainless steel. Following a brief description of bio-oil-derived chemicals and benefits of biochar applications in agriculture, this chapter explores upgrading technologies for the production of renewable fuels, providing a summary of companies using fast pyrolysis technology for drop-in fuel production from bio-oil.

Chapter "The Energy and Value-Added Products from Pyrolysis of Waste Plastics" presents advancements made in the pyrolysis of plastic waste for the generation of energy and value-adding coproducts. Although renewable fuel yields of 80 % with similar characteristics as diesel are possible under catalytic pyrolysis conditions, challenges still exist in converting plastic waste to renewable fuel, activated char for the remediation of industrial waste waters, heat storage, metal removal, etc. Based on a case study of the Kingdom of Saudi Arabia, these challenges are identified as high temperatures and retention times, catalyst cost, and performance improvements, which require improvement to make the pyrolysis of plastics more environmentally and economically favorable.

Chapter "Turning Food Waste into Biofuel" reviews waste to energy conversion of food waste, as the annually generated food waste amounts to ~1.3 billion tonnes or 1/3 of wastage of the overall human food production, representing a gross underutilization of energy-rich organic resources. Food waste is a valuable resource for bioenergy (methane, hydrogen, ethanol), bioplastics, and high-value bioproducts (organic acids and enzymes), but given the global challenges of meeting the future energy demands of the rapidly growing human population, this chapter places emphasis on techniques suitable for renewable fuel production (direct and microbial transesterification to biodiesel, production of ethanol via enzymatic pretreatment and microbial fermentation primarily by the yeast *Saccharomyces cerevisiae*, and microbial hydrogen and methane production). Review of life cycle analyses of bioethanol, biohydrogen, and methane production determined that greenhouse gas emissions are reduced by more than 50 %, but the processes are not sufficiently mature for incorporation into waste management. Particularly, evaluation of coproduct generation potential is required for economic sustainability.

Chapter "Solidification and Stabilization of Tannery Sludge" presents results on solidification and stabilization of tannery wastewater sludge from the leather industry, a process required to minimize adverse environmental impacts of the hazardous tannery agent chromium(III). Hazardous chromium(III)-containing wastewaters when treated produce hazardous primary and secondary sludge, with disposure in secure landfills. Increasing environmental safeguarding and space constraints in India make secure landfilling of this sludge is problematic. Data obtained for short- and long-term leaching tests show that chromium can be successfully encapsulated in a cement–lime sludge mixture, allowing the materials obtained to be utilized as alternatives to conventional construction materials, rather than requiring deposition in secure landfills.

Odor complaints from municipal solid waste management sites are a common complaint by nearby residents, and perception by the public often leads to problems in choosing sites. Therefore, odor management is as critical as management of solid wastes. Chapter "Odour Pollution from Waste Recovery Facilities" reviews odor compounds generated by MSW treatment using anaerobic digestion, incineration and refuse to fuel conversions, and odor control technologies. It summarizes odor measurement regulations in different countries, standards, and assessment strategies, i.e., how assessment needs to be conducted for different odor sources, e.g., point sources such as stacks and plumes, and area sources such as ponds. Odor treatment technologies discussed are biofilters packed with either organic or synthetic materials, wet scrubbers, and thermal or chemical oxidation, which are all equally effective. This chapter concludes that potential downstream environmental impacts of these odor control technologies require additional research.

As detailed, this book presents a solid background and method review for students and researchers in the municipal solid waste to energy conversion field. Additionally, synthesised details on current limitations to technologies and case study afford a great data reference for experts in the field of "solid waste management."

Townsville, Australia Obulisamy Parthiba Karthikeyan
Townsville, Australia Kirsten Heimann
Hong Kong Subramanian Senthilkannan Muthu

Contents

Prospects of Biomethanation in Indian Urban Solid Waste: Stepping Towards a Sustainable Future

Barkha Vaish, Abhijit Sarkar, Pooja Singh, Prabhat Kumar Singh, Chandan Sengupta and Rajeev Pratap Singh

Abstract Industrialization and urbanization together have a cumulative effect on generating significant amounts of urban solid waste which leads to increasing threats to the environment. India with a population of about 1.27 billion people alone generates about 0.2–0.5 kg of waste day^{-1} $capita^{-1}$ of which around 40–50 % is organic in nature. According to published reports, if these organic fractions of the waste are not treated properly and reach the landfill site; they can become a major source of greenhouse gas (GHG) emissions and causes leaching of harmful pollutants. These GHG and newly generated pollutants have been found to have detrimental effects on ground water, and create imbalances in the ecosystem. Therefore, '*need of the hour*' is to utilize the energy that is stored in the waste through different available technologies like composting, vermicomposting, fermentation and biomethanation etc. The process of biomethanation appears to be a more reliable and promising technology as it not only aims to solve the problem of organic solid waste, but also provides sustainable energy in the form of biogas. Moreover, when compared with other technologies, biomethanation is economic, eco-friendly and less labor intensive. Even though several research studies were conducted in the field of biomethanation, the process is still unpopular especially in developing countries due to lack of appropriate knowledge, treatment systems and due consideration by the government.

B. Vaish · P. Singh · R.P. Singh (✉)
Institute of Environment and Sustainable Development, Banaras Hindu University, Varanasi, India
e-mail: rajeevprataps@gmail.com; rps.iesd@bhu.ac.in

A. Sarkar
Laboratory of Applied Stress Biology, Department of Botany, University of Gour Banga, Malda 732103, West Bengal, India

P.K. Singh
Civil Engineering Department, Indian Institute of Technology, Banaras Hindu University, Varanasi, India

C. Sengupta
Laboratory of Microbial Research, Department of Botany, University of Kalyani, Kalyani, West Bengal, India

O.P. Karthikeyan et al. (eds.), *Recycling of Solid Waste for Biofuels and Bio-chemicals*, Environmental Footprints and Eco-design of Products and Processes, DOI 10.1007/978-981-10-0150-5_1

Keywords Industrialization · Urbanization · Urban solid waste · Greenhouse gas (GHG) · Biomethanation · Sustainable energy

1 Solid Waste Statistics: Global Perspective

Rapid global urbanization coupled with economic development is leading to the generation of enormous amounts of solid waste, hazardous to human health as well as to the environment. Population growth with declining job opportunities in rural areas has caused migration towards urban centers, which mainly contributes to urbanization (Sharholy et al. 2008). This leads to haphazard growth in urban cities which results in lack of nutrition, inadequate supply of water and sanitation, poor infrastructure, pollution, problems with solid waste management and eventually degradation of environment. Waste generation is mainly an urban problem. In many cities of the developing countries, municipalities cannot cope with the accelerated pace of waste generation. Likewise, in-lower middle income countries like India and Sub-Africa, waste collection rates are lower than 70 % (Kumar and Sharma 2014) while approximately more than 50 % of the waste collected is often disposed off carelessly in low-lying areas or thrown onto the streets and drains, which causes floods, becomes breeding place for insects and rodents adversely affecting residents who live near disposal sites (Chalmin and Gaillochet 2009). Also, the type and composition of municipal solid waste change according to the season and standard of living of particular city of a country. Therefore, the choice of technology and infrastructure depends on the composition and amount of waste generated and this underscores the importance of waste segregation at source.

This increasing population accompanied by an increase in GNI/capita of the developed and developing countries adds new waste into the waste stream that has to be managed every day. According to the reports, in 1900 the world had a total of 220 million urban residents, i.e. 13 % of the total population. According to a UN Report (2014), now about 2.9 billion people are residing in cities, i.e. 49 % of the world's total population. Adding to the problem, the approaching decades will witness more changes in the size and spatial distribution of world's population. With the continuing increase in urbanization, it is anticipated that by the year 2050, 2.5 billion more people will be added to the total urban population (UN 2014) which will be around 66 % by proportion of the total world's population. Of this population, 90 % of the total increase would be in developing countries like Asia and Africa. As societies became more affluent, the amount of waste production has risen tenfold.

Presently, the total amount of solid waste generated annually worldwide is approximately 4 billion tons (that includes municipal, industrial and hazardous waste) while the amount of municipal solid waste generated alone ranges from 1.6 to

2.0 billion tons (ISWA 2012). It is anticipated that by the year 2025, the total amount of waste will ascent from 1.3 billion tons of waste per day (TPD) to 2.2 billion TPD. Due to combined effects of population explosion and urban development, the rate of solid waste generation will double again and consequently the annual global cost will increase from $205 to $375 billion (World Bank 2012a, b). The per capita waste generation rate will also accelerate in the range of approximately 1.2–1.42 kg/person/day in the next coming fifteen years. For these reason, solid waste management is not only an environmental problem but also an economic problem for the country.

Generation of waste greatly vary from country to country as it depends on seasonal variations, cultural values, standard of living and consumption pattern of the area. Some countries generate more waste in comparison to another. Countries with lower earnings produce the least waste per capita as compared to the countries with higher incomes. The provisional figures provided by the World Bank (2012a, b) emphasized those high income countries like OECD account for generating relatively large amounts of solid waste i.e. around 572 million tonnes of solid waste/year and per capita values range from 1.1 to 3.7 kg per person per day with an average of 2.2 kg/capita/day compared to lower income countries like Sub-Saharan Africa which is 62 million tonnes of solid waste/year and per capita values ranging from 0.09 to 3.0 kg per person per day, with an average of 0.65 kg/capita/day. The major composition of waste in developing or low income countries like India and Bangladesh is basically organic in nature (Srivastava et al. 2014).

As stated in reports, India and China have disproportionately high urban waste generation rates as when related to their economic status. It is estimated that the per capita waste generated in India is about 0.4 kg/day with approximately 50–60 % compostable matter, whereas the wastes in high-income cities are more diverse with relatively larger shares of plastics and paper (Fig. 1) (Hoornweg and Bhada-Tada 2012). The estimated annual increase in per capita waste generation is about 1.33 % per year (Mohapatra (2006). With the existing population growth and limited disposal sites, discarding such large amounts of waste is a serious problem. In addition, organic waste when disposed off carelessly greatly increases greenhouse gas emissions which are major cause of climate change. In most Indian cities, MSW collection, segregation, transportation, processing and disposal is carried out by respective municipal corporations and state governments enforce regulatory policies.

In some cities like Mumbai, Chennai, Delhi, Bengaluru, Hyderabad and Ahmedabad garbage disposal is done by Public Private Partnerships (PPPs). The private sector has been involved in door-to-door collection of solid waste, street sweeping (in a limited way), secondary storage and transportation and for treatment and disposal of waste (Narendra et al. 2014). But, city leaders are faced with several challenges in their effort to streamline waste management services. A few of the

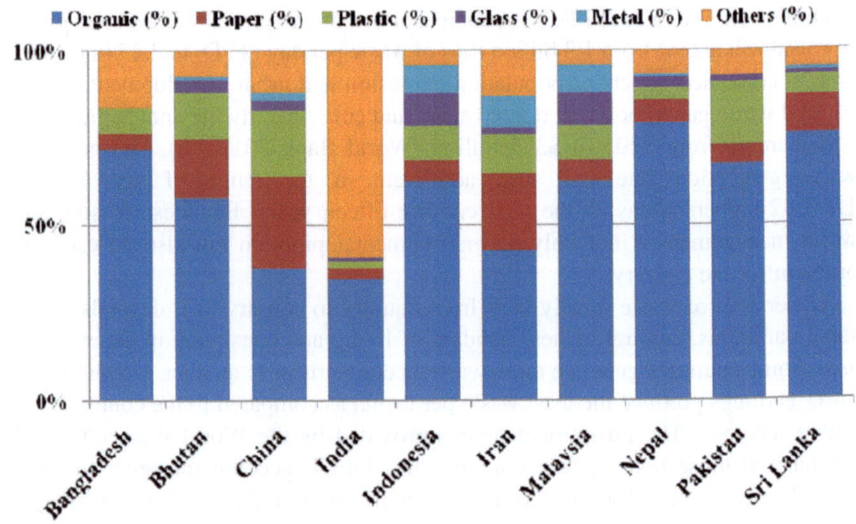

Fig. 1 MSW composition of some developing countries (Data adopted from—World Bank Report 2012a, b)

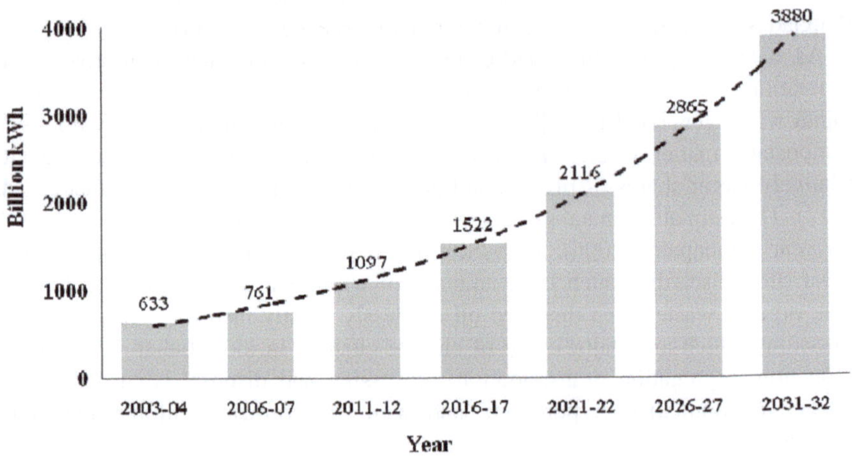

Fig. 2 Projected requirement of energy in India in near future (Data adopted from—Energy Statistics 2013)

pressing issues include rapidly increasing quantities and diverse characteristics of waste, the undesirable consequences of conventional methods of waste management, and failure to tap the resource value of waste (Figs. 2 and 3).

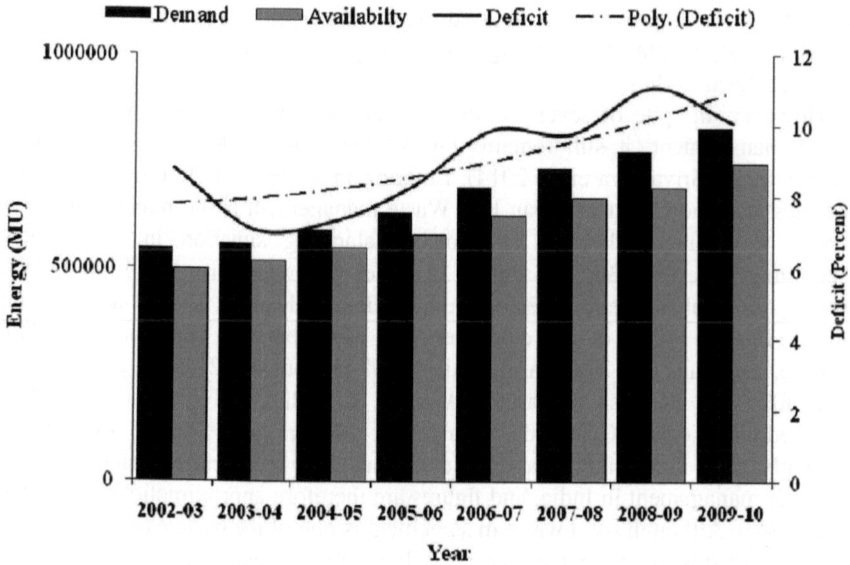

Fig. 3 Gap between demand and supply of energy in India (Data adopted from Central Electricity Authority 2012)

2 Solid Waste Management in India: Methods, Opportunities and Crisis

India, which is a land of physical, social, economical, climatic, geographical and linguistic diversity, with a population of about 1.21 billion (second largest after China) contains 17.5 % of the world's population (Vij 2012). The percentage population living in urban areas has increased from 17.35 % in 1951 to 31.2 % in 2011 (Census 2011). The current situation is that 1 out of every 3 person in India is living in urban cities of the country and it is projected that in the next 10 years as much as 50 % of India's population will live in cities (Khurshid and Sethuraman 2011). The level of urbanization directly contributes to the amount of waste generation and unscientific waste handling is the cause of health hazards and urban environment degradation. Municipal solid waste includes discarded material from households, industrial, commercial and institutional establishments and street sweepings which are collected by the municipal authorities for disposal (Jain 2007).

According to the Solid Waste Management Rules (2015) by the Ministry of Environment and Forests, it is the duty of every waste generator to segregate and store the waste in three different bags for biodegradable, non-biodegradable and hazardous waste, respectively and hand it over for proper disposal. Also, none of the waste generated will be allowed to be thrown in the streets. It would be the duty of the Ministry of Urban Development to coordinate with the state governments and union territories to take periodic review of the work done by the concerned authorities for improvement of solid waste management practices. The local municipal administrator is in-charge of ensuring implementation of these rules by all urban local bodies.

However, in spite of several efforts and heavily spending on municipal solid waste management, it still remains one of the most neglected areas of urban development (Srivastava et al. 2014). Littering in streets and choking of drains is the most common sight of urban life. Waste management is the major concern of India's governing authorities reaching an alarming situation in the country. According to reports, 36 % (8 out of 22) cities generate more than 1000 TPD of waste (Ahmedabad, Delhi, Greater Mumbai, Jaipur, Kanpur, Lucknow, Pune and Surat), 13.6 % (3 out of 22) cities generate waste between 500 and 1000 TPD (Indore, Ludhiana and Vadodara), while 50 % (11 out of 22) cities generate less than 500 TPD of waste (Agartala, Asansol, Chandigarh, Faridabad, Guwahati, Jamshedpur, Kochi, Kozhikode, Mangalore, Mysore and Shimla) (Vij 2012). Presently, there is no reliable national-level data on the technical or financial aspects of waste management in India, and figures are therefore approximations (Hanrahan et al. 2006). Although solid waste management is one of the basic essential services to be provided by municipal authorities in India, the present situation evidenced the overall lack of the waste management services.

2.1 Methods

Nowadays waste is seen as resource. It is necessary to harness the energy locked in the waste by applying appropriate waste to energy technologies. According to an analysis report by market analysts Frost and Sullivan, by the year 2013, Indian municipal solid waste to energy market could be growing at a compound growth rate of 9.7 % (EAI 2011). Hence, there is present need to find an appropriate waste to energy technology suitable for the different type of waste composition. Technologies for generating energy from biomass fall into two categories i.e.

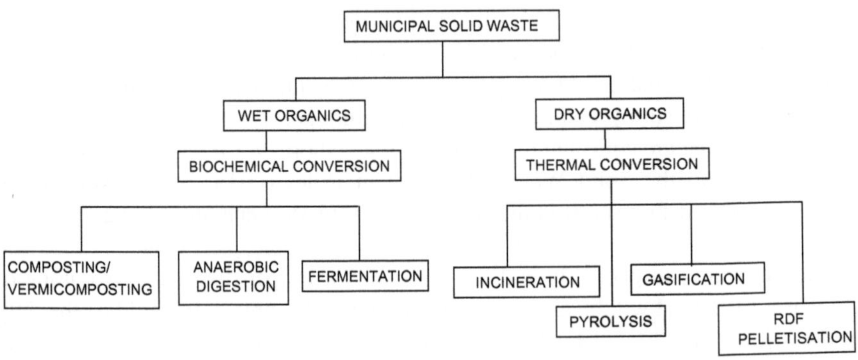

Fig. 4 Different technological options available for solid waste management

biochemical and thermochemical conversion (Fig. 4). In India, due to lack of proper management and resource availability, waste is mainly burnt in many instances (Karak et al. 2012). Waste composition is the main deciding factor for applying any technology available in market today. As more than 50 % of the waste is organic in nature in most of the cities of India, landfilling should be restricted to non-biodegradable waste only. 94 % of the waste is openly dumped in India, while only 5 % is composted and 1 % reaches other treatment techniques. However, a large numbers of technologies are available for treatment of solid waste that is discussed below.

2.1.1 Anaerobic Digestion/Biomethanation

Anaerobic digestion is a biochemical process that utilizes anaerobic bacteria to degrade organic fraction of waste in the absence of oxygen. This method is very useful for wastes containing a high percentage of moisture (>50 %) and C:N ratio of around 20–30 %. Anaerobic digestion plants operate with either single or multi stage under thermophilic (50–65 °C) or mesophilic (20–40 °C) conditions (Bolzonella et al. 2003). In the process of biomethanation, if assumption is made that the organic fraction is 50 % of the total waste, the total volume is reduced to around 35 % and the volume of organic fraction alone is reduced to around 70 % (Hartmann and Ahring 2005). After, digestion two end products are released namely biogas (mainly consisting of methane 55–60 % and carbon dioxide 30–45 %) and a bio-slurry that can be utilized as organic fertilizer. Biogas produced is scrubbed to improve its quality and can be directly utilized for household purposes or for power generation. Biogas has an energy content of about 20–25 MJ/m^3 (Braber 1995; Ambulkar and Shekdar 2004). As compared to incineration, the global warming potential of anaerobic digestion is quite low. In addition, this process emits around 0.2 kg CO_2/kWh_e of electricity generation. (Baldasano and Soriano 1999; Murphy and McKeogh 2004).

2.1.2 Composting/Vermicomposting

In India, the most common environmentally friendly and cost-effective technique is composting/vermicomposting, which is the biological decomposition of biodegradable waste through the joint activity of bacteria, micro-organisms and earthworms under aerobic conditions. At the end, a stable product that is relatively odour and pathogen free is achieved providing an excellent organic fertilizer (Tejada et al. 2008; Owamah et al. 2014). Under the National Scheme of Solid Waste Disposal (1975–1980), 10 mechanical composting plants were set up in cities with populations over 300,000 having processing capacities ranging from 150–300 tonnes of MSW per day (GoI 1995; Hoornweg et al. 2000). Currently, the major challenges faced by the composting/vermicomposting technology to be successfully implemented on large scales are low quality of waste, poor technical awareness, wrong selection of equipment, poor handling, low priority by the government and poor marketing efforts (Selvam 1996).

2.1.3 Incineration

Incineration of waste is one of the most common methods of waste disposal that have been adopted in different parts of the world which can reduce waste mass by 70 % and waste volume by 90 % respectively. At the end of the process, it provides steam for electricity generation and co-generation. But due to high organic content (>50 %) and low calorific value (<4.2 MJ/kg) of the waste generated (Unnikrishnan and Singh 2010), this process is not widely accepted in India.

Many unsuccessful attempts to start projects have been initiated by municipal authorities like in Lucknow, Rs. 76 crore wastes to energy plant for converting municipal solid waste (MSW) into electricity was closed in 2004. In 2005, Ahmedabad municipal standing committee planned to allow a private firm to set up a plant to generate power from solid waste collected in the city. The project was implemented in two phases. The first phase costing Rs. 10.94 crore and the second phase around Rs. 34.55 crore. Ahmedabad Municipal Corporation (AMC) is supposed to provide and deliver garbage to the plant site free of cost; the electricity generated will be sold to the AMC at Rs. 2.25 per unit plus 5 % annual escalation on a cumulative basis over the next 25 years. Moreover, environmental impacts of incineration plants are high as higher amounts of greenhouse gases (GHG's) and acidic gases (HCl, SOx, HF, NOx), volatile organic carbon (VOCs), polycyclic aromatic hydrocarbon (PAH) and heavy metal are emitted (Wikstrom and Marklund 2000; Wang et al. 2001; Liu and Liu 2005).

2.1.4 Gasification

Gasification involves the partial combustion of carbonaceous material at high temperature that is between 550 °C–900 °C to generate gas, after shredding of waste to reduce the size of particle (Belgiorno et al. 2003). During the gasification process, product gases such as CO_2 and H_2O are reduced to CO and H_2. The process may also generate some amount of methane and hydrocarbons and various contaminants such as small char particles, tars and ash (Kalyani and Pandey 2014). As reported by the Ministry of New and Renewable Energy (2015), the on-grid interactive power target was 100 MW but achieved only 45 MW, while for off-grid capacities in rural areas the target was 0.8 MWeq achieving 0.6 MWeq. For small gasifier plants in India (20-1000 kWe), capital costs are higher at around 900–1200$/kWe (IISc 2000). For production of electricity, the electrical demand load is around 20 % and carbon dioxide emission is around 114 g CO_2/kWhe in comparison to an incineration plant that emits 220 g CO_2/kWhe (Murphy and McKeogh 2004). High power generation efficiency can be achieved by using gasification-based power systems, which has an added advantage that CO_2 produced is in concentrated form and it is easier as well as cheaper to capture and sequester it, thereby preventing it from escaping into the atmosphere (Guan et al. 2010), mitigating the GHG emissions to some extent.

2.1.5 Pyrolysis

Pyrolysis is thermo-chemical decomposition of organic waste in an oxygen free environment that produces liquid, solid and gaseous end products, thereby reducing waste volume by 50–90 %. During pyrolysis, hydrocarbon present in the waste react at a temperature range of 450–500 °C in an oxygen-free environment generating end products in the form of pyrolysis gas, coke and char (fixed carbon + ashes) and tar (condensable gas) (Singh and Gu 2010). Electricity generation from pyrolysis is expected to cost around 0.07$/kWhe when co-fired (Uslu et al. 2008). Recently, Tamil Nadu has installed a 10-TPD capacity pyrolysis plant which is running successfully in Coimbatore. Likewise, a 10-TPD plant is running successfully in Masaipet (Andhra Pradesh) and Navapur (Maharashtra), Madhya Pradesh, Mathura (U.P.), Patna (Bihar). 5-TPD plants are installed in Bichhiwada (Rajasthan), Durg (Chhattisgarh) and Nagpur (Maharashtra).

2.1.6 Landfills/Sanitary Landfills (SLF)

Landfilling is the transfer of residual waste in a planned manner. Landfill/sanitary landfill is an engineered facility with a liner system at the base and sides to prevent leaching to the surrounding environment. It is equipped with leachate and landfill gas collection and treatment facility. It has a fenced buffer zone and a green belt area to serve as a barrier that helps in balancing adverse environmental impacts. Except for big metropolitan cities in India, there is no proper sanitary landfill in Class II and Class III cities for disposal of waste (Singh et al. 2011). Out of 22 surveyed cities by FICCI (2009), only 6 cities are equipped with proper landfills (Ahmedabad, Chandigarh, Jamshedpur, Mangalore, Surat and Vadodara). Guwahati, Indore and Jaipur are in the process of constructing sanitary landfills; and Agartala and Lucknow are considering construction of SLFs. The city of Lucknow has been sanctioned a project under the JNNURM for INR 42.92 crore, which would provide for two sanitary landfills and two composting units of capacities of 12 TPD each. With regards to taking initiatives on solid waste management, Gujarat emerges as one of the most active States as 3 cities of this State have already constructed sanitary landfills (Water and Sanitation Program, World Bank 2008).

2.2 *Opportunities*

India is growing rapidly both in terms of population and economy and so is the generation of waste. For disposal of waste, generally the most common practice that has been adopted in India is open dumping in low-lying areas (Biswas et al. 2010). Adverse impact in all aspects of environment and human health are the consequence. In the near future, amounts of solid waste are going to increase significantly (Jha et al. 2003; Ray et al. 2005; Rathi 2006; Gupta et al. 2007) and therefore proper and scientifically sound ways of disposing of waste should be considered so as to reach the goals of Integrated Solid Waste Management (ISWM) and sustainability. Sanitation and health conditions are pressing issues for the government but somehow the importance of municipal solid waste management (MSWM) is neglected. Incidents such as the plague episode of Surat and the dengue outbreak of West Bengal may be the eye opener for the government. It suggests that MSWM should not be neglected anymore and all issues of the environment from soil, water, air pollution to land crisis need to be adequately addressed. Therefore, due consideration should be given to this field as it is becoming more complicated with the rise in population and consumerism.

Although a large number of people are involved in this sector: waste pickers, informal rag pickers, municipal authorities, MSWM has a lower priority than sanitation, health and other issues. The scale of the problem is fairly unclear, as there is no authentic and reliable data available for waste generation quantities and disposal. Lack of investment and negligence by the government and policy makers are the main reasons for lack of progress in upgrading MSWM. Stakeholders involved in management of waste are: (a) the Ministry of Environment and Forests (MoEF) (b) the Ministry of Urban Development (MoUD) (c) Central and State Pollution Control Boards (d) the Department of Urban Development (e) State Level Nodal Agency (f) Urban Local Bodies (ULB) and (g) the Private Formal and Informal Sector. Of these, the most actively involved members are the informal sectors, rag pickers and non-governmental organizations (NGOs). The informal sector has been actively involved in door-to-door collection of waste, transportation, treatment and disposal (Vergara and Tchobanoglous 2012). Hence, it plays an important role in improving the management of solid waste but more attention needs to be given to strengthen ULB capacity to enter into contracts. With proper municipal solid waste management facilities, the Government of India, other ministries and nodal agencies involved in this sector have the opportunity to improve the living condition of urban and rural people, improve public health, conserve resources, mitigate GHG emissions and generate energy by adopting appropriate technology.

2.3 Crisis

Problems with present day municipal solid waste management are adversely affecting the environment as well as society. Due to the increase in complexity of waste and scarcity of land, disposal of waste is a common problem for developing countries like India. Uncontrolled dumping of waste without treatment at the dumpsites is everyday practice. It is very hard to find less filled landfill spaces or dump sites and locally unwanted land uses (LULUs) inside or on outskirts of the cities, as they are already carrying more waste than their capacities. Due to lack of proper segregation of waste at source, organic fraction of waste also reach the dump site and these are the main emitters of GHGs that ultimately lead to climate change (Annepu 2012).

Open burning of waste and fire from landfill sites release pollutants like carbon mono-oxide, dioxins and furans (carcinogenic), oxides of nitrogen (NOx) and sulphur dioxide into the environment (NEERI 2010). Searching for a satisfactory landfill space is a major concern for governing authorities and for urban local bodies (ULBs). Most of the waste produced in India is either openly burnt or finds its way to unsanitary landfills. Landfilling of unsegregated and untreated waste is equivalent to burying natural resources that could have been used for generation of energy and other useful products. Sorting waste at source and utilizing it properly by applying suitable technologies will help in minimizing the amount of waste reaching landfill sites significantly (Vaish et al. 2016c). Different technologies for treating different kind of waste are available in the market. Applying all the possible technologies in an integrated way can help to reach the goals of sustainability. Therefore, open dumping and unsanitary landfilling are not sustainable options and cannot be recommended for treating waste.

3 Energy Demand in India: Past, Present and Future

Increase in quantum and complexity of waste with the rising population, economy of the country and erratic changes in crude prices have demanded innovative ways to tackle the multiple issues. The energy requirement of the world heavily depends on fossil fuels such as coal, crude oil and gas providing almost 80 % of the global energy demands (Asif and Muneer 2007) while the remaining 20 % is supplied by renewable sources of energy. Likewise, India depends entirely on importation of crude oil from other countries, the prices of which are bound to rise as fossil fuels are becoming scarcer. India's economy is growing tremendously and the uncertainty about the supply of energy can have a negative impact on the functioning of economy. The production and consumption patterns of major energy source of India in the last five years i.e. from 2010–2014 are given in Tables 1 and 2 (BP 2015).

Table 1 Production pattern of major energy sources of India in the last five years (2010–2014)

	2010	2011	2012	2013	2014	Change in % (2014 over 2013)
Oil (Thousand barrels daily)	882	916	906	906	895	−1.3
Natural gas (Billion cubic meters)	5.8	46.1	40.3	33.7	31.7	−5.9
Coal (million tonnes oil equivalent)	217.5	215.7	229.1	228.8	243.5	6.4

Table 2 Consumption pattern of major energy sources of India in the last five years (2010–2014)

	2010	2011	2012	2013	2014	Change in % (2014 over 2013)
Oil (Thousand barrels daily)	3319	3488	3685	3727	3846	3.0
Natural gas (Billion cubic meters)	62.7	53.5	59.2	51.4	50.6	−1.5
Coal (million tonnes oil equivalent)	260.2	270.1	302.3	324.3	360.2	11.1

Although, India has improved in satisfying the energy requirement for different sectors of society but demand still outstrips supplies. According to the International Energy Outlook 2016, India and China are the world's fastest growing economies (of the non-OECD countries) over the last two decades. From the year 1990–2010, the economy of China rose by an average of 10.4 % per year and India's by 6.4 % per year. From the year 2010–2040, it is anticipated that the global energy demand will increase by 85 % and by the year 2040, both countries jointly will account for 34 % of the projected total world energy consumption (International Energy Outlook 2016) (Fig. 2).

During the 11th Five Year Plan of India, almost 55,000 MW of new generation capacity was created, yet overall an energy deficit of 8.7 % and peak shortages of 9.0 % remained (Energy Statistics 2013). An imbalance between demand and supply of energy persists across the country and as such, the country is facing a severe energy crisis (Fig. 3). Resources on which we are depending on like coal and petroleum are not sufficient for bridging the gap between demand and supply of energy (Vaish et al. 2016b).

The Indian energy market relies heavily on fossil fuels such as coal, petroleum, gas etc. Coal is the most abundant and important fuel for the Indian energy market and still the major source of electricity. Coal-fired generation currently provides two thirds of the generation capacity. Also, with a commitment to rural electrification, the Ministry of Power has accelerated the Rural Electrification Program with a target of 100,000 villages, by 2012 (Meisen 2010). Around 55 % of the Indian energy demands are fulfilled by the coal, with coal imports providing 22 % of total coal consumption requirements in 2006 (TERI 2013). It is estimated that the coal deficit in India will increase from around 50 million tonnes in the financial year (FY) 2011 to 400 million tonnes in the FY17 (Garg 2012). To combat these deficits, the Indian economy should shift its dependency from fossil fuel to renewable sources of energy. India is sanctified with vast resources of renewable energy such as solar, wind, biomass and hydro etc. In fact, the technical potential of these renewables exceeds the present installed generation capacity, i.e. 150,000 MW of exploitable renewable energy. Tapping India's wind, solar, biomass, and hydro could bring high quality jobs from a domestic resource. In recent years, clean and eco-friendly sources of energy are gaining importance, as they are easily available and affordable in many places (Detchon and Leeuwen 2014). The energy stored in biomass also incorporating solid organic waste could be used in generating green energy as this sector has a great future.

4 Biomethane Production from Waste: A Major Source of Energy

Energy generated from biomass produced by commercial, residential and the public sector can directly substitute fossil fuels as it is more effective in decreasing atmospheric CO_2. India is very rich in biomass. The current availability of biomass in India is estimated at about 500 millions metric tonnes per year. Studies sponsored by the Ministry has estimated surplus biomass availability at about 120–150 million metric tonnes per annum covering agricultural and forestry residues corresponding to a potential of about 18,000 MW (3500 MW from bagasse based cogeneration and 14,500 MW from surplus biomass). Currently, India has 537 MW commissioned and 536 MW under construction (MNRE). The facts reinforce the idea of a commitment by India to develop these resources for power production.

Technology to tap the energy stored in these biowastes is biomethanation. Biomethanation is the best option for households with feed materials to become self-sufficient for cooking gas and highly organic-enriched bio-manure. It provides the solution to protect households from problems of indoor air pollution, while saving on cost of refilling of LPG cylinders. The slurry produced at the end can serve as an organic fertilizer in their nearby agriculture fields. There could be a huge potential of installation of medium-size biogas-fertilizer plants in the country. During the demonstration phase, the Ministry is providing central financial assistance from 30–50 % of the cost (excluding cost of land) for a limited number of such projects. These projects are intended for implementation followed by an entrepreneurial model on Built, Own and Operate (BOO) and re-imbursement basis.

5 Biomethanation: A Window of Opportunity

In developing counties like India, the major portion of waste is organic in nature that is generated from agriculture, municipal, industrial, domestic sources etc. and their disposal is a serious ecological problem (Khalid et al. 2011). In a study by Troschinetz and Mihelcic (2009), the average of municipal solid waste generation rate in almost twenty three developing countries is about 0.77 kg/person/day. Accumulation of such large amount of waste is reaching critical levels in all affected countries, becoming a source of concern as they make their way to landfill sites or into open dumps. Due to high biodegradability of organic waste, emission of GHGs occur that are the cause of the changing climate (Bouallagui et al. 2003). Littering of waste on streets and in drains causes serious health and environmental problems. Recently, the organic fraction of waste has been identified as a valuable resource that can be transformed into organic manure and/or for the generation of electricity. A number of technologies can be applied for transforming the energy stored in the organic fraction of waste into useful forms of energy. These technologies are composting, vermicomposting, fermentation, biomethanation (anaerobic digestion). Of all these available technologies biomethanation is best suited and has a promising approach (Lee et al. 2009).

Biomethanation is an innovative and eco-friendly solution to many existing waste management and environmental problems as it significantly reduces the volume of solid organic fraction of waste, generates biogas and slurry (digestate) that is helpful in solving the crucial energy crisis, helps in reducing the indiscriminate use of inorganic fertilizers, and reduces GHG emissions thus mitigating the problem of climate change (Nixon et al. 2013; Abbasi et al. 2012a, b). Biomethanation of the organic fraction of waste is a therefore a win-win situation which provides environmental and economic benefits and helps in reaching the set goals of sustainability.

5.1 What Is Biomethanation?

Biomethanation (also called as anaerobic digestion) is an attractive waste management practice in which both energy and environmental problems could be solved. It is a biological process that converts organic matter into energy-rich biogas in the absence of oxygen by the help of anaerobic bacteria (Mata-Alvarez and Llabres 2000; Abbasi et al. 2012a, b; Surendra et al. 2014; Kalyani and Pandey 2014). At the end of the process biogas is generated that contains typically 40–60 % methane and the rest composition consists of mostly carbon dioxide with traces of other gases. This biogas can be utilized as combined heat and power or as fuel (Ferrer et al. 2011; González et al. 2011; Weiland 2010). In one of the study conducted by Saxena et al. (2009), it has been estimated that in a controlled environment of biomethanation 1 tonne of municipal solid waste will produce 2–3 times methane in 3 weeks compared to landfills that will produce the same amount of gas in 6–7 years (Ahsan 1999).

5.2 Biomethanation Process

Biomethanation is a multi-step process in which a number of microbial interaction takes place at every stage (Fig. 5). Three major groups of microorganisms are basically involved in anaerobic digestion of solid organic waste. These are

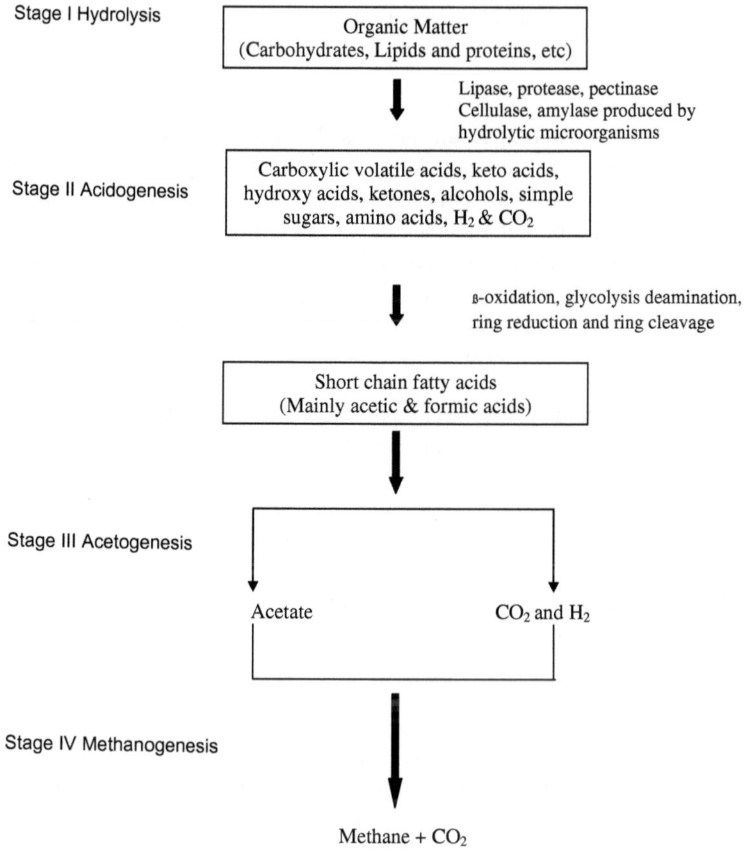

Fig. 5 Stages in biomethanation process

hydrolytic bacteria, anaerobic oxidizing bacteria as well as methanogenic bacteria (Hattori 2008; Thauer et al. 2008). Fermenting micro-organisms transforms biopolymers (lipids, protein, nucleic acids, carbohydrates, etc.) to soluble monomers such as long-chain fatty acids, glycerol, amino acids, purines, pyrimidines, mono-sugars, etc. and further to short chain fatty acids such as acetate, butyrate etc. Anaerobic oxidizing bacteria convert short chain fatty acid and alcohols to hydrogen, acetate, formate, and carbon dioxide by oxidation. At the end methanogenic bacteria convert these end products to methane and carbon dioxide (Angelidaki et al. 2011). Methane has a strong potential for replacing fossil fuels. The stages involved in biomethanation are discussed below:

Hydrolysis: It is a process in which complex polymeric organic compounds such as carbohydrates, lipids, and proteins are converted into soluble monomers such as amino acids and glycerols (Sevier and Kaiser 2002). Hydrolysis is widely regarded as the rate-limiting step of degradation of particulate organic matter (e.g., manure, sewage sludge, crop residues, etc. (Pavlostathis and Giraldo-Gomez 1991). Therefore, the overall rate of the process is determined by the hydrolysis rate of the complex substrate.

Fermentation: Fermentation is the process that uses microorganisms to convert sugar to acids, gases or alcohols. It occurs in absence of oxygen. The rate of fermentation depends on the concentration of cells and cellular components, microorganisms and different enzymes as well as temperature and pH. The process of fermentation takes much longer and a mixture of gases including CO and H_2S are produced. It is the anaerobic conversion of organic material such as amino acids, unsaturated fatty acids, glycerol etc. into alcohols and long chain organic acids, and fatty acids (Madigan et al. 2009).

Acetogenesis: In acetogenesis, acetate is synthesized by the reduction of CO_2 from organic acids such as propionate and butyrate and alcohols (ethanol) produced at the end of the fermentation process. Acetogenic bacteria are limited by the unfavorable energetics of the conversion processes (Schink and Stams 2006). Acetate formed by acetogenic bacteria can be either used directly by aceticlastic methanogens (Methanosarcina spp. and Methanosaeta spp.), or it can be degraded by syntrophic associations of bacteria (syntrophic acetate oxidizers) and hydrogen-consuming methanogenic archaea (Zinder and Koch 1984).

Methanogenesis: The process of methanogenesis takes place with the help of methanogenic bacteria belonging to the Archaea, phylum Euryarchaeota. The five orders of methanogens actively involved are: Methanobacteriales, Methanosarcinales, Methanomicrobiales, Methanococcales, and Methanopyrales. Recently a sixth order has been added Methanocellales and phylogenetically placed between the orders Methanosarcinales and Methanomicrobiales (Sakai et al. 2008). Methanogens are an important group of bacteria for the final step of biomethanation to form methane from acetate or CO_2, H_2, formate, alcohols and other methylated compounds (Thauer et al. 2008).

The main pathways known to be involved in methane formation are: aceticlastic methanogenesis where acetate is transformed to methane and CO_2 (Eq. 1). Secondly, hydrogenotrophic methanogenesis in which CO_2 is reduced to CH_4 (Eq. 2) and lastly methylotrophic methanogenesis where methylated C1 compounds such as dimethylsulfide, methanol, methylamines etc. are converted to methane (Eq. 3) (Deppenmeier 2002; Liu and Whitman 2008; Conrad et al. 2010).

$$CH_3COO^- + H^+ \longrightarrow CH_4 + CO_2 \quad (1)$$

$$4H_2 + CO_2 \longrightarrow CH_4 + 2H_2O \quad (2)$$

$$4CH_3OH \longrightarrow 3CH_4 + CO_2 + 2H \quad (3)$$

5.3 Operational Parameter of Biomethanation Process

Various operational parameters control the process of biomethanation by generating conditions which help these microorganisms to function properly (Demirel and Scherer 2008; Abbasi et al. 2012a, b). Presence of adequate quantities of nitrogen, micronutrients, and water is essential if an organic substrate is to undergo biomethanation and generate methane-rich biogas (Demirel and Scherer 2008; Takashima et al. 2011). These parameters are discussed below.

5.3.1 Composition of Urban Solid Waste

Characterization of waste is necessary to know with the changing trends in the composition of waste. Based on composition/characterization of waste, appropriate selection of waste processing technologies could be selected. In India, on an average 54 % compostable, 16 % recyclables and 30 % other types of waste are generated. The waste generated has a C/N ratio of 34 and a calorific value of about 1802 kcal/kg with a moisture content around 49 % (CPCB). Therefore, the waste generated if segregated properly could be used best for biomethanization.

Composition of the urban solid waste used as a feedstock is the most important determining factor in the process of biomethanation. The most significant types of solid wastes with considerable biomethanation potential are municipal solid waste (MSW), kitchen waste, garden waste, energy crops (maize, grass, sugarcane, etc.), etc. Feedstock preparation involves segregation of non-digestible parts and shredding of the waste to reduce particle size in order to increase the surface area for microorganisms to act upon.

5.3.2 pH

Regulating pH in the biomethanation process is a very important parameter as the microorganisms perform only at certain pHs and fluctuation in pH can be the inhibitor for the process. Optimum biogas production is achieved when the pH value of the feedstock is between 6.7 and 7.5 (Deublein and Steinhauser 2008; Daisy and Kamaraj 2011). During the initial period of digestion, large amounts of organic acids are produced and the pH of the mixture decreases. But because of digestion of nitrogen, the pH value increases. When the methane gas production stabilizes, the pH remains between 7.2 and 8.2 (Verma 2002). Reduction in pH due to formation of organic acid and increase in pH due to digestion of nitrogen inhibits the further digestion of the feedstock (Abbasi et al. 1991; Fricke et al. 2007).

5.3.3 Temperature

Different ranges of temperature is required for the certain groups of bacteria that fall into the category of thermophilic (50–65 °C), mesophilic (20–40 °C) and psychrophylic (<1.2 °C) (Cowan and Talaro 2009). The most common temperature range at which large biomethanation plants operate is generally mesophilic (i.e. 20–40 °C) at around 35 °C and the ideal thermophilic temperature is 55 °C (Monnet 2003; Suryawanshi et al. 2010). Eliyan (2007) demonstrated that, when compared to mesophilic temperature, thermophilic temperature range is much more efficient but it is very difficult to control the process and requires an extra input of energy. Therefore, the mesophilic temperature range is preferred as it maintains the energy balance.

5.3.4 Loading Rate

The design of anaerobic reactors is based on the loading rate and therefore it is another important parameter when digestion is carried out in continuous mode. It describes the amount of volatile solids to be fed into the digester each day typically expressed as weight of organic matter (volatile solids or COD) per bed volume of reactor in a certain period of time (Fannin and Biljetine 1987). The actual loading rate depends on the types of wastes fed into the digester, because the type of wastes determine the level of biochemical activity that will occur in the digester (Mattocks 1984). For soluble and easily degradable substrates, such as sugars and soluble starches, the acidogenic reactions can be much accelerated at high loadings. Loading rate is applicable to both types of reactors i.e. dry and wet anaerobic reactors. But generally for processing of municipal solid waste dry type of anaerobic reactors are used. In a study by Kiely (1998), food-to-microbes (F/M) ratio was also affected by the loading rate (Igoni et al. 2008). Overloading of the digestion plant can lead to failure of the process while inadequate mixing may cause significant rise of the volatile fatty acid concentration and decrease in pH.

5.3.5 Retention Time

It is the duration at which the feedstock and microorganism must be kept together in a digester tank to attain the desired level of degradation. Therefore, the shorter the retention time, the more efficient the reactor design (Abbasi and Nipaney 1993). There are two types of retention times: hydraulic retention time (HRT) and solid retention time (SRT). Hydraulic retention time denotes the time by which substrate (organic material) is retained in the digester for anaerobic degradation while solid retention time is the duration by which microorganisms reside in the digester tank. To enhance digester efficiency, we should aim to reduce HRT and increase the SRT. In other words more quantities of microorganisms should be present compared to feedstock.

For instance, in treatment of wastewater, the design is normally based on a loading rate of 10 kg-COD $m^{-3}d^{-1}$, with retention times of 8–24 h while in treatment of slurries retention time exceeds the doubling time of the slowest growing organisms present (m > D). As a consequence, in CSTR (Continuously Stirred Tank Reactor), hydraulic retention times (HRTs) are typically in the range 15–30 days at mesophilic conditions and 10–20 days at thermophilic conditions (Angelidaki et al. 2011).

5.3.6 Reactor Design

A reactor is basically the system which is primarily used for organic solid waste digestion by which two useful by-products are released-biogas and slurry. For different solid organic wastes different reactor designs are required. The anaerobic digesters used in India, China and other developing countries basically contain a large chamber of about 1000 L (1 m^3) or more. In a fixed-dome digester of China, biogas is collected under the fixed dome and pushes slurry to the overflow tank. Another type of biogas digesters is the floating dome biogas digester. To maintain the SRT > HRT, biomass should be retained in the tank for a long time and to achieve this anaerobic fixed reactor, continuous flow stirred tank reactor (CFSTR) and fluidized bed reactors are used.

With the introduction of anaerobic filters by Young and McCarty (1969), one after another reactors designs were introduced by different scientists such as downflow fixed film reactors, fluidized bed reactor, CSTR, diphasic/triphasic reactor, and anaerobic sequencing batch (ASB) biomethanation potential reactors, anaerobic baffled reactors etc. (Sutton and Huss 1984; Bachmann et al. 1985; Lettinga and Pol 1991; Fongastitkul et al. 1994; Sakar et al. 2009; Bai et al. 2009; Singh and Srivastava 2011; Wang et al. 2012). The main objective of all the designs of the reactors was to minimize HRT and F/M ratio, maximize SRT and enhance the digester loading.

6 Biomethanation Potential

The biomethanation potential is different for different types of feedstock. The economy of existing biogas plants depends on the methane potential of the feedstock used. These are of two types: Theoretical potential and practical potential that is discussed below:

Theoretical Potential: It is the expected maximum methane generation from the feedstock used following a stoichiometric equation. The specific theoretical methane potential (Bo, th) is usually expressed as 1 CH_4/g-VS, may then be calculated as per (Eq. 4): assuming 22.4 as the volume of 1 mol of gas under standard conditions (i.e., 273 K and 1 atm. pressure):

$$Bo, th = \frac{(n/2 + a/8 - b/4)\, 22.4}{12n + a + 16b} \frac{(STP\, l CH_4)}{g - VS} \quad (4)$$

An easy way to calculate the theoretical methane potential is:

Bo, th = 0.415 carbohydrates + 0.496 proteins + 1.014 lipids + 0.373 acetate + 0.530 propionate (Buendia et al. 2015)

This is an alternative way to obtain an estimation of the theoretical potential of methane.

Practical Potential: While the theoretical methane potential provides an estimation of the biogas production from any type of waste, the practical potential is close to the actual amount of biogas production from any kind of waste. The value obtained by the practical potential is always lower than the theoretical potential of methane. Estimation of the practical potential of methane is very difficult and depends on number of factors such as heterogeneous waste, activity of microorganisms, toxic substances, deficient in nutrients etc. Hashimoto et al. (1981) described the practical methane potential as "the ultimate volume of methane formed from a specific amount of waste (in either weight or COD) in a batch experiment for indefinite degradation time, that is, until the methane production ceases".

It is possible to achieve 100 % conversion specifically for water soluble feedstock materials. For highly particulate organic matter 30–60 % conversion is achieved while 40–50 % for cattle manure and 55–65 % for swine manure (Moller 2003). However, in Indian conditions highest conversion process efficiencies were found to be around 70–74 % (Rao and Singh 2004) because of high moisture content and lack of expertise.

7 Policies/Incentives for Such Facilities in India

In India, the policies dealing with the management of solid waste are inadequate and most of the municipalities have not been able to provide the satisfactory level of services due to a number of reasons (Gupta et al. 1998; Kansal 2002; Siddiqui et al. 2006). Biomethanation plants deal with a number of issues like lack of technological advancement, sustainable planning and insufficient funding by the government. However, now the Ministry of New and Renewable Energy (MNRE) has initiated promoting the innovations in waste to energy projects. MNRE has started a National Biogas and Manure Management Programme which is a Central Sector Scheme that provides incentives for setting up of Family Type Biogas Plants mainly for rural and semi-urban or households. The Ministry of New and Renewable Energy is trying to implement the National Biogas and Manure Management Programme (NBMMP) in all States and UTs of the country.

Approximately, 47.5 Lakh biogas plants have already been installed in the country up to 31st March, 2014. The government has set a target of installing

1,10,000 biogas plants by the year 2014–15. Some of the common types of anaerobic digesters that are used in India are:

The ARTI Compact Biogas Plant

ARTI (Appropriate Rural Technology Institute), which designed the plant in 2003 for treating organic waste at the household level of around 1–2 kg of food waste per day. It is simple low-cost, floating drum design that applies a wet digestion process. The smaller tank that holds the gas is inverted over the larger tank which holds the feedstock. The plant can be used equally well for urban households. Currently around 2000 such plants are used in Maharashtra, India, in both urban and rural households (WRAPAI 2009).

The BIOTECH Plants

BIOTECH is a nodal agency of the Ministry of Non-conventional Energy sources situated in Kerala, South India, which has been engaged in developing biogas plant using feedstock like cooked food waste, vegetable waste, waste water from kitchen waste. It has developed decentralized reactor plants from market waste for street lightening and household purposes. Like ARTI, the BIOTECH plant is also a floating drum design that includes a recirculation loop to optimize moisture content. To increase the retention time there is a baffle in the middle of the tank and orthogonal to the flow direction. The baffle holds back unsuspended solid waste so that liquid can flow. This leads to increased retention time and therefore improved degradation of solid waste. The digestate flows into the effluent tank from where it is used to flush the feedstock thereby avoiding the need of fresh water for flushing (Heeb 2009).

BIMA Digester

The BIMA (Biologically Induced Mixing Arrangements) digester is one of the most common types of digester being widely accepted in India. It is divided into three separate sections that are the main chamber, upper chamber and central tube to which the feed pipe is connected. The central tube is used for pre-hydrolysis of the substrate. Most of the gas is produced in the main chamber which in turn displaces an equal amount of feedstock into the upper chamber that builds a level of difference and thus a gas pressure is created in the main chamber. When the required level difference is achieved, the valve is opened to release the gas pressure in the gas connecting tube. Thus, the feedstock flows back with a high velocity into the main chamber.

BIMA digester is placed in Ludhiana, Punjab under the supervision of the Chemical Engineering Department of the Indian Institute of Technology, Roorkee for treating cattle dung. It is also installed in Koyambedu Wholesale Market Complex, Chennai for treating organic waste of around 30 metric tonnes of perishable wastes per day and will generate on an average of 2375 M^3 of gas (www.cdmchennai.gov.in).

Deenbandhu and KVIC models are equally common for digestion of organic waste in Indian condition.

A family type biogas plant generates biogas from organic substances such as cattle dung, and other bio-degradable materials such as kitchen wastes, garden wastes, night soils etc. (www.mnre.gov.in). In India, Gujarat is the only state that is successfully running waste to electricity plants through anaerobic digestion by M/S Kanoria chemicals Ltd., Ankleshwar that generates 2 MW of power. 4800 m^3 of biogas is generated by M/S Anil Starch Products Ltd through the anaerobic digestion process (Kalyani 2003). Similarly, Maharashtra is also taking initiative in utilizing waste and converting it to energy by installing many pilot projects at Mumbai, Pune, Nasik etc. (Ramachandra 2006). The Ministry of New and Renewable Energy (MNRE), Government of India has subsidized three demonstration projects of anaerobic digestion at Hyderabad (6.6 MW), Vijaywada (6 MW) and Lucknow (5 MW) (Annepu 2012).

8 Conclusion

Sustainability of cities is at stake in developing countries as cities are harnessing environmental resources at a furious pace, taking their ecological footprint far beyond the limit. Rapidly growing economies demands for energy consumption and eventually increase in CO_2 emission is observed that causes climate change (Vaish et al. 2016). Environmental pollution is widespread leading to degradation of the environment. Urban poverty, lack of water supply, energy constraints, sanitation and solid waste management are the common problems and the key parameters in the struggle for survival. It is a timely demand for us to understand our basic requirements, retaining our environment and reducing the dependency on the resources so as to reach the concept of sustainability. The most common problem is the enormous increase of organic solid waste in most developing countries. With an almost 3-fold increase in MSW generation by 2021, the situation may reach critical dimensions (Talyan et al. 2008). In order to achieve the goals of sustainability, it is necessary to establish a well harmonized solid waste management system with the objectives that solutions take care of the environment as well as fulfill the needs of the poor.

Biomethanation is a solution capable of solving many existing problem in developing countries such as solid waste management, GHG emissions, soil degradation due to injudicious use of inorganic fertilizers and the energy crisis. Biomethanation is a reliable, eco-friendly and affordable process. Currently, the important issue to be solved is to accelerate the development of biomass to energy conversion so as to relieve the pressure on resources and the environment. Moreover, it is the best practice and has number of advantages over other renewable sources of energy. Methane gas generated at the end of the process has capacity to reduce the reliance on imported fuels (DEFRA 2011). Therefore, biomethanation is an integral part of the solution to many existing problems.

Recommendations:

- Optimization of biogas reactors, feedstock used and pre-treatment processes should be promoted.
- A shared programme of encouraging biomethanation between industry, government, agencies, and others should be carried.
- Community participation should be encouraged, as it is a paramount in an approach to reach the goal of sustainability.
- Further research on its impact on human health and environment, and its prospects for the future should be encouraged.
- The research findings should be disseminated to the public easily and more effectively.
- Financial incentives and tariffs should be provided by the government to build strong and sustainable biogas plants (Saini et al. 2012).
- Granting of permits and incentives should be quicker and easier for the applicant. Inconsistencies in data recording, definitions, collection methods, and seasonal variations should be avoided for further planning.
- As the city population is added every year and the economic profile of the city changes, the magnitude of waste and the resources requirement to manage it will also increase.
- Given their financial limitations and competing demand of other services, urban local bodies may find it challenging to raise and sustain additional allocations for this sector. Thus waste minimization through the process of biomethanation seems the only sustainable way to manage existing and future quantities of waste.

Acknowledgment Authors are thankful to Department of Science and Technology (P-45/18) for providing fund and the Head, Dean and Director, Institute of Environment and Sustainable Development, Banaras Hindu University for providing necessary facilities. Ms. Pooja Singh is thankful to Council of Scientific and Industrial Research for Senior Research Fellowship.

References

Abbasi SA, Nipaney PC (1993) Modelling and simulation of biogas systems economics. Ashish Publishing House, New Delhi, p 356

Abbasi SA, Nipaney PC, Panholzer MB (1991) Biogas production from the aquatic weed pistia (Pistia stratiotes). Bioresource Technol 37(3):211–214

Abbasi T, Tauseef SM, Abbasi SA (2012a) Anaerobic digestion for global warming control and energy generation—an overview. Renew Sust Energ Rev 16:3228–3242

Abbasi T, Tauseef SM, Abbasi SA (2012b) Biogas energy. Springer Verlag, New York, p 169

Ahsan N (1999) Solid waste management plan for Indian megacities. Indian J Env Prot 19(2):90–95

Alvarez MJ, Llabres SMP (2000) Anaerobic digestion of organic solid waste. An overview of research achievements and perspectives. Bioresource Technol 74:3–16

Ambulkar AR, Shekdar AV (2004) Prospects of biomethanation technology in the Indian context: a pragmatic approach. Resour Conserv Recy 40(2):111–128

Angelidaki I, Karakashev D, Batstone DJ, Plugge CM, Stams AJM (2011) Biomethanation and its potential. In: Chapter sixteen, Methods in enzymology, vol 494. ISSN 0076-6879, doi:10.1016/B978-0-12-385112-3.00016-0

Annepu RK (2012) Sustainable solid waste management in India. Columbia University, Department of Earth and Environmental Engineering, New York

Asif M, Muneer T (2007) Energy supply, its demand and security issues for developed and emerging economies. Renew Sust Energ Rev 11:1388–1413

Bachmann A, Beard VL, McCarty PL (1985) Performance characteristics of the anaerobic baffled reactor. Water Res 19(1):99–106

Buendia IM, Fernandez FJ, Villasenor J, Rodriguez L (2015) Squeezing wastes in a wastewater treatment plant. Book: Waste to Energy. Ed. S. Syngellakis. WIT Press Southampton, Boston

Bai L, Shi Y, Sun X (2009) A review on expanded granular sludge bed (EGSB) reactor treating organic wastewater. Environmental Science and Management 34(1):93–97

Baldasano JM and Soriano C (1999) Emission of greenhouse gases from anaerobic digestion processes. Comparison with other MSW treatments. In: Proceedings of the second international symposium on anaerobic digestion of solid wastes, Barcelona 274–277

Belgiorno V, De Feo G, Della RC, Napoli RMA (2003) Energy from gasification of solid wastes. Waste Manage 23(1):1–15

Biswas AK, Kumar S, Babu SS, Bhattacharyya JK, Chakrabarti T (2010) Studies on environmental quality in and around municipal solid waste dumpsite. Resour Conserv Recy 55(2):129–134

Bolzonella Innocenti L, Pavan P, Traverso P, Cecchi F (2003) Semi-dry thermophilic anaerobic digestion of the organic fraction of municipal solid waste: focusing on the start-up phase. Bioresource Technol 86(2):123–129

Bouallagui H, Cheikh RB, Marouani L, Hamdi M (2003) Mesophilic biogas production from fruit and vegetable waste in tubular digester. Bioresource Technol 86:85–89

BP (2015) http://www.bp.com/en/global/corporate/about-bp/energy-economics/energy-outlook/country-and-regional-insights/india-insights.html

Braber K (1995) Anaerobic digestion of municipal solid waste: a modern waste disposal option on the verge of breakthrough. Biomass Bioenerg 9(1–5): 365–376

Census (2011) Provisional population totals, India

Central Electricity Authority (2012) Report of the Working Group on Power for Twelfth Plan (2012–17) Government of India, Ministry of Power, New Delhi. http://cea.nic.in/reports/others/planning/irp/report_working_group12.pdf

Chalmin P and Gaillochet C (2009) From waste to resource. an abstract of world waste survey, Cyclope, Veolia Environmental Services, Edition Economica, France

Conrad R, Klose M, Claus P, Enrich-Prast A (2010) Methanogenic pathway, 13C isotope fractionation, and archaeal community composition in the sediment of two clear- water lakes of Amazonia. Limnol Oceanogr 55(2):689–702

Cowan MK, Talaro KP (2009) Microbiology: a systems approach. McGraw-Hill, pp 869

Daisy A, Kamaraj S (2011) The impact and treatment of night soil in anaerobic digester: a review. J Microb Biochem Technol 3:043–50

DEFRA (Department of Environment, Food and Rural Affairs, UK) (2011) Anaerobic digestion strategy and action plan: a commitment to increasing energy from waste through anaerobic digestion. Executive summary

Demirel B, Scherer P (2008) The roles of acetotrophic and hydrogenotrophic methanogens during anaerobic conversion of biomass to methane: a review. Rev Environ Sci Biotechnol 7(2):173–190

Deppenmeier U (2002) The unique biochemistry of methanogenesis. Prog Nucleic Acid Res Mol Biol 71:223–283

Detchon R, Leeuwen RV (2014) Bring sustainable energy to the developing world. Comment. Nature 508:309–311

Deublein D, Steinhauser A (2008) Biogas from wastes and renewable resources: an introduction. John Wiley & Sons, Germany

EAI (2011) Energy Alternatives India. http://www.eai.in/ref/ae/wte/wte.html

Eliyan C (2007) Anaerobic digestion of municipal solid waste in thermophilic continuous operation. A M. Sc thesis in Environmental Engineering and Management, Asian Institute of Technology, School of Environment, Resource and Development, Thailand
Energy Statistics (2013) Central statistics office, national statistical organisation, ministry of statistics and programme implementation, government of India. www.mospi.gov.in
Fannin KF, Biljetine R (1987) Reactor designs. In: Chynoweth DP, Isaacson R (eds) Anaerobic digestion of biomass. Elsevier Applied Science, Ltd., London
Ferrer I, Garfí M, Uggetti E, Ferrer-Martí L, Calderon A, Velo E (2011) Biogas production in low-cost household digesters at the Peruvian Andes. Biomass Bioenerg 35(5):1668–1674
Fongastitkul P, Mavinic D, Lo K (1994) A two-phased anaerobic digestion process: concept, process failure and maximum system loading rate. Water Environ Res 66(3):243–254
Fricke K, Santen H, Wallmann R, Hüttner A, Dichtl N (2007) Operating problems in anaerobic digestion plants resulting from nitrogen in MSW. Waste Manage 27(1):30–43
Garg P (2012) Energy scenario and vision 2020 in India. J Sust Energ Environ 3:7–17
González C, Buenrostro O, Marquez L, Hernández C, Moreno E, Robles F (2011) Effect of solid wastes composition and confinement time on methane production in a dump. J Environ Prot 2(10):1310–1316
Government of India (GoI) (1995) Urban solid waste management in India. Report of the High Power Committee, Planning Commission
Guan G, Fushimi C, Tsutsumi A, Ishizuka M, Matsuda S, Hatano H, Suzuki Y (2010) High-density circulating fluidized bed gasifier for advanced IGCC/IGFC—Advantages and challenges. Particuology 8(6):602–606
Gupta S, Krishna M, Prasad R, Gupta S, Kansal A (1998) Solid waste management in India: options and opportunities. Resour Conserv Recycl 24:137–154
Gupta S, Choudhary N, Alappat BJ (2007) Bioreactor landfill for MSW disposal in Delhi. In: Proceedings of the international conference on sustainable solid waste management, Chennai, India, pp 474–481
Hanrahan D, Srivastava S, Ramakrishna AS (2006) Improving management of municipal solid waste in India overview and challenges. Environment Unit South Asia Region, The World Bank, New Delhi, pp 38–62
Hashimoto AG, Varel VH, and Chen YR (1981) Ultimate methane yield from beef cattle manure: effect of temperature, ration constituents, antibiotics and manure age. Agr Wastes 3:241–256
Hattori S (2008) Syntrophic acetate-oxidizing microbes in methanogenic environments. Microbes Environ 23:118–127
Hartmann H, Ahring BK (2005) Anaerobic digestion of the organic fraction of municipal solid waste: influence of co-digestion with manure. Water Res 39(8):1543–1552
Heeb F (2009) Decentralised anaerobic digestion of market waste. Case study in Thiruvananthapuram, India. Swiss Federal Institute of Aquatic Science and Technology, Duebendorf
Hoornweg D, Laura T, Lambert O (2000) Composting and its applicability in developing countries. Working paper series, No. 8. Urban Development Division, The World Bank, Washington, DC
Hoornweg D and Bhada-Tada P (2012) What a waste: a global review of solid waste management, No. 15, World Bank
http://www.cdmachennai.gov.in/mmc.html
International Energy Outlook (2016) U.S. Energy Information Administration. http://www.eia.gov/forecasts/ieo/world.cfm
ISWA (International Solid Waste Association) (2012) Globalization and Waste Management
Igoni AH, Ayotamuno MJ, Eze CL, Ogaji SOT, Probert SD (2008) Designs of anaerobic digesters for producing biogas from municipal solid-waste. Appl Energ 85:430–438
IISc Indian Institute of Science (2000–2004) Final report on the MNES sponsored project advanced biomass gasification. Combustion gasification and propulsion Lab, Department of Aerospace Engineering, Indian Institute of Science

Jain AK (2007) Sustainable development and waste management. Environews, Newsletter ISEB India 13(1)
Jha MK, Sondhi OAK, Pansare M (2003) Solid waste management—a case study. Indian J Environ Prot 23(10):1153–1160
Kalyani K (2003) Alternative energy supply option from household waste gasification process-a feasibility analysis on Rajkot city, Dehradun, University of Petroleum and Energy Studies, College of Management & Economic Studies
Kalyani AK, Pandey KK (2014) Waste to energy status in India: a short review. Renew and Sust Energ Rev 31:113–120
Kansal A (2002) Solid waste management strategies for India. Indian J Env Prot 22(4):444–448
Karak T, Bhagat RM, Bhattacharyya P (2012) Municipal solid waste generation, composition, and management: the world scenario. Critical Rev Environ Sci Technol 42(15):1509–1630. doi:10.1080/10643389.2011.569871
Khalid A, Arshad M, Anjuma M, Mahmooda T, Dawson L (2011) The anaerobic digestion of solid organic waste. Waste Manage 31:1737–1744
Khurshid S and Sethuraman S (2011, July 31) In: New India all roads lead to city. Hindustan Times, New Delhi, pg 10
Kiely G (1998) Environmental engineering, International edn. Irwin, McGraw-Hill, Boston, p 979
Kumar A, Sharma MP (2014) Estimation of GHG emission and energy recovery potential from MSW landfill sites. Sust Energ Technol Assess 5:50–61
Liu Y, Whitman WB (2008) Metabolic, phylogenetic, and ecological diversity of the methanogenic archaea. Ann N Y Acad Sci 1125:171–89
Lee M, Hidaka T, Hagiwara W, Tsuno H (2009) Comparative performance and microbial diversity of hyperthermophiclic and thermophilic co-digestion of kitchen garbage and excess sludge. Bioresource Technol 100:578–585
Lettinga G, Pol LWH (1991) UASB-process design for various types of wastewaters. Water Sci Technol 24(8):87–107
Liu Y, Liu Y (2005) Novel incineration technology integrated with drying, pyrolysis, gasification, and combustion of MSW and ashes vitrification. Environ Sci Technol 39:3855–3863
Madigan MT, Martinko JM, Dunlap PV, Clark DP (2009) Brock biology of microorganisms, 12th edn. Pearson Benjamin Cummings, San Francisco
Mattocks R (1984) Understanding biogas generation. Technical Paper No. 4.Volunteers in Technical Assistance, Virginia, USA, p. 13
Mohapatra PK (2006) Biotechnology for Waste Management. Textbook for Biotechnology. I.K. Intern Publishing House. 264–314
Meisen P (2010) Overview of sustainable renewable energy potential of India, Global energy network institute (GENI)
Ministry of Environment and Forest (2015) http://www.moef.nic.in/sites/default/files/SWM%20Rules%202015%20-Vetted%201%20-%20final.pdf
Moller HB (2003) Methane productivity and nutrient recovery from manure. Technical University of Denmark, PhD thesis, BioCentrum-DTU
Monnet F (2003) An introduction to anaerobic digestion of organic wastes, Final report, Remade Scotland
Murphy JD, McKeogh E (2004) Technical, economic and environmental analysis of energy production from municipal solid waste. Renew Energ 29(7):1043–1057
Narendra KG, Swamy C, Nagadarshini KN (2014) Efficient garbage disposal management in metropolitan cities using VANETs. J Clean Energ Technol 2(3)
National Environmental Engineering Research Institute (NEERI) (2010) Air quality assessment, emissions inventory and source apportionment studies. Central Pollution Control Board (CPCB), Mumbai, New Delhi
Nixon JD, Dey PK, Ghosh SK, Davies PA (2013) Evaluation of options for energy recovery from municipal solid waste in India using the hierarchical analytical network process. Energy 59:215–223

Owamah HI, Dahunsi SO, Oranusi US, Alfa MI (2014) Fertilizer and sanitary quality of digestate biofertilizer from the co-digestion of food waste and human excreta. J Waste Manage 34:747–752
Pavlostathis SG, Giraldo-Gomez E (1991) Kinetics of anaerobic treatment: a critical review. Crit Rev Environ Control 21(5–6):411–490
Ramachandra T (2006) Management of municipal solid waste. Capital Publishing Company, New Delhi
Rao MS, Singh SP (2004) Bioenergy conversion studies of organic fraction of MSW: kinetic studies and gas yield- organic loading relationships for process optimization. Bioresource Technol 95:173–185
Rathi S (2006) Alternative approaches for better municipal solid waste management in Mumbai, India. J Waste Manage 26(10):1192–1200
Ray MR, Roychoudhury S, Mukherjee G, Roy S, Lahiri T (2005) Respiratory and general health impairments of workers employed in a municipal solid waste disposal at open landfill site in Delhi. Int J Hyg Envir Heal 108(4):255–262
Saini S, Rao P, Patil Y (2012) City based analysis of MSW to energy generation in India, calculation of state-wise potential and tariff comparison with EU. Procedia Soc Behav Sci 37:407–416
Sakai S, Imachi H, Hanada S, Ohashi A, Harada H, Kamagata Y (2008) *Methanocella paludicola* gen. nov., sp. nov., a methane-producing archaeon, the first isolate of the lineage 'Rice Cluster I', and proposal of the new archaeal order Methanocellales ord. nov. Int J Syst Evol Microbiol 58:929–936
Sakar S, Yetilmezsoy K, Kocak E (2009) Anaerobic digestion technology in poultry and livestock waste treatment—a literature review. Waste Manage Res 27(1):3–18
Saxena R, Adhikari D, Goyal H (2009) Biomass based energy fuel through biochemical routes: a review. Renew Sustain Energy Rev 13:167–178
Schink B, Stams AJM (2006) Syntrophism among prokaryotes. In: Dworkin M, Falkow S, Rosenberg E, Schleifer KH, Stackebrandt E (eds) The prokaryotes: an evolving electronic resource for the microbiological community. Springer-Verlag, New York, pp 309–335
Selvam P (1996) A review of Indian experiences in composting of municipal solid wastes and a case study on private sector on private sector participation. In: Conference of recycling waste for agriculture: the rural-urban connection. Washington, DC, USA, pp 23–24
Sevier CS, Kaiser CA (2002) Formation and transfer of disulphide bonds in living cells. Nat Rev Mol Cell Biol 3(11):836–847
Sharholy M, Ahmad K, Mahmood G, Trivedi RC (2008) Municipal solid waste management in Indian cities—a review. Waste Manage 28:459–467
Siddiqui T, Siddiqui F and Khan E (2006) Sustainable development through integrated municipal solid waste management (MSWM) approach: a case study of Aligarh district. In: proceedings of the national conference of advanced in mechanical engineering, New Delhi, pp 1168–75
Singh J, Gu S (2010) Biomass conversion to energy in India—a critique. Renew Sust Energ Rev 14:1367–1378
Singh M, Srivastava RK (2011) Sequencing batch reactor technology for biological wastewater treatment: a review. Asia-Pacific J Chem Eng 6(1):3–13
Singh RP, Tyagi VV, Allen T, Ibrahim MH and Kothari R (2011) An overview for exploring the possibilities of energy generation from municipal solid waste (MSW) in Indian scenario. Renew Sust Energ Rev 15(9):4797–4808
Surendra KC, Takara D, Hashimoto AG and Khanal SK (2014) Biogas as a sustainable energy source for developing countries: Opportunities and challenges. Renew Sust Energ Rev 31:846–859
Srivastava V, Singh P, Ismail SA, Singh RP (2014) Urban solid waste management in the developing world with emphasis on India: challenges and opportunities. Rev Environ Sci Biotechnol. doi:10.1007/s11157-014-9352-4
Suryawanshi P, Chaudhari A, Kothari R (2010) Thermophilic anaerobic digestion: the best option for waste treatment. Crit Rev Biotechnol 30(1):31–40
Sutton PM, Huss DA (1984) Anaerobic fluidized bed biological treatment: pilot to full-scale demonstration. In: Proceedings of the water pollution control federation conference

Takashima M, Shimada K, Speece RE (2011) Minimum requirements for trace metals (iron, nickel, cobalt, and zinc) in thermophilic and mesophilic methane fermentation from glucose. Water Environ Res 83(4):339–346

Troschinetz AM and Mihelcic JR (2009) Sustainable recycling of municipal solid waste in developing countries. Waste Manag 29(2):915–23

Talyan V, Dahiya RP, Sreekrishnan TR (2008) State of municipal solid waste management in Delhi, the capital of India. Waste Manage 28:1276–1287

Tejada M, Garcia C, Gonzalez JL, Hernandez MT (2008) Use of organic amendment as a strategy for saline soil remediation: influence on the physical, chemical and biological properties of soil. Soil Biol Biochem 38:1413–1421

Thauer RK, Kaster AN, Seedorf H, Buckel W, Hedderich R (2008) Methanogenic archae: ecologically relevant differences in energy conservation. Nat Rev Microbiol 6:579–591

TERI (2013) The Energy Research Institute Governance in Coal Mining: Issues and Challenges. TERI-NFA Working Paper No. 9

United Nation Report (2014) World urbanization prospects

Unnikrishnan S, Singh A (2010) Energy recovery in solid waste management through CDM in India and other countries. Resour Conserv Recy 54(10):630–640

Uslu A, Faaij APC, Bergman PCA (2008) Pre-treatment technologies, and their effect on international bioenergy supply chain logistics. Techno-economic evaluation of torrefaction, fast pyrolysis and pelletisation. Energy 33(8):1206–1223

Vaish B, Singh P, Kothari R, Srivastava V, Singh PK, Singh RP (2016) The Potential of Bioenergy Production from Marginalised Lands and Its Effect on Climate Change. Climate Change and Environmental Sustainability 4(1):7–13

Vaish B, Singh P, Srivastava V, Singh PK, Singh RP (2016b) Municipal Solid Waste Management in India: Present Status and Energy Conversion Opportunities. Book: Emerging Energy Alternatives for Sustainable Environment. Eds. D P Singh, Richa Kothari and V V Tyagi. ISBN: 9788179934111

Vaish B, Srivastava V, Singh P, Singh A, Singh PK, Singh RP (2016c) Exploring untapped energy potential of urban solid waste. Energ. Ecol. Environ. DOI 10.1007/s40974-016-0023-x

Vergara SE, Tchobanoglous G (2012) Municipal solid waste and the environment: a global perspective. Ann Rev Environ Resour 37:277–309

Verma S (2002) Anaerobic digestion of biodegradable organics in municipal solid wastes. Columbia University, Master's thesis, Applied Science Columbia University, New York, NY, USA, pp 56

Vij D (2012) Urbanization and solid waste management in India: present practices and future challenges. In: Procedia—social and behavioral sciences. international conference on emerging economies—prospects and challenges (ICEE-2012), 37, pp 437–447

Wang KS, Chiang KY, Tsai CC, Sun CJ, Tsai CC, Lin KL (2001) The effects of FeCl3 on the 842 distribution of the heavy metals Cd, Cu, Cr, and Zn in a simulated multimetal incineration system. Environ Int 26:257–263

Wang X, Yang G, Feng Y, Ren G and Han X (2012) Optimizing feeding composition and carbon–nitrogen ratios for improved methane yield during anaerobic co-digestion of dairy, chicken manure and wheat straw. Bioresour Technol 120:78–83

Weiland P (2010) Biogas production: current state and perspectives. Appl Microbiol Biot 85 (4):849–860

Wikstrom E, Marklund S (2000) Secondary formation of chlorinated dibenzo-p-dioxins, dibenzofurans, biphenyls, benzenes, and phenols during MSW combustion. Environ Sci Technol 34:604–609

World Bank (2008) Secured landfills: the bucket at the end of the solid waste management chain

World Bank (2012a) http://www.worldbank.org/en/news/feature/2012/06/06/report-shows-alarming-rise-in-amount-costs-of-garbage

World Bank (2012b) What a waste: a global review of solid waste management. Urban development series knowledge papers

WRAPAI (ed) (2009) Document and data management, appendix S 06-energy research. Waste Refinery Australia Project Association Incorporated, Australia

www.wds.worldbank.org/external/default/WDSContentServer/WDSP/IB/2012/09/06/000425962_20120906103012/Rendered/PDF/722190WSP0Box30ured0Landfills0Oct08.pdf

Young JC, McCarty PL (1969) The anaerobic filter for waste treatment. J Water Poll Cont Fed 41(5):160–173

Zinder SH, Koch M (1984) Non-acetoclastic methanogenesis form acetate: acetate oxidation by a thermophilic syntrophic coculture. Arch Microbiol 138:263–273

Status and Prospects of Municipal Solid Waste to Energy Technologies in China

Suyun Xu, Hongfu He and Liwen Luo

Abstract As the world's largest developing country, China creates considerable quantities of municipal solid waste (MSW) every day, which is one of the most serious urban pollution sources. Waste to energy can not only reduce greenhouse gas emission from simple piling of solid waste, but also can generate energy to cope with the increasing demand on fossil fuel. So far, landfill gas-fired power generation, MSW incineration and anaerobic digestion are the primary waste to energy technologies successfully applied in China. In recent years, MSW incineration power generation technologies have undergone rapid development with the demand for a low carbon economy and the encouragement of national policies. The distribution and operation status of various waste to energy facilities built in China are assessed. Meanwhile, the limitations and potential development trend of landfill, incineration and anaerobic digestion are discussed. In addition, a series of preferential policies and regulations to encourage the expansion of MSW to energy is presented.

Keywords Anaerobic digestion · Incineration · Landfill gas-fired power generation · Municipal solid waste · Waste to energy · Solid waste management

1 Introduction

As the largest developing country in the world, China creates considerable quantities of municipal solid waste (MSW), which is one of the most serious urban pollution sources. China is faced with serious environmental and administrative challenges caused by MSW management. MSW to energy is a novel eco-friendly renewable energy resource and has attracted the attention of both national and local governments with various preferential policies. Not only can MSW to energy contribute to a significant reduction in greenhouse gas emissions caused by storeyards and landfills,

S. Xu (✉) · H. He · L. Luo
School of Environment and Architecture, University of Shanghai for Science and Technology, Shanghai 200093, China
e-mail: xusy@usst.edu.cn; xusuyun1127@gmail.com

O.P. Karthikeyan et al. (eds.), *Recycling of Solid Waste for Biofuels and Bio-chemicals*, Environmental Footprints and Eco-design of Products and Processes, DOI 10.1007/978-981-10-0150-5_2

but it can also generate clean energy to offset the increasing energy requirements. Waste-to-energy (WTE) incineration recovers energy from MSW and produces electricity and/or steam for heating, which is recognized as a renewable source of energy and is playing an increasingly important role in MSW management in China. Anaerobic digestion is a method engineered to decompose organic matter by a variety of anaerobic microorganisms under oxygen-free conditions. The end product of anaerobic digestion includes biogas (60–70 % methane) and an organic residue rich in nitrogen. This technology has been successfully implemented in the treatment of agricultural wastes, food wastes, and wastewater sludge due to its capability of reducing chemical oxygen demand (COD) and biological oxygen demand (BOD) from waste streams and producing renewable energy (Chen et al. 2008).

China's MSW to energy development has the characteristics of late starting, large scale and rapid growth, so it urgently needs to present and analyze the development status and challenges of MSW to energy technologies in China in view of latest situations. In recent years, central and local governments have made great efforts to improve MSW management in China. New regulations and policies have been issued, urban infrastructure has been improved, and commercialization and international cooperation have been encouraged. Nevertheless, China still falls behind developed countries in MSW technologies. Therefore, more efforts on developing efficient MSW disposal technologies are in demand based on the assessment of MSW management. In this chapter, the distribution and operation status of various waste to energy facilities (i.e., landfill, incineration and anaerobic digestion) in China will be assessed, as well as their limitations and potential development trend. In addition, a series of preferential policies and regulations to encourage the expansion of MSW to energy is presented.

2 Current Status of MSW Production and Management in China

2.1 MSW Generation and Characteristics

The sustained growth of the Chinese economy, with its rapid urbanisation and improved living standards, has generated a large amount of MSW and a significant rise in total energy consumption. From 1980s to 2000s, MSW production in China was expanding rapidly (Huang et al. 2006). The quantity of MSW collected and transported in 1981 was 26.1 million tons; in 2002, 136.5 million tons of MSW was handled, which was 3.2 times more than 1981. Nevertheless, the growth rate was slowing down in recent ten years, with annual waste production growing less than 10 % per year, resulting 172.4 million tons of MSW being handled in 2013. Figure 1 shows the situation of MSW management in China from 2004 to 2013 (China 2003–2014).

MSW management is a systematic project that includes collection, transportation, transferring, treating and recycling. Chinese MSW management began in the

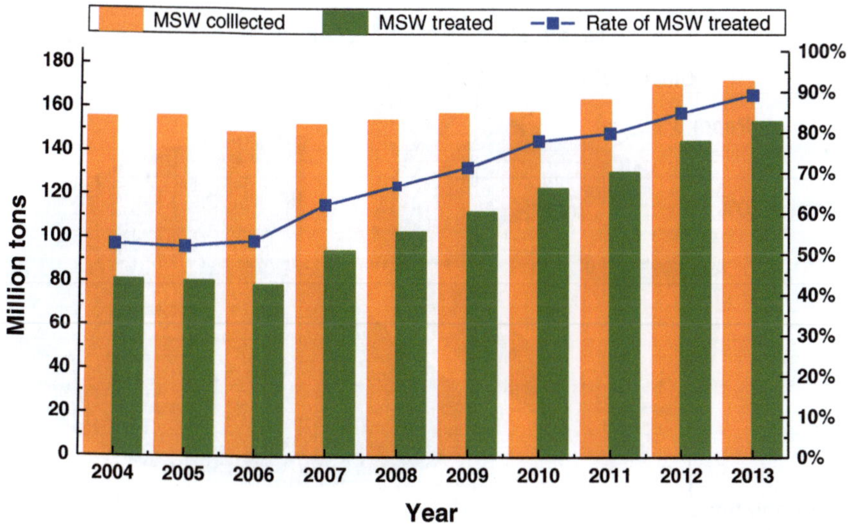

Fig. 1 The situation of MSW management in China from 2004 to 2013 (China 2003–2014)

late 1980s; prior to the adoption of this management, MSW was placed in open dumps. While the disposal rate of MSW was less than 2 % before 1990, the level of MSW disposal was gradually improved in the 1990s. By 1999, there were 696 MSW disposal plants with the disposal rate of 63.4 %, and less than 200 harmless treatment plants with a harmless treatment rate of 20.3 % in 668 Chinese cities.

In China, the MSW is generally divided into organic matter, inorganic matter, paper, fiber, timber bamboo, plastic, rubber, glass and metal (Li et al. 2001). The respective composition of MSW after entering the municipal recycling system is shown in Fig. 2. Characteristics of waste generation from several OECD countries are listed in Table 1. It can be seen that the the proportion of organic waste reaches as high as 59 %, which leads to the high moisture content of MSW, typically around 20–30 % in the U.S. and European countries (Cheng et al. 2007; Hu and Cheng 2013).

The composition of MSW is influenced mainly by the city size, geography circumstance, habit and living standard of the residence and fuel type people use. The quantity of paper, plastics, and glass is relatively small when compared with western countries, because most of the recyclable paper and glass are collected before entering the MSW management system. The calorific values (3000–6700 kJ/kg) of Chinese MSW are typically less than those of the developed countries (8400–17,000 kJ/kg), which are mainly composed of sorted organic wastes (Cheng and Hu 2010; Thipse et al. 2001). The low calorific value of the mix collected MSW is mainly caused by food remnants, resulting in incineration difficulties and applicable landfill sites. Nevertheless, the proportion of plastics is increasing, which is related to the increased use of plastic package materials (Huang et al. 2006).

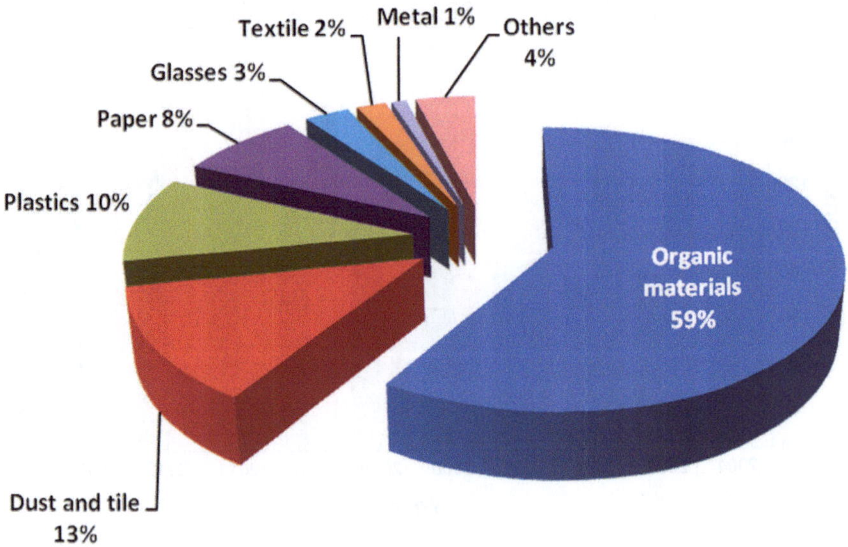

Fig. 2 The characteristic of MSW in China 2008, in dry weight, % (China 2014)

Table 1 OECD waste generation and disposal data (World Bank 2005; OECD library)

Country	Year	% of MSW					
		Organic materials	Paper and cardboard	Plastic	Glass	Metals	Textile and others
USA	1999	23	38	11	5	8	15
Japan	1999	34	33	13	5	3	12
Korea	2000	25	26	7	4	9	29
Germany	2000	23	41	3	22	8	3
France	1997	29	25	11	13	4	18
Denmark	2000	33	21	0.5	5	2	38
Australia	2000	50	22	7	9	5	8
Mexico	2000	52	14	4	6	3	20
Turkey	1997	64	6	3	2	1	24

2.2 *Source-Separated Collection and Transportation of MSW*

Utilizing source-separated collection is one of the key steps in MSW management. Source-separated collection begins at the source of MSW and involves the whole process of collection, transportation, disposal and recycling. The source-separated collection enables better waste minimization, resource utilization and hazardous waste disposal.

The principle underlying MSW source-separated collection requires that large volume wastes and hazardous wastes are first separated, and then the remaining MSW are classified in detail. The classification method of MSW applied in the collection system of eight megacities are presented in Table 2. As can be seen, all cities have followed relevant national standards. Although several differences existed among these cites, most of the cities share a similar philosophy: recyclables should be separated beforehand, while hazardous waste such as batteries and light tubes should be separated during the collection. Residents in Beijing, Xiamen, Shenzhen and Hangzhou are encouraged to classify waste as kitchen waste, recyclables, hazardous waste and other waste; while Shanghai is encouraging the use of a four-category classification: recyclables, hazardous waste, wet waste and dry waste.

It is worth pointing out that during the past years, Chinese cities have been consistently adjusting and refining the classification according to their different city conditions. For example, at the end of 20th century, sanitary landfill was the primary MSW disposal method in Shanghai, thus the separation of hazardous waste, such as battery and old lamp tube, from the buried waste is necessary. After that, the

Table 2 MSW source-separated classification in the eight cities, China. Adapted from Tai et al. (2011)

No	City	MSW source-separated classification	Current conditions
1	Beijing	Residential waste (RW): recyclables, kitchen waste, other waste Catering waste: recyclables, kitchen waste, other waste Institutional waste (IW): recyclables, other waste Village waste: ash, compostable waste, recyclables, hazardous waste, other waste	Partly implemented
2	Shanghai	RW: hazardous waste, recyclables, wet waste and dry waste IW: hazardous waste, recyclables, other waste Public places: recyclables, other waste	Partly implemented
3	Guangzhou	RW: recyclables, hazardous waste, bulky waste, other waste IW: plastic bottles, paper, retort pouch (TetraPak), other waste	Partly implemented
4	Shenzhen	RW: kitchen waste, non-kitchen waste, bulky waste, hazardous waste Commercial areas, road and public places: recyclables, non-recyclables, bulky waste, hazardous waste	No implementation
5	Hangzhou	RW: dry waste, wet waste, kitchen waste, non-kitchen waste Road and public place: recyclables, non-recyclables	Hardly implemented
6	Nanjing	Recyclables, non-recyclables, hazardous waste	No implementation
7	Xiamen	Recyclables, hazardous waste, other waste	No implementation
8	Guilin	Recyclables, kitchen waste, hazardous waste and other waste	No classified containers

development of incineration increased rapidly; because the fused glass would impose a hazardous effect on the incinerator, and the private sectors are not willing to recycle glass due to the low economic profit, Shanghai started to separate the glass from the household scraps. Thus for the service area of incineration, the MSW is classified as combustible materials, glass and hazardous waste; while in other service areas, waste is classified as organic waste, inorganic waste and hazardous waste. In 2014, Shanghai started to implement the "Method for MSW Classification and Reduction" and adopted a four-category classification: recyclables, hazardous waste, wet waste and dry waste.

3 Overview of Waste to Energy Technologies Status in China

Traditional integrated MSW management plans have focused on decreasing the amount of material that must be disposed of via incineration or landfilling. More recently, "zero waste" strategies have come to the fore, emphasizing prevention and materials recovery but also sharpening the focus on energy recovery as an approach for securing additional environmental benefits, including reductions in land use and emissions. Figure 3 displays an integrated sustainable solid waste management hierarchy, with environmental efficacy declining from top to bottom. The hierarchy of waste management advocates that the best approach to waste management is to first and foremost try to reduce waste generation and separate potential recyclables at source to improve the quality of materials for reuse, including organics for composting or anaerobic digestion. The waste hierarchy leads to reduced quantities requiring transfer and disposal; extends landfill lifetimes; reduces Greenhouse Gas generation; provides valuable recovered resources (e.g., methane gas and compost).

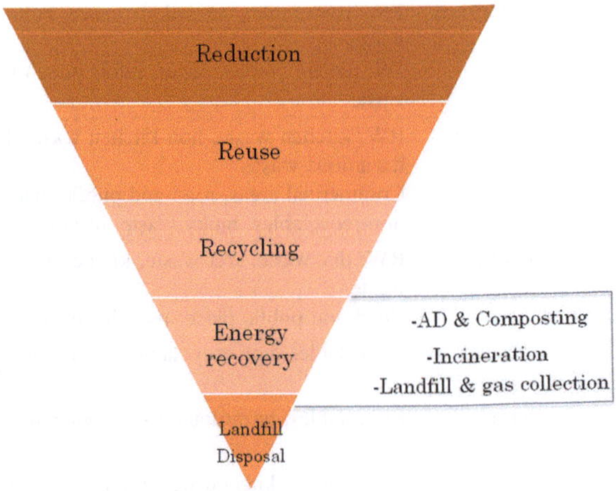

Fig. 3 Environmental hierarchy for solid waste management, revised from Themelis (2013)

In fact, waste-to-energy (WTE) technologies come in different forms, offer a variety of outputs, and are in various stages of development, but they have two common objectives: to both manage waste and generate energy. Digestion-based waste-to-energy technology can be deployed to extract useful energy from biodegradable organic materials and from landfill gas that is captured to reduce pollutant and greenhouse gas emissions (Kaufman et al. 2010). Incineration and advanced thermal conversion of the residual waste after recycling and composting represent environmentally sound MSW management options. Conventional combustion-based processes transform solid wastes into heat for direct use or further conversion into steam and electricity, while advanced conversion processes convert solids into gaseous or liquid fuels offering broader utility.

3.1 Landfill Gas-Fired Power Generation

3.1.1 Status of MSW Landfill in China

Landfill, incineration and composting are the three primary MSW disposal methods (Wang et al. 2009; Tian et al. 2013). In the late 20th century, landfill was recommended as the only option for waste management, because landfill is cost effective and easy to implement. However, it may lead to secondary pollution problems such as water and air pollution and soil contamination. Nowadays because of the change in the characteristics of waste, and the implementation of the 3R principle of waste management, i.e., waste reduction, reuse and recycle, composting and sanitary landfill are common and more suitable management practices (Tinmaz and Demir 2006). Nevertheless, waste disposal at landfill is still widely

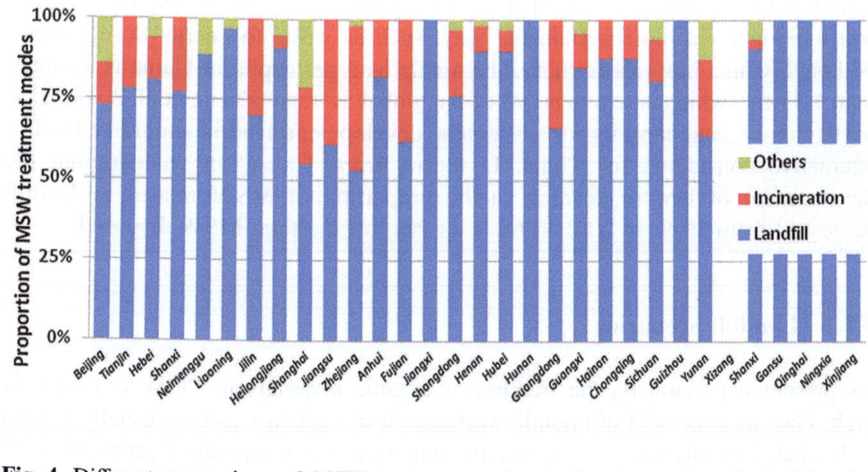

Fig. 4 Different proportions of MSW treatment modes in the perspective of administrative division in the year of 2013 (China 2014)

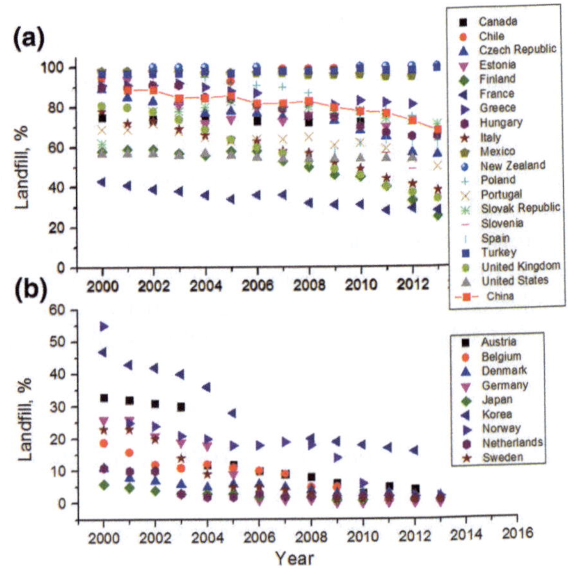

Fig. 5 Percentage of MSW disposed in landfill in China and OECD countries from 2000 to 2013 (OECD library)

applied in many cities in China, handling almost over 70 % of the total waste as shown in Fig. 4. Composting accounts for less than 10 %, which is classified into sub-type of "others" in Fig. 4. Composting, costly to implement and maintain, has become an unpopular method. Furthermore, the fertiliser, being of low nutrient contents and containing certain heavy metals, can only be used as a soil modifier.

As presented in Figs. 4 and 5, most of the Organization for Economic Cooperation and Development (OECD) countries tend to decrease the ratio of MSW disposed in landfill. Especially for Germany, all landfill sites were shut down in 2005. Other developed countries with limited land resource, i.e., Austria, Belgium and Sweden etc., are also vigorously decreasing MSW disposal capacity and looking for a more sustainable way, such as waste to energy development. Although China has vast territory, the arable land per capita in China is no more than 1.33 m^2, which is less than 1/2 of the global average (Zheng et al. 2014). Due to the infinite land capacity for sustainable development, landfill should be the last alternative considered for China. In recent years, China was encouraging the development of landfill, leading to the proportion of MSW treated in landfill decreased from 89 % in 2001 to 68 % in 2013 of the total MSW disposed.

3.1.2 Landfill Scenarios

As presented previously, the fraction of organic materials in MSW is relatively high. Poor management of organic waste (such as kitchen waste) is usually related to leachate percolation, causing serious sanitary and environmental problems such as unpleasant odors, the risk of explosion in landfill areas, as well as groundwater contamination (Mor et al. 2006). Necessary measures are required to cut down the emission of landfill gas (LFG) and leachate so as to prevent pollutions (MOHURD 2004, 2010). Larger sanitary landfill can reduce the cost of land requirement and environmental assessments, and can also be equipped with better pollution control facilities. Every sanitary landfill has designed carefully considered pollution prevention systems, such as leachate collection and treatment systems, gas collection and treatment systems, flood control systems and permeation prevention systems. For example, the Laogang Landfill Site in Shanghai has a daily capacity of 4900 tons, with an estimated service period of 45 years, which is equipped with leachate treatment and LFG collection facilities.

Yang et al. (2013) has summarized the landfilling technologies used in China into four scenarios. A brief overview of their main technical differences and the existing landfill situations in China, are summarized in Table 3. The open dump (Scenario 1) represents a dump site with no measures to control leachate or gas (e.g., lining and cover systems) and no regular landfilling operation processes (e.g., Waste compaction). Such sites mainly exist in remote, under-developed regions. Since the isolation of waste from the environment is not well managed, residual oxygen within the waste body may induce considerable aerobic degradation of the waste. In a sanitary landfill with LFG collection (Scenario 2), liner and cover systems have been introduced as well as leachate collection and treatment systems, but there is no LFG extraction system. In such a system, the LFG can migrate through the top cover layer. This type of landfill exists in most small Chinese cities. Compared with Scenario 2, the sanitary landfill (Scenario 3) has an LFG collection and flare system. According to the current Chinese standard of landfill management (MOHURD 2004), LFG must be collected effectively and be flared if it cannot be utilized for energy recovery. Hence, Scenario 3 may be a favorable replacement for Scenarios 1 and 2 in the near future. The sanitary landfill with LFG utilization (Scenario 4) represents the most advanced landfilling technology presently used in China. In this case, LFG is collected and used for energy recovery (usually for electricity production) with all thc othcr measures being the same as those described for Scenario 2. Scenario 4 has now been applied to a number of big cities.

3.1.3 Use of Combined Heat and Power (CHP) Systems

Depending on the fuel and power generation option, extensive pretreatment may be required to remove siloxane, hydrogen sulfide, and other constituents with potential to cause corrosion, erosion, environmental control, and odor problems. Further cleaning and purification are necessary to achieve the quality required for injection of pipeline-quality renewable fuel in natural gas delivery systems.

Alternatively, combined heat and power (CHP) system can realize in situ utilization of LFG by converting it to electricity and heat. CHP is commonly used in European countries as it can increase the energy recovery efficiency of LFG. Fueled by electric industry deregulation, environmental concerns, unease over energy

Table 3 Gas and leachate generation in the four different landfilling technology scenarios (revised from Yang et al. 2013)

	Liner and cover	Leachate generation volume (L, m^3/t)	Leachate collection efficiency (%)	Methane conversion rate	LFG collection efficiency (%)	LFG treatment	Methane oxidation efficiency of cover layer (%)
Scenario 1	No	4.3 + 0.25	0	0.8	0	No	0
Scenario 2	Yes	2 + 0.25	40	1	0	No	10
Scenario 3	Yes	2 + 0.25	40	1	20	Flare	10
Scenario 4	Yes	2 + 0.25	40	1	50	Electricity production	10

security, and many other factors, interest in CHP cogeneration technologies for distributed heat and power generation has been growing. Methane can replace natural gas as the fuel source in CHP systems. However, such facilities need to be strategically placed at or near customer facilities to efficiently supply the heating needs.

3.2 *MSW Incineration*

3.2.1 Potential of MSW as Biofuel

As a fuel, MSW poses a number of challenges. It is produced on a distributed basis, and its composition is highly variable, including a mix of organic and inorganic constituents. Hazardous and toxic waste stream components pose health and safety risks. Low energy density and high moisture, chlorine, and ash content lead to handling, combustion, slagging and fouling, corrosion, and byproduct management issues (Cheng and Hu 2010). Incineration of materials with heating capacities less than 3344 kJ/kg is probably not economical.

Lightly processed, post-recycling MSW received at mass-burn WTE plants has a heating value in the range of 10,000–12,500 kJ/kg; while high-intensity processing refuse derived fuel (RDF) is more amenable to firing in fluidized-bed combustor (FBC) units and advanced thermal conversion systems, with heating values roughly ranging from 12,500 to 15,000 kJ/kg, offers the potential for high-rate co-firing in pulverized coal plants. Mechanical, magnetic, thermal, biological, and other techniques may be applied to isolate and process combustible fractions. A typical system configuration of the MSW presorting process is illustrated in Fig. 6, which consists of three major units: shredding, air classification, and screening (Chang et al. 1998). After screening, the lightest portion in the MSW with the size greater than 100 mm can be exactly identified as the fluff-RDF; while the outputs with particle size larger than 100 mm and between 25 and 100 mm can be used as alternative fuels in the incineration facilities.

After adding certain amount of combustion-supporting agent, sulfur fixing agent and antiseptic, RDF made from MSW can be packed as cubes or pellets for easy storage

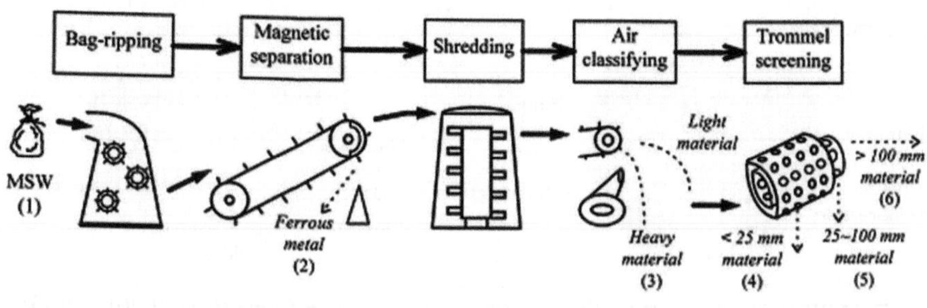

Fig. 6 Representative system of the MSW presorting process (Chang et al. 1998)

and transportation (Wei et al. 2009). It is not only convenient for storage and transportation, and can significantly improve the fuel performance, control pollution (Hernandez-Atonal et al. 2007; Jiang et al. 2008), has received widespread attention in waste energy reuse. Of course, installing and operating fuel-processing systems at the plant site imposes energy and cost penalties. Centralized manufacturing of higher-grade fuels offers potential economies of scale, while source-based production creates opportunities to reduce hauling costs and facilitate long-distance trade.

3.2.2 Incinerator Types and Process

The incinerator is the core of MSW incineration process, which cost accounts for approximately 50 % of the MSW incineration power plant. The technologies of its craft and design have a direct influence on MSW disposal effects and economic benefits, as well as a direct impact on the subsequent treatment of flue gases. There are various incinerators such as stoke grate incinerators, fluidised bed incinerators, rotary kiln furnaces and pyrolysis gasification furnaces.

Figure 7 shows the comparison of different MSW incinerators in China (CAEPI 2011). Stoke grate incinerators and fluidised bed incinerators predominate, while pyrolysis furnaces and rotary kiln furnaces are only adopted on a small scale. At present, most incineration facilities adopting mechanical stoke grate technologies are located in the more economically developed cities of eastern coastal areas, especially in the provincial capital and the sub-provincial cities. By contrast, incineration facilities using fluidised bed technologies are predominantly located in small and medium cities, as well as the large cities in the middle and western regions of China that are economically less developed. Comparatively, the cost of investment and operation of fluidised beds are relatively low. Furthermore, coal as the auxiliary fuel for fluidised beds, is abundant in central and western China. The facility costs for stoke grate incinerators vary between US$98 million and US$164 million per thousand tonnes daily of treatment capacity; while the costs for fluidised beds are merely half of that (CAEPI 2014). The operation and maintenance of stoke grate incinerators and fluidised beds are both costly.

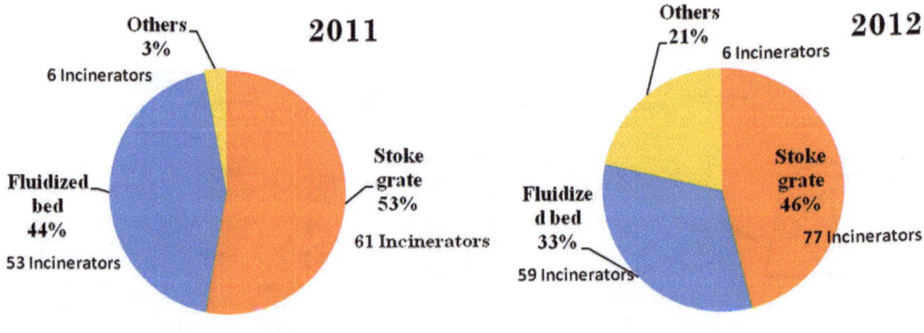

Fig. 7 MSW incinerators and total power generation capacity in China (CAEPI 2012, 2013)

It must be pointed out that the market share of stoke grate becomes increasingly higher than that of the fluidised bed. Total incineration capacity and total power generation capacity of stoke grate incinerator are 26.2, 31.5 and 32.6 % respectively from 2011 to 2013, while counterparts of fluidised bed incinerator are 11.3, 11.1 and 14.0 % respectively. There are 21 plants adopting stoke grate among the 28 newly built MSW incinerators in the year of 2013. The reason for this tendency is that the technology of fluidised bed is not as mature and stable as stoke grate; therefore, all advantages, such as complete combustion of native MSW through mixed unsorted collection, less dioxin emission, and no additional investment in sewage treatment etc., are not as good as expected. Due to this reason, the Chinese government is prone to support promotion of stoke grate at present, which is best illustrated by the first selection of "3A Selection of MSW Incineration Plants" in China lasting a whole year from March, 2012 to March, 2013. All of the MSW incineration plants adopting fluidised bed technology were unconditionally excluded for the selection, which represented the government's attitude and hindered the development of fluidised bed technology to a certain extent.

A MSW incineration power system generates electricity by driving turbines with high temperature steam produced by the incineration of MSW, as shown in Fig. 8. After transportation in closed trucks, MSWs were poured into a storage pool to ferment for approximately three days. The characteristics of MSW in China are unsorted coupled with low calorific values (3000–5000 kJ/kg) and high moisture rates (45–65 %), which organic ingredients account for 40–60 % of dry weight.

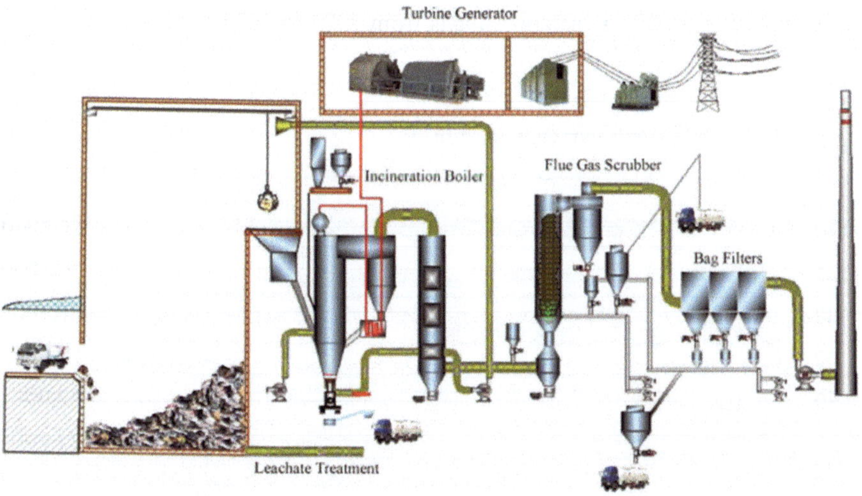

Fig. 8 Schematic diagram of MSW incineration and power generation process (Zheng et al. 2014)

This fermentation procedure could reduce the materials' humidity and increase their heating values. MSWs were then burned in incineration boilers to heat water to generate steam, which is the driving force of turbine generators. The flue gases and solid residues generated during the MSW incineration process should be treated accordingly to avoid secondary environmental pollution, especially the flue gases which contain significant amounts of dioxins, particulate matters, heavy metals, sulfur dioxide, and hydrochloric acid. The flue gases are first sent into a flue gas scrubber to remove acidic material, after which bag filters are used to remove dust particles so that the gas can meet the final emission standards. Fly ash, one of the flue gas residues, is a hazardous substance and should be dealt with in accordance with hazardous material waste laws.

3.2.3 Current Status and Development Trend of Waste Incineration in China

(1) **Rapid development of large scale waste incineration projects**
Although MSW incineration power generation in China is a recent development, the MSW incineration power generation technologies have undergone rapid development with the demand for a low carbon economy and the encouragement of national policies. Since the inception of the first incineration power plant in 1988, the number of MSW incineration plants has increased to 166 by the end of 2013, and the MSW incineration capacity has increased to 158488 tonnes/day. Figure 9 shows the daily processing capacity and the number of MSW incineration plants from 2003 to 2012 (China 2014).

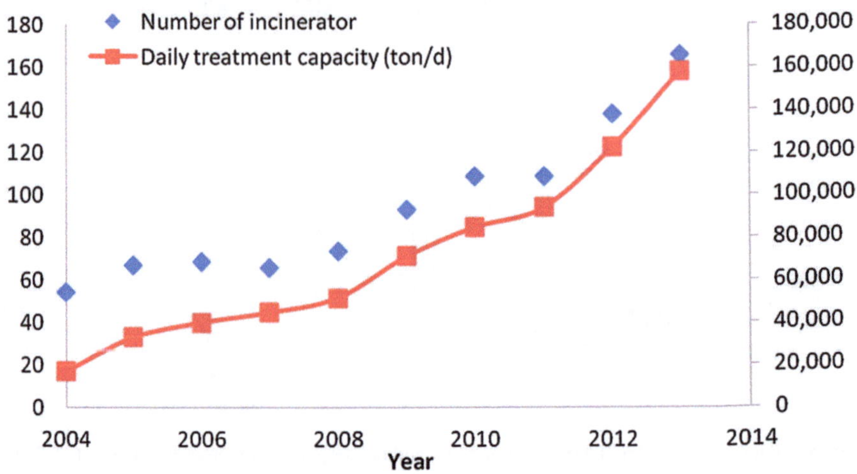

Fig. 9 The daily processing capacity and the number of MSW incineration facilities in China from 2004 to 2013 (China 2003–2014)

Table 4 MSW incineration power plants of significance in China (Zheng et al. 2014)

Year	Incineration capacity (ton/d)	Name	Installed capacity (MW)	Electricity parallel to grid (million kWh)	Type of incinerator
1988	300	Shenzhen Qingshui River MSW incineration plant	2 × 0.5	–	Stoke grate
2002	1000	Shanghai Pudong MSW incineration plant	2 × 8.5	100	Stoke grate
2011	2000	Shandong Jinan second MSW incineration plan	2 × 18	270	Stoke grate
2013	3000	Beijing Lujiashan MSW incineration plant	2 × 30	310	Stoke grate

In China, the development of MSW incineration power generation project tends to be large scale. Some significant MSW incineration power plants are shown in Table 4. In 2013, the Lujiashan incineration plant (Asia's largest MSW incineration power generation project) underwent trial operation and eventually supplied 320 million kWh of electricity to the grid per year. The first driving factor is the difficulty in selection a proper site for incineration plant, due to the "Not-In-My-Back-Yard" effect. Thus it is reasonable to build a large scale plant once the location is determined, which treatment capacity is up to 2000–3000 tonnes, sometimes even high up to 5000 tonnes. In addition, because MSW incineration requires massive funding and complex technologies, scale economy could ensure enterprises to obtain better economic benefits. More importantly, a larger scale MSW incineration tends to use more advanced technologies and will ultimately be better for the environment. Therefore, the Chinese government is vigorously promoting the construction of large incineration plants (Fig. 10).

As shown in Fig. 4, more MSW is treated by incineration in the eastern coastal areas of China, such as the provinces of Shanghai, Jiangsu, Fujian and Guangdong. Actually, more than 2/3 of the incineration power plants are concentrated in the Yangtze and Pearl River Delta regions; the rest are located in the middle and western provincial capital cities. Obviously, we can conclude that MSW incineration facilities currently operating in China are primarily located in cities that are economically developed and densely populated.

(2) **Disputes on waste incineration technology tend to be more rational**

"Whether a white cat or black cat, a good cat catches mice". In recent years, controversy over waste incineration technology has never stopped, one of the focus of the debate is entangled in the selection of stoke grate or fluidized bed.

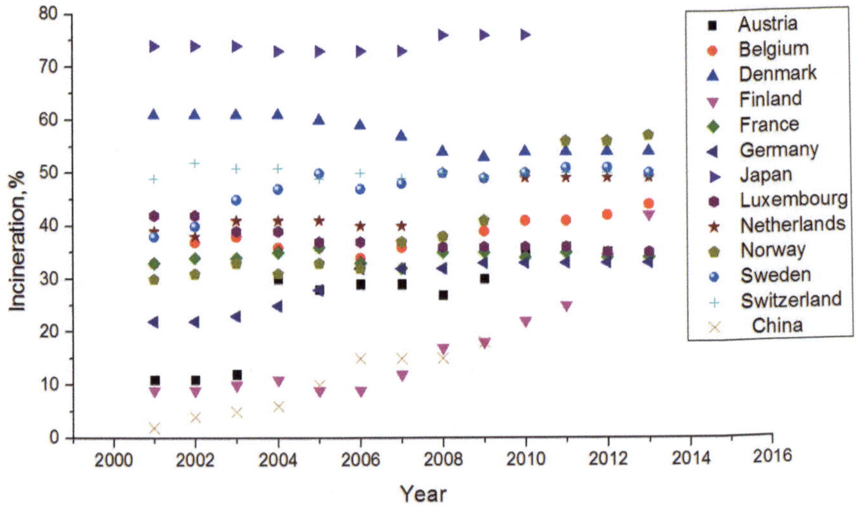

Fig. 10 The ratio of MSW treated by incineration in several OECD countries from 2001 to 2013 (OECD library)

As knowledge developed and practice experience accumulated, a rationale is gradually emerging for the selection of these two technologies.

Stoke grate has been developed for nearly a hundred years, which is more mature and stable than fluidised bed, thus currently dominating the Chinese market. In contrast, fluidised bed is being developed by domestic independent research with independent intellectual property rights to the technology, which gradually matures as improvement of more than ten years has been informed by practical experience. It is characterized by more fully and completely combustion, thus has good adaptability for the region acceptable to sludge and solid waste of low heat value.

(3) **Standards continue to improve**

In recent years, it is not rare to find that MSW incineration plants discharge pollutants beyond pollution limits. As the requirement of developing waste incineration technology and pollution control increased, the previous "Standard for pollution control on the municipal solid waste incineration (GWKB3-2001)" has been difficult to adapt to the new requirements. Revisions has been made and drafted to "Standard for pollution control on the municipal solid waste incineration (GB 18485-20)", which started to be implemented in July 2014. Furthermore, the provinces such as Beijing, Shanghai, Guangzhou have laid out local standards, which are stricter than the national standard. As per update of new pollution standards, some of the existing incinerators need to carry out technical renovation to improve the performance of end flue gas treatment facilities. In the long-term, it will be favorable for the development of the incineration industry in China.

3.3 Anaerobic Digestion

3.3.1 Advantages of Anaerobic Digestion Technology for MSW

With the increasing quantities of MSW, limited landfill areas and increased environmental impact awareness, alternative methods are being sought to treat municipal solid waste. Anaerobic digestion is attractive because it contributes to the solution of several important problems in China. Anaerobic digestion provides an environmentally friendly method for treating municipal solid waste. Landfills are currently the most common MSW disposal method in China (Fig. 4). In 2000, about 25 million m^3 of landfill leachate contaminated adjoining land and water sources while 17.6 million tons of undesirable CH_4 gas (Wang et al. 2001), which is a strong greenhouse gas, were produced by bio-waste deposited in landfills. In modern landfills, although improved liner and cover systems have been applied, some leakage is inevitable due to non-standard operation, leading to the releases of methane, volatile organic compounds, hazardous air pollutants, and leachate. Worldwide, societies are realizing that bio-waste needs to be disposed of in more environmentally friendly ways, such as waste digestion and methane gas recycling.

Anaerobic digestion will reduce land requirements for MSW disposal. 134.7 million tons of municipal solid waste was produced in China in 2001, and required about 500 million m^2 of land for disposal. With the increased urbanization in China, the cost of constructing and operating landfills will rapidly increase. Anaerobic digestion can substantially reduce the waste load on landfills. Meanwhile, anaerobic digestion can also produce useful fertilizer. An attractive option for treating the organic fraction of these wastes by anaerobic digestion is to compost the digestion sludge and then apply the stabilized residue on the soil as a fertilizer. Therefore, anaerobic digestion is the best technology for bio-waste disposal.

3.3.2 Status and Potential Assessment of Anaerobic Digestion Technology Development for MSW in China

Considerable success in using anaerobic technology for processing bio-waste is being reported by several recently constructed facilities in Europe. Although anaerobic digestion has been used in China for over 100 years, it is usually used to treat human and animal manure along with agricultural by-products. Compared with the large advanced anaerobic digestion plants in Europe, China still has a long way to go in developing effective bio-waste processing systems.

Introduction of Anaerobic Digestion Processes for MSW in China

Most anaerobic digestion plants used around the world can be divided into single phase, two-phase, and batch style plants. Batch reactors have economic advantages in developing countries, but their organic load rate is much lower than continuous feed systems and the reactors take up a larger area. Two-phase anaerobic digestion reactors have good shock load tolerance for the separation of acidification and methanation processes, but the technique is complex and relatively expensive. Single-phase reactors provide an acceptable result at less cost. Therefore, most recently built plants use single phase reactors. As listed in Table 5, most of the existing large-scale anaerobic digestion plants in China adopted single-phase digester. However, some plants setup one additional hydrothermal hydrolysis tank to facilitate efficient decomposition of biomass, such as the Suzhou Food Waste Treatment Plant (Jiangsu Province), Changchun Food Waste Treatment Plant (Jilin Province), and Shenzhen Municipal Organic Waste Treatment Plant.

New plants using high solid anaerobic digestion (with a total solid fraction of above 20 % compared to low solid plants with 4–8 % solid fraction) can improve energy production somewhat due to their high digestion efficiency and the process residues can be more easily dehydrated, which saves a great amount of water. Valorga and Dranco mode high solid anaerobic digestion plants have reactor volumes of less than 3300 m^3 and heights usually less than 25 m. However, high organic load rates and high salt and fat concentrations make high solid anaerobic digestion systems difficult to operate, so suitable control systems are needed for Chinese plants. Wet fermentation is the mainstream of the anaerobic treatment on MSW in biogas production currently with the advantage of high speed of fermentation, mature technology to construct and manage, and convenient input and output for the material.

Utilizing biogas to generate electricity via CHP system has become the major efficient way to use biogas. There is a large scale anaerobic digestion plant in Anyang, Henan designed to utilize biogas as vehicle fuel instead of electricity.

Factors Affecting MSW Digestion Efficiency

For anaerobic digestion technologies, the digestion process relies on anaerobic bacteria that break down organic materials into sugars, acids, and then gases, leaving behind liquid and solid residues. Decomposition occurs over years to decades in landfills and days to weeks in purpose-built digesters. However, the digestion efficiency is affected and limited by the characteristics of substrate.

There are great variations in the composition of MSW between developed countries and China. The disposal and treatment rates in China are much lower than those in developed countries. In China, organic matters in MSW is not sorted before disposal, typically, in landfills or in incineration plants, so most biogas plants are based on the digestion of animal manure and are built in the countryside using low efficiency and locally developed technology. Waste separation and collection are being developed in some big cities such as Beijing, Shanghai, and Shenzhen to enable the conversion of bio-waste into biogas containing methane and the separation of solid materials from anaerobic digestion processes for use as fertilizers. Furthermore, with the popularization of central heating and gas-fired heating systems, the proportion of coal residue in the MSW has decreased gradually while the proportion of organic waste has increased. Figure 1 shows that the main component of Chinese MSW is bio-degradable organic matters which includes kitchen waste,

Table 5 Large-scale anaerobic digestion plant for Food waste or separated MSW

Location	Substrate	Process	Scale (t/d)	Established time
Chongqing, Heishizi	Food waste	Anaerobic digestion	167, 1st stage 500, 2nd–3rd stages 1000, 4th stage	2012 2014 –
Lanzhou, Gansu	Food waste	Anaerobic digestion	200	2011
Ningbo, Zhejiang	MSW	Anaerobic digestion	200	2007
Sanming, Fujian	Food waste and waste oil	Anaerobic digestion	30	2009
Erdos, Inner Mongolia	Food waste	Anaerobic digestion	100	2010
Kunning, Yunnan	Food waste	Anaerobic digestion	200	2011
Beijing[a]	Food waste	Anaerobic digestion	150	2011
Dongcun, Beijing	Food waste	Anaerobic digestion	200	2012
Qingdao, Shandong	Food waste	Anaerobic digestion	200 600	2012 Under construction
Shenzhen[b]	Municipal organic waste	Hydrothermal hydrolysis-Anaerobic digestion	100	2011
Suzhou, Jiangsu[c]	Food waste	Hydrothermal hydrolysis-Anaerobic digestion	100, 1st stage 600, 2nd stage	2008 2012
Changchun, Jilin	Food waste	Hydrothermal hydrolysis-Anaerobic digestion	200	Under construction
Longgang, Shenzhen[d]	Food waste	Anaerobic digestion Two-stage CSTR	200	Under construction

(continued)

Table 5 (continued)

Location	Substrate	Process	Scale (t/d)	Established time
Anyang, Henan[e]	Biomass	Anaerobic digestion Two-stage CSTR	500	2010

[a]In Beijing Sanitation Group project, food waste is co-digested with faeces and sewage sludge
[b]The total capacity of Shenzhen Municipal Organic Waste (sludge) Treatment Plant is 500 t/d, including market vegetable garbage of 100 t, kitchen waste of 100 t/d, sewage sludge (water content 80 %) of 300 t. Wastes are pretreated with hydrothermal hydrolysis before feeding into anaerobic digester; after digestion, dehydrated cake is transferred into the composting system to produce organic fertilizer; odor generated in the entire system is processed by the biological deodorization system
[c]In Suzhou Food Waste Treatment Plant, the process of Hydrothermal hydrolysis—Anaerobic digestion is applied. Tsinghua University and Jiangnan university provide technical supports for thermal treatment and fermentation, respectively. The process of bio-diesel technology developed by our company, using a special catalyst to achieve a step catalytic production of bio-diesel oil
[d]The substrates of Shenzhen Longgang Food Waste Treatment Plant are food waste, sewage sludge and biodiesel production wastewater
[e]Anyang vehicle biogas project, is the first demonstration project for the industrial application of bio-gas, which is also the first application of biogas into vehicle. CSTR, continuous stirred tank reactor

hotel and restaurant waste, vegetable market waste, and garden waste (Jiang et al. 2007).

Since MSW includes many components, the ratio of carbon to nitrogen of the waste may not be suitable for anaerobic digestion, which will hinder efficient operation of the anaerobic digestion plant. Therefore, the various components in the waste stream must be collected separately so that the ratio of carbon to nitrogen can be adjusted to a suitable range by mixing the waste from different sources. In addition, the use of food residues as direct feedstuff is being limited by new regulations and transportation concerns in Chinese cities. Therefore, regulations are needed to facilitate separate collection systems for bio-waste with appropriate disposal technologies. If a bio-waste collection system can be set up and made compatible with existing collection systems, anaerobic digestion technologies for bio-waste disposal can be standardized. Since there are not yet any anaerobic digestion plants operating in any cities, an efficient design would be widely used throughout China.

Limitations to the Spread of Anaerobic Digestion Technology

(1) The capital investment for energy production systems may be somewhat higher for anaerobic digestion systems than for conventional systems, and electricity is fairly inexpensive in China, unlike in many other countries. As such, the development of anaerobic digestion technologies is currently limited.
(2) The costs of anaerobic digestion systems depend greatly on the local circumstances, including construction and labor costs, treatment capacity, the possibility of energy recovery, energy price, market, and taxes as well as the

energy purchase tariff, land prices, and the worth of the digested material. Although the equipment for high solid anaerobic digestion plants is expensive, the operating costs are relatively low considering their smaller sizes, higher digestion efficiencies and water savings.

With costs increasing for landfills and energy taxes on fossil fuels, anaerobic digestion should be encouraged as a renewable energy source. Anaerobic digestion will be a highly competitive alternative for the treatment of MSW in the near future.

4 The Prospects of MSW to Energy Technologies in China

Globally, more than 1 billion tons of post-recycling MSW continues to be disposed of in landfills each year, including more than 130 million tons in the U.S, while European, Asian, and other nations move forward with strong commitments to energy recovery. Global adoption of integrated resource management strategies could dramatically increase deployment of incinerators and advanced conversion technologies. This development would reduce landfilling and associated emissions of methane, while expanded landfill gas capture and energy production could further reduce the carbon footprint of waste management practices.

In recent years, China has issued a series of polices to promote the harmless treatment of MSW, including the development of WTE. The foremost influential policies are as follows,

(1) National 12th Five Year Plan (2011–2015)

- Reform garbage fee collection methods, appropriate increase garbage fee standards and the level of financial subsidies;
- Speed up the development of rural waste centralized treatment;
- Improve the recycling system of renewable resources; establish and improve waste separation and recovery system;
- Promote the resource utilization and harmless treatment of kitchen waste and other waste;
- In 2015, the rate of MSW harmless treatment in all cities rose to 80 %. It is clearly put forward that incineration technologies will be employed to handle more than 30 % of MSW, a huge increase over current practice (Fig. 11).

(2) Cleaner Production Law 2002
The Renewable Energy Law 2006. On February 28, 2005, the fourteenth conference of the National People's Congress (NPC) passed the "Renewable Energy Law of the People's Republic of China", effective on January 1, 2006, which is the first and only legislation providing strong legal protection, as well as phased support, for the development of renewable energy sources. As the cornerstone of the development of renewable energy, it boosted the proportion

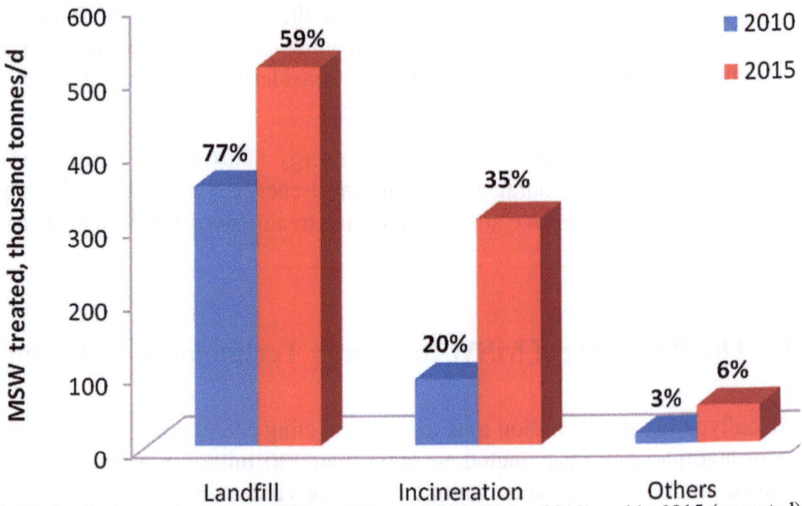

Fig. 11 Technologies used to treat MSW in China in 2010 (China 2010) and in 2015 (expected)

of renewable energy in the energy structure dramatically, particularly wind and photovoltaic energy. However, some imperfections in this law gradually appeared during the expansion of wind and solar energy. In order to provide better legal protection, it was amended on December 26, 2009 and the amendments affecting seven articles took effect on April 1, 2010. Owing to the prosperity later than wind and solar energy, MSW to energy could benefit more from the amendments, which ensure its benign development and avoid analogous problems occurring during the development of wind and solar energy. There are some amendments closely related to MSW to energy that should be pointed out. Firstly, the amendments reiterate purchase obligation (grid companies should purchase all the electricity generated by the MSW to energy generating entities), but put it within the overall framework of the national plan, suggesting that MSW to energy is viewed as an integral part of the total national energy resource. The amendments also established a renewable energy development fund financed by budgetary allocations from the central government and the collection of renewable energy fees charged throughout China, in which the amounts accumulated are slated for use as compensation to grid companies for the additional expenses they incur in purchasing renewable energy (as opposed to conventional energy), and for grid connection costs to facilitate the use of renewable energy, etc. (Renewable Energy Law of the People's Republic of China). Overall, the release of the law provides a broad path for the development of MSW to energy.

The economics of WTE incineration plants are extremely site-specific, depending on tipping fees, MSW characteristics, environmental regulations, byproduct management practices, and many other factors. WTE incineration installations often benefit from the investment and production tax credits granted to renewable energy sources. The economic viability of anaerobic digestion to energy (ADTE) installations is strongly influenced by policy drivers. Policies requiring control of air pollutant and greenhouse gas emissions from landfills, agricultural operations, and wastewater treatment plants improve economics by reducing the incremental cost of adding generating capacity. Depending on site-specific circumstances, these projects may also yield revenue streams in the form of marketable renewable energy certificates and carbon credits.

Market mechanisms are playing an increasingly prominent role in the Chinese economic system, and are also starting to be utilized to address certain environmental protection problems. In the area of MSW management, a handful of enterprises have entered the field; in late 1999, for example two private MSW enterprises were founded in Beijing. But generally speaking, outside of a few cities, the harnessing of market mechanisms is still not a significant part of MSW management in China, leaving local governments nationwide with the responsibility for and burden of MSW management. Typically, collection and transportation capability, as well as treatment plants, are paid for through direct government investment, and MSW management operations consistently draw on government subsidies. This leaves public sector MSW management entities without any profit incentive to rationalize operations and management.

China's economy is still underdeveloped, and in this context China must search for a least-cost path to achieve optimized MSW management through the promotion of a more sustainable municipal ecological model, and the use of economic incentives to improve MSW management. The primary method for resolving the growing MSW problem is to use the tools of environmental economics to design a rationalized system of MSW generation, collection, transportation, treatment and final disposal.

References

Chang Y-H, Chen WC, Chang N-B (1998) Comparative evaluation of RDF and MSW incineration. J Hazard Mater 58(1–3):33–45

Chen Y, Cheng J, Creamer K (2008) Inhibition of anaerobic digestion process: a review. Bioresour Technol 99(10):4044–4064

Cheng H, Hu Y (2010) Municipal solid waste (MSW) as a renewable source of energy: current and future practices in China. Bioresour Technol 101(11):3816–3824

Cheng H, Zhang Y, Meng A, Li Q (2007) Municipal solid waste fueled power generation in China: a case study of waste-to-energy in Changchun city. Environ Sci Technol 41(21):7509–7515

China, NBSO (2003–2014). In: NBSO China (ed) *China statistical year book 2003–2014*, China Statistical Press, Beijing

China Association of Environmental Protection Industry (CAEPI). China development report on urban domestic refuse treatment industry in 2011–2013. China Environmental Protection Industry, 2012–2014

Hernandez-Atonal FD, Ryu C, Sharifi VN, Swithenbank J (2007) Combustion of refuse-derived fuel in a fluidised bed. Chem Eng Sci 62(1–2):627–635

Hu Y, Cheng H (2013) Development and Bottlenecks of renewable electricity generation in China: a critical review. Environ Sci Technol 47(7):3044–3056

Huang Q, Wang Q, Dong L, Xi B, Zhou B (2006) The current situation of solid waste management in China. J Mater Cycles Waste Manage 8(1):63–69

Jiang J, Sui J, Wu S, Yang Y, Wang L (2007) Prospects of anaerobic digestion technology in China. Tsinghua Sci Technol 12(4):435–440

Jiang C, Huang L-W, Zhang X-M (2008) Pyrolysis analysis of RDF by TG-FTIR techniques. Environ Sci Technol 1:7

Kaufman SM, Krishnan N, Themelis NJ (2010) A screening life cycle metric to benchmark the environmental sustainability of waste management systems. Environ Sci Technol 44(15):5949–5955

MOHURD (Ministry of Housing and Urban-Rural Development of China) (2004). Technical Code for Municipal Waste Sanitary Landfill (CJJ 17-2004)

MOHURD (Ministry of Housing and Urban-Rural Development of China) (2010). Technical Code for Leachate Treatment of Municipal Solid Waste (CJJ 150-2010)

Mor S, Ravindra K, De Visscher A, Dahiya R, Chandra A (2006) Municipal solid waste characterization and its assessment for potential methane generation: a case study. Sci Total Environ 371(1):1–10

OECD library, Retrived on August 2015 from http://www.oecd-ilibrary.org/environment/data/oecd-environment-statistics/municipal-waste_data-00601-en

Tai J, Zhang W, Che Y, Feng D (2011) Municipal solid waste source-separated collection in China: a comparative analysis. Waste Manag 31(8):1673–1682

Themelis NJ (2013) Waste-to-Energy, Introduction. In: Kaltschmitt M, Themelis NJ, Bronicki LY, Söder L, Vega LA (eds) Renewable energy systems. Springer New York, New York

Thipse SS, Sheng C, Booty MR, Magee RS, Dreizin EL (2001) Synthetic fuel for imitation of municipal solid waste in experimental studies of waste incineration. Chemosphere 44(5):1071–1077

Tian H, Gao J, Hao J, Lu L, Zhu C, Qiu P (2013) Atmospheric pollution problems and control proposals associated with solid waste management in China: a review. J Hazard Mater 252–253:142–154

Tınmaz E, Demir I (2006) Research on solid waste management system: to improve existing situation in Corlu town of Turkey. Waste Manag 26(3):307–314

Wang W, Han F (2001) Study on methane generation from landfills in China. China Biogas, 19 (2):20–24. (in Chinese)

Wang L, Pen T, Huang C, Yuan H (2009) Management of municipal solid waste in the three Gorges region. Waste Manag 29(7):2203–2208

Wei X, Wang Y, Liu D, Sheng H, Tian W, Xiao Y (2009) Release of sulfur and chlorine during cofiring RDF and coal in an internally circulating fluidized bed. Energy Fuels 23(3):1390–1397

World Bank, Waste management in China: issues and recommendations, 2005. Retrieved on August 2015 from http://siteresources.worldbank.org/

Yang N, Zhang H, Shao LM, Lü F, He PJ (2013) Greenhouse gas emissions during MSW landfilling in China: influence of waste characteristics and LFG treatment measures. J Environ Manage 129:510–521

Zheng L, Song J, Li C, Gao Y, Geng P, Qu B, Lin L (2014) Preferential policies promote municipal solid waste (MSW) to energy in China: current status and prospects. Renew Sustain Energy Rev 36:135–148

GHGs Emissions and Sustainable Solid Waste Management

Suthirat Kittipongvises and Chongrak Polprasert

Abstract In the 21st century, global warming and climate change are among the greatest environmental challenges and humanitarian crisis. Globally, annual greenhouse gas (GHGs) emissions from solid waste disposal sites is estimated to be approximately a quarter of total anthropogenic methane emission. Integrated solid waste management, therefore, provides significant opportunities to control environmental pollution and minimize the negative impacts of global climate change. This chapter illustrates the current status of global GHGs emission in relationship with population growth and solid waste generation. Mathematic models used to quantify GHGs generated from the waste sector as the zero-order model (i.e., SWANA, German EPER and IPCC Default Method) and the first-order model (i.e., TNO, LandGEM, IPCC First-Order Decay; FOD) are explained including application to certain inventory in selected countries. Life Cycle Assessment (LCA), which involves the cradle-to-grave concept, environmental burden from global warming and selected case studies are described and applied to assess GHGs emissions from various solid waste management options such as recycling, composting, sanitary landfilling, anaerobic digestion, incineration, mechanical biological treatment (MBT), source reduction, and utilization and application of biochar. Existing solid waste management practices and innovative options to achieve GHGs mitigation and community adaptation including resiliency are presented. Lessons learned and best practices in solid waste management from Thailand (i.e., Bangkok Kamphaeng Sean West: Landfill Gas to Electricity Project) and from other countries (i.e., GHGs mitigation project: MBT plant in Gaobeidian, Hebei province, People's Republic of China; municipal solid waste composting project in Ikorodu, Lagos State, Federal Republic of Nigeria; and gasification, landfill gas and anaerobic digestion in Bali, Indonesia) are further discussed.

S. Kittipongvises (✉)
Center of Excellence on Hazardous Substance Management, Environmental Research Institute Chulalongkorn University, Bangkok, Thailand
e-mail: suthirat.k@chula.ac.th; suthirat.k@gmail.com

C. Polprasert
Faculty of Engineering, Thammasat University, Bangkok, Thailand

O.P. Karthikeyan et al. (eds.), *Recycling of Solid Waste for Biofuels and Bio-chemicals*, Environmental Footprints and Eco-design of Products and Processes, DOI 10.1007/978-981-10-0150-5_3

Keywords Climate change · Emission · Greenhouse gases (GHGs) · Integrated solid waste management · Life cycle assessment (LCA) · Sustainability

1 Introduction

1.1 Post-2015 Development Agenda: Population and Urbanization Growth, Global Climate Change and Solid Waste Problems

1.1.1 Rapid Population Growth and Associated GHG Emission

Currently, the global population has reached approximately 7.2 billion in 2014 and is projected to increase by more than 2 billion by 2050 (Fig. 1a). By comparison, the size of the rural population of the world remained essentially unchanged between 1994 and 2014. On the one hand, more than half of the global population now lives in the urban areas (Fig. 1b) (United Nations Department of Economic and Social Affairs, Population Division 2014).

The continuation of population and urbanization growth presents serious challenges for achieving all internationally agreed development goals. On this basis, according to the Millennium Summit of the United Nations (UN) in 2000, all UN member states and international organizations have committed their actions to help achieve the following Millennium Development Goals (MDGs) by 2015: *Goal 1* Eradicate extreme hunger and poverty, *Goal 2* Achieve universal primary education, *Goal 3* Promote gender equity and empower women, *Goal 4* Reduce child mortality, *Goal 5* Improve maternal health, *Goal 6* Combat HIV/AIDs, Malaria and other diseases, *Goal 7* Ensure environmental sustainability, and *Goal 8* Develop a

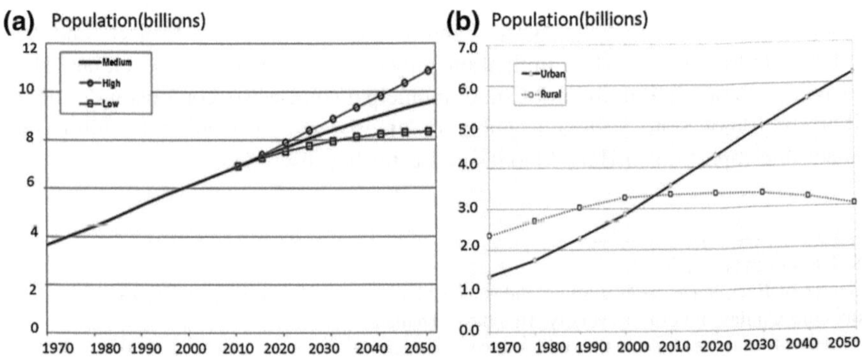

Fig. 1 **a** Projections of total global population, 1970–2050 (medium-, high- and low-fertility variants for 2015); **b** Urban and rural population of the world, 1970–2050 (United Nations Department of Economic and Social Affairs, Population Division 2014)

global partnership for development. However, as the 2015 deadline for the MDGs approaches, the international community is now actively engaged in various discussion forums to help define Sustainable Development Goals (SDGs) that will serve as the core of post-2015 global development agenda (UN 2014). There is a broad agreement on the need for close linkages between MGDs and integration of social, economic and environmental dimensions of sustainable development.

As previously detailed, increase in the number of the global population, combined with poverty, excessive resource consumption and wasteful production patterns cause or exacerbate environmental pollution and thus inhibit sustainable development. The environmental problems facing mankind in the 21st century are: global climate change, depletion of natural resources and destruction of our ecosystem. These crises are interrelated and connected to waste and waste management (Tanaka 2010). The details are given below:

Global climate change is one of the clearest manifestations of sustainability challenge of our time. Climate fluctuation is an issue that involves multiple interactions between global, social and human dimensions (Komiyama and Takeuchi 2006). The Intergovernmental Panel on Climate Change (IPCC WG2 2007) found global climate change to be the result of anthropogenic activities, particularly (i) carbon dioxide (CO_2) from fossil fuels burning, (ii) methane (CH_4) from agricultural activities, waste and wastewater management, (iii) nitrous oxide (N_2O) from agricultural activities such as fertilizer utilization, (iv) fluorinated gases (F-gases) from industrial progresses, etc. The results are higher greenhouse gas (GHGs) concentrations in the atmosphere which absorb heat from the Earth surface and consequently causing global warming and the associate frequent occurrence of extreme climate events. Globally, as shown in Fig. 2a, anthropogenic GHG emissions have risen since the late 19th century, with an increase approximately 35 % between 1990 and 2010 (EPA 2014; WRI 2014; FAO 2014).

Geologically, the majority of GHG emissions come from the following three regions: Asia, Europe, and the United States (Fig. 2b), which together accounted for 82 % of total global emissions in 2011 (Fig. 2b) (EPA 2014; WRI 2014). Furthermore, IPCC WG2 (2007) indicated that projections of total GHG emissions could increase by 25–90 % from 2000 to 2030.

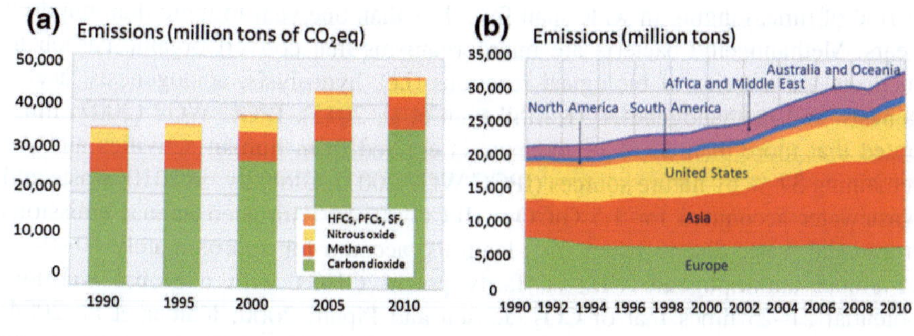

Fig. 2 **a** Type of global GHGs emissions, 1990–2010; **b** Global CO_2 emissions by region, 1990–2010 (EPA 2014; WRI 2014; FAO 2014)

The impacts of global climate change and extreme weather events such as storms, floods, droughts, and the rise of sea level are also observed (IPCC WG2 2007). It is acknowledged that climate change threatens biodiversity, ecosystem functions and the basic elements of life for people around the world such as: access to water, food production, human health, and use of land and the environment (Stern 2007). IPCC defined vulnerability of people as their propensity to be harmed by their exposure to hazards or stresses including climate stress (IPCC WG2 2007). Vulnerability to climate change, in a sense, will vary with adaptive capacity, social resilience or ability of a system to adjust to climate variability and extremes, and socio-economic pathway. Therefore, climate change issues should be of immediate concern and must be addressed at all levels in order to mitigate and for people to adapt to any changes that might occur.

1.1.2 Solid Waste and GHGs Emissions

In the rapidly urbanizing global society and expanding industries, solid waste management, particularly the issue of collection, management and disposal of solid wastes, represents the key challenges facing all the world's cities. Globally, the most recent global estimate (in 2012) of municipal solid wastes indicates approximately 1.3 billion tons is generated every year. This volume is expected to increase to 2.2 billion tons per year by 2025 (Hoornweg and Bhada-Tata 2012). The OECD countries make up almost half of the world's waste, while the people of Africa and South Asia generate the least amounts of waste being, 0.65 and 0.45 kg/capita/day, respectively (Table 1). In terms of disposal, landfill is the most popular method currently being used in the world (Hoornweg and Bhada-Tata 2012).

Given that climate change is a global concern, every waste management practice generates GHGs, both directly (i.e., emissions originate from the process itself) and indirectly through energy consumption (UNEP 2010a, b). The major GHGs from the waste sector are methane (CH_4) emissions from landfill site and, secondarily, CH_4 and nitrous oxide (N_2O) emissions from wastewater treatment operations (Bogner et al. 2007). By this logic, when solid waste is disposed in landfills, most organic matter will be degraded as a result of biological metabolism over a longer period of time, ranging in wide span from less than one year to more than hundred years. Methanogenic bacteria are microorganisms that convert organic carbon to methane via a series of biological reactions (i.e., hydrolysis, acidogenesis, acetogenesis, and methanogenesis) (Karthikeyan et al. 2015). IPCC WG2 (2007) illustrated that more than 63 % of methane is emitted from human activities and the remaining 37 % by nature sources (IPCC WG2 2007). Globally, in 2010, waste and wastewater accounted for 1.5 $GtCO_2e$ (IPCC 2014). Estimated annual emissions from solid waste disposal sites (i.e., landfills) account for approximately 10–19 % of global anthropogenic CH_4—a fairly potent GHGs with a global warming potential 21–25 times that of CO_2 (Jensen and Pipatti 2000; Kumar et al. 2004; Forster et al. 2007). Apparently, regarding to global GHGs emissions, CH_4 emission from solid waste disposal sites almost doubled during between 1970 and 2010.

Table 1 Global waste generation projections for 2025, by region (Hoornweg and Bhada-Tata 2012)

Region	Current available data			Projections for 2025			
	Total Urban population (millions)	Urban waste generation		Projected population		Project Urban waste	
		Per capita (kg/capita/day)	Total (tons/day)	Total population (millions)	Urban population (millions)	Per capita (kg/capita/day)	Total (tons/day)
Africa region	260	0.65	169,119	1152	518	0.85	441,840
East Asia and Pacific region	777	0.95	738,958	2124	1229	1.5	1,865,379
Europe and central Asia region	227	1.1	254,389	339	239	1.5	354,810
Latin America and the Caribbean region	399	1.1	437,545	681	466	1.6	728,392
Middle east and north Africa region	162	1.1	173,545	379	257	1.43	369,320
Organisation for economic co-operation and development	729	2.2	1,566,286	1031	842	2.1	1,742,417
South Asia region	426	0.45	192,410	1938	734	0.77	567,545
Total	2980	1.2*	3,532,252	7644	4285	1.4*	6,069,703

*Remark: Average value

Global waste emissions per capita also increased by approximately 5 % for the entire period 1970–2010 (IPCC 2014; JRC/PBL 2013).

In response to the problems of climate change, an important key focus of waste management activities is to reduce GHGs concentrations in the atmosphere. Particularly, climate benefits of waste management practices results from avoided landfill emissions, reduced raw material extraction, recovered materials and energy replacing virgin materials and fossil fuel energy sources, carbon storage due to recalcitrant materials in landfill disposal sites. Despite its importance, however, the direct measurement and monitoring of methane emission from solid waste disposal sites are still challenging tasks for concerned stakeholders.

1.2 Objectives and Scope

The purpose of this chapter is to highlight the relationship between waste and global climate change. Methods used to quantify GHGs generated from the waste sector including both zero-order model and first-order model are reviewed. LCA method which involves the estimation of GHGs emissions from various solid waste management options, lessons learned and innovative options to achieve GHGs mitigation from solid waste management in both Thailand and from other case countries are further illustrated.

2 Mathematic Models for GHG Generation from Solid Waste Disposal Sites

There are numerous models used to predict the amount of GHGs produced throughout the lifetime of disposal sites (USEPA 2005). Both zero-order and first-order model are presented in this chapter.

2.1 Zero-Order Model

In a zero-order model, GHGs formation from a certain amount of waste is assumed to be constant with time (i.e., methane generated from solid waste disposal sites remain steady over time). This implies that waste age and—waste type have no effects on the rate of waste decay or landfill gas production. This type of model is therefore used to estimate global and national emissions with the assumption that there is no major change in waste composition or the amount of solid waste landfilled. Examples of a zero-order model are the SWANA Zero-order model, German EPER model and IPCC Default method, as detailed below.

2.1.1 SWANA Zero-Order Model

The Solid Waste Association of North America (SWANA) has developed a zero-order model as shown in Eq. 1 (SWANA 1998).

$$Q = ML_o/(t_o - t_1) \quad \text{for } t_o < t < t_1 \tag{1}$$

where:

Q Methane generation rate in volume per time (m^3/year)
M Waste in the disposal site (mg)
L_o Methane generation potential (m^3/mg waste)
t_o Lag time
t_1 Time to endpoint of generation.

2.1.2 EPER Model

The German EPER model is a zero-order-model which can be described mathematically using Eq. 2 (Scharff and Jacobs 2006).

$$M_e = M \cdot BDC \cdot BDC_f \cdot F \cdot D \cdot C \tag{2}$$

where

M_e Amount of diffuse methane emission ($mgCH_4$/year)
M Annual amount of landfilled waste (mg)
BDC Proportion of biodegradable carbon (mg C in waste/mg waste)
BDC_f Proportion of biodegradable carbon converted into landfill gas (%)
F Fraction of methane in landfill gas
D Collection efficiency factor.

2.1.3 IPCC Default Method

The IPCC guidelines provide two methods for estimation of GHGs emissions from solid waste disposal sites: IPCC Default method (Tier 1) and First-order Decay (FOD) method (Tier 2) (see Sect. 2.2.3) (IPCC 1996; Pipatti and Svardal 2006).

Among the available methods, the default method is the simplest one for the estimation of GHGs emission from landfill. This method is based on 'mass balance calculation' which estimates the amount of CH_4 released from the disposal sites assuming that all the potential CH_4 are released during the same year the waste is disposed of (Eq. 3) (IPCC 1996):

$$\begin{aligned} CH_4 \text{ emission (Gg/yr)} = {} & (MSW_T \cdot MSW_F \cdot MCF \cdot DOC \cdot DOC_F \\ & \cdot F \cdot 16/12 - R) \cdot (1 - OX) \end{aligned} \tag{3}$$

where:

MSW_T Total municipal solid waste generated (Gg/yr), calculated from population (in thousand persons) · annual solid waste generation rate (Gg 10^{-3} persons yr^{-1})

MSW_F Fraction of solid waste disposed of at the disposal sites

MCF Methane correction factor (fraction). Three default values ranging from 1.0 to 0.4 are included, depending on the disposal site management and with 0.6 as general default value

DOC Degradable organic carbon (fraction) · (kgC/kgSW) · DOC value depends on the composition of waste. Equation to determine DOC is 0.4A + 0.17B + 0.15C + 0.3D; where A = Paper and textiles (% portion); B = Garden and park waste (% portion); C = Food waste (% portion); and D = Wood and straw waste (% portion)

DOC_F Fraction DOC dissimilated. IPCC default value is 0.77 (Bingemer and Crutzen 1987)

F Fraction of CH_4 in landfill gas (IPCC default is 0.5)

16/12 Conversion of C to CH_4

R Recovered CH_4 (Gg/yr)

OX Oxidation factor (fraction—IPCC default is 0).

Case Study: Estimation of CH_4 emissions from solid waste disposal sites in Panama Panama, located in Central America, provides a good case study for the quantification of GHGs emission from the waste sector because its solid waste management practices in data collection is typical of many developing countries, most of which are going to update their national GHGs inventory for the year 2000. According to Melissa et al. (2008), estimation of CH_4 emission used a MCF of 1, a DOC of 0.13, a DOC_F of 0.77, and a CH_4 content of the generated gas (F) of 0.5, as indicated in the IPCC guidance. Beside this, total waste generation from the 1994 inventory (374.1 Gg) was divided by total population in 1994 (2.57 million) to determine a total municipal waste generation rate of 145 kg/capita/year. Therefore, the default method for methane emission estimates in 1994 is 25 Gg. The IPCC default method has been widely applied in situations where detailed data are not available, but it may not provide a realistic estimate because it does not reflect the degradation profile of wastes over time.

2.2 First-Order Model

To estimate landfill gas generation, first-order models are the most commonly used currently.

These models consider both quantity and quality of waste in the disposal sites (i.e., age of waste, ability of waste to be degraded, moisture and carbon content). The first-order models always assume a direct relationship between carbon content of solid waste and exponential function of decay rate by time against generated landfill gas (Ozkaya et al. 2007). Examples of the first-order model are TNO, LandGEM, and IPCC.

2.2.1 TNO

The TNO model calculates landfill gas generation based mainly on amount of waste in the disposal site, degradation of organic carbon in the waste, and also degradation rate. The TNO model can be mathematically described as per Eq. 4.

$$\alpha_t = \varsigma 1.87 A C_o K_1 e^{-k_1 t} \tag{4}$$

where:

α_t Landfill gas production at a given time (m^3/year)
ς Dissimilation factor, 0.58
A Amount of waste in disposal site (m^3/kg C degraded)
C_o Amount of organic carbon in waste (kg C in waste/mg waste)
k_1 Degradation rate constant, 0.094.

2.2.2 LandGEM

The LandGEM model is used to estimate landfill gas generation for a given year from cumulative waste disposed of through that year (EPA 2005), as presented in Eq. 5.

$$Q_{CH4} = \sum\nolimits_{i=1}^{n} k L_o M_i \left(e^{-kt}\right) \tag{5}$$

where:

Q Methane generated in current year (m^3/year)
i 1 year time increment
n (year of the calculation)—(initial year of waste acceptance)
k Methane generation rate (1/yr)
L_o Methane generation potential (m^3/mg waste)
M_i Mass of solid waste disposed in place in specific time of t.

2.2.3 IPCC Fist-Order-Decay Method

Theoretically, the FOD method provides a time dependent GHGs emission profile that reflects the actual pattern of degradation process over the period of time. This method requires both historical disposals of waste and degradation rate in order to achieve an acceptable accurate result. It assumes that the *DOC* in waste decays slowly throughout a few decades, during which CH_4 and CO_2 are formed. In general, the rate of CH_4 production depends on the amount of carbon remaining in the disposal site. CH_4 emissions are highest in the first few years after deposition, and then gradually decline as the waste is degraded by anaerobic bacteria. The estimation of CH_4 emission from landfills using the FOD model can be done using Eq. 6 (IPCC 1996):

$$Q = L_o \cdot R \cdot \left(e^{-kc} - e^{-kt}\right) \tag{6}$$

where:

Q CH_4 generated in current year (m^3/yr)
L_o CH_4 generation potential (m^3/mg of refuse)
R Average annual waste acceptance rate during active life (Mg/yr)
k CH_4 generation rate constant (l/yr)
c Time since solid waste disposal site (SWDS) closure (yr)
t Time since SWDS opened (yr).

When estimating national figures, Eqs. 7 and 8 can be used for CH_4 estimation, according to the following details:

$$Q_{T,X} = k \cdot R_x \cdot L_o \cdot e^{-k(T-x)} \tag{7}$$

where:

$Q_{T,X}$ Amount of CH_4 generated in year T by the waste R_x (Mg)
x Year of waste input
R_x Amount of waste disposed in year x (Mg)
T Current year.

$$Q_T = \sum Q_{T,X} \tag{8}$$

where:

Q_T Total CH_4 emission in year T from waste disposed of in previous years (including year T).

Table 2 summaries the available mathematical models used to quantify the amount of methane emitted from solid waste disposal sites.

Case Study: Estimation of CH_4 emissions from solid waste disposal sites in Thailand According to Chiemchaisri et al. (2007), there were 425 solid waste disposal sites in the Kingdom of Thailand (95 landfills and 330 open dumps) in 2004. CH_4 emission was calculated based on the FOD method (Eq. 2) by assuming CH_4 production potential (L_o) of 170 m^3/ton, FOD rate constant (k) of 0.05 per year, and the total amounts of waste disposed to landfill and open dumps were 12,177 and 7704 ton/day, respectively. Therefore, the total CH_4 emissions were calculated to be 115.4 Gg in 2004.

Table 2 Mathematical models and their applications to estimate methane emission from solid waste disposal sites

Mathematic models for GHGs quantification	Examples of an application	Observation
Zero-order models	SWANA and EPER models are used by the United States and Germany	These models generate the rate of methane emission independent of the amount of waste remaining or the amount of landfill gases already produced
First-order models	TNO and LandGEM models are currently used by Denmark and the United States	First-order models consider the effect of age in methane generation. Methane emission is assumed to be in a steady, linear decrease over time proportional to the amount of organic material degraded in any given year and the remaining fraction of organic matter from previous years

Case Study: Estimation of CH_4 emissions from open dumps in Chennai, India A study done by Karthikeyan et al. (2012) found that the amount of methane emitted from the open dumps in Chennai, India, was approximately 33×10^6 m^3, 36×10^6 m^3 and 52×10^4 m^3 by using the FOD models as TNO, LandGEM and zero-order model, respectively. Of these, the total methane emission from the Chennai dumps contributed to 1.3 % of landfill methane emissions in India between 1986 and 2003.

In terms of accuracy, the first-order models are by far more accurate than the zero-model assuming that there is no major change in waste composition or the amount of solid waste landfilled from year to year. Coops et al. (1995) also concluded that the zero-order model was the most unreliable compared to other mathematic models. Under this scheme, zero-order model (i.e., IPCC Default method) may not provide realistic estimates because it is applied when national solid waste data is not available. In contrast, the first-order model method which sufficiently reflects the actual pattern of the waste degradation process over time based on both historical disposal information and degradation rate can provide more realistic estimate of GHGs emissions.

2.3 *Life Cycle Assessment for Solid Waste Management*

2.3.1 LCA Principle

LCA is an established environmental management tool since the 1990s for assessing environmental impacts of products (Frankl and Rubik 2000). LCA, often

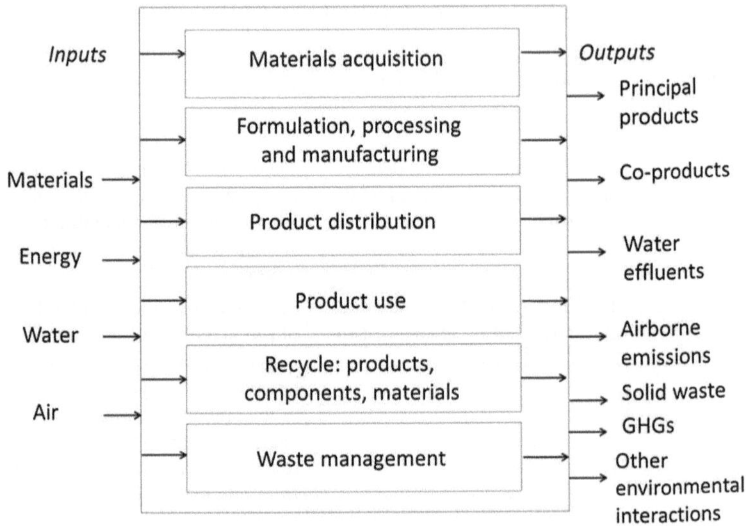

Fig. 3 The element of life cycle assessment (Adapted from SETAC 1993)

called from 'cradle to grave' analysis, is a tool for systematic evaluation of the environmental burdens associated with a process, product, or service system by quantifying energy and materials used through all stages of its life cycle and also wastes released to the environment. Figure 3 portrays environmental flows across the life cycle of a product in terms of the material inputs and pollution outputs to air, water and environment. A typical product's life cycle involves:

- Extracting a particular raw materials from natural ecosystems
- Refining resources into industrial feedstock
- Manufacturing and processing the product from these feedstocks
- Using and consuming the product by consumers, and
- Disposition of used product by 3Rs or disposal.

The first three phases (i.e., extraction, refining and manufacturing) are considered as the upstream phase in the product life cycle. The last phase, such as 3Rs, waste-to-energy, and landfill, is often termed the post-consumer or downstream phase. Basically, an LCA process has four major steps: (i) goal and scope definition (i.e., the products or services to be assessed are defined), (ii) life cycle inventory (i.e., energy, raw materials used and emissions to the environment are quantified), (iii) life cycle impact assessment (i.e., the effects of the resource use and emissions generated are grouped and quantified into a limited number of impact categories), and (iv) data interpretation (i.e., the results are reported and opportunities to reduce the impact of the products on the environment are evaluated) (ISO 1996; SETAC 1993; Frankl and Rubik 2000).

2.3.2 Application of LCA for Solid Waste Management

An LCA of materials management-related GHG sources encompasses emissions from the extraction and processing of raw materials, production, transportation, consumption, and end-of-life treatment. The USEPA (2006) illustrated that solid waste management decisions can potentially reduce GHG emissions by affecting one or more of the following GHG sources and sinks: (i) raw materials manufacturing, (ii) forest or soil carbon storage, and (iii) waste management. The sum of emissions across all steps in the life cycle represents net GHG emissions. Table 3 present the components of net emissions for various municipal solid waste management strategies.

As mentioned above, different waste management strategies have different implications for GHGs emissions and carbon sequestration. The following reviews of LCA studies of solid waste management options and practices are therefore presented.

LCA of Integrated Solid Waste Management in the United States of America

The USEPA (2006) investigated the net GHG implications of a waste management strategy by comparing net GHG emissions between baseline and alternative scenarios. For instance, a baseline scenario in which 10 tons of paper are produced, consumed, and landfilled can be compared with alternative scenarios in which 10 tons of paper are produced, consumed and recycled. Theoretically, the formula for net GHG emissions for each scenario of waste management is as follows:

$$\begin{aligned}\text{Net GHG emissions} &= \text{Gross manufacturing GHG emissions} \\ &\quad - (\text{Increase in carbon stocks} + \text{Avoided utility GHG emissions})\end{aligned} \tag{9}$$

At each point in the material LCA, the USEPA (2006) estimated of the total GHG emissions and sequestration resulting from source reduction and recycling of 21 single-material items: aluminium cans, steel cans, copper wire, glass, high-density polyethylene (HDPE), low-density polyethylene (LDPE), polyethylene terephthalate (PET), corrugated cardboards, magazines, newspaper, office paper, phonebooks, textbooks, dimensional lumber, medium-density fiberboard, carpet, personal computers, clay bricks, concrete, fly ash, and tires. A baseline scenario, in which the material is produced from the current mix of virgin and recycled inputs, but has not been disposed of or recycled, is considered to measure the GHGs impacts of source reduction and recycling. In this analysis, through source reduction, GHGs emissions throughout the life cycle of material (i.e., managing the post-consumer) are avoided. Additionally, when paper products are

Table 3 GHG mitigation potential from alternative solid waste management practices (USEPA 2006)

Management strategy	GHG sources and sinks		
	Manufacturing	Changes in forest or soil carbon storage	Waste management
Source reduction	Decrease in GHG emissions relative to the baseline of manufacturing	Increase in forest carbon sequestration (for organic materials)	No emissions/sinks
Recycling	Decrease in GHG emissions due to lower energy requirements (compared to manufacture from virgin inputs) and avoided process non-energy GHG	Increase in forest carbon sequestration (for organic materials)	Process and transportation emissions associated with recycling are accounted in the manufacturing stage
Composting	N/A	Increase in soil carbon storage	Compost machinery emissions and transportation emissions
Combustion	N/A	N/A	Non-biogenic CO_2, N_2O emissions, avoided utility emissions, and transportation emissions
Landfilling	N/A	N/A	CH_4 emissions, long-term carbon storage, avoided transportation emissions

reduced, carbon is then sequestered in forests, through reduced timber harvesting. In terms of recycling, GHGs emissions from producing an equivalent amount of raw material from virgin inputs in the manufacturing process are avoided. As presented in Table 4, GHGs reductions are calculated by taking the difference between (i) GHGs emissions from processing a raw material from 100 % recycled inputs, and (ii) GHGs emissions from processing an equivalent amount of raw material from 100 % virgin inputs. The minus values shown in column (f) of Table 4 indicate the extents of GHGs reduction from using the recycled products instead of the virgin inputs. The results showed that source reduction and using recycled products, especially aluminum cans, copper wire, carpet and tires, have a much higher potential to reduce GHGs.

Table 5 also presents the net GHGs emissions of municipal solid waste management in each practice. The results showed that source reduction represents an opportunity to reduce GHGs concentrations significantly. Recycling option is considered to be the second best opportunity to reduce the emissions of GHGs. The net emissions from composting and combustion are lower than other options (USEPA 2006).

Table 4 GHGs reductions from using recycled products metric tons of carbon equivalent (MTCE/ton or material recovered) (USEPA 2006)

(a)	(b)	(c)	(d)	(e)	(f) = b + c + d + e
Material	Recycled input credit: *Process energy*	Recycled input credit: *Transportation energy*	Recycled input credit: *Nonenergy*	Forest carbon sequestration	GHGs reductions from using recycled inputs
Aluminum cans	−2.92	−0.12	−0.66	0.00	−3.70
Steel cans	−0.48	−0.01	0.00	0.00	−0.49
Copper wire	−1.33	−0.02	0.00	0.00	−1.34
Glass	−0.03	0.00	−0.04	0.00	−0.08
HDPE	−0.34	0.00	−0.04	0.00	−0.38
LDPE	−0.42	0.00	−0.04	0.00	−0.46
PET	−0.40	0.00	−0.02	0.00	−0.42
Corrugated cardboard	0.00	−0.01	0.00	−0.83	−0.85
Magazines	0.00	0.00	0.00	−0.83	−0.84
Newspaper	−0.20	-0.01	0.00	−0.55	−0.76
Office paper	0.06	0.00	0.00	−0.83	−0.78
Phonebooks	−0.17	0.00	0.00	−0.55	−0.72
Textbooks	−0.01	0.00	0.00	−0.83	−0.85
Dimensional lumber	0.02	0.00	0.00	−0.69	−0.67
Medium density fiberboard	0.01	0.00	0.00	−0.69	−0.67
Mixed paper: Broad definition	−0.10	−0.03	0.00	−0.83	−0.96
Mixed paper: Residential definition	−0.10	−0.03	0.00	−0.83	−0.96
Mixed paper: Office paper definition	−0.08	−0.02	0.00	−0.83	−0.96
Carpet	−1.47	−0.02	−0.47	0.00	−1.96
Personal computers	−0.41	−0.01	−0.20	0.00	−0.62
Clay bricks	N/A	N/A	N/A	N/A	N/A
Concrete	0.00	0.00	0.00	0.00	0.00
Fly ash	−0.11	0.00	−0.12	0.00	−0.24
Tires	−1.75	0.00	0.00	0.00	−1.75

NA: Not applicable

Table 5 Net GHGs emissions of municipal solid waste management options (MTCE/ton) (USEPA 2006)

Material	Source of reduction*		Recycling*	Composting*	Combustion*
	Current mix of inputs	100 % virgin input			
Aluminum cans	−2.26	−4.28	−3.71	N/A	0.01
Steel cans	−0.88	−1.02	−0.50	N/A	−0.43
Copper wire	−2.01	−2.03	−1.35	N/A	0.00
Glass	−0.17	−0.19	−0.09	N/A	0.00
HDPE	−0.50	−0.55	−0.39	N/A	0.24
LDPE	−0.63	−0.65	−0.47	N/A	0.24
PET	−0.58	−0.60	−0.43	N/A	0.28
Corrugated cardboard	−1.63	−2.32	−0.96	N/A	−0.29
Magazines	−2.28	−2.36	−0.76	N/A	−0.05
Newspaper	−1.09	−1.39	−0.52	N/A	0.03
Office paper	−2.71	−2.79	−1.31	N/A	−0.70
Phonebooks	−1.49	−1.49	−0.49	N/A	0.03
Textbooks	−3.03	−3.11	−1.38	N/A	−0.70
Dimensional lumber	−0.42	−0.42	−0.54	N/A	−0.08
Medium density fiberboard	−0.47	−0.47	−0.54	N/A	−0.08
Food discards	N/A	N/A	N/A	−0.25	−0.25
Yard trimmings	N/A	N/A	N/A	0.01	0.00
Mixed paper: Broad definition	N/A	N/A	−1.06	N/A	−0.27
Mixed paper: Residential definition	N/A	N/A	−1.03	N/A	−0.25
Mixed paper: Office paper definition	N/A	N/A	−1.06	N/A	−0.29
Mixed metals	N/A	N/A	−1.44	N/A	−0.30
Mixed plastics	N/A	N/A	−0.42	N/A	0.26
Mixed recyclables	N/A	N/A	−0.83	N/A	−0.20
Mixed organics	N/A	N/A	N/A	−0.12	−0.12
Mixed waste (as disposed)	N/A	N/A	N/A	N/A	−0.15
Carpet	−1.10	−1.10	−1.97	N/A	0.10

(continued)

Table 5 (continued)

Material	Source of reduction*		Recycling*	Composting*	Combustion*
	Current mix of inputs	100 % virgin input			
Personal computers	−15.14	−15.14	−0.63	N/A	−0.06
Clay bricks	−0.09	−0.09	−0.01	N/A	−0.01
Concrete	−0.01	−0.01	−0.01	N/A	−0.01
Fly ash	−0.01	−0.01	−0.25	N/A	0.04
Tires	−1.10	−1.10	−0.51	N/A	

Remark: *Net emissions = net emissions minus landfilling
NA: Not applicable, or in the case of composting of paper, not analyzed

LCA of Integrated Solid Waste Management in the OECD Countries

ENV/EPOC/WGWPR (2010) investigated the mitigation opportunities for GHGs emissions resulting from implementing alterative municipal solid waste management practices among the OECD member countries in 2030. As shown in Table 6, the eight different scenarios of waste management were evaluated: (i) recycling, (ii) composting, (iii) anaerobic digestion, (iv) recycling and mechanical biological treatment (MBT), (v) landfill gas (LFG) collection, (vi) energy recovery from collected LFG, (vii) incineration, and (vii) source reduction.

To estimate the mitigation potential for each solid waste management scenario, ENV/EPOC/WGWPR (2010) multiplied the quality of waste managed by each management practice by material-specific GHGs emission factors. The difference between the baseline and the actual GHGs emissions for each practice is considered as the net GHGs reduction. For instance, consider the hypothetical situation where a country generates approximately 100 tons of municipal solid waste, landfilling 70 % and recycling 30 % of this waste in the baseline. Assume that landfilling releases 0.25 tons of CO_2e per ton of solid waste landfilled from landfill methane emissions, and recycling reduces GHGs by 0.5 tons of CO_2e per ton recycled by offsetting energy and emissions from product that would otherwise have been produced from virgin materials. Therefore, the emission reductions associated with switching to a scenario where 60 % of waste is recycled and 40 % of waste is landfilled with the following calculation: GHGs reduction from landfill = 100 × 0.25 × (40 %–70 %) = −7.5 tons CO_2e

GHGs reduction from recycling $= 100 \times (-0.5) \times (60\,\% - 30\,\%) = -15$ tons CO_2e
Total GHGs reduction $= -7.5 + (-15) = -22.5$ tons CO_2e

This calculation indicates that the change from landfilling to recycling option reduces about 7.5 tCO_2e from avoiding emissions produced by landfilling and also

Table 6 Alternative municipal solid waste management scenarios

Scenario	Description
Recycling	Recycling involves the reuse of a material in a production process that diverts material from the waste stream. Recycling a material can manufacture the same type or a different product
Composting	Composting is a process in which organic wastes are biodegraded by microorganism to produce compost or soil amendment
Anaerobic digestion	Anaerobic digestion technique is the decomposition of organic compounds by microorganisms in an oxygen-starved environment to produce a methane-rich gas or biogas
Recycling and MBT	MBT is a series of operations to sort solid waste prior to landfilling the biodegradable components and any remaining waste. The purpose of MBT is to separate waste which is not suitable for recycling, extract recyclable materials and produce a refuse-derived fuel (RDF) primarily for co-incineration
LFG collection	Landfill is a site for disposal of waste materials for all waste streams. LFG (i.e., CH_4) is generated as a result of anaerobic decomposition of organic materials
Incineration	Incineration is the thermal destruction that directly involves the combustion of waste stream
Source reduction	Source reduction is any activity that prevents the creation of solid waste at the point of solid waste generation

additional 15 tCO_2e from GHGs emissions savings directly associated with recycling to provide total emissions reductions of about 22.5 tCO_2e.

In this context, recognizing the potential for solid waste management to reduce GHGs emissions into the atmosphere, ENV/EPOC/WGWPR (2010) predicted change in GHGs emissions from each alterative waste management scenarios in 2030 compared to baseline emission of the OECD member countries (Fig. 4). The results found that MBT process with recycling, incineration with waste-to-energy (WTE), recycling, LFG energy recover, and source reduction strategies would significantly reduce the volume of GHGs released from the waste sector.

Apart from the eight different scenarios of waste management (Table 5), application of biochar could significantly contribute to mitigating GHGs emissions through several routes (Lehmann 2007a, b; Gaunt and Lehmann 2008; Roberts et al. 2010). For example, pyrolysis of biomass materials (a thermal decomposition of biomass in the absence of oxygen, which converts organic materials to produce condensable vapours, gases, and charcoal) offers the greater potential to sequester carbon in a stable form as biochar, generates renewable bioenergy, enhance agricultural productivity as fertilizers and soil amendments, and avoiding methane emissions from waste disposal sites. Several studies elaborated that both methane and nitrous oxide emissions were significantly suppressed (up to 50–92 %) when biochar was added to the soil (Van Zwieten et al. 2010; Clough and Condron 2010; Rondon et al. 2005). By combining these insights, LCAs have so far been increasingly used to systematically quantify climate change impacts of the

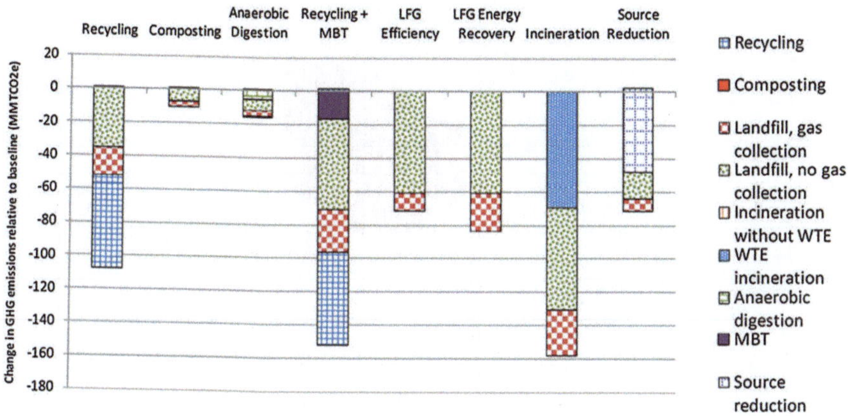

Fig. 4 Change in GHGs emissions relative to baseline municipal solid waste management practices based on the implementation of alternative waste management scenarios in 2030 in the OECD member countries (ENV/EPOC/WGWPR 2010)

pyrolysis-biochar system. Gaunt and Cowie (2009) found that the emissions abatement through utilization biochar as a soil amendment, expressed relative to 1 ton of biochar over a 10-year period, ranged between 0.1 and 0.5 ton carbon dioxide equivalent per ton biochar ($tCO_2e\ t^{-1}$). Additionally, when combined with the emissions abatement related to pyrolysis of biomass materials to produce biochar, the total emissions reduction ranged between 2.6 and 16 $tCO_2e\ t^{-1}$ biochar produced.

3 Case Studies

3.1 *Bangkok Kamphaeng Saen West: Landfill Gas to Electricity Project, Thailand*

Bangkok Kamphaeng Saen Landfill site is located in Kamphaeng Saen district, Nakhon Pathom province, approximately 75 km west of Bangkok. The project location is depicted in Fig. 5. Generally, there are about 3328 tons per day of municipal solid wastes with 60–70 % organic content and 0.7 t/m^3 compact density disposed at the landfill (Prime 2013).

Overall, the project activities comprise the capture and utilization of LFG (approximately 50 % CH_4 and 50 % CO_2) to generate electricity, which is subsequently connected to the national electrical grid. The excess LFG is sent to an enclosed ground flare (having more than 98 % of methane combustion efficiency) to minimize CH_4 emissions. The major source of GHGs emission from the project is CH_4 emission from decomposition of waste at the landfill site.

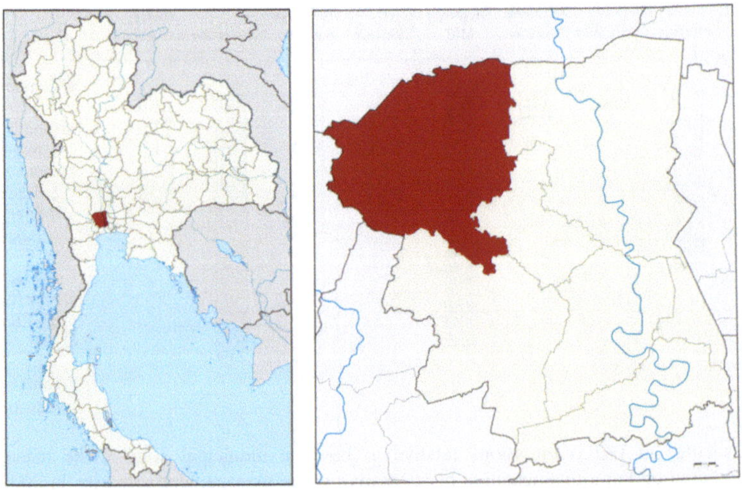

Fig. 5 Map of Kamphaeng Saen district, Nakhon Pathom province, Thailand

The construction of the entire LFG extraction pipe and collection system started in July 2009. In the LFG collection process, the horizontal collection pipes are installed in the waste pile as an integral part of tipping operation. The LFG then passed through the pre-treatment plant where both moisture and particles are removed prior to the power generating engine. There are currently eight power generator sets operable (each having 1500 round/min of rotating speed, 1063 kW of power, 400 V of voltage and 50 Hz of frequency). From October 2013 to October 2014, the total amount of CH_4 captured at normal temperature and pressure was approximately 3,158,908 m^3. The net amount of electricity fed to the grid was about 69,694 MWh (Bangkok Kamphaeng Saen West Landfill Gas to Electricity 2006).

According to climate mitigation relevance, based on the ACM0001 method*, namely "Consolidated baseline and monitoring methodology for LFG project activities" (UNFCCC 2015a, b, c), the total GHG emission reduction (Eq. 10) over the timeline of 8-year period is expected to be 1,726,214 tCO_2e (or 246,602 tCO_2e per year), as shown in Table 7.

$$ER_y = BE_y - PE_y - L \quad (10)$$

where:

ER_y Emission reduction in year y (tCO_2/year)
BE_y Baseline emission in year y (tCO_2/year)
PE_y Project emission in year y (tCO_2/year)
L Project leakage in year y (tCO_2/year).

Table 7 Estimation of GHGs emission reduction (tCO_2e) (Bangkok Kamphaeng Saen west landfill gas to electricity project PDD 2006)

Year	Estimate of project activity emissions	Estimate of baseline emissions	Estimate of leakage	Estimate of overall emission reductions
1	72	312,154	0	312,082
2	96	414,858	0	414,762
3	71	308,738	0	308,668
4	55	235,976	0	235,920
5	39	179,810	0	179,771
6	24	136,863	0	136,839
7	24	115,151	0	115,127
8	4	23,049	0	23,045
Total	384	1,726,599	0	1,726,214

Remark: ACM001 method is applicable to landfill gas capture project, where the baseline scenario is the partial or total atmospheric release of the gas and the project activities include situations such as:

(a) Captured gas is flared; and/or
(b) Captured gas is used to produce energy as electricity or thermal energy
(c) Capture gas is used to supply consumers through natural gas distribution network

3.2 GHGs Mitigation Project: MBT Plant in Gaobeidian, Hebei Province, People's Republic of China

In 2009, the German Federal Ministry of Environment granted a fund to a project developer to establish the MBT facility for GHGs mitigation in Gaobeidian City, approximately 80 km South-West of Beijing, Hebei, China, as depicted in Fig. 6.

The MBT plant involves both mechanical and biological treatment processes to stabilize 40,000 tons of municipal solid waste. The primary aim of mechanical treatment is to promote proper segregation of valuable wastes (i.e., paper, plastic, glass, etc.) and producing organic fertilizer. In this process, the waste feed stream is split into three fractions: coarse, medium, and fine. Both coarse and middle fractions are being passed to hand assorting stations (Fig. 7). After that, the coarse fraction is either crushed and fed to the waste stream again or directly disposed of at the landfill disposal site. Prior to final disposal, the fine fraction will undergo biological waste treatment that may be used as an amendment for landfill cover or as CH_4 oxidation layer. The biological treatment process is carried out partly under roof and in open air areas as an aerobic and actively ventilated stabilization (Kölsch et al. 2010).

As shown in Table 8, based on the AMS III.F method*, namely "Avoidance of CH_4 production from decay of biomass through composting" (UNFCCC 2015a, b, c), the net GHGs emission reduction is expected to be 83,814 tCO_2e over a 10 years period.

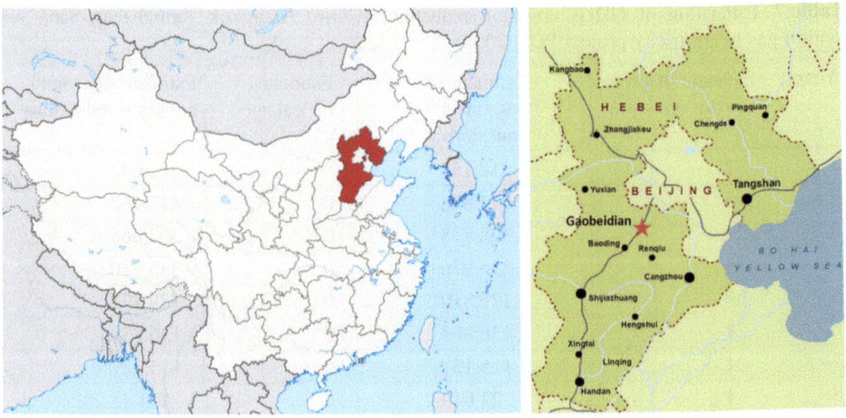

Fig. 6 Map of Gaobeidian city, Hebei province, People's Republic of China

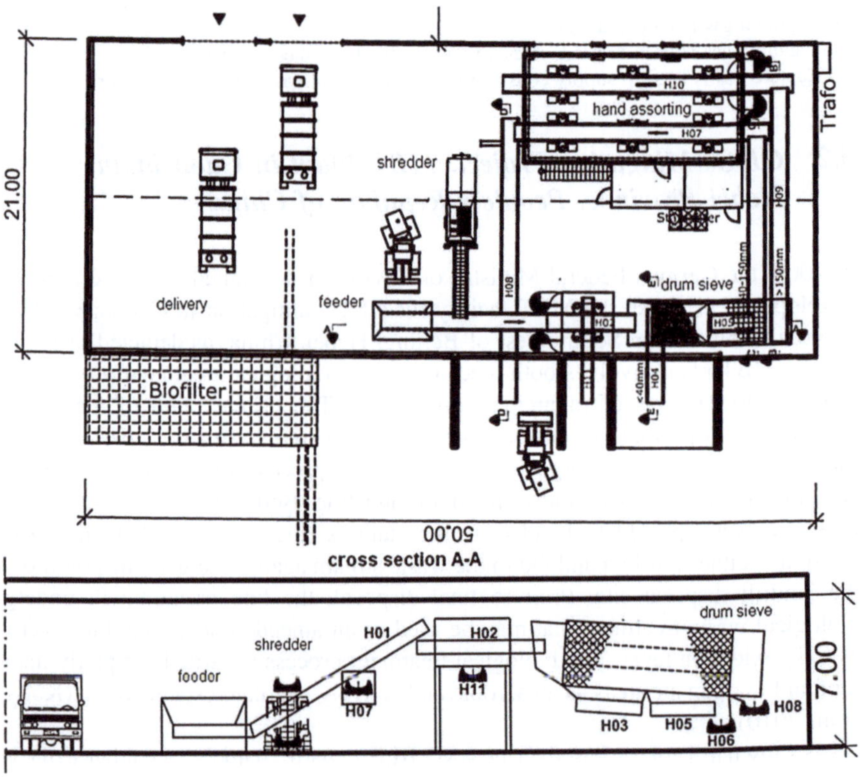

Fig. 7 Design of the MBT plant in Gaobeidian (Kölsch et al. 2010)

Table 8 Estimation of GHGs emission reduction from the MBT plant in Gaobeidian, China (tCO_2e) (Kölsch et al. 2010)

Year	Project emissions	Baseline emissions	Emission reductions
1	497	2339	1842
2	537	4376	3839
3	577	6156	5579
4	616	7714	7098
5	655	9081	8426
6	694	10,284	9590
7	732	11,345	10,613
8	770	12,283	11,513
9	807	13,114	12,307
10	845	13,852	13,007
Total	6730	90,544	83,814

Remark: The AMS III.F method comprises measures to avoid the emissions of CH_4 from biomass or other organic matter that would have otherwise been left to decay anaerobically in a solid waste disposal site, or in an animal waste management system. In the project activity, controlled biological treatment of biomass is introduced through one, or a combination, of the following measures:

(a) Aerobic treatment by composting and proper soil application of the compost
(b) Anaerobic digestion in closed reactors equipped with biogas recovery and combustion/flaring system

Additionally, this method is applicable under the following conditions:

(a) Where the solid waste would have been disposed and the methane emission occurs in absence of the proposed project activity
(b) In the case of projects co-composting wastewater, where the co-composting wastewater would have been treated anaerobically in the absence of the project activity
(c) Where the treatment of biomass through composting or anaerobic digestion takes place
(d) Where the residual waste from biological treatment or products from those treatments, like compost and slurry, are handled, disposed, submitted to soil application, or treated thermally/mechanically
(e) Where biogas is burned/flared or gainfully used
(f) And the itineraries between them (a, b, c, d and e), where the transportation of waste, wastewater, where applicable manure, compost/slurry/products of treatment or biogas occurs

3.3 Municipal Solid Waste Composting Project in Ikorodu, Lagos State, Federal Republic of Nigeria

The municipal solid waste composting site is located in Odogunyan, Ikorodu Local Government Council of Lagos State, Nigeria, as depicted in Fig. 8.

The composting site involves a biological process for organic matter decomposition of about 1500 tons of municipal solid waste per year by the unsheltered windrow system. The process of open windrow aerobic composting is considered a simple biological process in which organic fraction of wastes are directly converted into ammonia, CO_2, water vapour, and stable humus-like material called compost. In the windrow process, municipal solid wastes are unloaded in the shredding area,

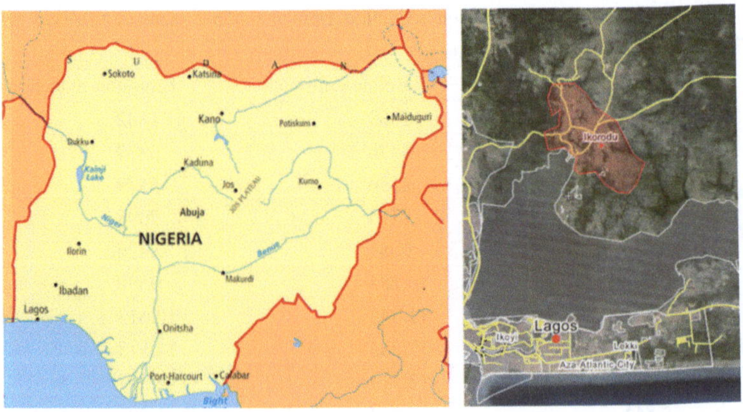

Fig. 8 Map of Ikorodu, Lagos state, Nigeria

where the wastes are chopped and shredded into pieces less that 7 cm for efficient composting. A discharge conveyor loads the shredded wastes in dump trucks for transport to the active compost site. Before composting, a tiller is used to dig a trench in the row for addition of water for optimal moisture content in the shredded material. To accelerate the degradation of organic waste, both dry and wet inoculants are added to the compost piles. When the desired C:N ratio and proper temperature and moisture contents have been achieved, the composting is completed. In this project, the composted product is tested for the presence of pathogens, heavy metals and soil nutrients according to the national and USEPA standards. Once the compost is satisfactory, the finished compost is either sold in bulk or packed in bags of 25 kg per each. The composting site has a capacity of 4800 bags (or approximately 120 tons) per day (Municipal Solid Waste Composting Project in Ikorodu, Lagos State PDD 2010).

According to the AM0025 method*, namely "Avoided emissions from organic waste through alternative waste treatment processes" (UNFCCC 2015a, b, c), the total GHG emission reduction (Eq. 6) over the timeline of 8-year period is expected to be 1,972,468 tCO_2e, as shown in Table 9.

3.4 *Gasification, Landfill Gas and Anaerobic Digestion (GALFAD) in Bali, Indonesia*

GALFAD project is located at a solid waste disposal site in Suwung, approximately 10 km from downtown Denpasar, the largest city and capital of the Bali island, Indonesia (Fig. 9). The Suwung landfill has been in operation since 2005. The project was split into two phases. In the first phase, the 4.8–2.8 MW gasification power plant and 2.0 MW anaerobic digestion were constructed. An additional

Table 9 Estimation of GHGs emission reduction from the municipal solid waste composting project in Ikorodu, Lagos State, Federal Republic of Nigeria (tCO_2e)

Year	Estimate of project activity emissions	Estimate of baseline emissions	Estimate of leakage	Estimate of overall emission reductions
1	1112	19,904	738	18,054
2	8291	200,372	4427	187,654
3	9393	255,473	4427	241,653
4	10,147	293,191	4427	278,617
5	10,667	319,203	4427	304,109
6	11,029	337,320	4427	321,864
7	11,285	350,097	4427	334,385
8	9557	299,378	3689	286,132
Total	71,481	2,074,938	30,989	1,972,468

Remark: The AM0025 method is applicable under the following conditions:

- The project activity involves one or a combination of the following waste treatment options for the fresh waste that in a given year would have otherwise been disposed of in a landfill:
  (a) Composting process in aerobic conditions
  (b) Gasification to produce syngas and its use
  (c) Anaerobic digestion with biogas collection and flaring and/or its use
- Mechanical or thermal treatment process to produce refuse-derived fuel (RDF)/stabilized biomass (SB) and its use. The thermal treatment process occurs under controlled conditions (up to 300 ° C). In case of thermal treatment process, the process shall generate a stabilized biomass that would be used as fuel or raw material in other industrial process. The physical and chemical properties of the produced RDF/SB shall be homogenous and constant over time
- Incineration of fresh waste for energy generation, electricity and/or heat

4.8 MW capacity—2.8 MW from gasification, 1.0 MW from anaerobic digestion and 1.0 MW from landfill gas recovery, were installed in the second phase.

To stabilize 500–1000 tonnes per day of municipal solid waste, of which about 60 % is organic, the GALFAD plant involves the process of gasification, landfill gas recovery and aerobic digestion, as briefly presented below:

Solid Waste Segregation Waste segregation is performed by using the combination of two rotary screens and manual sorting to separate waste to divide waste into wet (i.e., organic waste) and dry (i.e., paper, wood waste, etc.).

Pyrolysis Gasification Pyrolysis gasification is a combination of pyrolysis and gasification processes. The pyrolysis process involves the thermal decomposition of organic waste in an oxygen-free environment. The residue of the pyrolysis is fed into the gasification unit, in which this residue is broken down further into more useful combustible gases (syngas), such as carbon dioxide, methane, and hydrogen, through a water-gas shift reaction. The recovered energy is used to supply electricity to the local grid.

Landfill Gas Recovery System Landfill gas collection system of the Suwung solid waste disposal site consists of vertical wells drilled into the capped landfill cells,

Fig. 9 Map of Bali, Indonesia

connected together with medium density polyethylene piping. The waste in place at the Suwung landfill site is approximately 400,000 m^3, with an average depth of about 6–8 m.

Anaerobic Digestion The recovery biogas is extracted from the anaerobic digester. Organic waste is fed into digester in which anaerobic bacteria degrades organic matter into methane and carbon dioxide.

With respect to the greenhouse gas aspects, the GALFAD project could avoid methane emissions from disposal of untreated solid waste at landfills through the application of the gasification and anaerobic digestion technologies*. According to the AM0025 (Avoided emissions from organic waste through alternative waste treatment processes) and the ACM0001 methods (Consolidated baseline and monitoring methodology for LFG project activities) (UNFCCC 2015a, b, c), the total GHG emission reduction over the timeline of 8-year period is expected to be 863,962 tCO_2e (PT Navigat Organic Energy Indonesia Integrated Solid Waste Management Project PDD 2006), as shown in Table 10.

Table 10 Estimation of GHGs emission reduction from the gasification, landfill gas and anaerobic digestion (GALFAD) in Bali, Indonesia (tCO_2e)

Year	Annual estimation of emission reductions (tCO_2e)
1	49,033
2	79,912
3	96,857
4	114,688
5	121,533
6	129,417
7	158,571
8	113,951
Total	863,962

*Remark: The IPCC first-order-decay method was used to estimate methane emissions from anaerobic digestion and gasification

4 Summary and Conclusions

The significant increase in global population as a result of industrial revolution has caused rapid growth of urbanization and development. One of the challenges of our society is to address the excessive generation and environmentally safe disposal of solid wastes which in turn will increase the amount of GHGs emissions that global warming and subsequently climate change. In this chapter, the issues of solid wastes and GHGs management are presented, as summarized below:

- Currently, 1.3 billion tons of municipal solid wastes are being generated and the qualities are expected to increase to 2.2 billion tons per year by 2025. The OECD countries generate almost half of the total waste stream, whereas Africa and South Asia produce the least amount of solid wastes. Sanitary landfill is the method most popularly employed globally. The main GHGs from solid waste management are CH_4 from landfill, CH_4 and N_2O from wastewater treatment and CO_2 from waste incineration.
- Based on the available mathematic models, both the zero-order models (i.e., SWANA, German EPER, and IPCC Default method) and the first-order models (i.e., TNO, LandGEM, IPCC FOD) can be used to estimate amount of methane emitted from the waste sector. Because they realistically reflects the actual pattern of solid waste degradation process over the period of time, the first-order models are more accurate than the zero-order models.
- In the United States of America, the total GHGs emissions and sequestration resulting from source reduction and recycling of the following 21 single-material items are estimated: aluminum cans, steel cans, copper wire, glass, HDPE, LDPE, PET, corrugated cardboards, magazines, newspaper, office paper, phonebooks, textbooks, dimensional lumber, medium-density fiberboard, carpet, personal computers, clay bricks, concrete, fly ash, and tires. The amount of GHGs reductions are calculated by considering the differences between (i) GHGs emissions from manufacturing a raw material from 100 % recycled inputs, and (ii) GHGs emissions from manufacturing an equivalent amount of a raw material from 100 % virgin inputs. As a result, manufacturing with recycled materials such as aluminum cans, copper wire, carpet and tires has a high potential for GHGs emission reduction. Source reduction and recycling provide the best opportunity to reduce the GHGs emissions compared to other methods (i.e., composting and combustion).
- In the OECD member countries, as an alternative waste management practices, MBT with recycling, incineration with WTE, recycling, LFG energy recover, and source reduction strategies would significantly reduce GHGs emissions.
- The Bangkok Kamphaeng Saen West: Landfill Gas to Electricity Project captures and utilizes LFG from landfills to generate 69,694 MWh of electricity which is fed to the national grid. The total GHG emission reduction is expected to be 1,726,214 tCO_2e over the project timeline of 8 years.

- The GHGs Mitigation Project in Gaobeidian, Hebei province, People's Republic of China, employs MBT plant to promote proper segregation of valuable wastes and producing organic fertilizer. The net GHGs emission reduction is expected to be 83,814 tCO_2e over the project timeline of 10 years.
- The Municipal solid waste composting project in Ikorodu, Lagos State, Federal Republic of Nigeria, treats the municipal solid wastes by the process of open windrow aerobic composting. The net GHGs emission reduction is expected to be 1,972,468 tCO_2e over the project timeline of 8 years.
- GALFAD project in Bali, Indonesia, employs the process of gasification, landfill gas recovery, and aerobic digestion to treat 500–1000 tonnes per day of municipal solid waste. The total GHG emission reduction is expected to be 863,962 tCO_2e over the project timeline of 8 years.

As GHG become more regulated, there are still a number of areas that need attention to enhance waste management practices in mitigating the adverse effects of climate change. At present, lack of skilled human resources, technical capacity and enforcement of national policies in the implementation of solid waste management programs addressing the issue of climate change are considered to be the most important obstacles, especially in developing countries. The following recommendations can be drawn from above discussion.

- Allocating available budget for climate change mitigation and also enhancing the enforcement of existing solid waste legislation.
- Institutional strengthening and capacity building to promote environmental sound technologies for municipal solid waste management and climate change mitigation as well as create enabling conditions.
- Providing technical support and R&D on the concept of measuring, reporting and verification (MRV), carbon capture, LCA-type studies and GHGs mitigation relevant to changes in municipal solid waste management system at both local and national scales.
- Lifting the level of public awareness on the disposal of solid waste and its impact on global climate change.

References

Bangkok Kamphaeng Saen West Landfill Gas to Electricity (2006) Project design document. Sindicatum Carbon Capital Ltd., London, United Kingdom

Bingemer HQ, Crutzen PJ (1987) Production of methane from solid waste. J Geophys Res 87 (D2):2181–2187

Bogner J, Abdelrafie Ahmed M, Diaz C, Faaij A, Gao Q, Hashimoto S, Mareckova K, Pipatti R, Zhang T (2007) Waste management. In: Metz B, Davidson OR, Bosch PR, Dave R, Meyer LA (eds) Climate change 2007: mitigation. contribution of working group III to the fourth assessment report of the intergovernmental panel on climate change. Cambridge University Press, Cambridge, United Kingdom, New York, NY, USA

Chiemchaisri C, Juanga JP, Visvanathan C (2007) Municipal solid waste management in Thailand and disposal emission inventory. Environ Mon Ass 135:13–20. doi:10.1007/s10661-007-9707-1

Clough TJ, Condron LM (2010) Biochar and nitrogen cycle: introduction. J Environ Qual 39 (4):1218–1223

Coops O, Luning L, Oonk H, Weenk A (1995) Validation of landfill gas formation models. Proceedings Sardinia 95, 5th international landfill symposium, vol III. CISA, Environmental Sanitary Engineering Centre, Cagliari, Italy, pp 635–646

ENV/EPOC/WGWPR (2010) Greenhouse gas emissions and the potential for mitigation from materials management within OECD countries, Working group on waste prevention and recycling, Available at http://www.oecd.org/officialdocuments/publicdisplaydocumentpdf/publicdisplaydocumentpdf/?cote=ENV/EPOC/WGWPR%282010%291/FINAL&docLanguage=En

EPA (2005) Landfill gas emissions model (LandGEM), version 3.02. Eastern Research Group, USA

EPA (2014) Climate change indicators in the United States: global greenhouse gas emissions. Report, Available from https://www.epa.gov/climatechange/indicators. Updated May 2014 Access 28 Feb 2015

FAO (2014). FAOSTAT: emissions—land use. Report, Available at http://faostat3.fao.org/faostat-gateway/go/to/download/G2/*/E

Forster P, Ramaswamy V, Artaxo P, Berntsen T, Betts R, Fahey DW (2007) Changes in atmospheric constituents and in radiative forcing. In: Solomon S, Qin D, Manning M, Chen Z, Marquis M, Averyt KB, Tignor M, Miller HL (eds) Climate change 2007: the physical science basis. Contribution of working group I to the fourth assessment report of the intergovernmental panel on climate change. Cambridge University Press, Cambridge

Frankl P, Rubik F (eds) (2000) Life cycle assessments in industry and business, adoption patterns, applications and implications. Springer, Berlin, Heidelberg, New York

Gaunt J, Cowie A (2009) Biochar, greenhouse gas accounting and emissions trading. In: Lehmann J, Joseph S (eds) Biochar for environmental management: science and technology. Earthscan, UK

Gaunt JL, Lehmann J (2008) Energy balance and emissions associated with biochar sequestration and pyrolysis bioenergy production. Environ Sci Technol 42(11):4152–4158

Hoornweg D, Bhada-Tata P (2012) What a waste: a global review of solid waste management. World bank, urban development series knowledge papers, Available at http://wwwwds.worldbank.org/external/default/WDSContentServer/WDSP/IB/2012/07/25/000333037_20120725004131/Rendered/PDF/681350WP0REVIS0at0a0Waste20120Final.pdf

International Standardisation Organisation (ISO) (1996) Environmental management—life cycle assessment—principles and framework—ISO 14040. ISO, Paris

IPCC (1996). Report of the twelfth season of the intergovernmental panel on climate change, Mexico City, 11–13 September 1996

IPCC WG2 (2007) Climate change 2007: synthesis report. Intergovernmental Panel on Climate Change, Cambridge, UK, New York, USA

IPCC (2014) Annex V: expert reviewers, government reviewers and other scientific advisors of the IPCC WGIII fifth assessment report. In: Edenhofer O, Pichs-Madruga R, Sokona Y, Farahani E, Kadner S, Seyboth K, Adler A, Baum I, Brunner S, Eickemeier P, Kriemann B, Savolainen J, Schlömer S, von Stechow C, Zwickel T, Minx JC (eds) Climate change 2014: mitigation of climate change. contribution of working group III to the fifth assessment report of the intergovernmental panel on climate change. Cambridge University Press, Cambridge, United Kingdom, New York, NY, USA

IPCC WG2 (2007) Climate change 2007: impacts, adaptation and vulnerability. In: Parry ML, Canziani OF, Palutikof JP, van der Linden PJ, Hanson CE (eds) Contribution of working group

II to the fourth assessment report of the intergovernmental panel on climate change. Cambridge University Press, Cambridge, UK

Jensen JEF, Pipatti R (2000). CH_4 Emission from solid waste disposal. In: Good practice guidance and uncertainty management in national greenhouse gas inventories: IPCC. Available at http://www.ipccnggip.iges.or.jp/public/gp/bgp/5_1_CH4_Solid_Waste.pdf

JRC/PBL (2013) Emission database for global atmospheric research (EDGAR), Release Version 4.2 FT2010. European Commission, Joint Research Centre, Netherlands Environmental Assessment Agency, Available at http://edgar.jrc.ec.europa.eu

Karthikeyan OP, Kurian J, Esakku S (2012) Methane emission potential of open dumps in Chennai: a case study. Int J Environ Technol Manage 15(3/4/5/6):291–304. doi:10.1504/IJETM.2012.049229

Karthikeyan OP, Chidambarampadmavath K, Cirés Samuel, Heimann Kirsten (2015) Review of sustainable methane mitigation and biopolymer production. Crit Rev Environ Sci Technol 45 (15):1579–1610

Kölsch F, Ginter M, Fricke K (2010) German GHG mitigation lighthouse project MBT plant Gaobeidian (PR China). ISWA World Congress, HH

Komiyama H, Takeuchi K (2006) Sustainability science: building a new discipline. Sustain Sci 1:1–6

Kumar S, Gaikwad SA, Shekdar AV, Kshirsagar PS, Singh RN (2004) Estimation method for national methane emission from solid waste landfills. Atm Env 38:3481–3487. doi:10.1016/j.atmosenv.2004.02.057

Lehmann J (2007a) Bio-energy in the black. Front Ecol Environ 5(7):381–387

Lehmann J (2007b) A handful of carbon. Nature 447(7141):143–144

Melissa W, Jeffrey BC, Edgar S (2008) Estimating national landfill methane emissions: an application of the 2006 intergovernmental panel on climate change waste model in Panama. J Air Waste Manage Assoc 58(5):636–640. doi:10.3155/1047-3289.58.5.636

Municipal Solid Waste (MSW) Composting Project in Ikorodu, Lagos State PDD (2010) Project design document. International Bank for Reconstruction and Development as the Trustee for the Carbon Fund for Europe (CFE), Washington, USA

Ozkaya B, Demir A, Bilgili MB (2007) Neural network prediction model for the methane fraction in biogas from field-scale landfill bioreactors. Environ Model Softw 22:815–822

Pipatti R, Svardal P (2006) Solid waste disposal 2006, IPCC guidelines for national greenhouse gas inventories, vol 5, Chapter 3, pp 3.1–3.44

Prime (2013) 32 MW landfill gas project. Report, Available at http://www.primeroadgroup.com/32MWandfillGasProject.htm

PT Navigat Organic Energy Indonesia Integrated Solid Waste Management Project (GALFAD) PDD (2006) Project design document. PT Navigat Energy Indonesia (NOEI), Indonesia

Roberts KG, Gloy BA, Joseph S, Scott NR, Lehmann J (2010) Life cycle assessment of biochar systems: estimating the energetic, economic, and climate change potential. Environ Sci Technol 44:827–833

Rondon M, Ramirez JA, Lehmann J (2005) Greenhouse gas emissions decrease with charcoal additions to tropical soils. In: Proceedings of the 3rd USDA symposium on greenhouse gases and carbon sequestration in agriculture and forestry

Scharff H, Jacobs J (2006) Applying guidance for methane emission estimation for landfills. Waste Manage 26:417–429

Society of environmental toxicology and chemistry (1993) Guidelines for life—cycle assessment: a "code of practice", SETAC workshop in Sesimbra, Portugal 31 March–3 April. SETAC, Brussels

Stern N (2007) The economics of climate change—the stern review. Cambridge University Press, Cambridge

SWANA (1998) Comparison of models for predicting landfill methane recovery. Publication No. GR-LG 0075, The Solid Waste Association of North America, Dallas, TX

Tanaka M (2010) Sustainable society and municipal solid waste management. In: Agamuthu P, Tanaka M, Penerbit ITB (eds) Municipal solid waste management in Asia and the Pacific Islands, Indonesia

UNEP (2010a) Life cycle assessment, Available at http://www.unep.org/resourceefficiency/Consumption/StandardsandLabels/MeasuringSustainability/LifeCycleAssessment/tabid/101348/Default.aspx

UNEP (2010b) Waste and climate change global trends and strategy framework, Osaka/Shiga, Available at http://www.unep.or.jp/ietc/Publications/spc/Waste&ClimateChange/Waste&ClimateChange.pdf

UNFCCC (2015a) ACM0001: Flaring or use of landfill gas, CDM methodology, Available at http://cdm.unfccc.int/methodologies/DB/D44X8FH8SFCXREE6037AXJSBGGFVDO

UNFCCC (2015b) AMS-III.F.: Avoidance of methane emissions through composting, CDM methodology, Available at http://cdm.unfccc.int/methodologies/DB/7RF5DZ2T6T8F88BMPPHNDOATXD40Y0

UNFCCC (2015c) AM0025: Avoided emissions from organic waste through alternative waste treatment processes, CDM methodology, Available at http://cdm.unfccc.int/methodologies/DB/3S2TKBX3UY84WXOQWIO9W7J1B40FMD/view.html

United Nations (2014) The millennium development goals report, New York, Available at http://www.un.org/millenniumgoals/2014%20MDG%20report/MDG%202014%20English%20web.pdf

United Nations, Department of Economic and Social Affairs, Population Division (2014) Concise report on the world population situation in 2014, Reprinted with the permission of the United Nations

USEPA (2005) First-order kinetic gas generation model parameters for wet landfills. Available at http://nepis.epa.gov/Exe/ZyPDF.cgi?Dockey=P100ADRJ.PDF

USEPA (2006) Solid waste management and greenhouse gases: a life-cycle assessment of emissions and sinks, 3rd edn. USEPA, Washington, DC

Van Zwieten L, Kimber S, Morris S, Downie A, Berger E, Rust J, Scheer C (2010) Influence of biochars on flux of N_2O and CO_2 from Ferrosol. Aust J Soil Res 48(6–7):555–568

WRI (2014) Climate analysis indicators tool (CAIT) 2.0: WRI's climate data explorer, World Resources Institute. Accessed May 2014. http://cait.wri.org

Recent Advances of Anaerobic Digestion for Energy Recovery

Anthony Y.W. Ma, Billy K.W. Cheung, Katrina C.M. Kwok, Mingwei Cai and Patrick K.H. Lee

Abstract With climate change looming and the unsustainable supply of fossil fuels, the development of renewable and clean energy is urgently required. An often neglected source of clean energy is the organic material contained in waste and wastewater. Millions of tons of solid organic waste and wastewater are generated everyday worldwide. Instead of consuming energy, anaerobic digestion can be applied to treat the generated waste, thus achieving the objective of waste treatment for public health protection and also recovery of renewable methane for heat and power purposes. In this chapter, the benefits of anaerobic digestion will be introduced followed by a discussion on the mechanism and the typical design principles of anaerobic digestion systems. Some of the recent advancement of anaerobic digestion systems such as membrane bioreactors, fluidized bed reactors and co-digestion systems will be presented in the subsequent sections. The state-of-the-art molecular biological tools to monitor and diagnose the microbiology of anaerobic digestion systems will also be discussed. Lastly, the future outlook of anaerobic digestions will be addressed.

Keywords Anaerobic digestion · Waste treatment · Renewable energy · System design · Technology advancement · Molecular tools

Nomenclature

a.c.	Acoustic chemometrics
ABR	Anaerobic baffled reactor
ADM1	Anaerobic digestion model no. 1
AFBR	Anaerobic fluidized bed reactor

A.Y.W. Ma · B.K.W. Cheung · K.C.M. Kwok
Environmental Management Division, Hong Kong Productivity Council, Hong Kong, Hong Kong
e-mail: anthonym@hkpc.org

M. Cai · P.K.H. Lee (✉)
School of Energy and Environment, City University of Hong Kong, Hong Kong, Hong Kong
e-mail: patrick.kh.lee@cityu.edu.hk

O.P. Karthikeyan et al. (eds.), *Recycling of Solid Waste for Biofuels and Bio-chemicals*, Environmental Footprints and Eco-design of Products and Processes, DOI 10.1007/978-981-10-0150-5_4

AnMBR	Anaerobic membrane bioreactors
APBR	Anaerobic packed bed reactor
ASBR	Anaerobic sequencing batch reactor
CSTR	Continuous stirred tank reactor
DGGE	Denaturing gradient gel electrophoresis
EGSB	Expanded granular sludge bed
EN	Electronic nose
EPS	Extracellular polymeric substances
ET	Electronic tongue
FLU	Fluorescence spectroscopic
FW	Food waste
GC	Gas chromatographic
HPLC	High-performance liquid chromatographic
IR	Infrared spectroscopic
IWA	The international water association
LCFA	Long-chain fatty acid
MS	Mass spectrometry
NIR	Near infrared spectroscopic
OFMSW	Organic fraction municipal solid waste
PAT	Process analytical technology
PCR	Polymerase chain reaction
PFR	Plug-flow reactor
qPCR	Quantitative polymerase chain reaction
SHW	Slaughterhouse waste
SRB	Sulfate-reducing bacteria
TPAD	Two-phase anaerobic digester
T-RFLP	Terminal restriction-fragment length polymorphism
UAF	Upflow anaerobic filter
UASB	Upflow anaerobic sludge blanket
UV	Ultraviolet spectroscopic
VFAs	Volatile fatty acids
VIS	Visual spectroscopic
VS	Volatile solid
WWTP	Wastewater treatment plant

1 Introduction

The development of anaerobic digestion technology started in the beginning of the 19th century, although aerobic treatment and tertiary treatment were the mainstream treatment process after the Second World War. Nonetheless, anaerobic digestion of waste has been rapidly developed since the late 1960s and has been used to treat industrial wastewater as well as domestic wastewater for decades (Stronach et al. 1986; Speece 1996).

From a report on solid waste management conducted by the World Bank in 2013, it was estimated that cities currently generate roughly 1.3 billion tonnes of solid waste per year. With the current urbanization trends, this figure is expected to reach 2.2 billion tonnes per year by 2025, accounting for an increase of 70 % from the current level. Organic waste continues to be the largest component in municipal solid waste. The accumulation of solid organic waste is thought to be reaching critical levels in almost all regions of the world, becoming a pressing matter on public health, environmental quality, quality of life, and economic development. Anaerobic digestion can be considered as one of the oldest technologies for stabilization of wastes. There is now a growing interest in this technology to produce bioenergy as a result of increasing demand for energy coupled with the uncertainty surrounding fossil fuels cost. Bioenergy plays an important role in promoting renewable alternatives which is estimated to be the fourth largest energy resource in the world (Chen and Lee 2014).

Anaerobic digestion of wastes covers many aspects. In this chapter, the fundamental aspects including basic principles, microbiological processes, regime and limitation of anaerobic digestion on energy recovery will be introduced. Operational parameters such as acidic and alkaline conditions, occurrence of inhibitory compounds, together with the effect of temperature, are also considered. The design of anaerobic digestion reactor including fundamental design principles, performance enhancement by pretreatment, phase separation and co-digestion are reviewed, with special attention to technological advancement for improved methane recovery. Finally advanced molecular biological tools for system monitoring and the future outlook of anaerobic digestions will also be discussed.

1.1 Application of Anaerobic Digestion

Anaerobic digestion is applicable for a wide range of materials including municipal, agricultural and industrial wastes, and plant residues (Kalra and Panwar 1986; Gallert et al. 1998; Chen et al. 2008). It has a key role in residual waste stabilization for downstream processing. Recently, two new application areas, namely energy generation and production of value-added chemicals have drawn extensive interests (Batstone and Virdis 2014). One of the possible value-added chemicals is the production of volatile fatty acids (VFAs), which is a critical substrate for microorganism involved in the production of biodegradable plastics (Cai et al. 2009) and bioenergy (Lee et al. 2014).

1.2 Benefits of Anaerobic Digestion

Today, fossil fuels are the dominant energy sources meeting over 80 % of the world's energy demand in 2012 (International Energy Agency, France, 2013). The world energy demand was 5.5×10^{20} J in 2010. It is predicted to increase to 6.6×10^{20} J in 2020 and 8.6×10^{20} J in 2040 (Energy Information Administration, U.S. 2013). Nevertheless, fossil fuels are non-renewable and their reserves are limited. Moreover, tremendous amounts of greenhouse gases have been released from fossil fuel consumption driving the incentives of international communities to develop and utilize renewable energy. Of the renewable energy sources such as solar or wind power production, bioenergy becomes increasingly competitive on its own merits, primarily due to the extensive availability of biomass, biomass production technologies and infrastructure, and biomass being the sole feedstock for liquid fuels production. Biogas, a source of bioenergy, is a product of anaerobic digestion of organic substrates, which is one of the oldest processes used for the waste treatment and stabilization of sludge. The production of biogas through anaerobic digestion offers significant advantages over other processes of waste treatment such as (i) producing less residual solid generation in comparison to aerobic treatment, (ii) generating bioenergy in the form of biogas, (iii) yielding a digestate produced with high bioavailability as an improved fertilizer. The biogas formed is generally composed of 48–65 % methane, 36–41 % carbon dioxide, up to 17 % nitrogen, <1 % oxygen, 32–169 ppm hydrogen sulfide, and trace amounts of other gases (Rasi et al. 2007). Carefully designed and engineered anaerobic digestion of organic waste is therefore environmental beneficial in two ways:

(i) Generating of methane which is a kind of the greenhouse gases, in an enclosed reactor to prevent it from entering the atmosphere directly.
(ii) Displacement of energy from fossil fuels by clean bioenergy.

2 Mechanism of Anaerobic Digestion

2.1 Basic Principles of Anaerobic Digestion

Biogas formation is governed by microorganisms and the metabolic activities in the reactor. Typical anaerobic digestion of organic matters occurs in four steps, namely hydrolysis, acidogenesis, acetogenesis and methanogenesis in which a consortium of microorganisms including fermentative bacteria, acidogenic bacteria, acetogenic bacteria and methanogens are responsible for biogas production from organic materials such as carbohydrate, oils, fats and proteins. The carbohydrates, protein, oils and fats are firstly hydrolyzed into monomeric sugars, amino acids and fatty acids respectively by extracellular enzymes (amylase, lipase, proteolytic enzymes) produced by fermentative bacteria. Madigan proposed an approximate chemical formula for the mixture of organic materials as $C_6H_{10}O_4$ (Madigan et al. 2009).

The hydrolysis reaction can be written as:

$$C_6H_{10}O_4 + 2H_2O \rightarrow C_6H_{12}O_6 + H_2 \quad (1)$$

In acidogenesis, the hydrolyzed organic compounds denoted as $C_6H_{12}O_6$ are utilized by acidogenic bacteria or acid forming bacteria, which are a group of fast growing bacteria, and generate volatile fatty acids such as acetic acid, propionic acid, butyric acid and valeric acid as well as carbon dioxide, water and hydrogen. Generation of volatile fatty acids can be expressed as:

$$C_6H_{12}O_6 \Leftrightarrow 2CH_3CH_2OH + 2CO_2 \quad (2)$$

$$C_6H_{12}O_6 + 2H_2 \Leftrightarrow 2CH_3CH_2COOH + 2H_2O \quad (3)$$

In acetogenesis, the VFAs except acetic acid are utilized by acetogenic bacteria, which are a group of slow growing bacteria, to produce acetic acid and hydrogen. The acetogenesis can be written as:

$$CH_3CH_2COOH + 2H_2O \Leftrightarrow CH_3COOH + CO_2 + 3H_2 \quad (4)$$

Finally, methanogens utilize acetic acid, ethanol, methanol, hydrogen and carbon dioxide to form methane gas in methanogenesis. Methanogens utilizing acetic acid to produce methane are known as acetotroph while those utilizing hydrogen and carbon dioxide are known as hydrogenotroph. About 70 % of the methane are produced stoichiometrically via the acetate pathway and 30 % are produced via the hydrogen pathway (Siegrist et al. 2002; Madigan et al. 2009). The pathways for methanogenesis can be expressed as:

$$2CH_3CH_2OH + CO_2 \Leftrightarrow 2CH_3COOH + CH_4 \quad (5)$$

$$CH_3COOH + CO_2 \Leftrightarrow CH_4 + 2CO_2 \quad (6)$$

$$CH_3OH + H_2 \Leftrightarrow CH_4 + 2CO_2 \quad (7)$$

$$CO_2 + 4H_2 \Leftrightarrow CH_4 + 2H_2O \quad (8)$$

Figure 1 shows a scheme of the anaerobic digestion pathway, from long-chain organic compounds including proteins, carbohydrates and lipids to the final products, i.e., methane and carbon dioxide (Gujer and Zehnder 1983; Siegrist et al. 2002).

Methane production from anaerobic digestion process in waste treatment is generally limited by the rate of hydrolysis of suspended organic matters. Efficient pretreatment can enhance the ability of bacteria to access the suspended substrate and increase the methane yield. The objective of implementing different types of pretreatment in anaerobic digestion is to enhance the bioavailability of particular

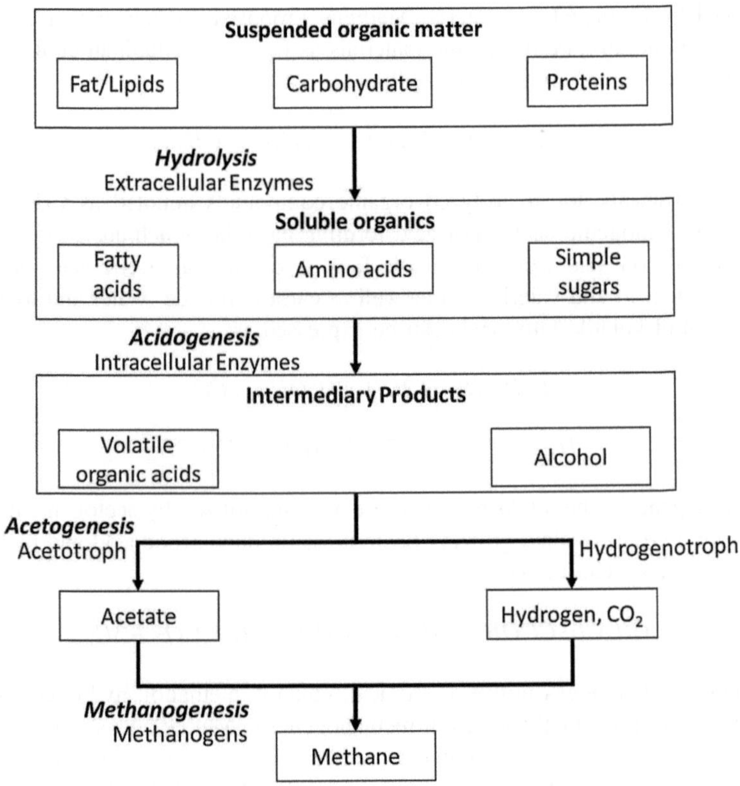

Fig. 1 Scheme of biodegradation steps of complex matter in anaerobic digestion

substrates so that enzymes can more efficiently hydrolyze the substrate. A number of pretreatment methods are discussed in Sect. 4.1. The overall organic matter stabilization can be improved through improvement from pretreatment technology.

Anaerobic digestion of waste is capable of recovering energy from a wide range of feedstock from different sources such as agricultural sector, industrial sector and municipal sector which needs to be (i) biodegradable, (ii) non-woody with low proportion of lignocellulosic material, and (iii) balanced in macro and micro nutrients (Kothari et al. 2014). Therefore, feedstock can range from readily biodegradable wastewater to complex high-solid waste. In order to obtain a higher yield of biogas, anaerobic co-digestion treatment, the simultaneous digestion of two or more substrates, is a feasible option to overcome the drawbacks of single substrate digestion and to improve the process efficiency. Figure 2 shows an overview of various feedstock from different sources. The choice of feedstock is influenced by various interrelated process factors such as reactor design and operation, quality of products, source and mass flux, economic considerations, bacterial physiology and specific purpose (Steffen et al. 2012).

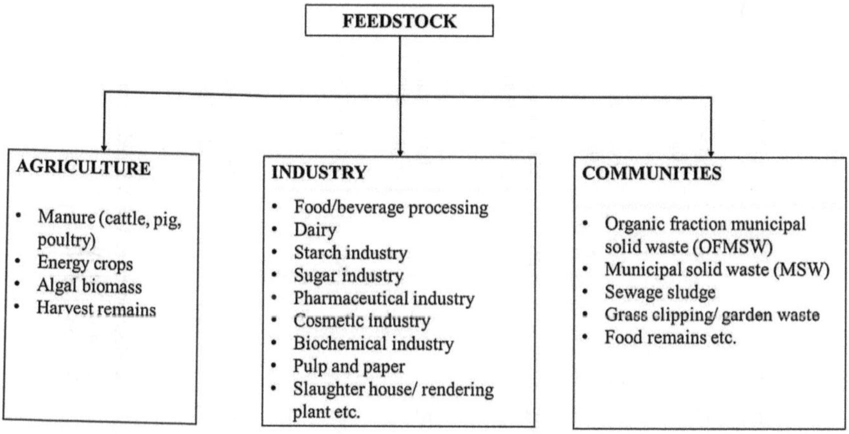

Fig. 2 Categorization of various feedstock from different sources

It should be noted that phase separation can be used for performance enhancement by optimizing the reactor configuration for the different stages of anaerobic processes in separate tanks whereby the conditions are optimized for specific groups of bacteria. The improvement of energy recovery by co-digestion and phase separation in anaerobic digestion are discussed in Sect. 4.2. Anaerobic digestion involves different groups of microorganisms which are highly sensitive to the environment. The operating parameters affecting anaerobic digestion are discussed in detail in Sect. 2.3.

Anaerobic digestion of waste serves the dual purpose of both energy recovery and waste management. Recently, it has been widely applied to municipal solid waste (MSW) to generate energy. Table 1 summarizes the high yield of methane production from anaerobic digestion of municipal solid waste. It should be noted that MSW is classified as a heterogeneous material in which the composition varies widely according to regional differences, climate, extent of recycling, collection frequency, season, cultural practices. Considering the biodegradability of OFMSW, the potential of anaerobic digestion of OFMSW are high as well as the methane yield, and co-digestion of MSW with sewage sludge is also becoming increasingly attractive.

2.2 *Microorganism and Microbiological Process*

A wide variety of microbial communities have been reported to be involved in the anaerobic digestion process (Fricke et al. 2007; Fantozzi and Buratti 2009). The microbial population distribution is highly dependent on the substrate and product concentration as well as on environmental conditions such as pH, temperature, hydrogen concentration etc. However, knowledge of the microorganism and

Table 1 Literature data of biomass with high yields of methane from municipal solid waste feeds

Feed	Bioreactor	Temp (°C)	HRT (days)	OLR (kg VSm^{-3} d^{-1})	CH_4 yield (m^3 kg^{-1} VS)	CH_4 PR (m^3 m^{-3} d^{-1})	VS_r (%)	Reference
HS-OF MSW 3–5.6 % TS VS = 82–87 % TS	CSTR Laboratory Plant	35	14–20	4	0.430	1.70	NR	Pauss et al. (1984)
HS-OF MSW 6.4 % TS VS = 89.9 % TS	CSTR 3 m^3	33–37	9–25	2.1–6.9	0.390	0.82–2.02	63–69	Cecchi et al. (1986)
SS-OF MSW VS = 88 % TS (Without paper, wood, plastic)	CSTR 3 m^3	35	25	2.1	0.399	NR	69	Mata-Alvarez et al. (1990)
SC-OF MSW 80:20 (% TS basis) VS = 88 % TS	CSTR 2.2 m^3	35	14.5	3.9	0.403	NR	71	Mata-Alvarez et al. (1990)
MS-OFMSW 35 % TS VS = 58.6 % TS	Pilot Plant 500 m^3 Valorga Process	37	15	13.7	0.230	NR	45	Valorga (1985)
Pre-composted MS-OFMSW VS = 43 % TS	CSTR 3 m^3	35	16.2	4.1	0.145	NR	27	Mata-Alvarez et al. (1990)

Remark
Temp Temperature, *HRT* Hydraulic retention time, *OLR* Organic loading rate, VS_a VS added, CH_4 *PR* Methane production rate, VS_r VS reduction
MS-OFMSW Mechanically sorted organic fraction of municipal solid waste, *HS-OFMSW* Hand sorted organic fraction of MSW
SS-OFMSW Source sorted organic fraction of MSW, *SC-OFMSW* Organic fraction of MSW from a separated collection, *NR* Not reported

microbiological processes involved is revealed gradually through the use of modern molecular techniques which complement traditional cultivation process and microscopic identification techniques (Merkel et al. 1999). The development of the modern molecular techniques on microorganisms in anaerobic digestion will be discussed in Sect. 5. The types of microorganism in the four distinct stages of anaerobic digestion are shown in Tables 2 and 3.

2.3 *Operational Parameters Affecting Anaerobic Digestion*

The ultimate methane production is influenced by a number of operational parameters in the anaerobic digestion reactor such as temperature, type of feedstock, pH level, retention time, C/N ratio, VFA concentration etc. Maximum methane production takes place when optimum range of these operational parameters is chosen. The optimum range of these parameters is reviewed in this section.

2.3.1 Temperature

Microorganisms in anaerobic digestion are very sensitive to temperature changes which affect hydrogen and methane production, and the decomposition of organic materials. There are three possible ranges of temperature in which the process can be carried out (psychrophilic, mesophilic and thermophilic) as shown in Table 4. Chae et al. studied the effects of temperature and temperature shock on the biogas yield from anaerobic digestion (Chae et al. 2008) and reported that methane content increased with increasing digestion temperature, but only to a small extent. Temperature shocks from 35 °C to 30 °C and 30 °C to 32 °C led to a drop in the biogas production rate. No lasting damage was observed from the digestion performance after recovery.

Thermophilic anaerobic digesters often manifest chronically higher VFAs concentration than those found in mesophilic anaerobic digesters (Kim et al. 2002). Therefore, the optimal conditions for anaerobic digestion to reduce energy consumption may be thermophilic hydrolysis/acidogenesis and mesophilic methanogeneis which is consistent with a two-phase anaerobic digestion process. This arrangement uses a mesophilic reactor as a polishing stage, eliminating the drawbacks of the thermophilic process. However, thermophilic conditions are applied in most of the large-scale centralized biogas co-digesters (Kothari et al. 2014). Digestion of organic urban wastes using thermophilic and mesophilic processes has also been studied by researchers and they found that thermophilic process is a more realistic and viable option as the added amount of heat required for thermophilic operations can be offset by the higher gas production yields and rates (Parkin and Owen 1986; De Baere 2000; Kim et al. 2002; Kuo and Lu 2004).

Table 2 Characteristics of typical hydrolytic bacteria and acidogenic bacteria in anaerobic digestion (Amani et al. 2010)

Type	Substrates	Products	Typical species	Reference
Hydrolytic bacteria	Proteins	Amino acids, sugars	*Clostridium sp., Proteus vulgaris, Peptococcus sp., Bacteroides sp., Bacillus sp., Vibrio sp.,*	Kim et al. (2009) Fang et al. (2009) Westlake et al. (1967) Ochoa-Reparaz et al. (2008) Chang et al. (2008) Parvez et al. (2008)
	Carbohydrates	Sugars	*Clostridium sp., Acetivibrio cellulolyiticus, Staphylococcus sp., Bacteroides sp.,*	Chong et al. (2009) Khan (1980) Ziagova et al. (2009)
	Lipids	Higher fatty acids, alcohols, amino acids	*Clostridium sp., Micrococcus sp.,*	Jo et al. (2008) Tuleva et al. (2009)
		Sugars	*Staphylococcus sp.,*	
Acidogenic bacteria	Amino acids	Valerate, isovalerate, propionate, butyrate	*Lactobacillus sp., Eschericia coli*	Kim et al. (2009) Li et al. (2003)
		Acetate, H_2, Higher fatty acids	*Staphylococcus sp., Bacillus sp., Pseudomonas sp., Micrococcus sp., Eubacterium limosum, Clostridium sp., Zymomonas mobiliz*	Kalyani et al. (2009) Hur and Rafii (2000) Fritsch et al. (2008) Wilkins (2009)
	Sugars	CO_2, H_2, formate, acetate, butyrate	*Eubacterium sp.,*	Kalyani et al. (2009)
		CO_2, H_2, formate, acetate, ethanol, lactate	*Eschericia coli*	
		Formate, acetate, ethanol, lactate	*Bifidobacterium sp.,*	Cheikhyoussef et al. (2009)
		Acetate	*Acetobacterium sp.,*	Bainotti et al. (1996)
	Fatty acids	Valerate, isovalerate, propionate, butyrate, acetate, H_2	*Clostridium sp.,*	
	Alcohols		*Syntrophomonas wolfei*	

Table 3 Characteristics of typical acetogenic bacteria and methanogens in anaerobic digestion (Amani et al. 2010)

Type	Substrates	Products	Typical species	Optimum growth condition	
				pH	Temp. (°C)
Acetogenic bacteria	Butyrate	Acetate	*Syntrophobacter wolinii*	NR	35–40
		H_2/CO_2, formate	*S. fumaroxidans*	NR	35–40
	Propionate	H_2/CO_2, formate	*Syntrophomonas wolfei*, *Pelotomaculum thermopropionicum*, *P. schinkii*	NR	35–40 50–60† 32–37
		Butyrate, acetate	*Smithella propionica*	NR	35–40
	H_2, CO_2	Acetate	*Clostridium aceticum*	NR	30–37
Methanogens	Acetate	CH_4, CO_2	*Methanothrix soehngenii*, *Methanosaeta concilii*, *Methanosarcina acetivorans*	7.4–7.8 7.1–7.5 6.5–7.5	35–40 35–40 35–40
	H_2, CO_2	CH_4	*Methanobacterium bryantii*, *M. thermoautotrophicum*, *M. alcaliphilum*, *Methanobrevibacter arboriphilus*, *Methanococus jannaschii*, *Methanolacinia paynteri*, *Methanospirillum hungatei*, *Methnanoplanus endosymbiosus*, *M. olentangyi*, *Methanothermus fervidus*	6.9–7.2 7.2–7.6 8.1–9.1‡ 7.8–8.0 5.0–7.0 6.6–7.2 NR 6.6–7.2 NR 6.5	37–39 65–70† 37 30–37 83§ 40 30–37 40 30–37 83§

(continued)

Table 3 (continued)

				Optimum growth condition	
	Formate, H_2, CO_2	CH_4, CO_2	*Methanobacterium formicicum,*	6.6–7.8	37–45
			Methanobrevibacter smithii,	7.0	37–39
			M. ruminantium,	7.0	37–39
			Methanococcus voltae,	6.5–8.0	35–40
			M. deltae,	NR	37
			M. maripaludis,	6.5–8.0	35–40
			M. thermolithoautotrophicus,	6.5–8.0	65†
			Methanoplanus limicola,	7.0	40
			Methanogenium cariaci,	6.2–6.6	20–25
			M. marisnigri,	6.8–7.3	20–25
			M. olentangyi,	NR	37
			M. tatii,	7.0	37–40
			M. thermophilicum,	6.5–7.2	55–60†
			M. bourgense,	6.3–6.8	35–42
			Methanocorpusculum aggregans	6.4–7.2	35–37

Remarks †Theromophilic; ‡Alkaliphilic; §Hyper-thermophilic
NR Not reported

Table 4 Classification of anaerobic digestion by operating temperature

Type	Operating temperature	Reference
Psychrophilic (or cryophilic)	10–20 °C	Sutter and Wellinger (1987)
Mesophilic	30–40 °C	Bolzonella et al. (2005), Zhang et al. (2014)
Thermophilic	55–70 °C	Buhr and Andrews (1977)
Ambient/seasonal temperature	Temperature changes in the surrounding environment (Typically 15–25 °C)	Yusuf and Ify (2011)
Hyperthermophilic	65–70 °C	Lee et al. (2009)

2.3.2 Acidic and Alkaline Conditions

Methanogens are extremely sensitive to pH, while fermentative microorganisms are generally less sensitive and can function in a wider range of pH between 4.0 and 8.5 (Hwang et al. 2004). A range of pH values suitable for anaerobic digestion has been reported by various researchers, but the optimal pH for methanogenesis has been found to be around 7.0 (Huber et al. 1982; Yang and Okos 1987; Khalid et al. 2011). The growth rate of fermentative bacteria is faster than those of methanogens, leading to the accumulation of acids in the digesters. Two main strategies for rectifying the low pH due to acid accumulation: (i) stopping the feed and allowing enough time for the methanogenic population to reduce the concentration of VFAs inside the system; and (ii) addition of bases to raise pH and provide additional buffering capacity. Another strategy suggested by Shah is that drastic reduction of pH could be prevented by the addition of another feed at a suitable ratio with the main feed (Shah et al. 2015) as practices in co-digestion.

2.3.3 Inhibitory Compounds

It is desirable to control inhibitory or toxic materials to achieve higher efficiency or a more economical operation of anaerobic digestion process performed by removal of toxic materials from waste stream or by dilution of the waste to below the toxicity threshold in the systems. Precipitation is commonly employed to remove the toxic materials from the systems.

Ammonia toxicity
Ammonia is an essential nutrient for the growth of microorganisms involved in anaerobic digestion but also acts as an inhibitor at high concentration. Fermentation of nitrogen-containing materials such as urea and proteins releases ammonia-nitrogen largely in the ionized form (NH_4^+). The toxic unionized form (NH_3)

increases with increasing pH in the system as the *pKa* value of ammonia is 9.3 (Koster and Lettinga 1984). Free ammonia is more toxic to methanogens than ionized ammonium (NH_4^+) because it is more readily diffusible through the cell membrane, causing proton imbalance, and/or potassium (K^+) deficiency, while ionized ammonium may just inhibit the methane synthesizing enzyme directly (Gerardi 2006). Another reason why ionized form of ammonia is less inhibitory than the free form is that the hydroxide ion produced can react with carbon dioxide to form bicarbonate, which increases the buffering capacity of the anaerobic reactor, making the process less susceptible to pH fluctuations when the production rates of acetogenic bacteria and methanogens differ.

Sulfide toxicity
A number of industrial wastes from petrochemical plants, tanneries, viscose rayon factories and coal gasification for electricity production generate sulfate-containing waste streams. Sulfidogens or sulfate-reducing bacteria (SRB) play a significant role in anaerobic digestion, which reduce sulfate to sulfide in the reactor under certain condition. Sulfide generated may be inhibitory to anaerobic digestion by (i) inhibiting methanogens, (ii) reducing rate of methanogenesis, and (iii) decreasing the quantity of methane produced by competing for the available carbon and/or hydrogen source. Inhibitory effect of sulfide in anaerobic digestion can be separated into two parts: competition for substrates between sulfate-reducing bacteria and methanogens directly and inhibition of methane formation by sulfide ions in the system. Competition between sulfate-reducing bacteria and methanogens in sulfate-containing waste streams for acetate as their common primary substrate can significantly affect the methane production efficiency.

The optimum conditions for anaerobic metabolic activity proposed by researchers are summarized in Table 5.

Table 5 Optimum conditions for anaerobic metabolic activity

Parameters	Optimum conditions	Reference
Temperature	Mesophilic range (35–40 °C) Thermophilic (50–65 °C)	Van Haandel and Lettinga (1994) Arsova (2010)
pH	6.3–7.8	Wang et al. (2012)
Carbon to nitrogen ratio (C/N ratio)	25–30	Ghosh and Pohland (1974)
Volatile fatty acid (VFA)	2000–3000 mg/L	Eastman and Ferguson (1981)
Organic loading rate (OLR) and nutrient concentration	Varies according to the substrate and inoculum	–

3 Design of Anaerobic Digestion Processes

Anaerobic digestion composes a broad family of processes which can be classified according to:

(a) their feedstock input mode: batch and continuous processes;
(b) single-step, double- or multiple steps; and
(c) geometry of the main treatment unit: vertical and horizontal unit.

3.1 Design Principles of Anaerobic Digestion System

Reactors of anaerobic digestion often operates under heterogeneous system whereby three phases namely: solid phase (sludge), liquid phase (wastewater) and gaseous phase (biogas) present simultaneously. The oldest and simplest type of anaerobic digester is not equipped with any mixing or heating, thus a long digestion period of 30–60 days is required. Some degree of natural mixing occurs inside the reaction tank due to bubbling of gas generated and thermal convection currents created from the digestion processes. Due to the lack of proper mixing, stratification usually occurs in four zones: (i) scum layer, (ii) supernatant layer, (iii) layer of digesting biosolid and (iv) layer of digested biosolid. A schematic diagram of this type of anaerobic digesters without mixing is given in Fig. 3. The accumulated biosolid at the bottom of the reaction tank is periodically discarded.

Different variations of anaerobic digestion have evolved over the years to improve the degradation performance including high-rate digestion and phase separated digestion. Figure 4 shows the schematic diagram of a typical high-rate digestion. The characteristic features of high-rate anaerobic digestion including

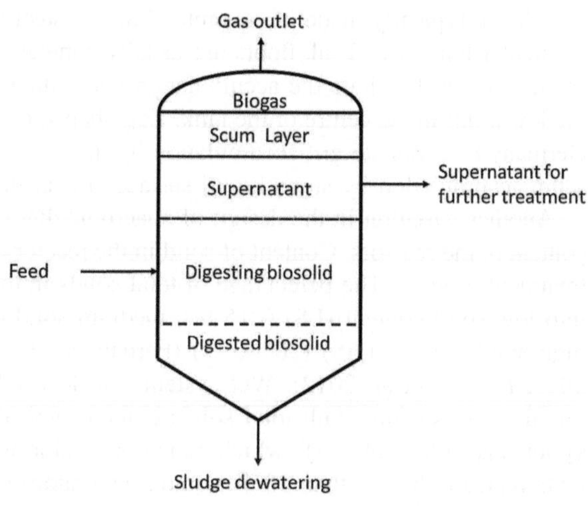

Fig. 3 Schematic diagram of low rate anaerobic digestion without mixing and heating

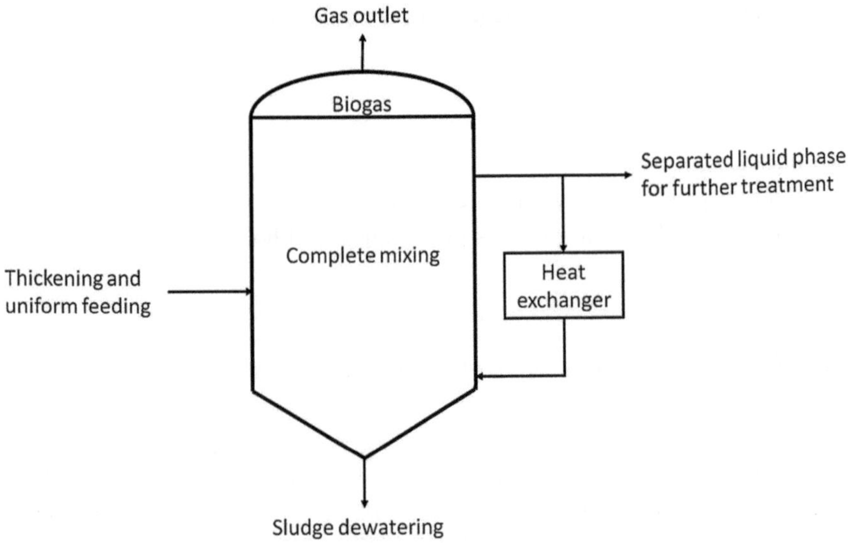

Fig. 4 Schematic diagram of high-rate anaerobic digestion

heating, auxiliary mixing, thickening and uniform feeding are introduced to the reactor design to create a uniform environment for microbial growth in order to improve stability and efficiency of biodegradation processes.

Determination of reaction tank volume is the first important consideration in designing an anaerobic digestion system. Various methods have been used for sizing of digestion tank including (i) per capita basis, (ii) solids loading, (iii) solids retention time, (iv) volatile solids destruction and (v) gas production (Turovskiy and Mathai 2006).

Anaerobic digestion reactors are mostly cylindrical or egg shaped. Vertical cylindrical digestion tanks are widely used in the United States, with diameter from 6 to 38 m, typically made of concrete although steel tank design are also common in smaller tank size. Tank floors are usually conical with slopes of varies between 1:3 and 1:6 to facilitate the accumulation and withdrawal of digested sludge from the low point in the centre of the tank. Egg-shaped digestion tanks are originated in Germany to eliminate grit accumulation by the steeply sloped bottom and to avoid scum accumulation by small liquid surface area at the top.

Another variation in the design of anaerobic digestion processes is on the solid content in the reactors. Content of solid in the reactor affects the reactor volume and treatment process. The percentage of total solids in the digester can be categorized into low solid content (LS) (<15 %), medium solid content (MS) (15–20 %) and high solid content (HS) (20–40 %) (Fernández et al. 2008; Cao and Pawlowski 2012; Raposo et al. 2012). Wet systems are low solid AD which are applied to liquid waste streams with total solids content typically less than 15 % while dry systems are high solid AD which handle stackable feedstock with total solid contents typically higher than 30 % without any addition of external liquids.

Single-stage low solids (SSLS) wet anaerobic digestion processes have been used for decades in the stabilization of sludge. The feedstock is conditioned to the appropriate solid content (10–15 %) by adding process water in the wet anaerobic digestion reactor with internal mixing to obtain homogeneity. The predominant reactor of wet anaerobic digestion is continuously stirred tank reactor (CSTR) with mechanical stirring to avoid stratification of the substrate inside the reactor. Short circuiting may be experienced in CSTR. Large amount of water consumption is needed to be mixed with the feedstock to obtain the low solid content, which can be acquired from treated supernatant.

High solid anaerobic digestion has been claimed to be more advantageous than low solid anaerobic digestion for several reasons, such as smaller reactor volumes, lower energy requirement for heating, higher biogas yield from undiluted wastes and less material handling (Duan et al. 2012). However, dry streams may suffer some drawbacks. They usually require proper preconditioning of the feedstock material, including substrate treatment and mixing with structure material, and special loading and unloading techniques. The content inside the digester may not be totally mixed, leading to lower methane yields than the wet systems. Different types of single-stage high solids (SSHS) dry anaerobic digestion processes have been developed and are in use commercially in Europe such as Dranco, Kompogas, and Valorga processes. The Dranco process developed in Belgium is a true dry-process for treatment of organic fraction of MSW, which is characterized by its design of feeding from the top, collection of digested biosolid at the bottom of the reactor and no internal mixing mechanism with total solid content at about 30–40 % (Cho et al. 2013). The Kompogas process developed in Switzerland takes place in plug flow in a horizontally cylindrical steel tank with total solid content at about 23 % (Hartmann and Ahring 2006). The Valorga process developed in France is a semi-dry mesophilic process in which mixing of waste with recycled process water takes place with total solid content of 30 % (Fernández et al. 2008). In additional to the improvement of reactor tank design in anaerobic digestion, the technological advancement in process design such as pretreatment, phase separation, co-digestion and biomass immobilization are discussed in Sect. 4.

4 Technology Advancement for Improved Methane Recovery

4.1 Pretreatment for Digestion Enhancement

Most researchers reported that the rate-limiting step for complex organic substrates is the hydrolysis step in an anaerobic digestion process (Valo et al. 2004; Izumi et al. 2010; Rafique et al. 2010; Bordeleau and Droste 2011; Fdez-Guelfo et al. 2011; Ma et al. 2011). Different pretreatments are utilized for anaerobic digestion such as mechanical (ultrasound, high pressure and lysis), thermal (<100 °C, >100 °C), chemical (ozonation, alkali, acids), microwave, ultrasonic, electric pulses, wet oxidation, freeze/thaw and biological treatment to increase the bioavailability of complex organic matters to microbes. Pretreatment methods to improve performance of anaerobic digestion have been the focus of many research studies over the last 30 years (Holm-Nielsen et al. 2009; Pilli et al. 2011) and the improvement of anaerobic digestion in terms of increasing methane generation and solid reduction are well known advantages of pretreatments.

The pretreatment effects are complex and generally linked to substrate characteristics and pretreatment mechanisms. Carlsson et al. have examined the effect of substrate pretreatment on anaerobic digestion (Carlsson et al. 2012), namely particle size reduction, solubilization, formation of refractory compounds, biodegradability enhancement and loss of organic materials for different substrate categories including wastewater treatment plant (WWTP) residues, organic waste from households, energy crops/plant residues, waste from food industry and manure. It is reported that thermal and ultrasonic pretreatments are predominantly applied on anaerobic digestion of WWTP residues, chemical and thermal pretreatment have been applied to less frequently studied substrates such as energy crops/harvesting residues, organic waste from food industry and manure, whereas mechanical and thermal pretreatments are commonly applied to organic fraction of municipal solid waste (OFMSW). Another focus on pretreatment methods is their ability to enhance anaerobic digestion process in terms of efficiency, energy balance, environmental sustainability as well as capital, operational and maintenance costs (Ariunbaatar et al. 2014).

4.1.1 Mechanical Pretreatment

Mechanical pretreatment is used to reduce both the particle size and crystallinity of lignocellulosic materials through a combination of chipping, grinding or milling processes, in order to increase the specific surface area and reduce the degree of polymerization of substrate (Sun and Cheng 2002). Smaller particles increase the surface area available to the microorganisms, resulting in increased bioavailability to bacteria and improved anaerobic degradability. Particle size reduction can accelerate the hydrolysis and acidogenesis processes as well as the production of soluble organic materials such as VFAs, resulting in a higher organic loading in the anaerobic digester. However excessive size reduction may result in higher solubilization and in turn excessive VFAs accumulation, leading to a decrease in methane production. The power requirement of mechanical pretreatment is relatively high depending on the final particle size and the substrate characteristics. In particular, the recalcitrant nature of cell walls of green waste makes mechanical pretreatment energy intensive (Izumi et al. 2010).

Ultrasonic disintegration is one type of mechanical pretreatments in which ultrasonic treatment acts to disrupt the cell structure and floc matrix of the substrate. There are two key mechanisms associated with ultrasonic treatment: (i) cavitation, which is favoured at a low frequency, and (ii) chemical reactions due to the formation of free radicals at a high frequency (Carrère et al. 2010). According to the studies by Show and co-workers, the optimal range of solid content for sonication lies between 2.3 and 3.2 % TS (Show et al. 2007). If the solid concentration of feedstock is too high, increased viscosity hinders cavitation bubble formation. The threshold specific energy ranges from 1000 to 16,000 kJ kg^{-1} TS with sludge as substrate although biogas production increases with energy input (Salsabil et al. 2009).

4.1.2 Thermal-Alkaline Solubilization Pretreatment

Alkaline treatment is one commonly used chemical treatment in anaerobic digestion in which there are two major reactions: (i) solvation and saphonication inducing swelling of solids to increase the specific surface area of the substrate; followed by (ii) simultaneous reactions of saponification and neutralization of various acids formed by degradation of the particulates leading to an increase in COD solubilization (Kim et al. 2003). Alkaline treatment is relatively effective in sludge solubilization, with the order of efficacy being NaOH > KOH > $Mg(OH)_2$ and $Ca(OH)_2$. Alkaline pretreatment by sodium hydroxide at relatively low dosage levels is effective in solubilizing municipal waste activated sludge at ambient temperature. Mouneimne et al. demonstrated that high concentration of Na^+ and K^+ may cause subsequent inhibition of anaerobic digestion (Mouneimne et al. 2003). Alkaline treatment is normally combined with thermal treatment. Waste solubilization and biodegradability improve with alkali dosage and temperature (Kim et al. 2003) Thermal-alkaline pretreatment usually proceeds at temperature lower than thermal hydrolysis alone and could result in a higher biogas production with a higher methane content.

4.1.3 Oxidative Pretreatment (Ozonation)

Oxidative pretreatment by ozonation is another chemical pretreatment method. It is the most widely used chemical method which does not lead to accumulation of salt and no chemical residues remain in the systems as compared to other chemical pretreatment methods (Carrère et al. 2010). Ozone is a strong oxidant which decomposes into radicals and reacts with organic substrates directly and indirectly. The direct reaction depends on the structure of the reactant whereas the indirect reaction is based on the hydroxyl radicals. Several studies have shown an optimal range of ozone dosage for the enhancement of anaerobic biodegradability such as 0.1 g O_3 g^{-1} COD (Weemaes et al. 2000), 0.2 g O_3 g^{-1} TSS (Yeom et al. 2002), and 0.15 g O_3 g^{-1} TSS (Bougrier et al. 2007). Ozonation has been combined with anaerobic digestion as a pretreatment (Weemaes et al. 2000; Yeom et al. 2002; Bougrier et al. 2007) or post-treatment with recycling back to the anaerobic digester (Battimelli et al. 2003; Goel et al. 2003).

Ariunbaatar et al. compared the efficiency of various pretreatment methods for enhancing the anaerobic digestion of OFMSW and food waste (FW) in terms of biogas production, VS reduction and COD solubilization as listed in Table 6.

Table 6 Efficiency of pretreatment for enhancing anaerobic digestion of OFMSW and FW

				Result			
Substrate	Pretreatment type	Pretreatment condition	Type of AD system	COD solubilization	VS reduction	CH_4 generation	Reference
OFMSW (source-sorted)	Mechanical	Disc screen	Thermophilic batch		80.6 %	338 ml CH_4/g VS	Davidson et al. (2007)
	Mechanical	Screw press	Thermophilic batch		63.2 %	354 ml CH_4/g VS	
	Mechanical	Shredder with magnetic separation	Thermophilic batch		63 %	289 ml CH_4/g VS	
OFMSW (source-sorted)	Mechanical	Disc screen	Thermophilic batch			428 ml CH_4/g VS	Hansen et al. (2007)
	Mechanical	Screw press	Thermophilic batch			461 ml CH_4/g VS	
	Mechanical	Shredder with magnetic separation	Thermophilic batch			487 ml CH_4/g VS	
OFMSW	Mechanical	Rotary drum	Thermophilic batch			457–557 ml CH_4/g VS	Zhu et al. (2009)
OFMSW (Synthetic)	Thermal hydrolysis process (THP)	Thermophilic pre-hydrolysis	Thermophilic continuous (2-stage)	81.5 % removal	95.7 %	2 times higher biogas production	Ueno et al. (2007)
OFMSW	Chemical	Alkaline	NA	11.5 % higher		150 ml CH_4/g VS	López Torres and Espinosa Lloréns (2008)

(continued)

Table 6 (continued)

				Result			
OFMSW	Thermal hydrolysis process (THP)	Pre-hydrolysis at 55 °C	Mesophilic continuous		47.5–71.6 %	299–418 ml CH_4/g VS	Schmidt and Ahring (1993)
OFMSW	Ultrasonic	Sonication at 20 kHz for 30-60 min	Mesophilic batch	60 % higher		24 % higher than untreated	Cesaro and Belgiorno (2014)
OFMSW (Synthetic)	Thermal hydrolysis process (THP)	Mesophilic and thermophilic pre-hydrolysis	Mesophilic and thermophilic continuous (2-stage)			341 ml CH_4/g VS	Escamilla-Alvarado et al. (2012)
OFMSW and corn stalk	Freeze/thaw	Freeze explosion followed by thermophilic pre-hydrolysis	Thermophilic			520 ml CH_4/g VS, 104 ml H_2/g VS	Kvesitadze et al. (2012)
FW	Thermal hydrolysis process (THP)	Semi-aerobic and anaerobic pre-hydrolysis	Mesophilic continuous	95 % removal		500 ml CH_4/g VS	Kim et al. (2000)
FW	Thermal hydrolysis process (THP)	Thermophilic pre-hydrolysis	Mesophilic		61.3 %	280 ml CH_4/g VS	Kim et al. (2004)
FW	Thermal hydrolysis process (THP)	Mesophilic pre-hydrolysis	Mesophilic continuous (2-stage)			9–13 % higher than mesophilic and thermophilic AD respectively	Verrier et al. (1987)
FW	Thermal hydrolysis process (THP)	Mesophilic pre-hydrolysis	Mesophilic continuous			546 ml CH_4/g VS, 65 ml H_2/g VS	Wang and Zhao (2009)

(continued)

Table 6 (continued)

				Result			
FW	Thermal hydrolysis process (THP)	Thermophilic pre-hydrolysis	Mesophilic continuous			464 ml CH_4/g VS, 205 ml H_2/g VS	Chu et al. (2008)
FW with polyactide	Thermal hydrolysis process (THP)	Hyper-thermophilic/thermophilic pre-hydrolysis	Thermophilic—Temperature phased AD			15–18 % higher than conventional thermophilic digester	Wang et al. (2011)
FW	Electric pulse	400 pulses with electroporation	Mesophilic continuous			20–40 % higher than untreated	
FW	Freeze/thaw	Frozen/thaw and pre-hydrolysis for 7 days	Mesophilic continuous	10 % higher		23.7 % higher than untreated	Stabnikova et al. (2008)
	Freeze/thaw	Frozen/thaw and pre-hydrolysis for 12 days	Mesophilic continuous	4 % higher		8.5 % higher than untreated	Carlsson and Anox Kaldnes(2008)

4.2 *Phase Separation and Co-digestion of Anaerobic Digestion*

4.2.1 Phase Separation

The prospects of phased anaerobic digestion of waste are extremely promising to achieve increased stability, higher loading capacities and greater process efficiencies than single-stage systems (Shuizhou and Zhou 2005). The advantages of two-phase anaerobic digestion (TPAD) have been extensively documented (Ghosh and Pohland 1974; Ghosh et al. 1985). Efficiency improvement of anaerobic digestion can be brought about by either digester design modification or advanced operating techniques.

Anaerobic digestion occurs in four steps, namely hydrolysis, acidogenesis, acetogenesis and methanogenesis as discussed above. The degradation process of organic matters in anaerobic digestion can be separated into two phases, (i) the "acid fermentation" phase or acidogenesis, leading to the production of intermediate products predominated by volatile organic acids such as acetic acid, propionic acid, butyric acid and valeric acid; and (ii) the "methane fermentation" phase or methanogenesis, resulting in the conversion of the intermediate products to stable end products mainly methane and carbon dioxide. The two phases in anaerobic digestion differ in bacterial populations, digestion rate, environmental requirements, degradation process and products. In two-phase anaerobic digestion system each phase can be controlled at the best environmental conditions in separate reactor.

Recently, various reactor configurations and substrates are being applied to two-phase anaerobic digestion as shown in Table 7. In order to accelerate the acidogenesis and methanogenesis processes in TPAD, the two separate reactors may be applied in various high rate anaerobic reactors such as upflow anaerobic sludge blanket (UASB)—UASB system (Fongsatitkul et al. 1995), continuous stirred tank reactor (CSTR)—upflow anaerobic filter (UAF) system (Held and Wellacher 2002), hybrid reactor (Yalcin et al. 2008), CSTR—anaerobic fluidized bed reactor (AFBR) system (Yu et al. 1999), two-phase plug-flow reactor (PFR) (Liu and Ghosh 1997; Liu 1998), and anaerobic packed bed reactor (APBR) (Tatara and Yamazawa 2004).

Two-phase anaerobic processes have been applied to treat many kinds of wastewater and solid wastes from difference sources such as distillery (Shin et al. 1992), landfill leachate (Agdag and Sponza 2005), coffee (Kida et al. 1994), cheese whey and dairy (Yilmazer and Yenigün 1999), starch (Demirel and Yenigün 2002), fruit and vegetable solid (Yu et al. 1999; Pavan et al. 2000), food (Shin et al. 1992), pulp and paper (Rintala and Puhakka 1994), olive mill (Borja et al. 1998), abattoir (Banks and Wang 1999), dye (Talarposhti et al. 2001), primary and activated sludge and solid (Bhattacharya et al. 1996).

Phase separation of anaerobic process has a number of major advantages (Shuizhou and Zhou 2005) including (i) isolation and optimization of potential rate-limiting steps; (ii) improvement of reaction kinetics and stability through pH

Table 7 Performance on two-phase anaerobic digestion with various kinds of feedstock

Feedstock	Reactor configuration	System HRT (days)	System organic loading rate (OLR) (kg COD/m^3.d)	Performance	CH_4 yield	Reference
Distillery wastewater	UASB-UASB	–	16.5–44.0	80 % COD removal	16.5 l/l COD. d	Shin et al. (1992)
Coffee waste	–	–	–	70 % COD removal	0.3 m^3/kg waste	Kida et al. (1994)
Cheese whey	CSTR-UAF	4–7	–	–	0.55 m^3/kg COD	Yilmazer and Yenigün (1999)
Abattoir waste	–	4–12	1.4–7.0	–	0.3 m^3/kg COD	Banks and Wang (1999)
Olive mill	–	–	2.3–2.4	–	0.36 m^3/kg COD	Borja et al. (1998)
Dye waste	APBR-APBR	3–5	0.25–1.0	90 % color removal	–	Talarposhti et al. (2001)
Fruit and vegetable	TPAD (mesophilic-thermophilic)	12	–	–	0.6 m^3/kg TVS	Pavan et al. (2000)
Primary and mixed primary-activated sludge	TPAD (mesophilic-mesophilic)	10	–	43 % VS removal	0.11 m^3/kg VS	Ghosh and Taylor (1999)

control, resistant to shock loading, selection of faster-growing microorganisms; and (iii) potential for detoxification in first phase. However, application of phase separation of anaerobic digestion has encountered certain barriers such as (i) disruption of syntrophic relationships; (ii) requirement of experience engineers and operators; (iii) uncertainty of linkage between reactor configuration and substrate types which determine the amenability of feedstock to two-phase anaerobic digestion.

4.2.2 Co-digestion

Mono-digestion (anaerobic digestion of a single substrate) usually suffers from its limitations in the cases of (i) low organic loads of sewage sludge; (ii) low organic loads and high nitrogen concentrations in animal manures; (iii) relatively high concentration of heavy metals in organic fraction of municipal solid waste (OFMSW); (iv) seasonal substrates such as crops and agro-industrial wastes; (v) potential inhibitors of methanogenic activity in slaughterhouse waste (SHW) such as the presence of high concentration of nitrogen and long-chain fatty acids (LCFA). Anaerobic co-digestion, i.e., simultaneous digestion of two or more substrates, is a feasible option to overcome the drawbacks of mono-digestion and to improve economic feasibility of anaerobic digestion (Mata-Alvarez et al. 2014).

Researchers found that the improvement of methane production is mainly a consequence of the increase in the organic loading rate (OLR) rather than synergisms between the primary substrate and co-substrate (Mata-Alvarez et al. 2011). Different kinds of mixtures can be considered and used in co-digestion as long as the blend ratio and types of co-substrate favor synergisms, dilute inhibitory compounds, optimize methane production and does not disrupt digestate quality. Typically, the decisions on the ratio between the primary substrates and co-substrates have been simplified to optimize the C/N ratio. The primary substrates like animal manures are characterized by high buffer capacities and a low C/N ratio while the co-substrates like agro-industrial waste and OFMSW are normally characterized as a high C/N ratio and low buffer capacity (Astals et al. 2012; Wang et al. 2012). However, the optimized combination in the mixture also requires consideration of other parameters such as macro and micronutrients equilibrium, pH and alkalinity, dilution of inhibitory compounds, amounts of biodegradable organics and dry matter (Hartmann et al. 2002).

4.3 Biomass Retention

Reactors of anaerobic process can be categorized according to how the biomass is retained in the system and the type of biomass in the system. Bacteria grow in the reactor liquid as flocculent or granular sludge in suspended growth reactors. Granular sludge exhibits higher activity rates and settling velocity that reduce the reactor volume required and allowing higher organic loading rates to the systems.

The most robust configurations for suspended growth anaerobic reactors are UASB (upflow anaerobic sludge blanket) and EGSB (expanded granular sludge bed).

4.3.1 UASB System

With the widespread industrial application of UASB reactors, increasing attention is focused on the granulation of anaerobic sludge (Fang 2000). Biomass is retained as granular matrix or blanket as suspension in the reactors. The advantages of granulation includes the establishment of a regular, thick and well-built microbial structure that is ready to operate with different transport phenomena, high biomass retention time leading to a high loading rate and better removal efficiencies, appropriate settleability, resistance to high OLR and toxicity shock (Speece 1996). The operation of UASB reactors may be limited by a number of factors including (i) inadequate retention of viable biomass for treating specific types of wastes that is not able to cultivate granular sludge, (ii) granule disintegration or wash-out of hollow granules, (iii) occurrence of fluffy granules, and (iv) scaling by inorganic precipitate.

4.3.2 EGSB System

EGSB systems are not equipped with an internal settler as in the conventional UASB, but with an advanced liquid-solid separation device. The main features of the EGSB reactors are: (i) high design organic loading rates; (ii) very small surface area; (iii) tall reactor system; and (iv) high upflow velocity. Engineering anaerobic sludge granules is a new area of research that targets at expanding the catabolic capabilities of the sludge.

4.3.3 Attached Growth Reactors

Attached growth reactors make use of either fixed film or carried media for the bacteria to grow and attach. Attached-growth systems comprise of fixed-film reactors and fluidized bed reactors involving immobilization of microbial biomass on inert media. In fixed film processes, bacteria reside on static support surface such as plastics rings, rocks, media modules or membrane modules. In fluidized bed processes, suspended carrier media such as sand, provide attachment surfaces in the reactors.

Hybrid anaerobic reactors are popular in recent development which take advantages of both suspended and attached growth processes in a single reactor. An example of hybrid anaerobic reactor design combine UASB as the lower section and upflow anaerobic filter as the upper section in a single reactor (Abdullah et al. 2005). The advantages of hybrid anaerobic reactors include (i) development of granular or flocculent sludge bed in the reactor, leading to an increased biomass inventory, (ii) suitability for treating wastes where granular sludge formation is difficult, and (iii) increasing process stability and removal efficiency.

4.3.4 Membrane Bioreactors

Efficient liquid-solids separation is the basis of any anaerobic high-rate reactor system for waste treatment. Anaerobic membrane bioreactors (AnMBRs) are emerging alternatives for UASB reactors. With the presence of the inert supportive media for bacterial growth, membrane bioreactors can achieve outstanding effluent quality (<20 ppm organics), and COD and solid removal (up to 99 and 100 % for domestic wastewater respectively) (Smith et al. 2012) with the advantages of (i) possible operation at approximately infinite SRT to reach very low effluent substrate concentrations, (ii) allowing the growth of slow-growing micro-organisms, and (iii) possible treatment of recalcitrant compounds. However, membrane bioreactors usually suffer the drawbacks of high pressure physical separation causing disruption to microbial communities and subjected to membrane fouling and scaling with typical precipitates such as calcium carbonate.

4.4 Reactor Configuration

Reactor configurations of anaerobic digester can be divided into conventional anaerobic digesters and high-rate anaerobic digesters. The first conventional anaerobic digester was used in 1881 to liquefy the solid components of sewage. In 1955, anaerobic contact process was developed to treat soluble organics and dilute wastewaters (Hassan et al. 2013). A variety of new bioreactor designs have been developed in recent years which facilitate a significantly high rate of reaction for the treatment of waste (Bouallagui et al. 2003; Mumme et al. 2010; Xing et al. 2010). High rate anaerobic reactors include completely mixed anaerobic digester, anaerobic contact process, anaerobic sequencing batch reactor (ASBR), anaerobic packed bed or anaerobic filter, anaerobic fluidized bed and expanded bed reactors, upflow anaerobic sludge blanket (UASB) reactor and anaerobic baffled reactors (ABR) (Barker et al. 1999). Through the development of innovative high-rate reactor designs, anaerobic treatment can now challenge the cost of aerobic treatment for many wastewater treatment applications (Malina Junior and Pohland 1992). Ward et al. reported that an anaerobic bioreactor should be designed in a way that allows a continuously high and sustainable organic loading rate with a short hydraulic retention time and has the ability to produce the maximum level of methane (Ward et al. 2008). Reactors can be classified into the following categories (i) batch and continuous process, (ii) single-phase, and (iii) multi-phase reactors. Reactor shape must also take into consideration, both mixing and heat transfer.

In addition to basic reactor design, mixing of the contents in anaerobic digesters are required to ensure efficient transfer of particulate organic material for active microbial biomass, to release gas bubbles trapped in the reactors and to prevent sedimentation of denser particulate materials. The mixing pattern may be intermittent, which is determined by the type of reactor, type of agitator used and the total solid contents of the feedstock (Burton and Turner 2003). Recirculation of biogas in the reactor or hydraulic mixing by recirculation of digestate with pump is commonly used to prevent the need of moving parts within the reactors. A certain degree of mixing is necessary but excessive mixing conditions can reduce biogas production (Gomez et al. 2006). It has been postulated that propionate-oxidizing bacteria and methanogenic archaea live in close proximity in granules with H_2 and formate as electron carriers. Excessive agitation can disrupt the granule structure, reducing the rate of oxidation of fatty acids and leading to digester instability (McMahon et al. 2001). Extracellular polymeric substances (EPS) are a combination of proteins and carbohydrates which are responsible for the formation of granules (Liu et al. 2004). An increase in mixing decreased the amount of EPS found, suggesting that minimal mixing produced larger anaerobic granules as greater quantities of EPS are required to maintain the granule structure (Ong et al. 2002). Mixing with biomass support media could be an important area in optimizing reactor configuration of anaerobic digester. Biomass support media provides an anchorage for the granular microbial communities and allows a high-shear type of mixing to increase solubility of COD without disruption to the microbial communities.

4.5 *Process Control and Monitoring*

Despite decades of academic and industrial research efforts, the complex anaerobic digestion processes are far from being understood in detail. Many anaerobic digestion plants are merely relying on a few simple-to-measure parameters mainly due to the conservative design of the over-sized reactors to guarantee process robustness, which gives a poor indication of the state of the biological process. Furthermore, unintentional organic loading, accidental addition of toxic substrates, process interruptions and lack of raw material quality control are believed to be one of the main limitations for effective process operation (Hjort-Gregersen et al. 1996; Holm-Nielsen et al. 2008; Nielsen and Angelidaki 2008; Kaparaju et al. 2009a, b). Introducing reliable monitoring and control technology would allow anaerobic digestion plants to be operated closer to their effective capacity limit instead of wasting reactor volume due to conservative design rules.

Process Analytical Technologies (PAT) is one of the recent advances in process monitoring in anaerobic digestion which allows complex bioconversion processes to be monitored and deciphered to a new level of reliability and effectiveness using spectroscopic and electrochemical measurement principles together with chemometric multivariate data analysis. Research efforts has been put in reviewing the potential application of PAT, Theory of Sampling (TOS) and chemometric data analysis within the field of anaerobic digestion monitoring (Madsen et al. 2011).

Table 8 Short-chained VFAs in sample matrices of manure and wastewater sludge

IUPAC nomenclature	Formula	CAS#	MW (Da)	bp (°C)	pK_a
Ethanoic acid	CH_3COOH	64-19-7	60.1	117.9	4.76
Propanoic acid	CH_3CH_2COOH	79-09-4	74.1	141.2	4.87
n-Butanoic acid	$CH_3CH_2CH_2COOH$	107-92-6	88.1	163.8	4.83
2-Methylpropanoic acid	CH_3CHCH_3COOH	79-31-2	88.1	154.5	4.84
n-Pentanoic acid	$CH_3(CH_2)_3COOH$	109-52-4	102.1	186.1	4.83
2-Methylbutanoic acid	$CH_3CH_2CHCH_3COOH$	116-53-0	102.1	177	4.80
3-Methylbutanoic acid	$CH_3CHCH_3CH_2COOH$	503-74-2	102.1	176.5	4.77

Table 9 Four main classes of reviewed analytical modalities collectively known as PAT in AD process monitoring

Main class	Reviewed analytical modalities
Spectroscopic	Fluorescence (FLU) Peck and Chynoweth (1992), Infrared (IR) Steyer et al. (2002), Near Infrared (NIR) Nordberg et al. (2000), Hannsson et al. (2002), Holm-Nielsen et al. (2007), Holm-Nielsen et al. (2008), Raman, Visual (VIS), Ultraviolet (UV) Redondo et al. (2008), Rudnitskaya and Legin (2008), Buczkowska et al. (2010)
Electro-chemical	pH, Redox potential, Electronic tongue (ET), Electronic nose (EN)
Chromatographic	GC, GC headspace, HPLC Pind et al. (2003), Boe et al. (2005), Diamantis et al. (2006)
Other	Acoustic chemometrics (a.c.) Nacke et al. (2005), Mass spectrometry (MS), Microwaves Lomborg et al. (2009), Titration Feitkenhauer et al. (2002), Lahav and Morgan (2004)

The use of multivariate sensor technologies and electrochemical arrays is encouraged as many studies have shown promising results in both laboratory-scale and pilot-scale. Many authors have suggested VFAs as control parameters as these acids are indicative of the activity of methanogenic consortia. A number of analytical methods have been developed for quantification of relevant VFAs for anaerobic digestion process monitoring. Short-chained VFAs commonly present in sample matrices such as manure and wastewater sludge are listed in Table 8 (Hill and Holmberg 1988; Christiansen et al. 1995; Nielsen et al. 2007). The application of numerous monitoring techniques for quantifying these parameters has been reported in the literature. An overview of available techniques is provided in Table 9.

4.6 *Mathematical Models*

Two most widely used models for anaerobic digestion are the Anaerobic Digestion Model no. 1 (ADM1) developed by a task group for the International Water Association (IWA) and Siegriest Model (Siegrist et al. 2002). The two models are constructed with different approaches: Siegriest model parameters are based on experiments, whereas the ADM1 uses review consensus (Batstone 2006).

Lyberatos and Skiadas (1999) gave an extensive review on modelling for anaerobic digestion. They pointed out that some important factors describing the behavior of anaerobic digesters should be evaluated and taken into account from a modelling point of view including digester startup conditions, degree of acclimation to the feedstock, hydraulic loading, organic loading, biogas production per unit volume, concentration of inhibitors, availability of nutrients, cation concentrations, types and concentration of solids in the feedstock. Anaerobic digestion is a complex system of biochemical and physical processes. Due to its complexity, it has traditionally been treated as a black box system and optimization has been based on experience or trial-and-error methods. As experiments of anaerobic digestion are expensive and time-consuming, modelling can provide a useful tool for process understanding and optimization (Kothari et al. 2014).

5 Advanced Molecular Biological Tools for System Monitoring

Anaerobic digestion is carried out by a mixture of different *Bacteria* and *Archaea* living in a microbial community. The microbial community is generally considered complex as hundreds of different types of organisms are involved in the process and these organisms are also interacting among themselves. In order to optimize the yield of biogas and for trouble shooting purposes in case of a process upset, a comprehensive view on the composition and metabolic functions of the organisms in the system is warranted. In the past decade, a number of advanced molecular tools targeting the DNA, RNA and proteins of microbial cells have become available to provide detailed biological information on the microbial community and it is now possible to move beyond the traditional 'black box' approaches of operating an anaerobic digester. An overview of these tools is described in this section.

5.1 Low-throughput Methods

Given the high microbial diversity present in anaerobic digestion, culture-dependent methods to analyze a microbial community are not practical and not feasible. Hence, culture-independent methods are required. In the 1990s and early 2000s, polymerase chain reaction (PCR) targeting the 16S rRNA gene of *Bacteria* and *Archaea* with universal primers followed by clone library and Sanger sequencing was a popular method to identify the organisms present in anaerobic digesters (Chouari et al. 2005). Typically, a few hundred clones are randomly picked and sequenced as the method is labor intensive and expensive. Because a relatively small number of clones can be analyzed, clone library method can only capture the dominant populations and the organisms that are present at a low relative abundance are usually not captured. Other methods that are suitable to identify the dominant populations include terminal restriction-fragment length polymorphism (T-RFLP) (Ike et al. 2010) and chemical or temperature denaturing gradient gel electrophoresis (DGGE) (Bialek et al. 2012). The aforementioned methods are semi-quantitative where the relative proportion of the taxa is determined. When absolute quantification is required to determine the concentration of a specific population in an anaerobic digester, quantitative PCR (qPCR) can be applied and qPCR has the advantage that the quantification range spans a few orders of magnitude, making it possible to quantify the low and high abundant organisms such as different methanogens (Goberna et al. 2010). The application of these molecular methods to a single sample can provide a snapshot of the microbial community, but when multiple samples at different time points and under different conditions are analyzed, the shift in composition of the microbial community can be revealed.

5.2 *High-throughput Methods*

The advent of sequencing technology in the past few years has revolutionized the ability to analyze microbial communities, providing both breadth and depth in coverage of information. Furthermore, the cost per DNA base has decreased and robotic instruments have automated many procedures in the lab, making the analysis less labor intensive. First, it was the emergence of the next-generation sequencing platforms by 454 Life Sciences that can generate a few hundred million bases per run and long read length up to 450 bp. Later, the sequencers developed by Illumina (Solexa) have further increased throughput to as much as a few hundreds gigabases per run with a shorter read length (~ 125 bp). With the Illumina platforms, the number of reads that can be obtained per sample to analyze the composition and structure of a microbial community can range from a few thousands to tens of thousands, which is substantially more than a clone library analysis (Sundberg et al. 2013). With this sequencing depth, both the dominant and minor members of the community can be identified, which represents substantial improvement over previous methods as the minor members could also be functionally important. In addition to targeting the 16S rRNA gene, high-throughput sequencing has also been applied to analyze functional genes such as the methyl coenzyme M reductase (*mcrA*) gene that is ubiquitously present in all methanogens for catalyzing the last step of methane generation (Ellis et al. 2012; Wilkins et al. 2015).

Targeting taxonomic and/or functional genes as biomarkers can identify the organisms present. However, in order to determine the metabolic functions of these organisms, shotgun metagenomic sequencing of the microbial community can be performed to determine the gene content present in these organisms. Hampered by the lower throughput of previous sequencing platforms, a gene-centric approach was usually taken in early metagenomic studies where the goal is to simply identify the metabolic functions present (Li et al. 2013; Wong et al. 2013). Recently, with the increase in the throughput of sequencers, acquiring a substantial quantity of reads per sample is possible, enabling a genome-centric approach in metagenomic sequencing where genes are placed in a genomic framework (Sekiguchi et al. 2015). A genome-centric approach offers the advantages that the metabolic capability of an organism and its interactions with other members in the community can be better deciphered.

Building on the metagenomic sequencing results, two complementary approaches, namely metatranscriptomics and metaproteomics, can be further applied to query the dynamics and expression of genes under different conditions in anaerobic digesters. Metatranscriptomics make use of high-throughput sequencing to analyze the expressed RNA (Zakrzewski et al. 2012), while metaproteomics utilize advanced mass-spectrometry to analyze the expressed proteins (Hanreich et al. 2013; Lü et al. 2014). Metagenomics are useful to determine what organisms are present and what biochemical functions these organisms possess. However, under what conditions these organisms are active and what metabolic functions are executed cannot be easily interpreted from the metagenomic data. Therefore, metatranscriptomics and metaproteomics are useful tools to provide detailed information on the activity of the organisms in a digester. The combination of metagenomics, metatranscriptomics and metaproteomics is generally referred to as 'omics' methods and these innovative molecular biology tools can help microbiologists and engineers to better diagnose anaerobic digesters.

6 Future Outlook

Climate change, waste treatment and renewable energy are pressing issues facing society in the 21st century. Anaerobic digestion of organic materials can address all these issues simultaneously. Given that anaerobic digestion is a mature technology, the deployment of this technology in large centralized scale or small decentralized scale is expected to gain widespread use in the near future. Further optimization and enhancement of the engineering coupled with knowledge in the microbiology will certainly further improve the robustness and performance of anaerobic digestion. Without a doubt, the outlook of anaerobic digestion is promising and this technology will play an important role in our society.

References

Abdullah AGL, Idris A et al (2005) A kinetic study of a membrane anaerobic reactor (MAR) for treatment of sewage sludge. Desalination 183(1–3):439–445

Ağdağ ON, Sponza DT (2005) Anaerobic/aerobic treatment of municipal landfill leachate in sequential two-stage up-flow anaerobic sludge blanket reactor (UASB)/completely stirred tank reactor (CSTR) systems. Process Biochem 40(2):895–902

Amani T, Nosrati M et al (2010) Anaerobic digestion from the viewpoint of microbiological, chemical, and operational aspects—a review. Environ Rev 18(NA):255–278

Ariunbaatar J, Panico A et al (2014) Pretreatment methods to enhance anaerobic digestion of organic solid waste. Appl Energy 123:143–156

Arsova L (2010) Anaerobic digestion of food waste: current status, problems and an alternative product. Columbia University, Berlin

Astals S, Nolla-Ardèvol V et al (2012) Anaerobic co-digestion of pig manure and crude glycerol at mesophilic conditions: biogas and digestate. Bioresour Technol 110:63–70

Bainotti AE, Setogaichi M et al (1996) Kinetic studies and medium improvement of growth and vitamin B12 production by acetobacterium sp. in batch culture. J Ferment Bioeng 81(4): 324–328

Banks CJ, Wang Z (1999) Development of a two phase anaerobic digester for the treatment of mixed abattoir wastes. Water Sci Technol 40(1):69–76

Barker DJ, Mannucchi GA et al (1999) Characterisation of soluble residual chemical oxygen demand (COD) in anaerobic wastewater treatment effluents. Water Res 33(11):2499–2510

Batstone DJ (2006) Mathematical modelling of anaerobic reactors treating domestic wastewater: rational criteria for model use. Rev Environ Sci Bio/Technol 5(1):57–71

Batstone DJ, Virdis B (2014) The role of anaerobic digestion in the emerging energy economy. Curr Opin Biotechnol 27:142–149

Battimelli A, Millet C et al (2003) Anaerobic digestion of waste activated sludge combined with ozone post-treatment and recycling. Water Sci Technol 48(4):61–68

Bhattacharya SK, Madura RL et al (1996) Volatile solids reduction in two-phase and conventional anaerobic sludge digestion. Water Res 30(5):1041–1048

Bialek K, Kumar A et al (2012) Microbial community structure and dynamics in anaerobic fluidized-bed and granular sludge-bed reactors: influence of operational temperature and reactor configuration. Microb Biotechnol 5(6):738–752

Boe K, Batstone DJ et al (2005) Online headspace chromatographic method for measuring VFA in biogas reactor. Water Sci Technol 52(1–2):473–478

Bolzonella D, Pavan P et al (2005) Mesophilic anaerobic digestion of waste activated sludge: influence of the solid retention time in the wastewater treatment process. Process Biochem 40(3–4):1453–1460

Bordeleau EL, Droste RL (2011) Comprehensive review and compilation of pretreatments for mesophilic and thermophilic anaerobic digestion. Water Sci Technol 63:291–296

Borja R, Banks CJ et al (1998) Anaerobic digestion of slaughterhouse wastewater using a combination sludge blanket and filter arrangement in a single reactor. Bioresour Technol 65 (1–2):125–133

Bouallagui H, Ben Cheikh R et al (2003) Mesophilic biogas production from fruit and vegetable waste in a tubular digester. Bioresour Technol 86(1):85–89

Bougrier C, Battimelli A et al (2007) Combined ozone pretreatment and anaerobic digestion for the reduction of biological sludge production in wastewater treatment. Ozone Sci Eng 29 (3):201–206

Buczkowska A, Witkowska E et al (2010) The monitoring of methane fermentation in sequencing batch bioreactor with flow-through array of miniaturized solid state electrodes. Talanta 81 (4):1387–1392

Buhr HO, Andrews JF (1977) The thermophilic anaerobic digestion process. Water Res 11(2): 129–143

Burton CH, Turner C (2003) Manure management treatment strategies for sustainable agriculture

Cai M, Chua H et al (2009) Optimal production of polyhydroxylalkanoates (PHA) in activated sludge fed by volatile fatty acids (VFAs) generated from alkaline excess sludge fermentation. Bioresour Technol 100(3):1399–1405

Cao Y, Pawlowski A (2012) Sewage sludge-to-energy approaches based on anaerobic digestion and pyrolysis: brief overview and energy efficiency assessment. Renew Sustain Energy Rev 16 (3):1657–1665

Carlsson M, Anox Kaldnes AD (2008) Electroporation for enhanced methane yield from municipal solid waste. In: ORBIT 2008, Wageningen, The Netherlands
Carlsson M, Lagerkvist A et al (2012) The effects of substrate pre-treatment on anaerobic digestion systems: a review. Waste Manag 32(9):1634–1650
Carrère H, Dumas C et al (2010) Pretreatment methods to improve sludge anaerobic degradability: a review. J Hazard Mater 183(1–2):1–15
Cecchi F, Traverso PG et al (1986) Anaerobic digestion of organic fraction of municipal solid wastes—digester performance. Sci Total Environ 56:183–197
Cesaro A, Belgiorno V (2014) Pretreatment methods to improve anaerobic biodegradability of organic municipal solid waste fractions. Chem Eng J 240:24–37
Chae KJ, Jang A et al (2008) The effects of digestion temperature and temperature shock on the biogas yields from the mesophilic anaerobic digestion of swine manure. Bioresour Technol 99 (1):1–6
Chang JJ, Chou CH et al (2008) Syntrophic co-culture of aerobic Bacillus and anaerobic Clostridium for bio-fuels and bio-hydrogen production. Int J Hydrogen Energy 33(19): 5137–5146
Cheikhyoussef A, Pogori N et al (2009) Antimicrobial activity and partial characterization of bacteriocin-like inhibitory substances (BLIS) produced by Bifidobacterium infantis BCRC 14602. Food Control 20(6):553–559
Chen HH, Lee AHI (2014) Comprehensive overview of renewable energy development in Taiwan. Renew Sustain Energy Rev 37:215–228
Chen Y, Cheng JJ et al (2008) Inhibition of anaerobic digestion process: a review. Bioresour Technol 99(10):4044–4064
Cho SK, Im WT et al (2013) Dry anaerobic digestion of food waste under mesophilic conditions: performance and methanogenic community analysis. Bioresour Technol 131:210–217
Chong ML, Sabaratnam V et al (2009) Biohydrogen production from biomass and industrial wastes by dark fermentation. Int J Hydrogen Energy 34(8):3277–3287
Chouari R, Le Paslier D et al (2005) Novel predominant archaeal and bacterial groups revealed by molecular analysis of an anaerobic sludge digester. Environ Microbiol 7(8):1104–1115
Christiansen N, Hendriksen HV et al (1995) Degradation of chlorinated aromatic compounds in UASB reactors. Water Sci Technol 31(1):249–259
Chu CF, Li YY et al (2008) A pH and temperature phased twostage process for hydrogen and methane production from food waste. Int J Hydrogen Energy 33:4739–4746
Davidson A, Gruvberger C et al (2007) Methane yield in source-sorted organic fraction of municipal solid waste. Waste Manag 27:406–414
De Baere L (2000) Anaerobic digestion of solid waste: state-of-the-art. Water Sci Technol 41 (3):283–290
Demirel B, Yenigün O (2002) Two-phase anaerobic digestion processes: a review. J Chem Technol Biotechnol 77(7):743–755
Diamantis V, Melidis P et al (2006) Continuous determination of volatile products in anaerobic fermenters by on-line capillary gas chromatography. Anal Chim Acta 573–574:189–194
Duan N, Dong B et al (2012) High-solid anaerobic digestion of sewage sludge under mesophilic conditions: feasibility study. Bioresour Technol 104:150–156
Eastman JA, Ferguson JF (1981) Solubilization of particulate organic carbon during the acid phase of anaerobic digestion. J Water Pollut Control Fed 53(3):352–366
Ellis JT, Tramp C et al (2012) Characterization of a methanogenic community within an algal fed anaerobic digester. ISRN Microbiol 2012:753–892
Escamilla-Alvarado C, Rios-Leal E et al (2012) Gas biofuels from solid substrate hydrogenogenic-methanogenic fermentation of the organic fraction of municipal solid waste. Process Biochem 47:1572–1587
Fang HHP (2000) Microbial distribution in UASB granules and its resulting effects. Water Sci Technol 42:201–208
Fang Y, Lu Y et al (2009) Improvement of alkaline lipase from proteus vulgaris T6 directed evolution. Enzyme Microbial Technol 44(2):84–88

Fantozzi F, Buratti C (2009) Biogas production from different substrates in an experimental continuously stirred tank reactor anaerobic digester. Bioresour Technol 100(23):5783–5789

Fdez-Guelfo LA, Alvarez-Gallego C et al (2011) The effect of different pretreatments on biomethanation kinetics of industrial organic fraction of municipal solid wastes (OFMSW). Chem Eng J 171:411–417

Feitkenhauer H, von Sachs J et al (2002) On-line titration of volatile fatty acids for the process control of anaerobic digestion plants. Water Res 36(1):212–218

Fernández J, Perez M et al (2008) Effect of substrate concentration on dry mesophilic anaerobic digestion of organic fraction of municipal solid waste (OFMSW). Bioresour Technol 99 (14):6075–6080

Fongsatitkul P, Mavinic DS, Lo KV (1995) A two-phase anaerobic digestion (UASB-UASB) process: design criteria and optimal system loading capacity. Can J Civ Eng 22(3):551–565

Fricke K, Santen H et al (2007) Operating problems in anaerobic digestion plants resulting from nitrogen in MSW. Waste Manag 27(1):30–43

Fricke K, Santen H et al (2007) Operating problems in anaerobic digestion plants resulting from nitrogen in MSW. Waste Manag 27(1):30–43

Fritsch M, Hartmeier W et al (2008) Enhancing hydrogen production of Clostridium butyricum using a column reactor with square-structured ceramic fittings. Int J Hydrogen Energy 33 (22):6549–6557

Gallert C, Bauer S et al (1998) Effect of ammonia on the anaerobic degradation of protein by a mesophilic and thermophilic biowaste population. Appl Microbiol Biotechnol 50(4):495–501

Gerardi MH (2006) Methane-forming bacteria. In: Gerardi MH (ed) Wastewater bacteria, edn. Wiley, New Jersey, USA, pp 161–163

Ghosh S, Pohland FG (1974) Kinetics of substrate assimilation and product formation in anaerobic digestion. J Water Pollut Control Fed 46(4):748–759

Ghosh S, Taylor DC (1999) Kraft-mill biosolids treatment by conventional and biphasic fermentation. Water Sci Technol 40(11–12):169–177

Ghosh S, Ombregt JP et al (1985) Methane production from industrial wastes by two-phase anaerobic digestion. Water Res 19(9):1083–1088

Goberna M, Gadermaier M et al (2010) Adaptation of methanogenic communities to the cofermentation of cattle excreta and olive mill wastes at 37 °C and 55 °C. Appl Environ Microbiol 76(19):6564–6571

Goel R, Tokutomi T et al (2003) Optimal process configuration for anaerobic digestion with ozonation. Water Sci Technol 48(4):85–96

Gomez X, Cuetos MJ et al (2006) Anaerobic co-digestion of primary sludge and the fruit and vegetable fraction of the municipal solid wastes—conditions for mixing and evaluation of the organic loading rate. Renewable Energy 31:2017–2024

Gujer W, Zehnder JB (1983) Conversion processes in anaerobic digestion. Water Sci Technol 15 (8–9):127–167

Hannsson M, Nordberg Å et al (2002) Early warning of disturbances in a laboratory-scale MSW biogas process. Water Sci Technol 45(10):255–260

Hanreich A, Schimpf U et al (2013) Metagenome and metaproteome analyses of microbial communities in mesophilic biogas-producing anaerobic batch fermentations indicate concerted plant carbohydrate degradation. Syst Appl Microbiol 36(5):330–338

Hansen TL, Jansen JC et al (2007) Effects of pre-treatment technologies on quantity and quality of source-sorted municipal organic waste for biogas recovery. Waste Manag 27:389–405

Hartmann H, Angelidaki I et al (2002) Co-digestion of the organic fraction of municipal waste with other waste types. London, UK

Hartmann H, Ahring B (2006) Strategies for the anaerobic digestion of the organic fraction of municipal solid waste: an overview. Water Sci Technol 53(8):7–22

Hassan SR, Zwain HM et al (2013) Development of anaerobic reactor for industrial wastewater treatment: an overview, present stage and future prospects. J Adv Sci Res 4(1):7–12

Held Christof, Wellacher Martin et al (2002) Two-stage anaerobic fermentation of organic waste in CSTR and UFAF-reactors. Bioresour Technol 81(1):19–24

Hill DT, Holmberg RD (1988) Long chain volatile fatty acid relationships in anaerobic digestion of swine waste. Biological Wastes 23(3):195–214

Hjort-Gregersen K, Chartier P et al (1996) Centralized biogas plants: economic potential. In: Biomass for energy and the environment, edn. Pergamon, Oxford, pp 1695–1700

Holm-Nielsen JB, Andree H et al (2007) Remove from marked records transflexive embedded near infrared monitoring for key process intermediates in anaerobic digestion/biogas production. J Near Infrared Spectrosc 15(2):123–135

Holm-Nielsen JB, Lomborg CJ et al (2008) On-line near infrared monitoring of glycerol-boosted anaerobic digestion processes: evaluation of process analytical technologies. Biotechnol Bioeng 99(2):302–313

Holm-Nielsen JB, Al Seadi T et al (2009) The future of anaerobic digestion and biogas utilization. Bioresour Technol 100(22):5478–5484

Huber H, Thomm M et al (1982) Methanococcus thermolithotrophicus, a novel thermophilic lithotrophic methanogen. Arch Microbiol 132(1):47–50

Hur H-G, Rafii F (2000) Biotransformation of the isoflavonoids biochanin A, formononetin, and glycitein by Eubacterium limosum. FEMS Microbiol Lett 192(1):21–25

Hwang MH, Jang NJ et al (2004) Anaerobic bio-hydrogen production from ethanol fermentation: the role of pH. J Biotechnol 111(3):297–309

Ike M, Inoue D et al (2010) Microbial population dynamics during startup of a full-scale anaerobic digester treating industrial food waste in Kyoto eco-energy project. Bioresour Technol 101(11): 3952–3957

Izumi K, Okishio Y et al (2010) Effects of particle size on anaerobic digestion of food waste. Int Biodeterior Biodegradation 64(7):601–608

Jo JH, Lee DS et al (2008) Statistical optimization of key process variables for enhanced hydrogen production by newly isolated Clostridium tyrobutyricum JM1. Int J Hydrogen Energy 33(19): 5176–5183

Kalra MS, Panwar JS (1986) Anaerobic digestion of rice crop residues. Agric Wastes 17(4): 263–269

Kalyani DC, Telke AA et al (2009) Ecofriendly biodegradation and detoxification of reactive red 2 textile dye by newly isolated Pseudomonas sp. SUK1. J Hazard Mater 163(2): 735–742

Kaparaju P, Ellegaard L et al (2009a) Optimisation of biogas production from manure through serial digestion: lab-scale and pilot-scale studies. Bioresour Technol 100(2):701–709

Kaparaju P, Serrano M et al (2009b) Effect of reactor configuration on biogas production from wheat straw hydrolysate. Bioresour Technol 100(24):6317–6323

Khalid A, Arshad M et al (2011) The anaerobic digestion of solid organic waste. Waste Manag 31 (8):1737–1744

Khan AW (1980) Degradation of cellulose to methane by a coculture of Acetivibrio cellulolyticus and Methanosarcina barkeri. FEMS Microbiol Lett 9(3):233–235

Kida K, Ikbal et al (1994) Anaerobic digestion of coffee waste by two-phase methane fermentation with slurry-state liquefaction. J Ferment Bioeng 77(3):335–338

Kim SW, Park JY et al (2000) Development of a modified three-stage methane production process using food wastes. Appl Biochem Biotechnol 84:731–741

Kim M, Ahn YH et al (2002) Comparative process stability and efficiency of anaerobic digestion; mesophilic vs thermophilic. Water Res 36(17):4369–4385

Kim J, Park C et al (2003) Effects of various pretreatments for enhanced anaerobic digestion with waste activated sludge. J Biosci Bioeng 95(3):271–275

Kim HW, Han SK et al (2004) Anaerobic co-digestion of sewage sludge and food waste using temperature-phased anaerobic digestion process. Water Sci Technol 50:107–114

Kim JY, Lee MY et al (2009) Production of gamma-aminobutyric acid in black raspberry juice during fermentation by Lactobacillus brevis GABA 100. Int J Food Microbiol 130(1):12–16

Koster IW, Lettinga G (1984) The influence of ammonium-nitrogen on the specific activity on pelletized methanogenic sludge. Agric Wastes 9:205–216

Kothari R, Pandey AK et al (2014) Different aspects of dry anaerobic digestion for bio-energy: an overview. Renew Sustain Energy Rev 39:174–195

Kuo WC, Lu HY (2004) A complete resource recovery of the thermophilic anaerobic digestion. In: 10th Congress on anaerobic digestion

Kvesitadze G, Sadunishvili T et al (2012) Two-stage anaerobic process for bio-hydrogen and biomethane combined production from biodegradable solid waste. Energy 37:94–102

Lahav O, Morgan BE (2004) Titration methodologies for monitoring of anaerobic digestion in developing countries—a review. J Chem Technol Biotechnol 79(12):1331–1341

Lee M, Hidaka T et al (2009) Two-phased hyperthermophilic anaerobic co-digestion of waste activated sludge with kitchen garbage. J Biosci Bioeng 108(5):408–413

Lee WS, Chua ASM et al (2014) A review of the production and applications of waste-derived volatile fatty acids. Chem Eng J 235:83–99

Li Z, Han J et al (2003) Nitrobenzocyclophosphamides as potential prodrugs for bioreductive activation: synthesis, stability, enzymatic reduction, and antiproliferative activity in cell culture. Bioorg Med Chem 11(19):4171–4178

Li A, Chu Y et al (2013) A pyrosequencing-based metagenomic study of methane-producing microbial community in solid-state biogas reactor. Biotechnol Biofuels 6(1):3

Liu T (1998) Anaerobic digestion of solid substrates in an innovative two-phase plug-flow reactor (TPPFR) and a conventional single-phase continuously stirred-tank reactor. Water Sci Technol 38(8):453–461

Liu T, Ghosh S (1997) Phase separation during anaerobic fermentation of solid substrates in an innovative plugflow reactor. Water Sci Technol 36(6):303–310

Liu YQ, Liu Y et al (2004) The effects of extracellular polymeric substances on the formation and stability of biogranules. Appl Microbiol Biotechnol 65:143–148

Lomborg CJ, Holm-Nielsen JB et al (2009) Near infrared and acoustic chemometrics monitoring of volatile fatty acids and dry matter during co-digestion of manure and maize silage. Bioresour Technol 100(5):1711–1719

López Torres M, Espinosa Lloréns MdC (2008) Effect of alkaline pretreatment on anaerobic digestion of solid wastes. Waste Manag 28(11):2229–2234

Lü F, Bize A et al (2014) Metaproteomics of cellulose methanisation under thermophilic conditions reveals a surprisingly high proteolytic activity. ISME J 8(1):88–102

Lyberators G, Skiadas IV (1999) Modelling of anaerobic digestion—a review. Global Nest Int J 1 (2):63–76

Ma J, Duong T et al (2011) Enhanced biomethanation of kitchen waste by different pretreatments. Bioresour Technol 102:592–599

Madigan M, Martinko J et al. (2009) Brock biology of micro-organisms, 12th edn

Madsen M, Holm-Nielsen JB et al (2011) Monitoring of anaerobic digestion processes: a review perspective. Renew Sustain Energy Rev 15(6):3141–3155

Malina Junior JF, Pohland FG (1992) Design of anaerobic processes for the treatment of industrial and municipal wastes. Lancaster, Pa. (EUA)

Mata-Alvarez J, Cecchi F et al (1990) The performances of digesters treating the organic fraction of municipal solid wastes differently sorted. Biol Wastes 33(3):181–199

Mata-Alvarez J, Dosta J et al (2011) Co-digestion of solid wastes : a review of its uses and perspectives including modeling. Crit Rev Biotechnol 31:99–111

Mata-Alvarez J, Dosta J et al (2014) A critical review on anaerobic co-digestion achievements between 2010 and 2013. Renew Sustain Energy Rev 36:412–427

McMahon KD, Stroot PG et al (2001) Anaerobic codigestion of municipal solid waste and biosolids under various mixing conditions—II: microbial population dynamics. Water Res 35:1817–1827

Merkel M, Manz W et al (1999) Population dynamics in anaerobic wastewater reactors: modelling and in situ characterization. Water Res 33(10):2392–2402

Mouneimne AH, Carrère H et al (2003) Effect of saponification on the anaerobic digestion of solid fatty residues. Bioresour Technol 90(1):89–94

Mumme J, Linke B et al (2010) Novel upflow anaerobic solid-state (UASS) reactor. Bioresour Technol 101:592–599

Nacke T, Brückner K et al (2005) New type of dry substances content meter using microwaves for application in biogas plants. Anal Bioanal Chem 383(5):752–757
Nielsen HB, Uellendahl H et al (2007) Regulation and optimization of the biogas process: propionate as a key parameter. Biomass Bioenergy 31(11):820–830
Nielsen HB, Angelidaki I (2008) Strategies for optimizing recovery of the biogas process following ammonia inhibition. Bioresour Technol 99(17):7995–8001
Nordberg Å, Hansson M et al (2000) Monitoring of a biogas process using electronic gas sensors and near-infrared spectroscopy (NIR). Water Sci Technol 41(3):1–8
Ochoa-Reparaz J, Mielcarz D et al (2008) F.31. Bacteroides fragilis regulates the induction of FoxP3 + CD4 + CD25 + treg cell populations in peripheral lymph nodes of mice with microflora-depleted guts. Clin Immunol 127, Supplement: S53
Ong HK, Greenfield PF et al (2002) Effect of mixing on biomethanation of cattle-manure slurry. Environ Technol 23:1081–1090
Parkin GF, Owen WF (1986) Fundamentals of anaerobic-digestion of wastewater sludges. J Environ Eng 112(5):867–920
Parvez S, Venkataraman C et al (2008) Toxicity assessment of organic pollutants: reliability of bioluminescence inhibition assay and univariate QSAR models using freshly prepared Vibrio fischeri. Toxicol In Vitro 22(7):1806–1813
Pauss A, Nyns EJ et al (1984) Production of methane by anaerobic digestion of domestic refuse. In: EEC Conference on anaerobic and carbohydrate hydrolysis of waste, Luxembourg, 8–10 May
Pavan P, Battistoni P et al (2000) Two-phase anaerobic digestion of source sorted OFMSW (organic fraction of municipal solid waste): performance and kinetic study. Water Sci Technol 41(3):111–118
Peck MW, Chynoweth DP (1992) On-line fluorescence-monitoring of the methanogenic fermentation. Biotechnol Bioeng 39(11):1151–1160
Pilli S, Bhunia P et al (2011) Ultrasonic pretreatment of sludge: a review. Ultrason Sonochem 18 (1):1–18
Pind PF, Angelidaki I et al (2003) A new VFA sensor technique for anaerobic reactor systems. Biotechnol Bioeng 82(1):54–61
Rafique R, Poulse TG et al (2010) Effect of thermal, chemical and thermo-chemical pretreatments to enhance methane production. Energy 35:4556–4561
Raposo F, De la Rubia MA et al (2012) Anaerobic digestion of solid organic substrates in batch mode : an overview relating to methane yields and experimental procedures. Renew Sustain Energy 16:861–877
Rasi S, Veiganen A et al (2007) Trace compounds of biogas from different biogas production plants. Energy 32(8):1375–1380
Redondo R, Machado VC et al (2008) On-line monitoring of gas-phase bioreactors for biogas treatment: hydrogen sulfide and sulfide analysis by automated flow systems. Anal Bioanal Chem 391(3):789–798
Rintala JA, Puhakka JA (1994) Anaerobic treatment in pulp- and paper-mill waste management: a review. Bioresour Technol 47(1):1–18
Rudnitskaya A, Legin A (2008) Sensor systems, electronic tongues and electronic noses, for the monitoring of biotechnological processes. J Ind Microbiol Biotechnol 35(5):443–451
Salsabil MR, Prorot A et al (2009) Pre-treatment of activated sludge: effect of sonication on aerobic and anaerobic digestibility. Chem Eng J 148(2–3):327–335
Schmidt JE, Ahring BK (1993) Effects of magnesium on thermophilic acetate degrading granules in upflow anaerobic sludge blanket (UASB) reactors. Enzyme Microbial Technol 15:304–310
Sekiguchi Y, Ohashi A et al (2015) First genomic insights into members of a candidate bacterial phylum responsible for wastewater bulking. Peer J 3:e740
Shah FA, Mahmood Q et al (2015) Co-digestion, pretreatment and digester design for enhanced methanogenesis. Renew Sustain Energy Rev 42:627–642
Shin HS, Bae BU et al (1992) Anaerobic digestion of distillery wastewater in a two-phase UASB System. Water Sci Technol 25(7):361–371

Show KY, Mao T et al (2007) Optimisation of sludge disruption by sonication. Water Resour 41 (20):4741–4747

Shuizhou K, Zhou S (2005) Application of two-phase anaerobic degradation in industrial wastewater treatment. Int J Environ Pollut 23(1):65–80

Siegrist H, Vogt D et al (2002) Mathematical model for meso- and thermophilic anaerobic sewage sludge digestion. Environ Sci Technol 36(5):1113–1123

Smith AL, Stadler LB et al (2012) Perspectives on anaerobic membrane bioreactor treatment of domestic wastewater: a critical review. Bioresour Technol 122:149–159

Speece RE (1996). Anaerobic biotechnology for industrial wastewater

Stabnikova O, Liu XY et al (2008) Digestion of frozen/thawed food waste in the hybrid anaerobic solid-liquid system. Waste Manag 28:1654–1659

Steffen R, Szolar O et al (2012) Feedstocks for anaerobic digestion

Steyer JP, Bouvier JC et al (2002) On-line measurements of COD, TOC, VFA, total and partial alkalinity in anaerobic digestion processes using infra-red spectrometry. Water Sci Technol 45 (10):133–138

Stronach SM, Rudd T et al (1986) Anaerobic digestion processes in industrial wastewater treatment. Heidelberg, Germany

Sun Y, Cheng J (2002) Hydrolysis of lignocellulosic materials for ethanol production: a review. Bioresour Technol 83:1–11

Sundberg C, Al-Soud WA et al (2013) 454 pyrosequencing analyses of bacterial and archaeal richness in 21 full-scale biogas digesters. FEMS Microbiol Ecol 85(3):612–626

Sutter K, Wellinger A (1987) Güllengrube als biogasanlage. FAT-Berichte 304:1–8

Talarposhti AM, Donnelly T et al (2001) Colour removal from a simulated dye wastewater using a two-phase anaerobic packed bed reactor. Water Res 35(2):425–432

Tatara M, Yamazawa A et al (2004) High-rate thermophilic methane fermentation on short-chain fatty acids in a down-flow anaerobic packed-bed reactor. Bioprocess Biosyst Eng 27(2): 105–113

Tuleva B, Christova N et al (2009) Isolation and characterization of trehalose tetraester biosurfactants from a soil strain Micrococcus luteus BN56. Process Biochem 44(2):135–141

Turovskiy IS, Mathai PK (2006) Wastewater sludge processing. New York, US

Ueno Y, Tatara M et al (2007) Production of hydrogen and methane from organic solid wastes by phase-separation of anaerobic process. Bioresour Technol 98:1861–1865

Valo A, Carrère H et al (2004) Thermal, chemical, and thermo-chemical pretreatment of waste activated sludge for anaerobic digestion. J Chem Technol Biotechnol 79(11):1197–1203

Valorga (1985) Waste recovery as a source of methane and fertilizer—The Valorga process. In: 2nd Annual international symposiumvon industrial resource management Philadelphia, U.S.A

Van Haandel A, Lettinga G (1994) Anaerobic sewage treatment—a practical guide for regions with a hot climate. New York

Verrier D, Roy F et al (1987) Two-phase methanization of solid vegetable wastes. Biol Wastes 22:163–177

Wang X, Zhao Y (2009) A bench scale study of fermentative hydrogen and methane production from food waste in integrated two-stage process. Int J Hydrogen Energy 34:245–254

Wang F, Hidaka T et al (2011) Co-digestion of polylactide and kitchen garbage in hyper-thermophilic and thermophilic continuous anaerobic process. Bioresour Technol 112:67–74

Wang X, Yang G et al (2012) Optimizing feeding composition and carbon-nitrogen ratios for improved methane yield during anaerobic co-digestion of dairy, chicken manure and wheat straw. Bioresour Technol 120:78–83

Ward AJ, Hobbs PJ et al (2008) Optimisation of the anaerobic digestion of agricultural resources. Bioresour Technol 99(17):7928–7940

Weemaes M, Grootaerd H et al (2000) Anaerobic digestion of ozonized biosolids. Water Res 34 (8):2330–2336

Westlake WS, Horler DF et al (1967) The effect of sodium on the fermentation glutamic acid by Peptococus aerogens. Biochem Biophys Res Commun 26(4):461–465

Wilkins MR (2009) Effect of orange peel oil on ethanol production by Zymomonas mobilis. Biomass Bioenergy 33(3):538–541

Wilkins D, Lu XY et al (2015) Pyrosequencing of mcrA and archaeal 16S rRNA genes reveals diversity and substrate preferences of methanogen communities in anaerobic digesters. Appl Environ Microbiol 81(2):604–613

Wong MT, Zhang D et al (2013) Towards a metagenomic understanding on enhanced biomethane production from waste activated sludge after pH 10 pretreatment. Biotechnol Biofuels 6(1):38

Xing W, Ngo HH et al (2010) Enhancement of the performance of anaerobic fluidized bed bioreactors (AFBBRs) by a new starch based flocculant. Sep Purif 72:140–146

Yalcin AO, Orhan I et al (2008) Anaerobic treatment of a chemical synthesis-based pharmaceutical wastewater in a hybrid upflow anaerobic sludge blanket reactor. Bioresour Technol 99(5): 1089–1096

Yang ST, Okos MR (1987) Kinetic study and mathematical modeling of methanogenesis of acetate using pure cultures of methanogens. Biotechnol Bioeng 30(5):661–667

Yeom IT, Lee KR et al (2002) Effects of ozone treatment on the biodegradability of sludge from municipal wastewater treatment plants. Water Sci Technol 46(4–5):421–425

Yilmazer G, Yenigün O (1999) Two-phase anaerobic treatment of cheese whey. Water Sci Technol 40(1):289–295

Yu J, Ji M et al (1999) A three-phase fluidized bed reactor in the combined anaerobic/aerobic treatment of wastewater. J Chem Technol Biotechnol 74(7):619–629

Yusuf MOL, Ify NL (2011) The effect of waste paper on the kinetics of biogas yield from the co-digestion of cow dung and water hyacinth. Biomass Bioenergy 35(3):1345–1351

Zakrzewski M, Goesmann A et al (2012) Profiling of the metabolically active community from a production-scale biogas plant by means of high-throughput metatranscriptome sequencing. J Biotechnol 158(4):248–258

Zhang W, Lang Q et al (2014) Anaerobic digestion characteristics of pig manures depending on various growth stages and initial substrate concentrations in a scaled pig farm in Southern China. Bioresour Technol 156:63–69

Zhu B, Gikas P et al (2009) Characteristics and biogas production potential of municipal solid wastes pretreated with rotary drum reactor. Bioresour Technol 100:1122–1129

Ziagova M, Kyriakou G et al (2009) Co-metabolism of 2,4-dichlorophenol and 4-Cl-m-cresol in the presence of glucose as an easily assimilated carbon source by Staphylococcus xylosus. J Hazard Mater 163(1):383–390

Anaerobic Digestion of Solid Waste: A Focus on Microbial Community Structures

Surakasi V. Prasad and Kalluri K. Meher

Abstract Anaerobic digestion (AD) is one of the significant strategy for the management of solid organic waste. It is a biological process that degrade the organic matter in the absence of oxygen with ultimate products being CO_2 and CH_4. Solid waste has to be treated, mechanically or chemically or biologically prior to fed into the anaerobic digesters for an efficient treatment. Solid wastes with lignocellulosic and hemi cellulosic materials are difficult to degrade and need proper pre-treatment. The anaerobic digester should follow optimum parameters such as; temperature 37 °C for mesophilic and 55 °C for thermophilic digestion, pH 6.5–8.0, hydraulic retention time (HRT) of about 35–40 days for mesophilic and 15 days for thermophilic digestion, feed C/N ratio 30/1 for a successful degradation of waste material and biogas production. A balanced active bacterial and methanogenic *Archaeal* population in the AD is most important factor that influence the stable digestion of the waste material. Molecular techniques based on 16S rDNA gene and other functional gene markers such as *McrA*, *Pct*, *nif* are handy to monitor the treatment process. The most advanced next generation DNA sequencing platforms have been serving to identify the community structure and playing an important role in assigning the microbial communities involved to their function. These techniques further helps in rapid bioaugmentaion of AD for the stable operation of digestion process.

Keywords Solid waste · Anaerobic digestion · Pre-treatment · Microbial communities · Methanogens · Bioaugmentation

---

S.V. Prasad (✉)
Biorem, Prolog Biologicals Pvt. Ltd, 139, Phase V, Jeedimetla,, Hyderabad, Telangana, India
e-mail: prasadbior@gmail.com

K.K. Meher
Calorifique Renewable Energies (I) Pvt Ltd, Hyderabad, Telangana, India

O.P. Karthikeyan et al. (eds.), *Recycling of Solid Waste for Biofuels and Bio-chemicals*, Environmental Footprints and Eco-design of Products and Processes, DOI 10.1007/978-981-10-0150-5_5

1 Introduction

An unprecedented increase in the amount of solid waste (SW) was concomitant in developed and rapidly growing developing countries in the past three decades. Most of the SW is being disposed in the landfills. According to the 2013 EPA data, the United States generated 251 million tons of MSW, about 67 % of which were disposed in landfills (EPA report). In China, about 190 million tons of MSW were produced annually, nearly 90 % of which were disposed by landfills (Zhang et al. 2010a, b). In India, about 31.5 Million tonnes of MSW were produced and disposed mostly in landfills. In Greece, the main destination for MSW is landfills (Nikolaou et al. 2010). However, with unprecedented changes in the world economics, and the emergence of real estates business, the cost of land has tremendously increased near cities for landfills. With the enhancement of public environmental consciousness, safe disposal and recycling of SW has become utmost important. On the other side, unorganized decomposition of organic SW causing direct impact on green concept.

Handling of the organic fraction of the MSW are most challenging task. Biodegradation of organic fraction occurs principally through the action of aerobic microorganisms. Ultimately, complete oxidation of the carbonaceous organic materials by aerobic microbes result in the production of carbon dioxide (CO_2) and water (H_2O). Whereas anaerobic microorganisms degrade the organic matter in the absence of oxygen with ultimate products being CO_2 and methane (CH_4).

With better and advanced understanding of the biology of anaerobic digestion (AD) and its core mechanism, it is now widely used to treat a number of solid organic wastes such as MSW, sugar mills, pulp and paper mills, tanneries, slaughter houses, meat packing, cotton and wool processing, antibiotics, food processing industries etc., apart from the organic fraction of the MSW.

Among the biological treatment processes, anaerobic treatment is the most attractive method of solving the twin problems of energy consumption and pollution in a cost effective manner. With the revolutionary advancement in the DNA sequencing technologies, understanding of microbial community structure is simplified. The rapid diagnosis of microbial community structure and its dynamics of AD will help in maintaining the smooth operation of AD. In this chapter, SW preparation for anaerobic digestion, anaerobic degradation, importance of microbial communities and socioeconomic benefits with few case studies were discussed.

2 Substrate Pre-treatment

The ability to make biogas out of many organic substrates is one of the main advantages of anaerobic digestion. Organic waste in the form of liquids are easily treated in the ADs without much pre-treatment. However, solid waste substrates can be very slow to break down because of the following limitations:

- They contain chemicals that inhibit the growth and activity of the microorganisms,
- They create physical problems like floating, foaming or clumping and block impellers and pipes in biogas plants, or
- Their molecular structure is poorly accessible to microorganisms and their enzymes due to their highly crystalline structure and low surface area.

Sometimes all these problems occur simultaneously hence, pre-treatment of substrate(s) can be used to overcome some of these problems.

The main substrate sources of methane production in AD, are carbohydrates, lipids and fats and proteinaceous materials. The agricultural biomass contains most complex substrates, such as starch, cellulose, hemi-cellulose and lignins. While starch (α-1-4 linked D-glucose) is relatively easy to break down biologically, cellulose (β-1-4 linked D-glucose) and hemi-cellulose (a polymer of various sugars and uronic acids) are difficult and slow to break down. It is generally believed that lignin is difficult to degrade by anaerobic bacteria and may even inhibit the degradation of other substances like cellulose. Breaking down this lignocelluloses complex is the key to biogas production (Noike et al. 1985).

In recent years, various pre-treatment technologies have been developed to increase the availability of simple organic matter to the microbial communities in ADs (Table 1), particularly lignocellulosic material. These pre-treatment technologies are aimed to:

- make AD faster
- potentially increase biogas yield
- make use of new and/or locally available substrates
- prevent processing problems such as high electricity requirements for mixing or the formation of scum layers.

Table 1 Pre-treatment Principles and techniques

No.	Principle	Technique
1	Physical	Mechanical
		Thermal
		Ultrasound
		Electrochemical
2	Chemical	Alkali
		Acid
		Oxidative
3	Combined process	Stream
		Extrusion
		Thermo chemical
4	Biological	Microbial
		Enzymatic

2.1 Mechanical Pre-treatment

2.1.1 Knifes and Hammer Milled

It is carried out by knifes and hammer mills for particle size reduction (1–2 mm) and increasing the specific surface area of the biomass. Particle size reduction gives greater possibility for enzymatic attack and the rate of degradation; it can also reduce the digesting slurry viscosity thus mixing easier and can reduce the problems of floating layers. A major disadvantage is that machines can be damaged by inert materials (stones or pieces of metal) in the substrate and equipment repairs can be very expensive. Biomass with less than 15 % moisture can only be used for mechanical pre-treatment. The energy demand increases with higher moisture content. Batch tests at laboratory scale reported that about 10 % higher gas yield with knife milled hay to 0.5 mm compared to 20–30 mm (Menind and Normak 2010). Similarly knife-milled sisal fibers from 100 to 2 mm achieved an approximately 20–25 % higher gas yield (Mshandete et al. 2006). Milling pre-treatment is recommended for very bulky substrates to ease processing. Hammer mills are relatively easy and cheap to operate. Menardo et al. (2011) showed that mechanical pre-treatment of barley and wheat straw, increased methane yield, but not for maize stalks or rice straw.

2.1.2 Extrusion

The biogas substrates are fed into the extruder and conveyed by screw along a tube, where it is exposed to high pressure, temperature and shear forces causing the tough fibers to break. The sudden drop in pressure as the substrate leaves the extruder also help substrate breakdown. Depending on the final consistency required, the substrate can be placed under a pressure of up to 300 bar at temperatures from 60 to 300 °C. Extrusion effectively breaks and opens the cell structure of biomass which results in faster methane production, which in turn facilitates higher organic loading rates (Rotter et al. 2011).

2.1.3 Thermal Pre-treatment

In thermal pre-treatment, the substrate is heated (typically 125–190 °C) under pressure and hold at that temperature for up to 1 h. The presence of heat and water disrupts the hydrogen bonds that hold together crystalline cellulose and the lignocelluloses complexes. Thermal pre-treatment is often carried out with chemicals or in combination with mechanical shearing. Thermal pre-treatment is only effective up to a certain temperature. The maximum temperature varies with different substrates and using batch AD tests has been found to be 175 °C for sludge (52 % increase in methane production) (Distefano and Ambulkar 2006), 190 °C for crops, and 160 °C for brewers' spent grains. However, these values are dependent on pre-treatment retention time. Many studies (Distefano and Ambulkar 2006) show that thermal (including thermo-chemical or thermo-mechanical) pre-treatment only increases biogas yield up to a certain temperature, above which biogas production decreases. Therefore, the trick with all pre-treatment involving high temperatures is to find the optimum conditions that break down the substrate. Thermal pre-treatment is particularly well suited to locations where there is a supply of waste heat, for example from a nearby factory or power plant.

2.2 *Chemical Pre-treatments*

Chemical pre-treatment is conducted using a range of different chemicals, mainly acids and bases of different strengths under different conditions.

2.2.1 Alkali Pre-treatment

Alkali pre-treatment of lignocellulosic materials with lime or sodium hydroxide (NaOH) causes partial lignin solubilization. Alkali treatment of agricultural crop residue is effective for AD. Small-scale batch tests, with alkali (6 % NaOH) pre-treated straw for 3 weeks at ambient temperatures increased biogas yield significantly (He et al. 2008). Pre-treatment and methanization of fallen leaves using 3.5 % NaOH had shown about 20 % increase in the methane yield (Liew et al. 2011). It is important to note that alkali pre-treated substrates have high pH values and the above experiments were carried out using small scale batch digesters, but during continuous fermentation, alkali pre-treatment leads to salt build up and increased pH values. The high salt concentration and the resulting effect on the ammonium-ammonia balance inhibits methanization (Chen et al. 2008). The pH increase due to alkali pre-treatment might be beneficial for substrates with low pH or high lipid content (Beccari et al. 2001) with olive oil mill effluent. This pre-treatment technology is economically unattractive due to the high costs of alkali (Chang et al. 1997), but it may be useful for acidic and lignin rich substrates that could otherwise not be anaerobically digested.

2.2.2 Acid Pre-treatment

Acid pre-treatment does not disrupt lignin but is thought to work by breaking down hemicellulose and disrupting ether bonds between lignin and hemicellulose (Knappert et al. 1981). Acid pre-treatment is typically used in combination with heat.

2.2.3 Oxidative Pre-treatment

Oxidative pre-treatment with hydrogen peroxide or ozone affects lignocellulose in a similar way to alkaline pre-treatment as it can also break down lignin. Rice straw was successfully pre-treated with hydrogen peroxide and ammonium at room temperature and found 100 % increase in the biogas production (Song et al. 2012). However, one possible disadvantage is that introducing more oxygen into the system increases the proportion of CO_2 in the biogas produced. This pre-treatment is also not carried out at large scale presumably or partly due to high costs.

2.2.4 Thermo-Chemical Pre-treatment

Different kinds of bases and acids can be used at temperatures from 60 to 220 °C in thermo-chemical pre-treatment. Batch AD of pre-treated (160 °C, 3 % H_2SO_4 and 20 min retention time) cassava yielded 57 % higher gas yield when compared to untreated cassava. Batch tests with sunflower stalks pre-treated with different chemicals and temperatures (Monlau et al. 2012) found that pre-treatment with heat alone was not very effective, but pre-treatment with H_2O_2 or NaOH (4 % total solids) did increase methane yield by about one third at 55 °C (rather than 30 or 80 °C). They found that this pre-treatment solubilised lignin. Pre-treatment with HCl at 170 °C increased methane yield by around 20 % and solubilised hemicellulose but not lignin. Although thermo-chemical pre-treatment has been tried at pilot scale several times, large-scale thermo-chemical pre-treatment of substrates for biogas production is not found.

2.3 *Biological Pre-treatment*

The advantage of biological pre-treatments over chemical or thermal pre-treatment is that biological pre-treatment can take place at low temperature without using chemicals. Disadvantage is that they can be slower than non-biological method.

2.3.1 Anaerobic Microbial Pre-treatment

In the anaerobic pre-treatment, the first step of anaerobic digestion i.e., hydrolysis and acid production is separated from acetogenesis and methanogenesis. The pH value of the first digester should lie between 4 and 6, which inhibits methane production and causes volatile fatty acids to accumulate (Deublein and Steinhauser 2010; Thauer 1998). In general, cellulose, hemicellulose and starch-degrading enzymes work best between pH 4 and 6 at temperatures from 30 to 50 °C, so the pre-acidification step increases the degradation rate by creating an optimal environment for these enzymes. Two-stage continuous AD of household waste at a hydraulic retention time (HRT) of approximately 30 days has yielded about 21 % more biogas (Liu et al. 2006).

Another positive effect of this pre-treatment method is on the methane concentration in the biogas. In addition to H_2 and volatile fatty acids, CO_2 is formed during the pre-acidification step. Due to the low pH, most of the carbonate is in the form of CO_2, which is volatile and is released out of the digester. This means there is less CO_2 in the gas phase of the methanogenesis step, and therefore a higher CH_4 concentration is obtained. Two stage AD of grass silage produced biogas with 71 % methane as compared with 52 % methane content in a single-stage continuous AD (Nizami et al. 2012). Another advantage of two-stage digestion is that the microorganisms of the first stage are less sensitive to chemicals such as phenol, ammonia, etc. than the microorganisms of the second stage, and many inhibiting chemicals can be broken down in the first stage.

Overall, two-stage digestion is useful for a range of different substrates and higher investment costs for an additional reactor are typically offset by faster digestion rates (due to optimized pH and temperature for the hydrolytic enzymes) and the added stability of feeding with a constant pH. In addition, higher gas methane yields might lead to lower gas upgrading costs.

2.3.2 Aerobic Microbial Pre-treatment

Aerobic microbial pre-treatment can be carried out with naturally occurring mixed cultures. Aerobic organisms produce cellulose, hemicellulose and/or lignin degrading enzymes rapidly and in large amounts, and these solubilise the substrate.

In the integrated aerobic-anaerobic pre-treatment process, the organic fraction of municipal SW is fed into an aerated leach bed reactor. The leachate (hydrolysis juice) is collected and fed into an anaerobic digester. The remaining undegraded solid fraction is disposed by composting. This process is similar to the anaerobic leach bed system (Lehtomäki et al. 2008), but with aerobic conditions in the leach bed reactor.

In leach bed reactors (aerobic or anaerobic) there are no processing problems due to fibers or large chunks in the anaerobic digester. In general, the advantage of an aerobic process is that it is considerably faster, but the disadvantage is that a lot of the organic matter that could be degraded to methane is instead degraded to CO_2 if the pre-treatment phase is too long. It is possible to combine the two processes, for example with micro aeration in an anaerobic pre-treatment reactor increase methane yields significantly (Jagadabhi et al. 2009).

2.3.3 Fungal Pre-treatment

Many fungi, particularly white-rot fungi, are known for their ability to remove environmental pollutants from solid and liquid waste (Reddy 1995). Fungal pre-treatment has been used to detoxify coffee/cherry husks for anaerobic digestion (Jayachandra et al. 2011). There has also been some research on fungal pre-treatment of waste to increase biogas yields (Wagner et al. 2014). It is not clear what effect fungal pre-treatment has on biogas yields, because although white-rot fungi can delignify substrates, they also remove some of the organic matter that could be used for anaerobic digestion. Fungal pre-treatment has not been carried out at large scale for anaerobic digestion.

2.3.4 Enzyme Addition

Enzymes that break down biomass are already present in anaerobic digesters as they are produced by the microorganisms of AD. To enhance this breakdown, a mixture of enzymes can be added, and may include cellulose-, hemicellulose-, pectin- and starch-degrading enzymes. Enzyme additives can be applied in three different ways: by direct addition to a single-stage anaerobic digester, by addition to the hydrolysis and acidification vessel (first stage) of a two-stage system or by addition to a dedicated enzymatic pre-treatment vessel.

Effect of 25 different commercially available enzyme preparations including enzyme mixtures marketed to biogas plants as well as pure enzymes normally marketed to other industries. They found that the effect of enzymatic pre-treatment on biogas yield from sludge and manure was minimal and speculated that this was because the enzymes were being degraded by the native microorganisms. Some of the enzyme products increased the biogas yield by around 10 % in grass silage and green waste silage. Enzyme products for biogas plants are offered by several different companies, but some enzymes have a relatively high price for a limited increase in biogas yield. Hence enzyme dosage for biogas production with increased methane concentration is unlikely economically feasible.

3 Anaerobic Digestion

The application of the anaerobic digestion has evolved from the treatment of mainly sewage sludge and manure toward the use of more energy-rich waste mixtures with higher methane potentials, such as slaughterhouse, food industry, and household wastes; glycerol; and frying fats (Deublein and Steinhauser 2010). Anaerobic digestion of waste is an environmentally and economically beneficial process in which the biological degradation of organic wastes results in the production of CH_4 as a carbon-neutral energy source (Zitomer et al. 2008). The anaerobic degradation of organic matter mainly proceeds as hydrolysis, fermentation, acetogenesis, and methanogenesis of organic matter where various microorganisms are involved in each step (Zinder et al. 1984). The organic substrates may increase the stress on the operational system, because these substrates may result in a less stable process (Salminen and Rintala 2002; Edström et al. 2003). An improved understanding of the microbial communities and their function during the different aspects of AD may help to optimize biogas production, and molecular biology techniques offer possible tools.

3.1 *Hydrolysis*

A diverse number of bacteria take part in the hydrolysis and fermentation steps; the oxidation of intermediate fermentation products to acetate is performed by either hydrogen- or formate-producing acetogens (Stams and Plugge 2009). Hydrolysis of the polymerized mostly insoluble organic compounds, like carbohydrates, proteins and fats yields soluble monomers and dimers, i.e., monosaccharides, amino acids, and fatty acids. Several extracellular enzymes from the group of hydrolases (amylases, proteases, and lipases) produced by appropriate strains of hydrolytic bacteria help the hydrolysis. Decomposable polymers, that is, cellulose and cellulo-cottons are difficult to decompose, which is considered to be a rate limiting step of hydrolysis. About 50 % of the organic fraction of solid waste undergoes biodegradation. The remaining part of the compounds remains in their primary state because of the lack of enzymes participating in their degradation (Parawira et al. 2008). The bacterial community associated with ADs of different substrates and tretment systems are listed in Table 2 (Karthikeyan and Visvanathan 2013).

The biodegradation rate depends on parameters such as size of particles, pH, production of enzymes, diffusion, and adsorption of enzymes on the particles of waste. Some industrial operations overcome this limitation by the use of chemical reagents to enhance hydrolysis. The application of chemicals to enhance the first step has been found to result in a shorter digestion time and provide a higher methane yield. Diverse group of bacteria are reported to perform the hydrolysis (Smith and Bryant 1979). *Clostridia* from *Fermicutes* are most prominently identified in several anaerobic microbial community studies (Leven et al. 2007; Krause et al. 2008; Cardinali-Rezende et al. 2009; Patil et al. 2010) which are capable of degrading proteins, lipids, and polymeric carbohydrates. *Bacteroidetes*, and *Flavobacteria* are next most abundant groups which degrade complex organic matter.

3.2 *Fermentation and Acidogenesis*

During this stage, the monomers undergo the acid and alcohol fermentation process and convert water-soluble chemical substances, including hydrolysis products, to short-chain organic acids (C1 to C5 compounds such as formic, acetic, propionic, butyric, and pentanoic), alcohols (methanol, ethanol), aldehydes, carbon dioxide, and hydrogen. Amino acids and peptides will form the decomposition of proteins, which may be a source of energy for anaerobic microorganisms. This process may be divided into two types: hydrogenation and dehydrogenation. The basic pathway of transformations passes through acetates, CO_2, and H_2, whereas other acidogenesis products play an insignificant role. As a result of these transformations, methanogens may directly use the by-products as substrates and energy source. Bacterial species from several phylogroups such as Fermicutes, Bacteroidetes, Proteobacteri, are widely found to hydrolyse the complex organic compounds.

Table 2 Bacterial community distribution in single- and two-stage bioreactors

Bacterial community (similarity percentage)	Arachaeal community (similarity percentage)	Organic substrate and conditions (Reference)
Uncultured Thermotogae bacterium clone (100)	*Methanothermobacter thermautotrophicus* (96)	Simulated food waste; two stage system; thermophilic condition (Yabu et al. 2010)
Anaerobaculum mobile (100)	*Methanosarcina barkeri* (99)	
Thermacetogenium sp. (98)	Uncultured *Methanosarcina* sp. KT19 (99)	
Coprothermobacter proteolyticus (100)	*Methanosarcina thermophila* (99 %)	
Clostridium caenicola (98)		
Tepidanaerobacter syntrophicus (99)		
Syntrophomonas wolfei (97)		
Clostridium sufflavum (96)		
Clostridium populeti (96)		
Clostridium thermopalmarium (100)		
Leuconostoc mesenteroides (99)		
Desemzia incerta (99)		
Bacteroidales bacterium 28bM (96)		
Mycobacterium elephantis (99) and		
uncultured bacterium		
Bacillus thermoterrestris (93)	*Methanoculleus thermophilicus* (97)	Cattle manure; single-stage system; thermophilic condition (Chachkhiani et al. 2004)
Bacteroides sp. (92)	*Methanosarcina thermophila* (100) and uncultured archaeon	
Clostridium sp. FCB90-3 (96)		

(continued)

Table 2 (continued)

Bacterial community (similarity percentage)	Arachaeal community (similarity percentage)	Organic substrate and conditions (Reference)
Clostridium hydroxybenzoicum (97)		
Eubacterium limosum (92)		
Desulfotomaculum thermobenzoicum (92)		
Zoogloea sp. (99)		
Sulfide-oxidizing bacterium N9-1 (98)		
Pseudomonas halodenitrificans (97) and uncultured bacterium		
Psychrobacter sp. (91 and 85)	Methanobacteriaceae (74)	Mixed waste; single-stage system; mesophilic condition (Supaphol et al. 2011)
Arcobacter sp. (96)	*Methanothermobacter thermautotrophicus* (74)	
Lactobacillus sp. (98)	Uncultured Methanosaeta (90)	
Bacillus sp. (77)	Uncultured Methanosarcinaceae (96)	
Brevibacterium sp. (90)	*Methanobrevibacter* (93)	
Comamonas denitrificans strain 2B7 (90)	*Methanosaeta concilii* (94) and uncultured archaeon clone	
Streptomyces sp. (87)		
Anaerovorax sp. (87)		
Thiobacillus denitrificans (89)		
Nocardioides sp. (96)		
Symbiobacterium sp. (82)		
Shewanella algae (92) and		
uncultured *Clostridium* sp. (81)	*Methanobacterium beijingense*	Molasses waste; two-stage system; mesophilic condition (Park et al. 2010)[a]
Costridium butyricum	*Methanothrix soehngenii*	
Clostridium leptum	*Methanobacterium formicicum*	
Clostridium drakei	*Methanobacterium aarhusense*	

[a]Study results from W-ABC processes in which first stage is Hydrogen production and second stage is Methane production

Levén et al. (2007) studied the effect of process temperature on the AD of organic household waste and reported a dominance of Thermotogae and Clostridia in their thermophilic reactors, while *Bacteroidetes* and *Chloroflexi* were the main phyla in the mesophilic reactors (Sundberg et al. 2013). The acetateformed during acetogenesis are due to a number of different microbes, e.g., *Syntrophobacter wolinii*, a propionate decomposer and *Sytrophomonos wolfei*, a butyrate decomposer will highly limit the AD process since they grow in symbiosis else the process will be incomplete. Other acid formers are *Clostridium* spp., *Peptococcus anerobus*, *Lactobacillus*, and *Actinomyces* etc. However the most unpleasant gases such as ammonia, hydrogen sulphide etc. also will form in acidogenesis phase.

Among the products of acidogenesis, ammonia and hydrogen sulfide which give an intense unpleasant smell to this phase of the process should also be mentioned. The acid phase bacteria belonging to facultative anaerobes use oxygen accidentally introduced into the process, creating favorable conditions for the development of obligatory anaerobes of the following genera: *Pseudomonas*, *Bacillus*, *Clostridium*, *Micrococcus*, or *Flavobacterium* (Nikolaou et al. 2010; Conrad 1999).

3.3 Acetogenesis

In this process, the acetate bacteria including those of the genera of *Syntrophomonas* and *Syntrophobacter* convert the small chain fatty acids into acetates and hydrogen which may be used by methanogenic Archaea (Schink 1997). Bacteria like *Methanobacterium suboxydans* account for decomposition of pentanoic acid to propionic acid, whereas *Methanobacterium propionicum* accounts for decomposition of propionic acid to acetic acid. As a result of acetogenesis, hydrogen is released, which exhibits toxic effects on the indigenous microorganisms of process. Therefore, a symbiosis is necessary for acetogenic bacteria with autotrophic methane bacteria using hydrogen (Schink 1997). Acetogenesis is a phase which depicts the efficiency of biogas production, because approximately 70 % of methane arises in the process of acetates reduction. Consequently, acetates are a key intermediate product of the process of methane digestion. In acetogenesis phase, approximately 25 % of acetates are formed and approximately 11 % of hydrogen is produced in the wastes degradation process (Schink 1997). Homoacetogenesis, i.e. acetate formation from carbon dioxide and dihydrogen via the acetyl-CoA pathway, out competes methanogens for common substrate, H_2, which contributed greatly to acetate production (Ye et al. 2014). Butyrate degradation for hydrogen production under conditions suppressing methanogenesis was evaluated in continuously fed-tank reactors operated at 55 °C and started up with digested manure as inoculum (Siriwongrungson et al. 2007)

Finally, methane formation is mainly derived from acetate and H_2/CO_2 by methanogenic *Archaea*. A balanced interaction between the microorganisms in this degradation chain is crucial for the continuous transformation of the intermediates formed and subsequently an efficient biogas production. Methanogens are especially important for the obligatory syntrophic interactions driving the acetogenic proton reduction needed for growth on, for example, fatty acids and alcohols (Stams and Plugge 2009).

3.4 Methanogenesis

Finally, methanogenic organisms consume the acetate, hydrogen, and some of the carbon dioxide to produce methane using the by-products produced from Acetogenesis and partly Acidogenesis. A comparison of the free energetics of hydrolysis of ATP (−31.8 kJ/mol) and those of methane formation from the substrates hydrogen and carbon dioxide, formate, methanol, methylamines, carbon monoxide and acetate leads to the conclusion that only small amounts of energy is available to these organisms (Surakasi et al. 2007). Most of these substrates contain one carbon (Table 3). The methanogens are classified based on their substrates as acetotrophic which uses acetate as substrate, hydrogenotrophic which uses H_2/CO_2 as a substrate, and methylotrophic which uses methyle group as a substrate. Methanogenesis has not been observed with complex organic matters. Methanogens cannot degrade more complex molecules such as glucose since it requires a complex and specialized metabolic machinery that is lacking in methanogens (Zinder 1993). Acetate is one of the most important substrates for methanogenic *Archaea*, because over 70 % of biomethane comes from processing of acetic acid which will be processed to methane and carbon dioxide. The Methanogenic Archaea include *Methanobacterium, Methanobacillus, Methanococcus and Methanosarcina*. *Methanosarcina* spp., *Methanothrix* spp and *Methanosaeta* are considered to be important in AD both as acetate and H_2/CO_2 consumers.

Methanoculleus thermophilicus is dominant hydrogenotrophic *Archaea* that uses four moles of H_2 and one mole of CO_2 to form one mole of CH_4. *Methanosarcina*

Table 3 Stoichemoetric reactions of methanogenic *Archaea*

Reaction	$\Delta G^{0'}$ (KJ/molCH_4)
4 H_2 + CO_2 → CH_4 + $2H_2O$	−35.6
4 Formate → CH_4 + $3CO_2$ + $2H_2O$	−130.1
Methanol + H_2 → CH_4 + H_2O	−112.5
4 Methanol → $3CH_4$ + CO_2 + $2H_2O$	−104.9
4 Methylamine + $2H_2O$ → CH_4 + CO_2 + $4NH_4^+$	−75.0
4 Trimethylamine + $6H_2O$ → $9CH_4$ + $3CO_2$ + $4NH_4^+$	−74.3
2 Dimethylsulphide + $2H_2O$ → $3CH_4$ + CO_2 + H_2S	−73.8
2 Dimethylamine + $2H_2O$ → $3CH_4$ + CO_2 + $2NH_4^+$	−73.2
4 2-Propanol + CO_2 → CH_4 + 4 Acetone + $2H_2O$	−36.5
Acetate → CH_4 + CO_2	−31.0

and Methanobacteriales are postulated to be the dominant hydrogenotrophic methanogens at both mesophilic and thermophilic conditions of organic waste, because of the low sludge retention time of 3–5 days, compared to conventional anaerobic digesters, which have a SRT of about of 20–30 days (Appels et al. 2008).

From the above stoichiometric reactions it can be seen that theoretically the biogas produced would contain 50 % methane and 50 % carbon dioxide. However, acetogenesis typically produces some amount of hydrogen, and for every four moles of hydrogen consumed by hydrogenotrophic methanogens a mole of carbon dioxide is converted to one mole of methane. Substrates other than sugar, such as fats and proteins, can yield larger amounts of hydrogen leading to higher typical methane content for these substrates. Furthermore, hydrogen and acetate can be biochemical substrates for a number of other products as well. Therefore, the overall biogas yield and methane content will vary for different substrates, biological consortia and digester conditions. Typically, the methane content of biogas ranges from 50–70 %.

3.5 *Allied Microbiological Conversions*

Several microbiological communities with specialized biochemical reactions find their substrate and thus niche, which plays an important role in the complete degradation of feed that come from the pre-treated solid waste. These biochemical reactions may co-exist with basic methanogenesis processes or can be separated depending on the characteristics of the effluent for ex:anammox reaction can be separated from methanogenesis in a dedicated digester where the /N ration of the effluent is low. Many of the allied biochemical reactions are discussed below.

3.5.1 Sulphate Reduction

Micro-organisms (most frequently bacteria) are often integrally involved in the chemical alteration of minerals. The dissolution of sulphide minerals under acidic conditions, the precipitation of minerals under anaerobic conditions, the adsorption of metals by bacteria and the formation and destruction of organo-metallic complexes are all examples of indirect micro-organism participation. Where minerals are available as soluble trace elements, they serve as specific oxidizing substrates or are electron donors/acceptors in oxidation-reduction reactions and they may be directly involved in cell metabolic activity.

Reduction of sulphate ions to hydrogen sulphide is effected by specialized strictly anaerobic bacteria of the genera *Desulfovibrio* and *Desulfotomaculum*. Sulphate-reducing bacteria are physiologically and phylogenetically highly diverse (Castro et al. 2000); they oxidize a wide variety of low-molecular weight compounds (short-chain fatty acids, alcohols, alkanes, aromatic compounds, acetate) to CO_2. In natural systems the specific requirement of SRB, i.e., the low molecular weight compounds, is met by the fermentative activity of other anaerobic bacteria on complex organic substrates. Sulphate behaves as an alternative electron acceptor to support anaerobic respiration. Sulphate reducing bacteria convert sulphate (SO_4^{2-}) or sulphite (SO_3^{2-}) to sulphide (S^{2-}). This is the major natural process for the conversion of sulphate ion. The substrates are either partially oxidized for ex: to acetate or fully oxidized to carbon dioxide. The formation of biogenic sulphide is the first step in biotechnological processes directed at the removal and recovery of sulphur or heavy metals. Lactate is used by the SRB during anaerobic respiration to produce acetate (Cork and Cusanovich 1979).

$$2CH_3CHOHCOO^- + SO_4^- = 2CH_3COO^- + 2HCO_3^- + H_2S$$

3.5.2 Dissimilative Sulphate Reduction Process

Sulphate is activated by reaction with ATP, forming adenosine phosphosulphate

$$(APS)\ SO_4^{2-} + ATP \rightarrow APS + Pp_i$$

Hydrogenase splits molecular hydrogen, and the electrons contained therein are used to reduce the sulphur atom of APS, releasing sulphite (SO_3^{2-}). This reaction involves an intermediate electron carrier, cytochrome c3.

$$APS + H_2 \rightarrow SO_3^{2-} + AMP + H_2O$$

Using more electrons derived from molecular hydrogen, sulphite is reduced to hydrogen sulphide:

$$SO_3^{2-} + 6\,H^+ + 6\,e^- \rightarrow H_2S + H_2O + 2\,OH^-$$

3.5.3 Nitrate Reduction

In anaerobic environments denitrification is the key dissimilative pathway for nitrate reduction. Denitrifying bacteria metabolize nitrogenous compounds with the assistance of the molybdenum-containing enzyme, nitrate reductase, in the reverse way that nitrifying bacteria does: they turn nitrogen oxides (NO_2 and NO_3) back into nitrogen gas or nitrous oxides (N_2, N_2O, or NO) for energy generation. These gases then volatize, and return back into the atmosphere.

$$NO_3^- \rightarrow NO_2^- \rightarrow NO \rightarrow N_2O \rightarrow N_2\text{ gas}$$

The enzyme *nitrate reductase* is synthesized only when O_2 is repressed, anoxic conditions are obligatory for most denitrifying bacteria.

Dissimilative denitrification is desirable in sewage treatment and in bioremediation as it aids in converting organic nitrogen to clean nitrogen gas that escapes to the atmosphere. This method allows for very clean disposal of nitrogenous pollutants.

3.5.4 Anammox Process

Anaerobic ammonium oxidation (Anammox), is an important microbial process of the nitrogen cycle (Arrigo 2005). In this biological process, nitrite and ammonium are converted directly into dinitrogen gas (Reimann and Keltjens 2015) and is used for the removal of nitrogen from ammonia rich (>50 mg/lit) wastewater streams originating from municipal, industrial, agricultural activities. Nevertheless, oxygen is still required for the production of nitrite by ammonia-oxiding bacteria. However, in partial nitrification/anammox systems, oxygen demand is greatly reduced because only half of the ammonium needs to be oxidized to nitrite instead of full conversion to nitrate. The autotrophic nature of anammox bacteria and ammonia-oxidizing bacteria guarantee a low yield and thus less sludge production (Hu et al. 2013). Additionally, anammox bacteria easily form stable self-aggregated biofilm (granules) allowing reliable operation of compact systems characterized by high biomass concentration and conversion rate up to 5–10 kg N m^{-3} (von Loosdrecht 2008). Overall, it has been shown that efficient application of the anammox process in wastewater treatment results in a cost reduction of up to 60 % (Siegrist et al. 2008; van Dongen et al. 2001) as well as lower CO_2 emissions.

Anammox is a combination of two separate treatment steps: a partial nitrification process (Sharon) followed by Anammox. About 50 % of the ammonium is converted to nitrite in the Sharon process. This conversion takes place in a single, completely mixed reactor without biomass retention at average process temperatures between 30 and 40 °C and retention times of 1–2 days.

$$NH_4^+ + HCO_3^- + 0.75\ O_2 \rightarrow 0.5\ NO_2^- + 0.5\ NH_4^+ + CO_2 + 1.5\ H_2O$$

To ensure that only 50 % of the ammonium is converted to nitrite, the oxygen supply is limited. The reactor will be operated either under alternating oxic and anoxic conditions, or continuously with limited oxygen supply. In the latter case, oxygen consumption by the nitrifying organisms will also generate the anoxic process conditions required for the second treatment step—the Anammox process.

The ammonium-nitrite mixture produced in the Sharon process is converted under anoxic conditions to nitrogen gas with ammonium as electron donor. The conversion of ammonium and nitrite to nitrogen gas is described by the following formula:

$$NO_2^- + NH_4 = N_2 + 2\,H_2O$$

The bacteria catalysing the reaction are autotrophic and facilitate the conversion without the use of COD or the addition of external carbon sources.

Bacteria that perform the anammox process belong to the phylum Planctomycetes. Currently, five anammox genera have been discovered: *Brocadia, Kuenenia*, *Anammoxoglobus*, *Jettenia* (all fresh water species), and *Scalindua* (marine species) (Jetten et al. 2009).

3.5.5 Iron Reduction

Ferric iron, Fe (III) is one of the important electron acceptor for micro-organisms in the AD of solid waste. The iron metal is one of the most abundant and potential electron acceptor for organic matter decomposition. Further it is becoming increasingly apparent that microbial metal reduction may be manipulated to aid in the remediation of polluted environments and waste streams contaminated with metals and certain organics.

A great diversity of micro-organisms can metabolize sugars or amino acids with Fe (III) reduction; however it is a trivial side reaction in the metabolism of these organisms. The primary products of the metabolism of the fermentative Fe (III)-reducing micro-organisms are typical fermentation acids, alcohols, and H_2. Micro-organisms that can completely oxidize sugars and amino acids to carbon dioxide with Fe (III) as the sole electron acceptor are unknown.

$$\text{Acetate}^- + 8\,\text{Fe (III)} + 4\,H_2O \rightarrow 2\,HCO_3^- + 8\,\text{Fe(II)} + 9\,H^+$$

Desulphuromonas acetoxidan is known primarily for its unique ability to couple the oxidation of acetate to the reduction of Sulphate. However *D. acetoxidans* can also oxidize acetate with Fe (III) as the electron acceptor. Formate is oxidized to carbon dioxide whereas lactate and pyruvate are incompletely oxidized to carbondioxide and acetate:

$$\text{formate}^- + 2\,\text{Fe (III)} + H_2O \rightarrow H\,CO_3^- + 2\,\text{Fe(II)} + 2\,H^+$$
$$\text{lactate}^- + 4\,\text{Fe (III)} + 2\,H_2O \rightarrow \text{acetate}^- + HCO_3^- + 4\,\text{Fe(II)} + 5\,H^+$$
$$\text{pyruvate}^- + 2\,\text{Fe (III)} + 2\,H_2O^- \rightarrow \text{acetate}^- + HCO_3^- + -2\text{Fe(II)} + 3\,H^+.$$

Formate oxidation coupled to Fe (III) reduction is a potentially important process in anaerobic environments if formate replaces H_2 as an important fermentation product in Fe (III)-reducing environments.

Several *Desulfovibrio* species also oxidize H_2 with the reduction of Fe (III) at rates comparable to those observed with other Fe (III) reducers, but no net cell growth occurs. The minimum threshold for H_2 uptake in *D. desulphuricans* is lower with Fe (III) serving as the electron acceptor than with sulphate, suggesting that under conditions of limiting electron donor availability, *Desulfovibrio* species will preferentially reduce Fe (III). A wide variety of monoaromatic compounds can be completely oxidized to carbon dioxide with Fe (III) serving as the sole electron acceptor *G. metallireducens* is the only aromatic-oxidizing, Fe (III)-reducing organism. Contaminants such as toluene, p-cresol, and phenol are the aromatics oxidized by *G. metallireducens*.

Long-chain fatty acids are another important component of organic matter that is metabolized in ADs. Enrichment cultures that can oxidize long-chain fatty acids have been established, but no isolates have been purified and the pathways for oxidation of long-chain fatty acids in Fe (III)-reducing environments have not been elucidated. In acidic conditions of AD, elemental sulphur can serve as an electron donor for Fe (III) reduction. *Thiobacillus ferrooxidans, Thiobacillus thiooxidans, and* the thermophile, *Sulfolobus acidocaldarius,* reduce Fe (III) by the following reaction:

$$S^0 + 6\ \text{Fe (III)} + 4\ H_2O \rightarrow HSO^{-4} + 6\ \text{Fe (II)} + 7\ H^+.$$

Initial studies indicated that neither *T. thiooxidans* nor *T. ferrooxidans* could conserve energy to support growth from this reaction. The *S. acidocaldarius* can reduce the Fe (III) at thermophilic conditions (Gold 1992).

4 Operating Parameters in AD Process

The growth rate of the microorganisms is of paramount importance in the AD process. A variety of factors affect the rate of digestion and biogas production. Design of the AD is one of the important factors that influence the complete degradation of solid wastes. Dry and wet type of anaerobic reactors have been employed for anaerobic decomposition of different types of organic substrates. Karthikeyan and Viswanathan (2013) reviewed the applicability of dry and wet type ADs for different types of organic substrates with merits and demerits. The operating parameters of the digester must be controlled so as to enhance the microbial activity and thus increase the anaerobic degradation efficiency of the system. Some of these parameters are discussed below.

4.1 Temperature

It is the most important one. Anaerobic bacterial communities can endure temperatures ranging from below freezing to above 57.2 °C, but they thrive best at mesophilic temperatures of about 36 °C and thermophilic 55 °C. Bacterial activity and thus the biogas production falls off significantly between mesophilic and thermophilic temperature ranges and gradually from 35 to 0 °C. Although digesters operated in the mesophilic range must be larger to accommodate a longer period of decomposition within the reactor (residence time), the process is less sensitive to any changes in operating conditions. To optimize the digestion process, the digester must be kept at a consistent temperature, as rapid changes will upset bacterial activity.

4.2 pH

Anaerobic bacteria, especially the methanogens, are sensitive to the acid concentration within the digester and their growth can be inhibited by acidic conditions. An optimum pH value for AD lies between 6.5 and 8.0. During digestion, the two processes of acidification and methanogenesis require different pH levels i.e. 4.0–6.0 and 6.5–8.0 for optimal process control. The retention time of the digestate affects the pH value and in a batch reactor acetogenesis occurs at a rapid pace. Acetogenesis can lead to accumulation of large amounts of organic acids resulting in pH below 5. Excessive generation of acid inhibit methanogens. The pH reduction can be controlled by the addition of lime or by recycled filtrate obtained during residue treatment. In fact, the use of recycled filtrate can even eliminate the lime requirement. As digestion reaches the methanogenesis stage, the concentration of ammonia increases and the pH value can increase to above 8. Once methane production is stabilized, the pH level stays between 7.0 and 7.5.

4.3 Waste Composition

The wastes treated by AD may comprise a biodegradable organic fraction, and an inert combustible fraction. The biodegradable organic fraction includes kitchen scraps, food residue, and grass etc. The combustible fraction includes slowly degrading lignocellulosic organic matter. Finally, the inert fraction contains stones, glass, sand, metal, etc. The removal of the inert fraction prior to digestion is important. The volatile solids (VS) in organic wastes are measured as total solids minus the ash content, obtained by complete combustion of the feed wastes. The volatile solids comprise the biodegradable volatile solids (BVS) fraction and the refractory volatile solids (RVS). The knowledge of the quantity of BVS fraction in waste helps in better estimation of the biodegradability of waste, biogas generation, organic loading rate and C/N ratio. Waste with high VS and low non-biodegradable matter, or RVS, is best suited to AD treatment. The composition of wastes affects the yield and biogas quality as well as the compost quality.

4.4 C/N Ratio

The relationship between the amount of carbon and nitrogen present in organic materials is represented by the C/N ratio. A C/N ratio of 20/1 to 30/1 is best. A high C/N ratio is an indication of rapid consumption of nitrogen by methanogens and results in lower gas production. On the other hand, a lower C/N ratio causes ammonia accumulation and an increase in pH values exceeding 8.5, which is toxic to methanogenic *Archaea*. Optimum C/N ratios of the digester materials can be achieved by mixing materials of high and low C/N ratios, such as organic SW mixed with sewage or animal manure.

4.5 Organic Loading Rate

Organic loading rate (OLR) is a measure of the biological conversion capacity of the AD system. Charging the system above its sustainable OLR, results in low biogas yield due to accumulation of inhibiting substances such as fatty acids in the digester slurry. In such a case, the feeding rate to the system must be reduced. OLR is a particularly important control parameter in continuous systems. Many plants have reported system failures due to overloading (Varma 2002). A general observation is that OLR is double that in high solids than low solids systems.

4.6 Residence Time

The required residence time for completion of the AD varies with differing technologies, process temperature, and waste composition. The residence time for wastes treated in mesophillic digester range from 15 to 40 days. Lower residence times are required in digesters operated in the thermophilic range. A high solids reactor operating in the thermophilic range has a residence time of 14 days (Personal Communication with M. Lakos, May 2001).

4.7 Mixing

The kind of mixing equipment and amount of mixing varies with the type of reactor and the solids content in the digester. The purpose of mixing in a digester is to blend the fresh material with digestate containing microbes. Furthermore, mixing prevents scum formation and avoids temperature gradients within the digester and release the trapped biogas in the deep layers of digestate. However excessive mixing can disrupt the microbes and their activity so slow mixing is preferred.

5 Molecular Biology of Anaerobic Digestion

A diverse number of Bacteria and Archaea take part in the hydrolysis and fermentation steps of AD. The oxidation of intermediate fermentation products to acetate is performed by either hydrogen- or formate-producing acetogens (Stams and Plugge 2009), and in methane formation by methanogenic *Archaea*. In this degradation chain a balanced interaction among the microbial population is crucial for the continuous transformation of the intermediates formed and subsequent biogas production. However, the roles and interactions of specific microorganisms within the biogas-producing communities are very complex (Levén et al. 2007; Cheon et al. 2008; Lee et al. 2008). A deeper understanding of the microbial community structure and functional dynamics of AD is therefore vital to improve process performance.

5.1 Analysis of Microbial Community Structure

Use of culture-dependent techniques helped in identifying key populations capable of carrying out specific metabolic processes in anaerobic digestion. However, majority of microorganism in these systems have not been cultured in the laboratory (Amann et al. 1995). Our understanding of the microbial community and metabolism of anaerobic digester remain incomplete if restricted to culture dependent techniques. Moreover, a complex microbial web can exhibit characteristics that each of its component organism do not have when studied in isolation. Over the last one and half decades the application of culture-independent methods to anaerobic digesters together with other techniques such as imaging, isotope labeling and chemical analyses have provided insight into community composition and function of dominant population. In the recent past several marker genes have been employed to study the community structure. Advancements of molecular biology techniques, helped in linking of these community structure with their metabolic attributes have also been achieved.

Molecular tools have also improved our understanding of how feed stock, reactor configuration and operational conditions influence microbial community structure and dynamics (Talbot et al. 2008; Warner et al. 2011; Sundberg et al. 2013; Li et al. 2013).

5.2 The Marker Genes

5.2.1 16S rDNA Analysis

Since the Woesien revolution, the 16S rDNA has become a major tool for microbial ecological studies (Whitman et al. 2001). Norma Pace and his colleagues developed a technique to retrieve rRNA gene sequences and identify organisms in natural habitats without the need for cultivating them (Pace 1997; Hugenholtz et al. 1998). Universal distribution and regions with various degrees of sequence conservation of 16S rRNA gene made it a good taxonomic tool. Several AD microbial community studies that include, cloning and sequencing of 16S rRNA gene have been reported (Wani et al. 2006; Surakasi et al. 2007; McHugh et al. 2003; Cheon et al. 2008; Klocke et al. 2008). There are variations in the gene sequence from *Bacteria* and *Archaea,* hence kingdom 16S rRNA gene stands very handy marker to establish the overall community of the AD. However, requires other specific marker genes to understand the functional attributes of the microbes present in the AD.

5.2.2 McrA

Methanogenesis requires reduction of the methyl group of methyl-coenzyme M to CH_4 by the enzyme methyl-coenzyme M reductase (MCR), involving a nickel-containing factor F430 (Pramanik and Kim 2013). All known genomes of the methanogenic archaea encode at least one copy of the *mcr*BDCGA operon, which is composed of two alpha (*mcr*A), beta (*mcr*B), and gamma (*mcr*G) subunits (Luo et al. 2002). Therefore, the presence of this enzyme is a reliable diagnostic indicator of methanogenesis in diverse environments (Reeve et al. 1997; Luton et al. 2002; Steinberg and Regan 2009; Palacio-Molina et al. 2013). Strategies can be developed by combine the analysis of differential gene expression of *mcr* alpha sub unit and the traditional approaches to monitor the performance of bio-digesters on real time basis. Several studies have established that the presence and transcription of the gene for the alpha sub-unit of MCR (*mcrA*) can be used to detect the presence, abundance and/or activity of methanogens in natural and engineered environments (Springer et al. 1995; Luton et al. 2002; Juottonen et al. 2008; Gagnon et al. 2011; Kampmann et al. 2012). Several studies demonstrated that the methane flux correlated with the abundance of *mcrA* in AD (Freitag and Prosser 2009; Freitag et al. 2010; Traversi et al. 2012).

5.2.3 Pct

Syntrophic acetogenic bacteria are an important guild because they are essential for maintaining efficient and stable AD operation. However, this guild is poorly understood due to difficulties to culture them in a pure form. Syntrophic acetogenesis is an important step responsible for converting a number of acidogenesis products, including propionate, butyrate, isopropionate, isobutyrate, valerate, isovalerate, and ethanol to the substrates of methanogenesis, i.e., acetate, H_2, and CO_2. The oxidation of propionate in syntrophy is particularly important because nearly 30 % of the electrons generated from complex substrates flow through propionate during AD. Propionate-CoA transferase gene (pct) is one that is used to investigate syntrophic acetogenic bacterial diversity and distribution (Li et al. 2013).

5.2.4 Other Marker Genes

Apart from the 16S rDNA many other genes were also used to find the different metabolic groups in different habitats. Functional genes as phylogenetic markers have several advantages because they not only support interpretation of the phylogenetic diversity but also enable detailed studies of particular microbial guilds, such as sulfate-reducing bacteria, nitrogen fixing bacteria etc. with respect to their distribution, population dynamics, and in situ metabolic activities.

The *fhs* gene and the *acsB* gene, which encode the formyl-tetrahydrofolate synthetase and the acetyl-CoA synthase, respectively, of the homoacetogenesis pathway, have also been proved as useful markers in investigating homoacetogenesis in anaerobic environment (Leaphart and Lovell 2001; Gagen et al. 2010). The gene that codes for subunit of the dinitrogenase (*nif* genes) was used to estimate abundances of N_2-fixing bacteria (Ueda and Carmichael 1995). Methane monooxygenase and methanol dehydrogenase genes were used to analyze methanotrophic communities (Khmelenina et al. 2000). Dissimilatory sulfate reductase coding genes were used to analyze the sulfate reducing bacterial populations. Several other protein coding genes also were used for phylogenetic analysis of prokaryotes, For example; *rpo*B which encodes the β subunit of RNA polymerase (Peixoto et al. 2002; Dahllof et al. 2000), *rec*A which encodes the DNA repair protein *RecA* (Marechal et al. 2000), and *gyrB* the structural gene for the β subunit of DNA gyrase (Yamamoto et al. 1999) etc. A reduced expression of *mcrA* gene is associated with high VFA concentration and lower pH values, gradually resulting in poor methane concentration in the biogas which also indicates a possible dominance of SRB over methangens. Similarly a low propionate-CoA transferase gene (*pct*) count indicates poor acetogenesis which may result in accumulation of VFA. The diagnosis will help to sense the coming disturbances in the digestion so the necessary measure can be implemented to avoid the sickening of AD.

5.3 Molecular Techniques in Microbial Community Dynamics

Having identified several nucleic acid probes, several advancements have been evolved to study the microbial communities, their metabolic activities and syntrophic associations which helps to smooth running of the AD.

5.3.1 Cloning/Shotgun Cloning Methods

Restriction fragmented community DNA can be shotgun-cloned into lambda bacteriophages (or into bacterial artificial chromosome (BAC) vectors which are able to "hold" larger size fragments), and then screened for the presence of rRNA genes.

The advantage of such libraries is that they are also sources of genes other than those encoding for rRNA. This is a laborious procedure, as rRNA genes will only constitute a small fraction of the total clones requiring screening of thousands of clones.

The simplest way to obtain phylotypes from the environment is through the use of the polymerase chain reaction (PCR) (Saiki et al. 1988). The DNA is extracted form a mixed microbial population, and primers directed at universally conserved regions of the 16S rRNA gene. The resulting population of rDNA are then cloned and sequenced. The different 16S rDNA clones can be analyzed phylogenetically by comparison to the databases of the known 16S rRNA genes (Surakasi et al. 2007; Wani et al. 2006). Thus a semi-quantitative census or community analysis of the organisms present in a habitat can be obtained without culturing them. Use of such methodology is quite common in present day ecological studies. However, there are biases in every step of the procedure. For example: an organism from which DNA is not extracted by the procedure will not be included in the census (Suzuki et al. 1998). So care must be taken.

The reverse transcriptase utilizes universal or group-specific primers to make single-stranded DNA that is complementary to rRNA, and then PCR is used to make duplex ribosomal DNA for cloning. The resulting community profile will offer some reflection of the most metabolically active organisms, because cells that produce more RNA (i.e. those that are metabolically more active) will be represented to a greater extent in the clone library than metabolically inactive cells.

5.3.2 Electrophoresis Techniques

The polymorphism in the community genes specific to either 16S rDNA or any functional genes can be identified via gel electrophoresis. The genes amplified using specific primers can be digested using the restriction enzymes and can be anylyzed using electrophoresis, which is called restriction fragment length polymorphism (RFLP) (Surakasi et al. 2007). Such methods will give preliminary insights on the complexity of the community. Denaturant gradient gel electrophoresis (DGGE) is another method that is widely used to analyze the specifically amplified genes. It is based on the differing mobility on a gel of denatured DNA-fragments of the same size but with different nucleic acid sequences, thus generating band patterns that directly reflect the genetic biodiversity of the sample. The number of bands corresponds to the number of dominant species. The most important application of DGGE is monitoring dynamic changes in microbial communities, especially when many samples have to be processed. There are multiple applications of DGGE related to anaerobic digestion processes: studies on mesophilic and thermophilic reactors (Lapara et al. 2000), urban solid waste (Silvey et al. 2000), communities in a continuous stirred tank reactor (CSTR) (Ueno et al. 2001) etc. An alternative approach is the generation of Single Strand Conformation Polymorphism (SSCP) patterns; a technique that has been employed to study anaerobic digesters both on laboratory- and industrial scales (Bouallagui et al. 2004; Delbès et al. 2001). However these techniques are low throughput methods, hence the coverage of microbial community will be limited.

5.3.3 Metagenomics

The more recent development of 'next-generation sequencing' (eg: Roche 454 and Illumina sequencing platforms) has made it possible to efficiently deep-sequence microbial communities in complex biological samples without the time-consuming cloning procedure. The technique has so far been used for the sequencing of metagenomes from a number of biogas reactors (Schlüter et al. 2008; Werner et al. 2011; Lee et al. 2012). However this technology suffers from several limitations such as homo-polymers errors which lead to over estimation of number of phylotypes. (Carvalhais et al. 2012; Reeder and Knight 2009). Pyrosequencing techniques is limited to genus level identification due to its limitations in the base read length of the 16S rRNA gene (250–500 bp). The Illumina MiSeq platform is becoming increasingly popular for 16S rRNA gene amplicon sequencing because it can generate longer paired end reads and up to ten times more sequences per run (Caporaso et al. 2012). However, complexity of the whole community will be huge.

5.3.4 Metatranscriptomics

Traditionally microbial community gene expression has been screened on microarrays which are time consuming and expensive to build, but cannot detect novel genes as the data set is designed based on the known gene sets (Mutz et al. 2012). Metatranscriptomics involves the sequencing of the reverse transcribed mRNA extracted from a microbial community and provides a way to measure in situ gene expression (Su et al. 2011). The method can detect novel genes also. This method avoids the level of complexity seen in the metagenomics by focusing only on active participants in the AD.

5.3.5 Metaproteomics

Metaproteomics is the characterization of expressed proteins during the anaerobic digestion. In metaproteomics, proteins are extracted from a mixed microbial community sample, followed by fractionation, separate using liquid chromatography and detection with tandem mass spectrometry (MS/MS) (Langley et al. 2012). Metaproteomics can be used to identify the distribution of metabolic activities among a community and how populations cooperate or compete (Hettich et al. 2013). A combined metagenomics and metaproteomics study revealed that the population of methanogens was a very minor percentage of total community in the metagenomic data, while metaproteomics data showed that key enzymes of methanogenesis were highly expressed.

5.3.6 Metabolomics

Metabolomics is analysis of quantitative and qualitative measure of all low molecular weight molecules involved in metabolic reactions those participate in growth and natural function of microbial community. Metabolic fluxes are not regulated by gene expression alone since one transcript can direct the production of multiple proteins, post-translational modifications can alter the location and function of proteins, and the chemical environment can effect protein function.

5.3.7 Microscopic Methods

MAR-FISH

The meta-omics approach provides direct evidence to link microbial community to specific metabolic processes. However, approaches that allow us to measure substrate uptake by specific populations and visualize the spatial organization of the community further enhance our understanding of the processes. Fluorescence in situ hybridization (FISH) is widely applied in diverse and complex microbial systems. Micro-autoradiography (MAR) is one technique that uses radioactive isotopes to study the in situ uptake of specific substrates (Talbot et al. 2008; Okabe et al. 2004).

While revolutionary methodologies have evolved in the recent past, diagnostic methodologies, however, need to be transformed to simple tools and techniques to aid the rapid analysis of microbial diversity of ADs at low cost.

6 Anaerobic Digestion Technology in Solid Waste Management and Economics

Many factors affect the performance of an anaerobic digester such as feed stock, reactor design, operational conditions and substrate utilization etc. Characterization of solid waste and perfect pre-treatment of the feed stock is most important step in the AD. An optimal designing of anaerobic digester is possible only after thorough understanding of the feed stock nature. Designing of anaerobic digester and operational conditions together play in important role in the fine functioning of AD. Substrate utilization by a balanced microbial population is up most importance for any anaerobic digester. Though there are several studies explains the microbial community of the digester, maintaining these communities in a right combination even during the shock loads is challenging. Several microbes those found in the digester could not be cultured in the laboratory, since most of the microbes grow in syntrophy or by depending on the metabolite of the other *Bacteria*. However, commercial supply of such microbial cultures that bioaugment the digester will greatly reduce the startup times and accepts the shock loads. Here we discuss two case studies where the solid waste has been successfully treated in the anaerobic digester with economic benefits.

6.1 Case Study-1

SW from vegetable oil industry: De-oiled castor cake biomethantion: India produces a number of non-edible oil seeds such as zetropa, neem, mahua, sal, karanj and castor etc. Oils are extracted from these seeds by crushing or by solvent extraction. These oils are not edible due to the presence of toxic alkaloids but are used in manufacturing lubricants, paints and soaps. Some of these oils are used in making biodiesel. After oil extraction about 70–80 % of the seed remains as de-oiled cake. These de-oiled cakes contain about 5–8 % of residual oil and are not used as livestock feed due to the presence of toxic substances. At present it is used as manure due to its NPK value in horticultural crops or burnt directly as fuel in boilers. The cake has a calorific value comparable to other fuels because of the presence of carbohydrates, proteins and fats.

In the present case, AD of de-oiled castor cake is studied for the techno-economic feasibility. Castor is grown worldwide as a cash crop. Global castor seed production is around 1.4 million tons per year. India is the major castor seed producer with 0.6 MT per annum. Almost all the seed is used for oil extraction. This result in the generation of about 0.4 Mt of de-oiled cake. The chemical analysis of de-oiled cakes has 94–96 % TS of that 91 % are VS indicating rich organic matter. The cakes are composed of good amounts of protein, carbohydrates and fat.

Pilot-scale: These studies were conducted in a 1 m^3 size fixed dome digester made with PVC material. The active culture volume of the digester was 800 L and the rest was head space for gas collection. The feed pipe was provided from the top portion of the digester and submerged about 0.7 m into the digesting slurry. An effluent pipe was provided on the side wall of the digester; about 0.3 m from the digester top portion, care was taken that air will not enter through this into the digester while effluent is drained out. About 40 L of effluent were drained once a day through the effluent pipe after closing the gas valve on the digester. Then equal volume of the feed slurry was fed into the digester along with the specially enriched microbial consortia that can bioaugment in the system rapidly. Feeding was done mechanically, after feeding gas valve was opened. Feed preparation, feeding and re-circulation carried out using a 0.5 HP motor and valves (1, 2, 3 and 4) on the pipes. The digester slurry was recirculated for 5 min every 12 h to mix the digester contents. Biogas produced was measured using a wet type gas flow meter. Feed slurry was prepared in a separate feed tank mixed thoroughly with the help of pump for about 10 min and then fed to the digester. The pilot trials were conducted for about 6 months. The results given in the table are an average of 15 days after digester stabilization.

De-oiled cakes come in the form of flakes. Daily feed material was soaked in water for 12 h. Before feeding it to the digester, it was mixed thoroughly with the pump for about 10 min resulting in a homogeneous feed slurry.

The plant was operated in semi-batch mode at a 20 day HRT with 4.5 % TS (~2 kg castor cake/day) in feed slurry. The results obtained in pilot-scale trials (Table 4) are in conformity with lab-scale studies.

The digester effluents and dried digested sludge are very good organic manures. Digested effluents consists of N, P, K (0.13, 0.25 and 0.17 %) and micronutrients (Cu-0.02 %, Zn-0.24 %, Mn-0.05 %, Fe-0.65 %, Mg-0.24 %). Due to the dissolved state, the nutrients are readily absorbed by the roots. Whereas the NPK value of the digested sludge was 1.95, 0.29 and 0.13 %. Horticultural studies showed that the effluent and digested sludge were more effective fertilisers than the castor cake.

Techno-economic feasibility of castor cake biomethanation over a 6 months study period; expenditure considered geed material.

Total castor cake used: 360 kg.
Cost of castor cake: $10 \times 360 = 60$ USD.

Value of Biogas

Biogas produced: 163 m^3.
1 m^3 of biogas with 70 % methane: 0.55 kg LPG Biogas produced (163 m^3) = 89.65 kg LPG
Value of biogas produced as LPG = 245 USD.
Manure value: recoverable digested sludge after drying is about 125 kg only @ of 0.5 USD/kg of dried digested sludge. Dry solid manure value is 10 USD Recoverable liquid effluent (6 months) 5000; L the value is about 40 USD @Rs. 0.5/lit value of liquid manure is Rs. 2500/- (Table 5).

Table 4 Performance of pilot-scale plant

S. No.	Particulars	Data
1	Biogas production(L/day)	817
2	Biogas yield (L/kg/day)	453
3	Methane content (%V/V)	70
4	Gas production rate (L/L/day)	1.0
5	Total solids degradation (W/W)	72.8
6	Volatile solids degradation (W/W)	79.5
7	Influent slurry pH	4.5
8	Effluent slurry pH	7.6 ± 0.2
9	Effluent VFA (mg/L)	~2614–3200
10	Temperature (°C)	Room temperature 30 ± 1

Table 5 Economics of castor cake biomethanation with respect to LPG

S. No.	Input cost (USD)		Output cost (USD)	
1	Cake	56	Value of biogas—as LPG replacement	245
2	Water	16	Value of solid manure	8
3	Electricity	3.4	Value of effluent	40
4	Manpower	100		
5	Maintenance	24		
	Total	183.4		293

Profit = 293 − 183.4 = 110 USD/6 months and 220 USD/year.

Pilot-plant cost

Digester 1 m^3 = 64.5 USD; Feed tank 0.25 m^3 = 16 USD
Piping = 8 USD
0.5 HP motor = 48.4 USD 000/-
Valves = 16 USD
Others = 8 USD
Total = 160 USD
Payback period is about 10 months.

A 225 m^3 pilot scale biogas plant on castor cake was erected. The plant is in operation successfully since 2010 and is being managed by M/s. Green Leaf Technologies Ltd. Pune which has obtained the sub-license for commercialization purpose. Biogas produced was used for thermal application as a fuel in the boiler in place of diesel. Biogas replaced nearly 200 L of diesel required daily for boiler operation.

The performance of the industrial-scale digester is given below:

- Rate of gas production (v/v): 1:1
- Biogas yield (L/kg): 425–450
- Biogas calorific value: 7500 kcal/ m^3
- Methane %: 77 %.

The industrial-scale biogas plant was arranged in clusters of 10 m^3 size and maintained the conditions as mentioned in the pilot-scale plant. Biogas yield was 425–450 L/kg of cake per day and methane content was 75–77 %. The calorific value of the biogas produced was 7500 kcal/m^3, whereas biogas produced with Gobar is 4500 kcal/m^3 of gas. This might be due to the higher methane percentage in biogas produced with castor cake. The higher gas yields are due to high organic solids content in the feed, whereas cattle dung is a predigested one hence its organic content is lower than in oil cakes. The gas produced has replaced 200 L of diesel that was being used as furnace fuel prior to the biogas plant installation. Sludge is about 25–30 % of the cake used per day and is separated dried and used as manure.

6.2 Case Study-2

Two stage anaerobic digestion of kitchen waste

Step-1 Food waste collection: Food waste (pre-cooked and cooked and left overs) from kitchen, and dining areas was collected and brought to the biogas plant site once in 24 h.

Step-2 Feed preparation: Food waste was screened for non-degradable material and pre-treated (masticated in a feed crusher along with water) and made into a homogeneous slurry. This automatically entered into the primary digester while the food crusher is in operation. This operation is manual. The present system is designed to handle 500 kg of food waste per day. Approximately 400–450 L of water is used for 500 kg of food waste per day and it take about 2–3 h time.

Step-3 Primary digester: In this digester, the food waste in the form of slurry is degraded by specially developed hydrolytic and acetogenic bacteria into volatile fatty acids, alcohols, aldehydes, carbon dioxide and hydrogen. The pH of this slurry is acidic to about 4–5. Here the HRT is 5 days.

Step-4 Stabilization pond: Digesting slurry from primary digester is sent to the stabilization tank by gravity feed and is mixed with the digested slurry from the secondary digester before feeding it to the secondary digester. The pH of the slurry is about 6.5.

Step-5 Secondary digester: Neutralized and stabilized feed slurry from primary digester is fed to the secondary digester using a submersible pump. The commercial methanogenic consortia were applied to the digester and helped in the rapid stabilization of the AD. The HRT of this digester is about 30 days. Digested slurry from this digester flows out simultaneously when the feed slurry is fed into it. A gas dome is placed on top portion of the secondary digester for collecting the biogas produced. Biogas is collected into a gas balloon when the gas dome is full with biogas.

Step-6 Gas balloon: The total capacity of this balloon is 20 m^3. It is connected to the biogas dome through a pipe. Excess biogas from gas dome was collected into this balloon. This balloon is placed in a separate room. Total biogas storage system is designed to accommodate total biogas produced from 500 kg of food waste.

Step-7 Gas pressure tank: In order to provide sufficient gas pressure at the gas stoves, biogas from the balloon is pumped into the pressure tank using a gas blower. When sufficient pressure ($\sim$3 bar) is reached, the blower switches off. Blower operation is based on the gas usage in the kitchen. Biogas is supplied to the kitchen stoves using pipes.

Economics of the food waste biogas plant:

- Plant design is to handle 500 kg of food waste/day
- 12.5 kg of kitchen waste = 1 m^3 biogas (with 60 % methane)
- 500 kg of kitchen waste = 40 m^3 of biogas
- 1 m^3 biogas = 0.54 kg of LPG
- 40 m^3 of biogas = 21.6 kg LPG
- 1 m^3 biogas = 5500 kcal heat energy
- 40 m^3 of biogas is equal to 21 kg LPG/day
- Value of 21 kg of LPG = Rs. 2100/day
- Daily 1000 L of digested effluent are obtained. It is good organic manure with plant nutrients like micro and macro nutrients (NPK). This can be used for growing plants and in agriculture.
- Total expected revenues from biogas plant per day = ~Rs. 2100/-
- **Monthly revenue: Rs. 63,000/-**
- Per year: ~Rs. 756,000/-
- Total monthly expenditure on biogas plant: Rs. 17,500/-
- Electricity: Rs. 1000/-
- Operator salary: Rs. 10,000/-
- Water: Rs. 1500/-
- Miscellaneous expenditure: Rs. 5000/-
- Net benefit: 63,000–17,500 = Rs. 45,500
- Per year: Rs. 546,000/- lakhs
- **Payback period = ~5 years.**
- Project cost + interest @ 11 %/net benefit
- Rs. 20 lakh + interest @ 11 %/5.46 lakh.

7 Summary and Conclusion

Disposal of SW in open landfills pose major environmental and ecological problems besides occupying a large area of land for their storage/disposal. Anaerobic digestion is a sustainable treatment process for organic solid waste management. With a rapid civilization of third world countries and globalization, SW management using anaerobic digestion will become a mainstream technology in the near future. The digested residue or sludge can be considered quite stable organic manure with a very slow turnover of several decades for adequate soil conditioning. In this way the natural imbalance in CO_2 can be compensated by restoring or creating organic-rich soils. The removal of CO_2 constitutes another benefit that could help place AD among the most sustainable technologies in the field of solid waste treatment.

Environment consciousness of the public is forcing many governments to look for safe disposal and recycling of solid waste. The digestion of SW starts with hydrolysis, fermentation and acedogensis, followed by acitogenesis and methanogenesis with a great co-ordination among the different micorbial communities.

However, the pre-treatment of the SW will greatly enhance the digestion of solid waste. Culture-dependent methods for investigating the microbial community structure has left us with limited knowledge. Culture-independent methods of studying microbial communities in the AD has been simplified with the advent of several molecular techniques such as, 16S rDNA and other gene markers for phylogeny studies, electrophoresis methods for identification of polymorphism in community genes, next generation sequencing for meta-omics studies, microscopy techniques associated with molecular marker genes for investigating community dynamics and qualitative PCR methods for finding the active microbial species in the digester. The combination of molecular tools, such as DGGE, gene sequencing, and FISH, with microbial activity tests seems to be essential for a better characterization of anaerobic biomass present the AD. These techniques have helped us to understand community dynamics, but needs simplification and regular usage while running the digesters to pre-determine likely future imbalances of the microboal population and to help take the necessary precautions. The commercial availability of microbial populations is very much necessary to fortify the digester with, when shock loads are expected to occur or for rapid bioaugmentation of the digester. Since many of the microbes are obligate anaerobes, cultivation in the laboratory and commercial availability are challenging.

References

Amann RI, Ludwig W, Schleifer KI (1995) Phylogenetic identification and in situ detection of individual microbial cell with out cultivation. Microbiol Rev 59:143–155

Beccari M, Majone M, Papini MP, Torrisi L (2001) Enhancement of anaerobic treatability of olive oil mill effluents by addition of Ca(OH)2 and bentonite without intermediate solid/liquid separation. Water Sci Technol 43:275–282

Bouallagui H, Torrijos M, Godon JJ, Moletta R, Cheikh RB, Touhami Y, Delgenes JP, Hamdi M (2004) Microbial monitoring by molecular tools of a two-phase anaerobic bioreactor treating fruit and vegetable wastes. Biotechnol Lett 26:857–862

Caporaso JG, Lauber CL, Walters WA, Berg-Lyons D, Huntley J, Fierer N, Owens SM, Betley J, Fraser L, Bauer M, Gormley N, Gilbert JA, Smith G, Knight R (2012) Ultra-high-throughput microbial community analysis on the Illumina HiSeq and MiSeq platforms. ISME J 6:1621–1624

Cardinali-Rezende J, Debarry RB, Colturato LF, Carneiro EV, Chartone-Souza E, Nascimento AM (2009) Molecular identification and dynamics of microbial communities in reactor treating organic household waste. Appl Microbiol Biotechnol 84:777–789

Carvalhais LC, Dennis PG, Tyson GW, Schenk PM (2012) Application of metatranscriptomics to soil environments. J Microbiol Methods 91:246–251

Castro HF, Williams NH, Ogram A (2000) Phylogeny of sulfate-reducing bacteria(1). FEMS Microbiol Ecol 31:1–9

Chachkhiani M, Dabert P, Abzianidze T, Partskhaladze G, Tsiklauri L, Dudauri T, Godon JJ (2004) 16S rDNA characterisation of bacterial and archaeal communities during start-up of anaerobic thermophilic digestion of cattle manure. Bioresour Technol 93:227–232

Chang VS, Burr B, Holtzapple MT (1997) Lime pretreatment of switchgrass. Appl Biochem Biotechnol 63–65:3–19

Chen Y, Cheng JJ, Creamer KS (2008) Inhibition of anaerobic digestion process: a review. Bioresour Technol 99:4044–4064

Cheon J, Hidaka T, Mori S, Koshikawa H, Tsuno H (2008) Applicability of random cloning method to analyze microbial community in full-scale anaerobic digesters. J Biosci Bioeng 106:134–140

Conrad R (1999) Contribution of hydrogen to methane production and control of hydrogen concentration in methanogenic soils and sediments. FEMS Microbiol. Ecol. 28:193–202

Cork DJ, Cusanovich MA (1979) Continuous disposal of sulphate by bacterial mutualism. Dev Ind Microbiol 20:591–602

Dahllof I, Baillie H, Kjelleberg S (2000) rpoB-based microbial community analysis avoids limitations inherent in 16S rRNA gene intraspecies heterogeneity. Appl Environ Microbiol 66:3376–3380

Delbes C, Leclerc M, Zumstein E, Godon JJ, Moletta R (2001) A molecular method to study population and activity dynamics in anaerobic digestors. Water Sci Technol 43:51–57

Deublein D, Steinhauser A (2010) Biogas from waste and renewable resources: an introduction edn. WILEY-VCH Verlag GmbH & Co. KGaA, Mörlenbach

Distefano TD, Ambulkar A (2006) Methane production and solids destruction in an anaerobic solid waste reactor due to post-reactor caustic and heat treatment. Water Sci Technol 53:33–41

Edstrom M, Nordberg A, Thyselius L (2003) Anaerobic treatment of animal byproducts from slaughterhouses at laboratory and pilot scale. Appl Biochem Biotechnol 109:127–138

Freitag TE, Prosser JI (2009) Correlation of methane production and functional gene transcriptional activity in a peat soil. Appl Environ Microbiol 75:6679–6687

Freitag TE, Toet S, Ineson P, Prosser JI (2010) Links between methane flux and transcriptional activities of methanogens and methane oxidizers in a blanket peat bog. FEMS Microbiol Ecol 73:157–165

Gagen EJ, Denman SE, Padmanabha J, Zadbuke S, Al Jassim R, Morrison M, McSweeney CS (2010) Functional gene analysis suggests different acetogen populations in the bovine rumen and tammar wallaby forestomach. Appl Environ Microbiol 76:7785–7795

Gagnon N, Barret M, Topp E, Kalmokoff M, Masse D, Masse L, Talbot G (2011) A novel fingerprint method to assess the diversity of methanogens in microbial systems. FEMS Microbiol Lett 325:115–122

He Y, Pang Y, LiuY Li X, Wang K (2008) Physicochemical characterization of rice straw pretreated with sodium hydroxide in the solid state for enhancing biogas production. Energ Fuels 22:2775–2781

Hugenholtz P, Goebel BM, Pace NR (1998) Impact of culture independent studies on the emerging phylogenetic view of bacterial diversity. J Bacteriol 180:4765–4774

Jagadabhi PS, Kaparaju P, Rintala J (2009) Effect of micro-aeration and leachate replacement on COD solubilization and VFA production during mono-digestion of grass-silage in one-stage leach-bed reactors. Bioresour Technol 101:2818–2824

Jayachandra T, Venugopal C, Anu appaiah KA (2011) Utilization of phytotoxic agro waste: coffee cherry husk through pretreatment by the ascomycetes fungi Mycotypha for biomethanation. Energ Sustain Dev 15:104–108

Jetten MS, Niftrik L, Strous M, Kartal B, Keltjens JT, Op den Camp HJ (2009) Biochemistry and molecular biology of anammox bacteria. Crit Rev Biochem Mol Biol 44:65–84

Juottonen H, Tuittila ES, Juutinen S, Fritze H, Yrjala K (2008) Seasonality of rDNA- and rRNA-derived archaeal communities and methanogenic potential in a boreal mire. ISME J 2:1157–1168

Kampmann K, Ratering S, Baumann R, Schmidt M, Zerr W, Schnell S (2012) Hydrogenotrophic methanogens dominate in biogas reactors fed with defined substrates. Syst Appl Microbiol 35:404–413

Karthikeyan OP, Visvanathan C (2013) Bio-energy recovery from high-solid organic substrates by dry anaerobic bio-conversion processes: a review. Rev Environ Sci Biotechnol 12:257–284

Khmelenina VN, Eshinimaev B, Kaliuzhnaia MG, Trotsenko IuA (2000) Potential activity of methane and ammonia oxidation by methanotropic communities from soda lakes of the southern Transbaikal. Mikrobiologiia 69:553–558

Klocke M, Nettmann E, Bergmann I, Mundt K, Souidi K, Mumme J, Linke B (2008) Characterization of the methanogenic Archaea within two-phase biogas reactor systems operated with plant biomass. Syst Appl Microbiol 31:190–205

Knappert D, Grethlein H, Converse A (1981) Partial acid hydrolysis of poplar wood as a pretreatment for enzymatic hydrolysis. Biotechnol Bioeng Symp 11:67–77

Krause L, Diaz NN, Edwards RA, Gartemann KH, Kromeke H, Neuweger H, Puhler A, Runte KJ, Schluter A, Stoye J, Szczepanowski R, Tauch A, Goesmann A (2008) Taxonomic composition and gene content of a methane-producing microbial community isolated from a biogas reactor. J Biotechnol 136:91–101

Langley SR, Dwyer J, Drozdov I, Yin X, Mayr M (2012) Proteomics: from single molecules to biological pathways. Cardiovasc Res 97:612–622

LaPara TM, Nakatsu CH, Pantea L, Alleman JE (2000) Phylogenetic analysis of bacterial communities in mesophilic and thermophilic bioreactors treating pharmaceutical wastewater. Appl Environ Microbiol 66:3951–3959

Leaphart AB, Lovell CR (2001) Recovery and analysis of formyltetrahydrofolate synthetase gene sequences from natural populations of acetogenic bacteria. Appl Environ Microbiol 67: 1392–1395

Lee MH, Oh KH, Kang CH, Kim JH, Oh TK, Ryu CM, Yoon JH (2012) Novel metagenome-derived, cold-adapted alkaline phospholipase with superior lipase activity as an intermediate between phospholipase and lipase. Appl Environ Microbiol 78:4959–4966

Lee MY, Cheon JH, Hidaka T, Tsuno H (2008) The performance and microbial diversity of temperature-phased hyperthermophilic and thermophilic anaerobic digestion system fed with organic waste. Water Sci Technol 57:283–289

Lehtomaki A, Huttunen S, Lehtinen TM, Rintala JA (2008) Anaerobic digestion of grass silage in batch leach bed processes for methane production. Bioresour Technol 99:3267–3278

Leven L, Eriksson AR, Schnurer A (2007) Effect of process temperature on bacterial and archaeal communities in two methanogenic bioreactors treating organic household waste. FEMS Microbiol Ecol 59:683–693

Li H, Xu X, Chen H, Zhang Y, Xu J, Wang J, Lu X (2013) Molecular analyses of the functional microbial community in composting by PCR-DGGE targeting the genes of the beta-glucosidase. Bioresour Technol 134:51–58

Liew LN, Shi J, Li Y (2011) Enhancing the solid-state anaerobic digestion of fallen leaves through simultaneous alkaline treatment. Bioresour Technol 102:8828–8834

Liu D, Zeng RJ, Angelidaki I (2006) Hydrogen and methane production from household solid waste in the two-stage fermentation process. Water Res 40:2230–2236

Gold T (1992) The deep, hot bio-sphere. Proc Natl Acad Sci USA 89:6045–49

Luo HW, Zhang H, Suzuki T, Hattori S, Kamagata Y (2002) Differential expression of methanogenesis genes of Methanothermobacter thermoautotrophicus (formerly Methanobacterium thermoautotrophicum) in pure culture and in cocultures with fatty acid-oxidizing syntrophs. Appl Environ Microbiol 68:1173–1179

Luton PE, Wayne JM, Sharp RJ, Riley PW (2002) The mcrA gene as an alternative to 16S rRNA in the phylogenetic analysis of methanogen populations in landfill. Microbiology 148: 3521–3530

Marechal J, Clement B, Nalin R, Gandon C, Orso S, Cvejic JH, Bruneteau M, Berry A, Normand P (2000) A recA gene phylogenetic analysis confirms the close proximity of Frankia to Acidothermus. Int J Syst Evol Microbiol 50(Pt 2):781–785

McHugh S, Carton M, Mahony T, O'Flaherty V (2003) Methanogenic population structure in a variety of anaerobic bioreactors. FEMS Microbiol Lett 219:297–304

Menardo S, Airoldi G, Balsari P (2011) The effect of particle size and thermal pre-treatment on the methane yield of four agricultural by-products. Bioresour Technol 104:708–714

Menind A, Normak A (2010) Study on grinding biomass as pre-treatment for biogasification. Presented at the international scientific conference, biosystems engineering 2010, 13. Estonian Research Institute of Agriculture, Tartu, Estonia, 24 May 2010, pp 155–164

Monlau F, Barakat A, Steyer JP, Carrere H (2012) Comparison of seven types of thermo-chemical pretreatments on the structural features and anaerobic digestion of sunflower stalks. Bioresour Technol 120:241–247

Mshandete A, Bjornsso L, Kivaisi AK, Rubindamayugi MST, Mattiasson B (2006) Effect of particle size on biogas yield from sisal fibre waste. Renew Energ 31:2385–2392

Mutz KO, Heilkenbrinker A, Lonne M, Walter JG, Stahl F (2012) Transcriptome analysis using next-generation sequencing. Curr Opin Biotechnol 24:22–30

Nikolaou A, Giannis A, Gidarakos E (2010) Comparative studies of aerobic and anaerobic treatment of MSW organic fraction in landfill bioreactors. Environ Technol 31:1381–1389

Nizami A-S, Orozco A, Groom E, Dieterich B, Murphy JD (2012) How much gas can we get from grass? Appl Energ 92:783–790

Noike T, Endo G, Chang JE, Yaguchi J, Matsumoto J (1985) Characteristics of carbohydrate degradation and the rate-limiting step in anaerobic digestion. Biotechnol Bioeng 27:1482–1489

Okabe S, Kindaichi T, Ito T, Satoh H (2004) Analysis of size distribution and areal cell density of ammonia-oxidizing bacterial microcolonies in relation to substrate microprofiles in biofilms. Biotechnol Bioeng 85:86–95

Pace NR (1997) A molecular view of microbial diversity and the biosphere. Science 276:734–740

Palacio-Molina SL, Oropeza-Navarro R, Balagurusamy N (2013) Quantitative analysis of *mcrA* trascripts and its correlation with methanogenic activity in two natural wetlands. Oral Presented at the II Congreso de Bioquímica y Biología Molecular de Bacterias, Cuatrocienegas, Coahuila

Parawira WJSRJ, Mattiasson B, Björnssona L (2008) Energy production from agricultural residues: high methane yields in pilot-scale two-stage anaerobic digestion. Biomass Bioenerg 32:44–50

Park MJ, Jo H, Park D, Lee DS, Park JM (2010) Comprehensive study on a two-stage anaerobic digestion process for the sequential production of hydrogen and methane from cost-effective molasses, Int. J. Hydrogen Energy 35:6194–6202

Patil SS, Kumar MS, Ball AS (2010) Microbial community dynamics in anaerobic bioreactors and algal tanks treating piggery wastewater. Appl Microbiol Biotechnol 87:353–363

Peixoto RS, da Costa Coutinho HL, Rumjanek NG, Macrae A, Rosado AS (2002) Use of rpoB and 16S rRNA genes to analyse bacterial diversity of a tropical soil using PCR and DGGE. Lett Appl Microbiol 35:316–320

Reddy CA (1995) The potential for white-rot fungi in the treatment of pollutants. Curr Opin Biotechnol 6:320–328

Reeder J, Knight R (2009) The 'rare biosphere': a reality check. Nat Methods 6:636–637

Reeve JN, Nolling J, Morgan RM, Smith DR (1997) Methanogenesis: genes, genomes, and who's on first? J Bacteriol 179:5975–5986

Reimann JJM, Keltjens JT (2015) Sustaining life on planet earth: metalloenzymes mastering dioxygen and other chewy gasesedn. Springer

Rotter VS, Lehmann A, Marzi T, Mohle E, Schingnitz D, Hoffmann G (2011) New techniques for the characterization of refuse-derived fuels and solid recovered fuels. Waste Manag Res 29:229–236

Saiki RK, Gelfand DH, Stoffel S, Scharf SJ, Higuchi R, Horn GT, Mullis KB, Erlich HA (1988) Primer-directed enzymatic amplification of DNA with a thermostable DNA polymerase. Science 239:487–491

Salminen EA, Rintala JA (2002) Semi-continuous anaerobic digestion of solid poultry slaughterhouse waste: effect of hydraulic retention time and loading. Water Res 36:3175–3182

Schink B (1997) Energetics of syntrophic cooperation in methanogenic degradation. Microbiol Mol Biol Rev 61:262–280

Schluter A, Bekel T, Diaz NN, Dondrup M, Eichenlaub R, Gartemann KH, Krahn I, Krause L, Kromeke H, Kruse O, Mussgnug JH, Neuweger H, Niehaus K, Puhler A, Runte KJ, Szczepanowski R, Tauch A, Tilker A, Viehover P, Goesmann A (2008) The metagenome of a biogas-producing microbial community of a production-scale biogas plant fermenter analysed by the 454-pyrosequencing technology. J Biotechnol 136:77–90

Siegrist H, Salzgeber D, Eugster J, Joss A (2008) Anammox brings WWTP closer to energy autarky due to increased biogas production and reduced aeration energy for N-removal. Water Sci Technol 57:383–388

Silvey P, Pullammanappallil PC, Blackall L, Nichols P (2000) Microbial ecology of the leach bed anaerobic digestion of unsorted municipal solid waste. Water Sci Technol 41:9–16

Siriwongrungson V, Zeng RJ, Angelidaki I (2007) Homoacetogenesis as the Alternative Pathway for H2 Sink During Thermophilic Anaerobic Degradation of Butyrate Under Suppressed Methanogenesis.ˮ Water Res. 41:4204–10

Smith CJ, Bryant MP (1979) Introduction to metabolic activities of intestinal bacteria. Am J Clin Nutr 32:149–157

Song Z, Yang G, GUO Y, Zhang T (2012) Comparison of two chemical pretreatments of rice straw for biogas production by anaerobic digestion. BioResources 7:3223–3236

Springer E, Sachs MS, Woese CR, Boone DR (1995) Partial gene sequences for the A subunit of methyl-coenzyme M reductase (mcrI) as a phylogenetic tool for the family Methanosarcinaceae. Int J Syst Bacteriol 45:554–559

Stams AJ, Plugge CM (2009) Electron transfer in syntrophic communities of anaerobic bacteria and archaea. Nat Rev Microbiol 7:568–577

Steinberg LM, Regan JM (2009) mcrA-targeted real-time quantitative PCR method to examine methanogen communities. Appl Environ Microbiol 75:4435–4442

Su C, Lei L, Duan Y, Zhang KQ, Yang J (2011) Culture-independent methods for studying environmental microorganisms: methods, application, and perspective. Appl Microbiol Biotechnol 93:993–1003

Sundberg C, Al-Soud WA, Larsson M, Alm E, Yekta SS, Svensson BH, Sorensen SJ, Karlsson A (2013) 454 pyrosequencing analyses of bacterial and archaeal richness in 21 full-scale biogas digesters. FEMS Microbiol Ecol 85:612–626

Supaphol S, Jenkins SN, Intomo P, Waite IS, O'Donnell AG (2011) Microbial community dynamics in mesophilic anaerobic co-digestion of mixed waste. Bioresour Technol 102: 4021–4027

Surakasi VP, Wani AA, Shouche YS, Ranade DR (2007) Phylogenetic analysis of methanogenic enrichment cultures obtained from Lonar Lake in India: isolation of Methanocalculus sp. and Methanoculleus sp. Microb Ecol 54:697–704

Suzuki M, Rappe MS, Giovannoni SJ (1998) Kinetic bias in estimates of coastal picoplankton community structure obtained by measurements of small-subunit rRNA gene PCR amplicon length heterogeneity. Appl Environ Microbiol 64:4522–4529

Talbot G, Topp E, Palin MF, Masse DI (2008) Evaluation of molecular methods used for establishing the interactions and functions of microorganisms in anaerobic bioreactors. Water Res 42:513–537

Thauer RK (1998) Biochemistry of methanogenesis: a tribute to Marjory Stephenson. Marjory Stephenson prize lecture. Microbiology 144(Pt 9):2377–2406

Traversi D, Villa S, Lorenzi E, Degan R, Gilli G (2012) Application of a real-time qPCR method to measure the methanogen concentration during anaerobic digestion as an indicator of biogas production capacity. J Environ Manage 111:173–177

Ueda H, Carmichael GR (1995) Formation of secondary pollutants during long-range transport and its contribution to air quality in east Asia. Terr Atmos Oceanic Sci 6:487–500

Ueno Y, Haruta S, Ishii M, Igarashi Y (2001) Changes in product formation and bacterial community by dilution rate on carbohydrate fermentation by methanogenic microflora in continuous flow stirred tank reactor. Appl Microbiol Biotechnol 57:65–73

van Dongen U, Jetten MS, van Loosdrecht MC (2001) The SHARON-Anammox process for treatment of ammonium rich wastewater. Water Sci Technol 44:153–160

Varma S (2002) Anaerobic digestion of biodegradable organics in municipal solid wastes. Columbia University

von Loosdrecht MC (2008) Innovative nitrogen removal. In: Henze M, van Loosdrecht MCM, Ekama GA, Brdjanovic D (eds) Biological wastewater treatment: principles, modelling and design. IWA Publishing, London, pp 139–155

Wagner AO, Schwarzenauer T, Illmer P (2014) Improvement of methane generation capacity by aerobic pre-treatment of organic waste with a cellulolytic Trichoderma viride culture. J Environ Manage 129:357–360

Wani AA, Surakasi VP, Siddharth J, Raghavan RG, Patole MS, Ranade D, Shouche YS (2006) Molecular analyses of microbial diversity associated with the Lonar soda lake in India: an impact crater in a basalt area. Res Microbiol 157:928–937

Warner JR, Reeder PJ, Karimpour-Fard A, Woodruff LB, Gill RT (2011) Rapid profiling of a microbial genome using mixtures of barcoded oligonucleotides. Nat Biotechnol 28:856–862

Werner JJ, Koren O, Hugenholtz P, DeSantis TZ, Walters WA, Caporaso JG, Angenent LT, Knight R, Ley RE (2011) Impact of training sets on classification of high-throughput bacterial 16 s rRNA gene surveys. ISME J 6:94–103

Whitman WB, Boone DR, Koga Y, Keswani J (2001) Taxonomy of methanogenic archaea. In: Begey's manual of systematic bacteriology: 2nd edn 1:211–294. Springer, New York

Yabu H, Sakai C, Fujiwara T, Nishio N, Nakashimada Y (2010) Thermophilic two-stage dry anaerobic digestion of model garbage with ammonia stripping. J Biosci Bioeng 111:312–319

Yamamoto S, Bouvet PJ, Harayama S (1999) Phylogenetic structures of the genus Acinetobacter based on gyrB sequences: comparison with the grouping by DNA-DNA hybridization. Int J Syst Bacteriol 49(Pt 1):87–95

Ye R, Jin Q, Bohannan B, Keller JK, Bridgham SD (2014) Homoacetogenesis: A potentially underappreciated carbon pathway in peatlands Soil Biology & Biochemistry 68:385–391

Zhang Q, Tang L, Zhang J, Mao Z, Jiang L (2010a) Optimization of thermal-dilute sulfuric acid pretreatment for enhancement of methane production from cassava residues. Bioresour Technol 102:3958–3965

Zhang W, Yue B, Huang QF, Huang ZC, Zhang ZQ, Wang Q, Lin Y, Wang JA (2010b) Physical and chemical properties of land-filling pile and aged refuse in 5-year-old semi-aerobic and anaerobic landfills. Ying Yong Sheng Tai Xue Bao 21:763–769

Zinder SH, Anguish T, Cardwell SC (1984) Effects of temperature on methanogenesis in a thermophilic (58 degrees C) anaerobic digestor. Appl Environ Microbiol 47:808–813

Zitomer DH, Adhikari P, Heisel C, Dineen D (2008) Municipal anaerobic digesters for codigestion, energy recovery, and greenhouse gas reductions. Water Environ Res 80:229–237

Recycling of Livestock Manure into Bioenergy

Qigui Niu and Yu-You Li

Abstract The total amount of manure production increased fast in response to rapidly increasing demand for livestock production, which poses a strong public health threat due to the greenhouse gases (GHG) emissions and leachates without appropriate treatment. Anaerobic digestion is available technology for livestock manure treatment with new business opportunities and benefits for the society, such as bioenergy of CH_4 and nutrients are extra income, odor and pathogens are reduced and GHG emission are limited. The productions of chicken manure, cattle manure and pig manure were introduced with evaluation of potential energy production. The toxicity of ammonia/ammonium in the anaerobic digestion process was evaluated with microbial community dynamics. Operation conditions effects on methane production were analyzed, such as pH, organic loading rate (OLR) effects and the ammonia stripping pretreatment effects on the CH_4 conversion. The dynamic of functional archaeal and bacterial community were also conducted. *Methanosaeta* dominated in the steady stage of chicken manure thermophilic digestion but *Methanothermobacter* dominated in the inhibition stage and *Methanosarcina* thrived in the recovered stage. In contrast, under mesophilic conditions, *Methanosarcina* dominated in the steady stage while in the inhibition stage *Methanosaeta* and *Methanoculleus* thrived and lastly recovered to *Methanosaeta*. Poultry manure can be easily inhibited by ammonia compared to cattle manure and pig manure digestion since it has a high nitrogen content, which was more suitable for mesophilic digestion with evidence of process resilience in mesophilic digestion. Pre-treatment of ammonia stripping or co-digestion are the effective ways to generate a stable process.

Keywords Manure production · Manure treatment · Methane fermentation · Anaerobic digestion · Removal efficiency

Q. Niu (✉) · Y.-Y. Li
Department of Civil and Environmental Engineering, Graduate School of Engineering, Aobayama 6-6-06, Sendai 980-8579, Japan
e-mail: Niuqg@epl1.civil.tohoku.ac.jp

Y.-Y. Li
e-mail: yyli@epl1.civil.tohoku.ac.jp

O.P. Karthikeyan et al. (eds.), *Recycling of Solid Waste for Biofuels and Bio-chemicals*, Environmental Footprints and Eco-design of Products and Processes, DOI 10.1007/978-981-10-0150-5_6

Abbreviations

TAN	Total ammonia nitrogen
FA	Free ammonia nitrogen
VFA	Volatile fatty acid
CM	Chicken manure
TS	Total solid
VS	Volatile solids

1 Introduction

Environmental pollution and energy supply are two of the main problems in the fast developing society globally. In 2013, the world energy consume reached to 89,774 million ton oil/year, with 0.9 % increased than last year. Furthermore, by 2030, the world is projected to consume two-third more energy than today. The use of fossil fuel, associated pollution and the real need of safe energy supply have promoted technological development. It is worldwide accepted that renewable energy such as biogas from anaerobic digestion is environmental friendly and benefits energy security (Dorian et al. 2006). Meanwhile, the increase of intensive and mechanized livestock breeding industries with approximate populations of 1.43 billion cattle, 1.87 billion sheep and goats, 0.98 billion pigs, and 19.60 billion chickens worldwide (Robinson et al. 2014). Following the livestock feeding, a huge amount of manure (about 40 Mt cattle manure/day, 4.9 Mt pig manure/day and 2 Mt chicken manure/day) are produced, causing strong public health threats due to greenhouse gases (GHG) and leachates production without appropriate treatment. In recent years, anaerobic digestion has attracted considerable attention for livestock manure treatment leading to the conversion of organic waste into the renewable energy in the form of CH_4. Anaerobic digestion of livestock manure has many new business opportunities such as developing rural economies, improving farm incomes, mitigating climate change and proving alternative energy source and nutrients as extra income. Especially for large farms, many agricultural companies are interested in implementing anaerobic digestion for better manure management with energy recovery and socio-economic benefits.

Research on the manure digestion have increased exponentially in recent years. Anaerobic digestion has distinct advantages over conventional compost and direct fertilizer treatments without air pollution and toxic hazards. Anaerobic digestion of livestock manure has many advantages which allow for bioenergy recovery besides minimizing the waste:

- The production of biogas (CH_4, CO_2) that can be used as a green source of energy. This provides a low-cost energy source with valuable fertilizer production.
- Reduction of GHG emission of the farm. Reducing nitrous oxide emissions with low demand of alternative fuels.

- It can accommodate high COD loads, which adapts to remove and/or work in the presence of various toxic components provided that adaptation time is allowed for the anaerobic biomass.

Anaerobic digestion of manure is becoming a common technology in many countries with produced bio-methane converted to electrical and thermal energy or upgraded to vehicle fuel. However, in industrial plants it is still difficult to manage because of limitations in monitoring and controlling applications and lack of knowledge. To utilize the renewable energy potential of livestock manure through anaerobic digestion, process operation should be well controlled. Anaerobic digestion was usually conducted in mesophilic (35 °C) or thermophilic (55 °C) conditions with different dominance profiles of functional microbial groups at optimized temperature. Generally, manure digestion can generate 50–80 % of CH_4 and 20–50 % of CO_2, with trace NH_3(g) and H_2S(g) gas, especially in poultry manure digestion (Fig. 1). Livestock manure wastes being an important part of biomass resources shared 28 % of the worldwide biogas production compared to the biggest contributor agricultural wastes (59 %), which was higher than for municipal organic wastes (11 %) and municipal sewage and industrial organic wastes (2 %) (Maghanaki et al. 2013).

Methane fermentation is conducted by various specialized bacterial groups. The entire anaerobic fermentation process can be divided into four steps (Fig. 2): (1) Hydrolysis, (2) Acidogenesis, (3) Acetogenesis and (4) Methanogensis. Methanogens are the most important functional group for biogas production. To date, the known classes of methanogens are well-established with six orders with more than 110 species: *Methanobacteriales*, *Methanococcales*, *Methanomicrobiales*, *Methanosarcinales*, *Methanopyrales* and *Methanocellales*. Methanogens are abundant in a wide variety of anaerobic environments catalyzing the terminal step of food chain by converting methanogenic substrates into methane.

In this chapter, the main livestock productions and manure production are summarised. The potential bioenergy production of different manure types were evaluated and illustrated in case studies. Process control for long-term anaerobic

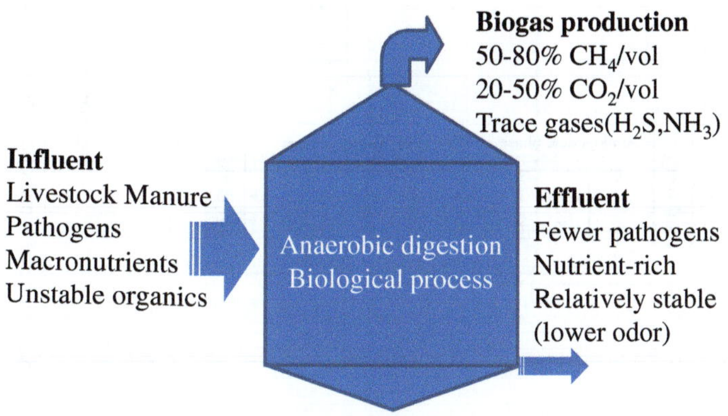

Fig. 1 Schematic of basic process of anaerobic digestion

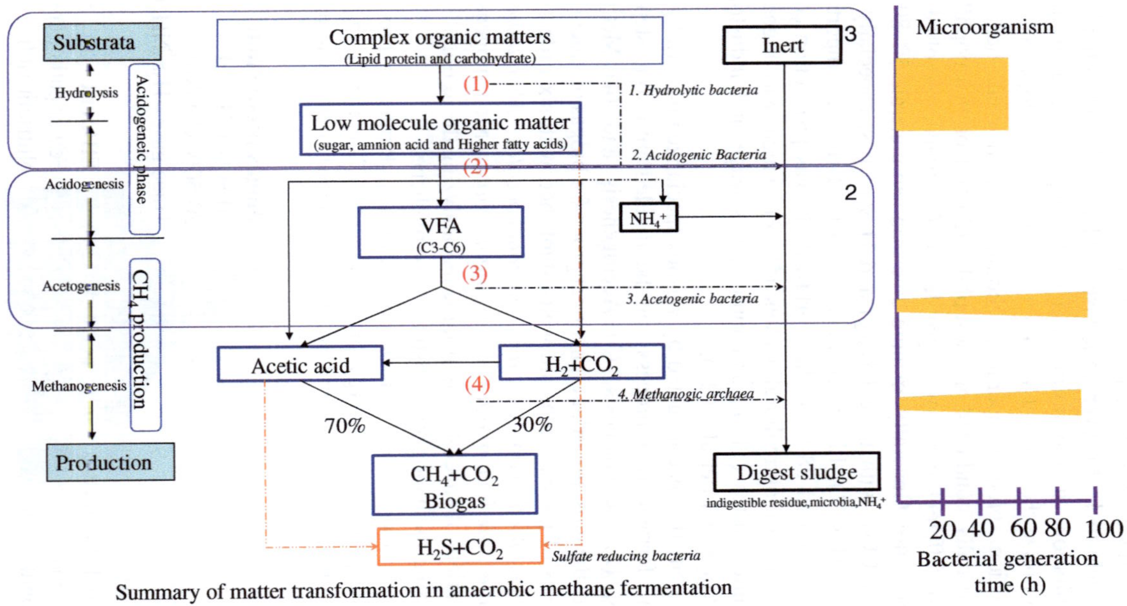

Fig. 2 Schematic of metabolic chains of anaerobic digestion with microorganisms

digestion was also evidenced, especially for chicken manure digestion. The most important inhibitor of livestock manure digestion following ammonia variation was evaluated. The control region of total ammonia/free ammonia concentration was proposed based on different manure stabilization processes. Moreover, process resilience of chicken manure digestion was also investigated. A fast recovery strategy of the inhibited process was developed.

2 Livestock Production and Livestock Manure Characteristics

2.1 Livestock and Livestock Manure

Livestock manure is an inevitable by-product of livestock production. Nowadays, in developed countries, demand for livestock products is stagnating with many production systems increasing in efficiency and environmental sustainability. However, in developing countries, the number of livestock increases year by year. Manure contains valuable material that can be used as a source of organic matter and fertilizer for crop and pasture production. Factors of animal species, diet, digestibility, protein and fiber content, animal age, housing, environment, and stage of production could affect manure properties.

Livestock manure production is hard to measure directly from storage volumes or counting loads spread. Theoretically, the larger the livestock production and meat consumption, the more manure production is generated. The global livestock distribution (head/km^2) is shown in Fig. 3. China and the USA are the largest producer of chicken manure, Brazil and India lead in cattle production, while China leads pig production (Fig. 3). Table 1 shows manure production and nutrients contents of the livestock manure. For calculating manure production, Eq. (1) was used in an approximate scale.

Estimate by equation:

$$M = Q_1 * a - Q_2 \tag{1}$$

M manure production
Q_1 production of livestock
a conversion coefficient
Q_2 fodder digestibility rate.

Quantity, composition and value of fresh manure are summarized in Table 1.

The top two countries of cattle production reached 200 million heads in 2013 were Brazil and India (Fig. 4). Most countries have an increased tend of production with the top six countries with production over 50 million heads were Brazil, India, China, USA, Ethiopia and Argentina, respectively (Fig. 4). Based on Eq. (1), the cattle manure produced can be calculated for approximation. The country of top pig

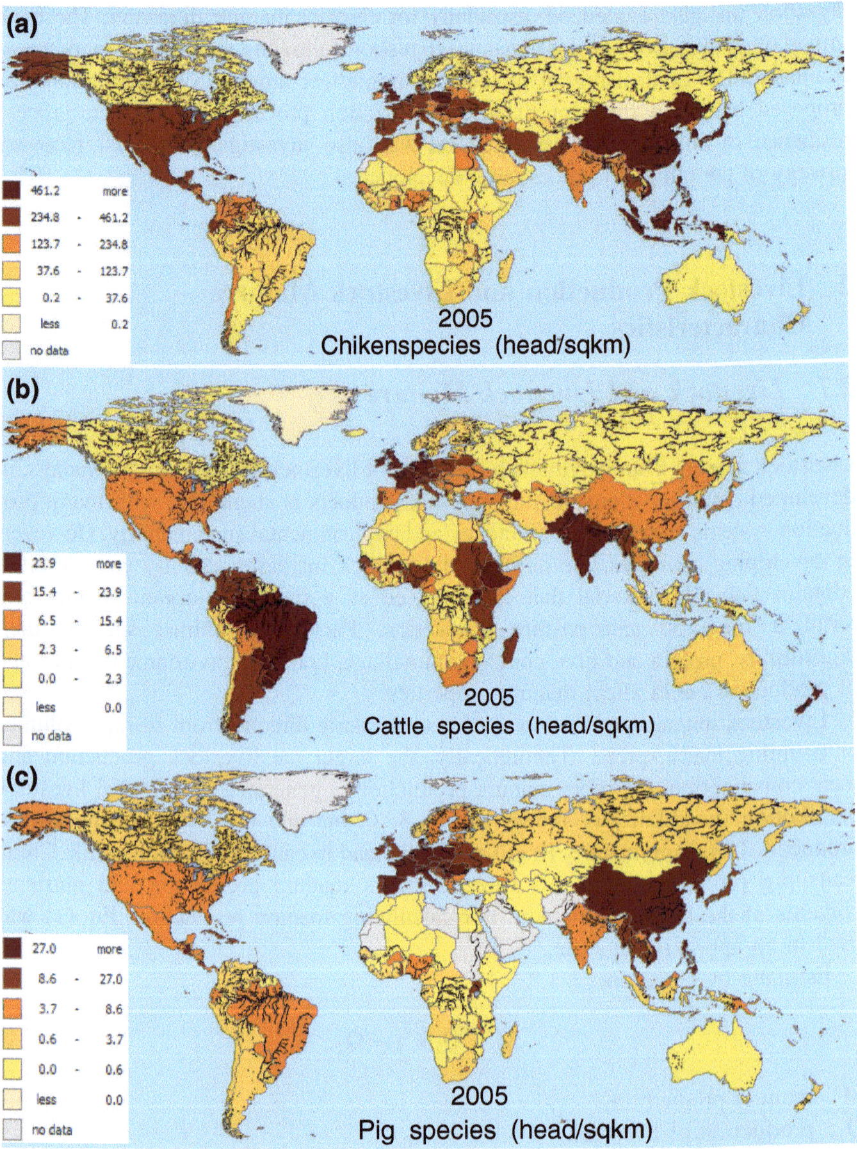

Fig. 3 Chicken, cattle, pig distribution worldwide in 2005 (head/km^2) (FAO 2013)

production is China with 9 times of USA which were the second large production. In 2013 year, China has 45 million pig production (Fig. 5). Poultry production encompasses a number of different species, including the chicken (eggs- "layers", "broilers"), turkeys, ducks, and so on, chicken occupied over than 80 % of the poultry production. China and USA are the top two countries with 6.5 billion and 2 billion heads of chicken produced in 2013 (Fig. 6).

Table 1 Quantity, composition and value of fresh manure

Animal	Tons excreted/year/ 1000 lbs. live weight	Lbs. (N) Nitrogen	Lbs. (P) Phosphate	Lbs. (K) Potash	Value/ton
Cow (beef or dairy)	12 tons	11.2	4.6	12	5.46
Finishing cattle	8.5 tons	14	9.2	10.8	7.14
Sheep and goats	6 tons	28	9.6	24	12.46
Swine	16	10	6.4	9.1	5.23
Poultry	4.5	31.2	18.4	8.4	13.54
Horse	8	13.8	4.6	14.4	6.42

Excreted by 1000 lb of live weight of various kinds of farm animals
http://www.paffa.state.pa.us/paffa2/files/specialrecords/manuretable.pdf

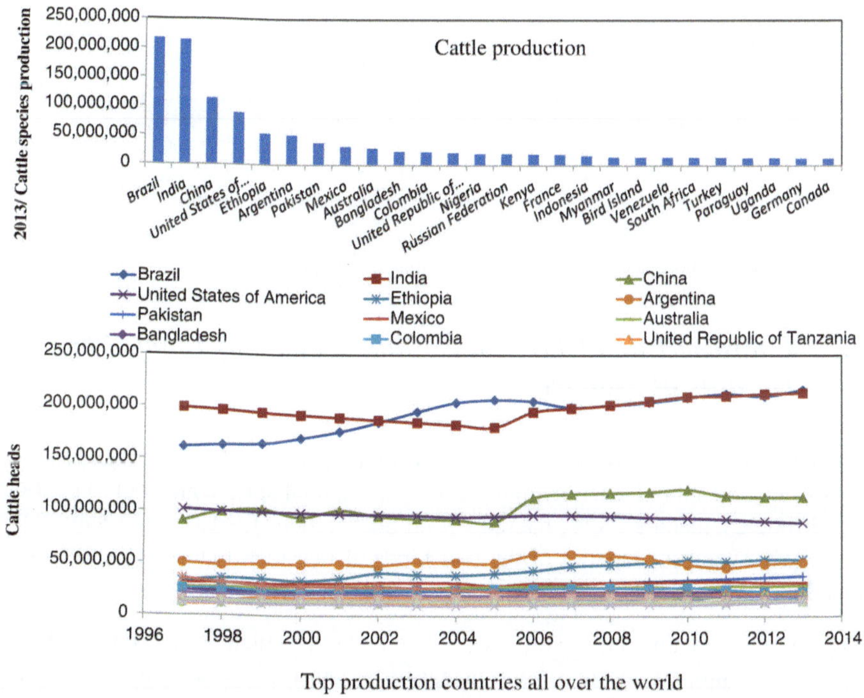

Fig. 4 Cattle production following time and the dominate production countries

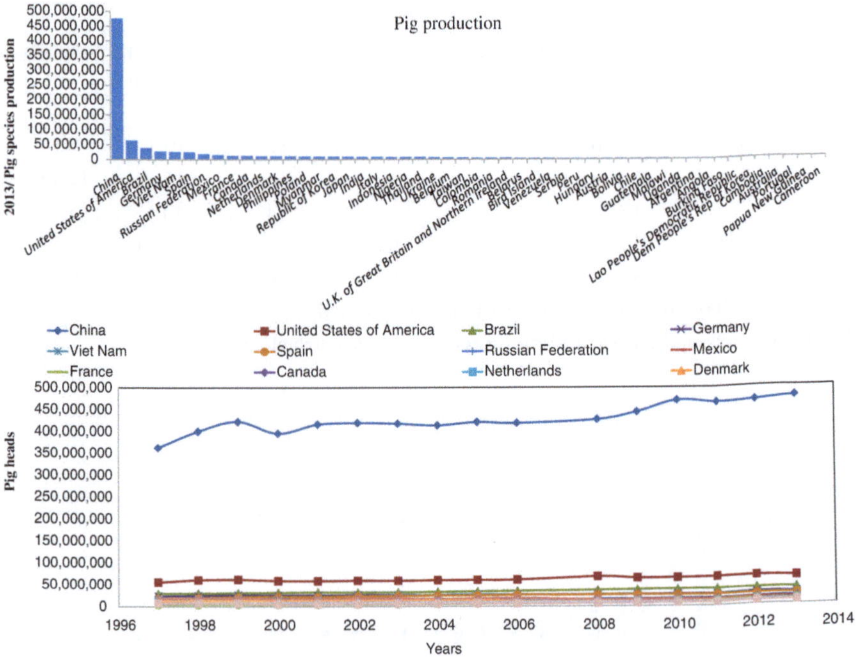

Top production countries all over the world

Fig. 5 Pig production following time and the dominate production countries

2.2 *Characteristic of Livestock Manure and Environmental Hazards of Manure*

The characteristic of livestock manure can be divided into 4 kinds: liquid, slurry, semi-solid and solid. Generally, pig manure have a total solid content (TS) of about 10 %, poultry manure 25 %, manures from beef feeders 12–13 % and dairy cows 13–14 %. The characteristics of the four kinds of livestock manure is summarized as follows:

Liquid contains the lowest TS ($\sim$4 % or less), which is easily treated with properly designed and managed by anaerobic digestion.

Slurry 4–10 % solids content can be handled as slurry. Pig manure typically contains between 2 and 6 % solids. Solid contents may increase to 8 to 12 %, resulting in thicker slurry.

Semi-solid 10–20 % solids content, handling characteristics vary by the type of solids present. Difficulties arise as these manures are too thick to pump, and too thin to scoop, therefore being usually diluted with water or special pumps are used to agitate and move.

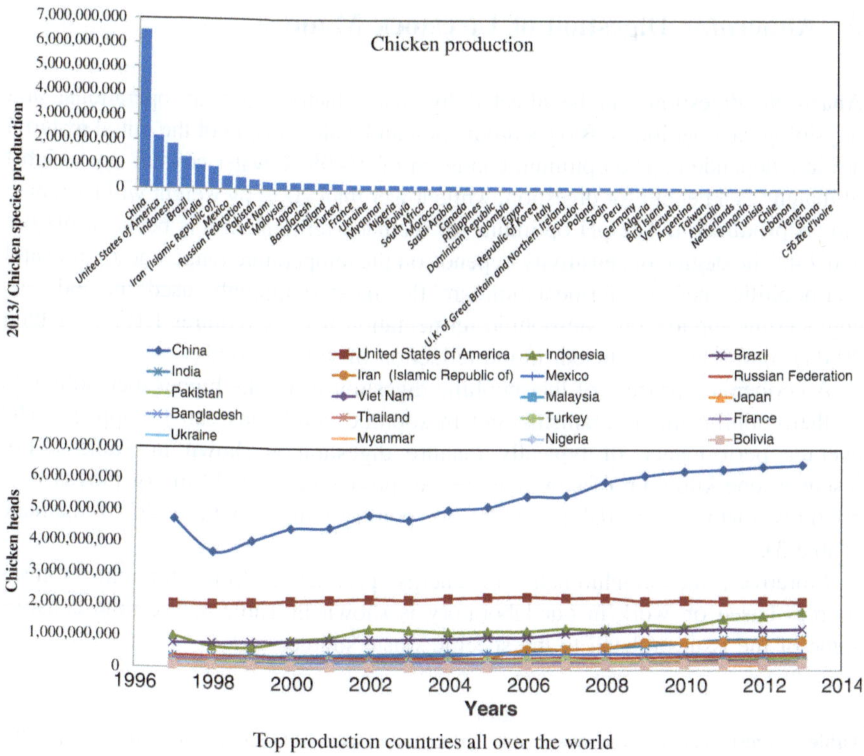

Fig. 6 Chicken production following time and the dominate production countries

Solid 20 % solids content (80 % moisture content) or more can be handled as a solid. It can be stacked and picked up with a fork- or bucket loader.

Livestock manure is rich in nitrogen and phosphorous, which are common sources of pollution. Livestock manure also contains bacteria, viruses, parasites and arsenic, the latter especially for the poultry manure, as arsenic is added to chicken feed to control parasites and promote growth. Copper, iron and zinc, which boost dioxin formation, are also present livestock manure, which in land applications can lead the second pollution. GHG emissions lead to air pollution and leachate from composting can cause soil and land pollution with a threat to public health. The hazards of manures are described as follows:

- Foul smelling gas/air pollution
- Inframicrobe, pathogen/soil, water pollution
- Ammonia hazard/water and soil pollution
- Organic acid hazard/soil and water pollution
- Water damage/soil damage.

3 Anaerobic Digestion of Livestock Manure

Anaerobic digestion can be affected by many factors such as operational and physiological conditions. Biogas production and water content of the initial material are interdependent. The optimum condition of 91–98 % water content (by weight) of manure is a benefit for digestion, while lower than 20 % by weight will terminate biogas production. The pH optimum for methane fermentation is between pH 6.7 and 7.4. The degree of sensitivity depends on the temperature range, mesophilic and thermophilic methane fermentation are the most commonly used methods for engineering application. Mesophilic fermentation usually requires HRT over than 20-day, but is not so efficient in the reduction of volatile solids.

A cooperated process of thermophilic digestion using the higher metabolic rate of thermophilic microorganisms and mesophilic conditions can be applied. The average performance of typically manure digestion is shown in Table 2. For instance, one kilogram of cow manure can produce 0.26–0.28 m^3/kg biogas with 50–60 % methane and 0.4–0.6 m^3/kg for poultry manure with 50–72 % methane (Table 3).

Moreover, the stoichiometry of energy production from different biomass sources based on work in our laboratory is shown in Table 4. We now consider some of the issues related to the specific waste sources.

Table 2 Fresh manure production and characteristics per 1000 kg live animal mass per day (Engineers 2003)

Parameter	Units	Dairy	Beef	Swine	Broiler	Turkey
Total manure	kg	86 ± 17	58 ± 17	84 ± 24	85 ± 13	47 ± 13
Urine	kg	26 ± 4.3	18 ± 4.2	39 ± 4.8	–	–
Total solids	kg	12 ± 2.7	8.5 ± 2.6	11 ± 6.3	22 ± 1.4	12 ± 3.4
Volatile solids	kg	10 ± 0.79	7.2 ± 0.57	8.5 ± 0.66	17 ± 1.2	9.1 ± 1.3
BOD5	kg	1.6 ± 0.48	1.6 ± 0.75	3.1 ± 0.71	–	2.1 ± 0.46
COD	kg	11 ± 2.4	7.8 ± 2.7	8.4 ± 3.7	16 ± 1.8	9.3 ± 1.2
pH	–	7 ± 0.43	7 ± 0.31	7.5 ± 0.57	–	–
Total Kjeldahl nitrogen	kg	0.45 ± 0.096	0.34 ± 0.073	0.52 ± 0.21	1.1 ± 0.24	0.62 ± 0.13
Ammonia nitrogen	kg	0.079 ± 0.083	0.086 ± 0.052	0.29 ± 0.10	–	0.08 ± 0.018
Total phosphorus	kg	0.094 ± 0.024	0.092 ± 0.0072	0.18 ± 0.10	0.30 ± 0.053	0.23 ± 0.093
Potassium	kg	0.29 ± 0.09	0.21 ± 0.063	0.29 ± 0.16	0.4 ± 0.064	0.24 ± 0.08
Calcium	kg	0.16 ± 0.069	0.14 ± 0.11	0.33 ± 0.18	0.41	0.63 ± 0.034
Magnesium	kg	0.071 ± 0.016	0.049 ± 0.015	0.075 ± 0.03	0.15	0.073 ± 0.007
Chloride	kg	0.13 ± 0.039	–	0.26 ± 0.052	–	–
Iron	g	12 ± 6.6	7.8 ± 5.9	16 ± 9.7	–	75 ± 28
Manganese	g	1.9 ± 0.75	1.2 ± 0.51	1.9 ± 0.74	–	2.4 ± 0.33

Table 3 Share of biogas volume produced from different livestock wastes (Babaee et al. 2013; Jain et al. 1981)

Raw material	Biogas (m^3/kg)	%CH_4	%CO_2	Volume of methane (m^3 kg)	OLR_{max}
Cow manure	0.26–0.28	50–60	34–38	0.14	4
Sheep manure	0.22–0.24	40–50	37.6	0.1	–
Poultry manure	0.4–0.6	50–72	30–50	0.27	3
Pig manure	0.22–0.73	50–60	40–50	–	5

Table 4 Calculation and stoichiometry of methane fermentation

	Organic	Gas production (Nm^3/kg − $VS_{degradation}$)	CH_4 (%)
Carbohydrate	$(C_6H_{10}O_5)_n + nH_2O \rightarrow 3nCH_4 + 3nCO_2$	0.83	50.0
Protein	$C_{16}H_{24}O_5N_4 + 14.5H_2O \rightarrow 8.25CH_4 + 3.75CO_2 + 4NH_4^+ + 4HCO_3^-$	0.764	68.8
Lipid	$C_{50}H_{90}O_6 + 24.5H_2O \rightarrow 34.75CH_4 + 15.25CO_2$	1.425	69.5
Cooking scrap	$C_{17}H_{29}O_{10}N + 6.5H_2O \rightarrow 9.25CH_4 + 6.75CO_2 + NH_4^+ + HCO_3^-$	0.881	57.8
Cattle manure	$C_{22}H_{29}O_{10}N + 6.5H_2O \rightarrow 9.25CH_4 + 6.75CO_2 + NH_4^+ + HCO_3^-$	0.970	56.0
Kitchen garbage	$C_{46}H_{73}O_{31}N + 14H_2O \rightarrow 24CH_4 + 21CO_2 + NH_4^+ + HCO_3^-$	0.888	53.3
Sewage sludge	$C_{10}H_{19}O_3N + 5.5H_2O \rightarrow 6.25CH_4 + 2.75CO_2 + NH_4^+ + HCO_3^-$	1.003	69.4
Chicken manure	$C_{7.5}H_{12.4}O_{4.8}NS_{0.13} + 4.15H_2O \rightarrow 3.7CH_4 + 2.8CO_2 + NH_4^+ + HCO_3^- + 0.13H_2S$	0.75	63.5

3.1 *Cattle Manure Digestion*

The concept of utilizing cattle manure in an anaerobic digester to generated CH_4 has been well established. The effect of organic loading and retention times on dairy manure fermentation have been reported in (Coats et al. 2011). The pilot-scale digestion exhibited an average yield of 0.09 mg VFA (as COD)/mgVS with an average effluent total VFA concentrations of 6398 mg VFA (as COD)/L. High concentration of VFA was generated at HRTs of less than 20 h in mesophilic condition, while at low organic loading rates, minimal net organic acid production may be realized in a continuously operated fermenter (Fig. 7).

As illustrated in Fig. 8, with an average of cattle manure digestion of $37.7\ m^3/m^3_{inf}$ (60.1 % of CH_4), the TS reduction efficiency reached to 32.3 % in the digester. The bio-methane can be converted into heat or electric energy since CH_4 has 891 kJ/mol (at standard conditions), which benefits for the farmers.

Co-digestion of cattle manure has been conducted to improve methane production, a ratio of 40:60 for wheat straw/cattle manure has been identified as the most suitable mixture for optimum biogas production at mesophilic condition

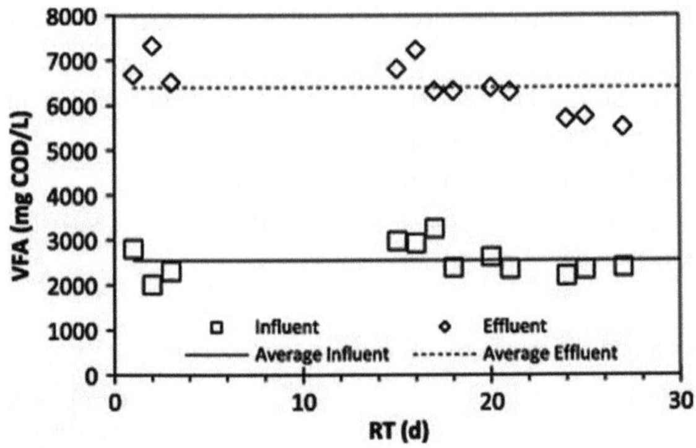

Fig. 7 Influent and effluent VFA concentrations in pilot-scale dairy manure digestion (Coats et al. 2011)

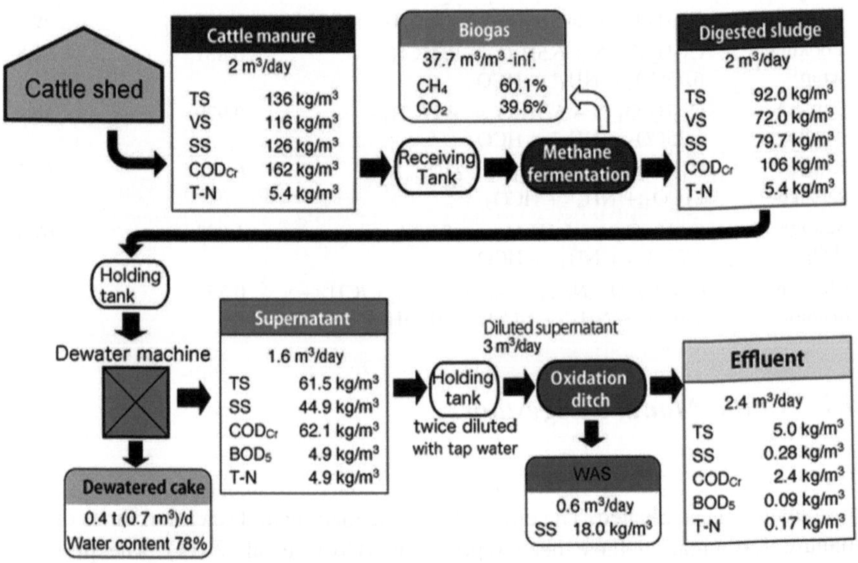

Fig. 8 Flow diagram for cattle manure fermentation (Li et al. 2015)

(Krishania et al. 2013). In thermophilic conditions, Liu et al. (2009) conducted co-digestion of garbage, screened swine and dairy cattle manure in batch experiments. The results showed that when the mixed waste (Garbage: SM: DCM = 1:19:12) was treated using a thermophilic UAF reactor, methanogens responsible for the methane production were *Methanoculleus* and *Methanosarcina* species with the dominated bacterial phylum being *Firmicutes*. Dairy cattle manure digestion obviously improved when co-digested with limited garbage (2–3 %), which has been proposed as a prospective method to treat cattle manure for the construction of a sustainable environment and society.

Table 5 Kinetic constants and maximum volumetric methane production rates for anaerobic fermentation of beef cattle manure (Chen et al. 1980)

T °C	K	μ_{max} day^{-1}	Correlation coefficient	θ_m day	V_{max} LCH_4/L/day	$\theta_m V_{max}$ day
30	1.086	0.284	0.995	3.52	1.36	7.19
35	0.870	0.326	0.987	3.06	1.68	5.93
40	0.856	0.382	–	2.62	2.04	5.04
45	0.764	0.456	0.994	2.19	2.49	4.11
50	0.798	0.55	0.955	1.81	2.95	3.44
55	0.761	0.518	0.987	1.93	2.84	3.61
60	0.856	0.618	–	1.62	3.20	3.12
65	0.856	0.313	–	3.19	1.67	6.15

The temperature effect on methane fermentation kinetics of beef-cattle manure were conducted by Chen et al. (1980) with kinetic constants shown in Table 5. The V_{max} of 3.20L/L/d were found at 60 °C but decreased to 1.67 L/L/d at 65 °C, which means that while thermophilic conditions can improve the degradation of cattle manure, excessively high temperatures may affect functional bacteria groups resulting in decreased biogas production efficiency.

Cattle manure still contains fibrous components like cellulose resulting in hydrolysis being the limited step of the digestion process. Conventional anaerobic digestions are operated as single-stage reactors under mesophilic or thermophilic conditions. Thermophilic pre-treatment results in higher degradation rates (shorter treatment times) and a better sanitation effect for cattle manure digestion. Increases in specific methane yields ranging from 24 to 56 % were obtained when cattle manure and its fractions (fibers and liquid) were pre-treated at 68 °C for periods of 36, 108, and 168 h, and subsequently digested at 55 °C. The two stage thermophilic (68/55 °C) fermentation process had a short HRT of 3 day at 68 °C and 12 day at 55 °C compared to the single stage HRT of 15 days (Nielsen et al. 2004). The 68 °C reactor generated 7–9 % of the total amount of methane of the two-stage system and maintained a volatile fatty acids (VFA) concentration of 4.0–4.4 g acetate/L.

3.2 *Pig Manure Digestion*

The biogas of pig manure digestion in a CSTR conducted by Hansen et al. (1998) achieved 188 mL CH_4/gVS at 37 °C, while digestion at 55 °C produced only 67 mL CH_4/gVS being inhibited by an FA content of 1.4 g/L which is twice that of 37 °C levels even under similar total ammonia concentration.

The co-digestion of pig manure with grass silage was investigated by Xie et al. (2011) as shown in Table 6. The highest specific CH_4 yields were 304.2 and 302.8 mL CH_4/gVS at pig manure/grass silage ratios of 3:1 and 1:1, respectively. The digestion systems failed at the ratio of 0:1. Lag phase times decreased

Table 6 Anaerobic digestion at different pig manure to grass silage ratios (Xie et al. 2011)

PM/GS ratio	1:0	3:1	1:1	1:3
Total methane production (mL)	7833	8517	8478	7484
Lag phase a, λ(d)	29.5 ± 0.3	28.1 ± 0.3	24.6 ± 0.3	21.3 ± 0.3
R_{max} (mL CH_4/d)	210 ± 4	287 ± 5	309 ± 7	280 ± 7
Specific methane yield (mL CH_4/gVS)	279.8	304.2	302.8	267.3
VFA yields (g/gVS)	0.25 ± 0.01	0.32 ± 0.02	0.34 ± 0.03	0.33 ± 0.02
pH	7.5–8.0	7.1–8.0	6.9–7.9	6.5–7.8
NH_4^+–N	1562–2368	1430–2240	1288–1850	1160–1330
Free NH_3 range (mg/L)	55–246	22–210	12–136	4–93
Free NH_3 at pH = 7.8	158	150	124	89
VS removal rate (%)	60.5	63.8	64.7	59.5

following the grass silage percentile increase. The daily methane yield was linearly correlated with the acetic acid concentration indicating that acetoclastic methanogenesis being the dominated methanogens. The experimental result recommend applying a pig manure/grass silage ratio of 1:1 in practice due to high specific methane yields and a short lag phase.

3.3 *Chicken Manure Digestion*

In our lab, two lab-scale thermophilic (a) and mesophilic (b) CSTR reactors have been developed with an average performance shown in Fig. 9 feeding with raw chicken manure (CM) and ammonia-stripped CM, where ammonia was removed by means of recycling biogas followed by gas washing in sulfuric acid to capture ammonia (Table 7) (Niu et al. 2013a, 2015). High solid methane fermentation techniques are valuable engineering applications in the fields of both renewable energy production and bio-waste degradation. However, high VS load results in high ammonia load, thus the high solid methane fermentation is more sensitive to ammonia concentration. Moreover, the relationship between operational performance and the dynamic transition of the archaeal and bacterial community remains poorly understood.

At total ammonia-nitrogen concentrations of <5000 mg/L, the process can deliver a steady performance. Due to the high organic-nitrogen content of CM, the process indicator of volatile fatty acid (VFA) accumulated correlates with ammonia concentration. A distinct rise in VFA accumulation combined with low methane production of 0.29 L/gVS$_{in}$ occurred at a total ammonia-nitrogen (TAN) of 4000 ~ 4500 mg/L. The biogas production and mass transfer in the thermophilic reactor are illustrated in Fig. 10 for the ammonia-stripped and raw CM. With

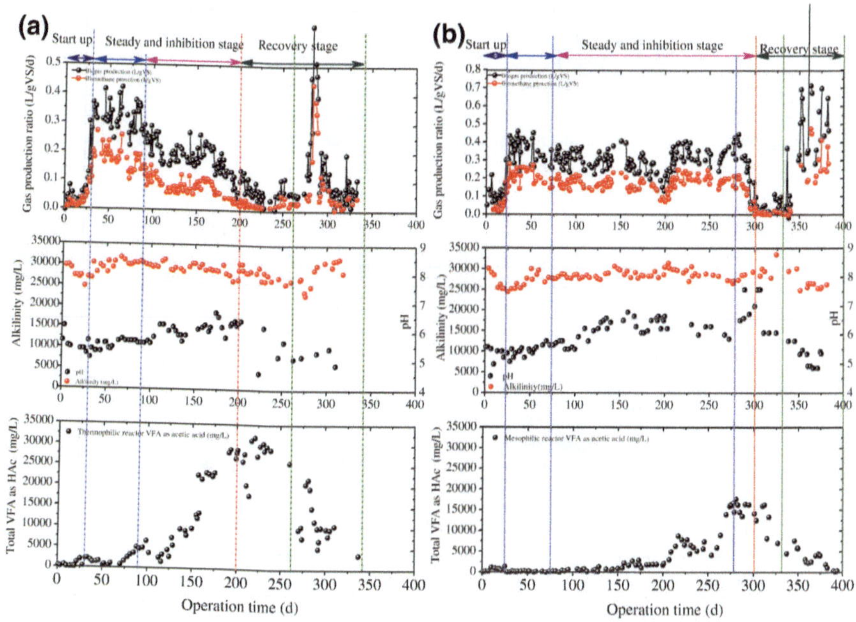

Fig. 9 Process performance comparison of thermophilic (**a**) and mesophilic reactor (**b**) (Niu et al. 2015)

Table 7 Characteristic of chicken manure

Biomass	Kinds	Total amount		TS mg/L	VS mg/L	T-N mg/L
		t/year	t/day			
Chicken manure (CM)	Raw CM	13	0.035	112	82.7	6450
	Ammonia stripped CM	–	–	89.3	61.3	3590
	∑/average	13	0.035	–	–	–

proposed design, feed with TS 10 % of CM 10 tons/day, a full scale reactor can produce 6000 m^3CH_4/d, which equals 1800 kWh/d of electric power.

The two reactors were fed with TS around 10 % with TVS 7–8 %. Methane production rates were similar in both reactors being stable at 0.25 L CH_4/gVS_{in} until day 75 when feeding with ammonia-stripped chicken manure, achieving over 70 % removal efficiency of the initial TCOD at TAN lower than 3500 mg/L (Fig. 10). Raw CM was fed from day 75 to investigate the effect of ammonia on CM fermentation. At day 76, initial accumulation of VFA was observed in the thermophilic reactor but not in the mesophilic reactor (Fig. 9). Meanwhile, the methane production rate in the thermophilic reactor decreased to 0.13 L/gVS_{in}, which is 35 % lower than that of the mesophilic reactor. Following a TAN concentration increase from 3500 to 4000 mg/L, the methane production rate decreased to 0.08 L/gVS_{in} in the thermophilic reactor, which is less than half of the steady

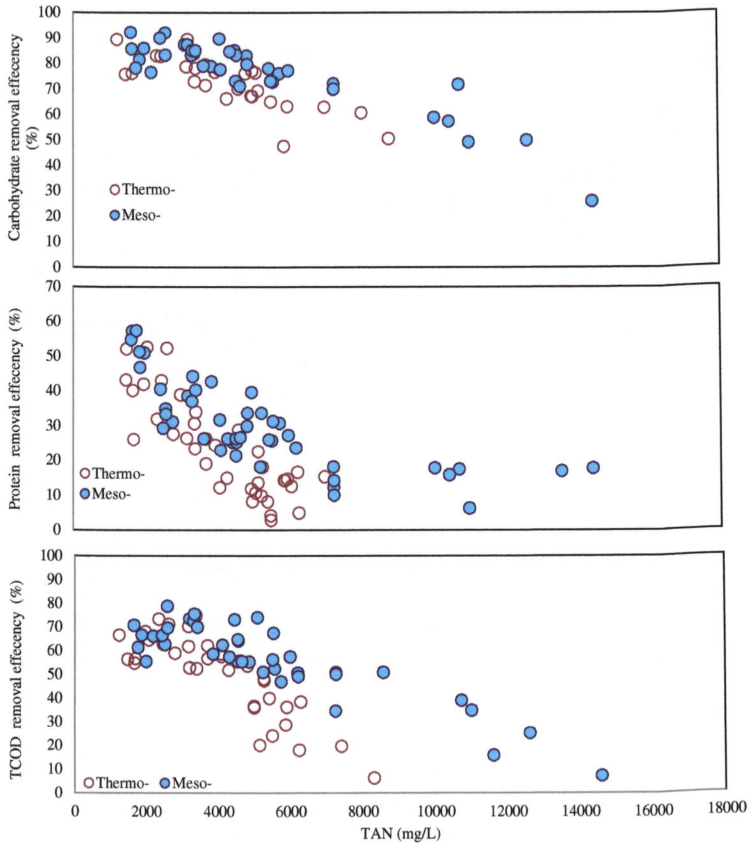

Fig. 10 TAN effects on the process removal efficiency (Niu et al. 2015)

stage (Fig. 9a). The protein removal efficiency decreased sharply from 40 to 20 % (Fig. 10). An inhibited steady stage was found with an average methane production rate of 0.07 L/gVS$_{in}$. At day 200, NH_4HCO_3 was artificially added in order to investigate the effects of ammonia overload on the performance and the shifts in the microbial community. Similar biogas production rates (0.74 m^3/kgVS) were observed in thermophilic and mesophilic reactors. The thermophilic reactor had a higher hydrolysis rate of ammonia production about 70.93 g/kgVS$_{degraded}$ in the steady stage (Fig. 11). The mesophilic reactor could afford a TAN concentration over 12,000 mg/L with a stable methane production rate, whereas the thermophilic reactor lost about half of the methane production activity at a TAN of 6000 mg/L. This shows that the mesophilic reactor has a higher tolerance for TAN than the thermophilic reactor. The mesophilic reactor successfully recovered from serious inhibition of TAN 16,000 mg/L. In contrast, recovery of the thermophilic reactor failed. The mesophilic reactor was dominated by aceticlastic methanogens compared to the thermophilic reactor.

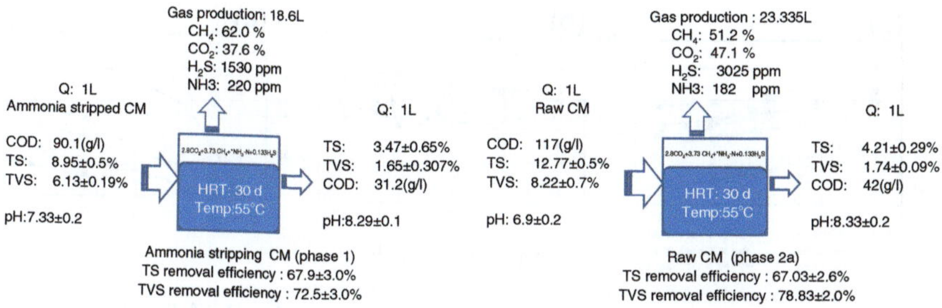

Fig. 11 The gas production and mass transfer in the thermophilic fermentation (Niu 2014)

Microbial communities were analysed in the steady stage, inhibition stage and recovery stage. The population profiles showed significant variations in the proportion of methanogens. The aceticlastic *Methanosarcinaceae* population increased with the degradation of acetate in both reactors in the steady stage. Canonical correspondence analysis (CCA) results showed that the archaeal communities' dynamics responded to the physico-chemical parameters of TAN (total ammonia nitrogen), VFA and free ammonia (FA). For thermophilic digestion, the steady stage was dominated by *Methanosaeta* which transferred to *Methanothermobacter* in the inhibition stage and *Methanosarcina* in the recovery stage. In contrast, under the mesophilic condition, *Methanosarcina* dominated in the steady stage while in the inhibition stage *Methanosaeta and Methanoculleus* thrived and with *Methanosaeta* dominance in the recovery stage. The CCA graph obtained for the methanogens in the two reactors showed a separation and shifts between different stages of archaeal communities (Fig. 12). Significantly shifts of hydrogenotrophic methanogens in the thermophilic reactor were proved following TAN variation. The dynamic results strongly indicate high resilience in the mesophilic reactor and lower resilience of both microbial community and function in the thermophilic reactor. The recovery stage of mesophilic reactor had the highest functional group diversity and higher gas production rate.

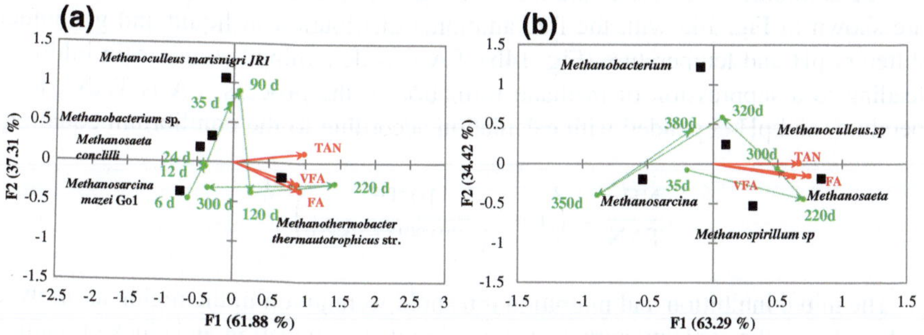

Fig. 12 CCA comparison of methanogens in the thermophilic (**a**) and mesophilic reactor (**b**)

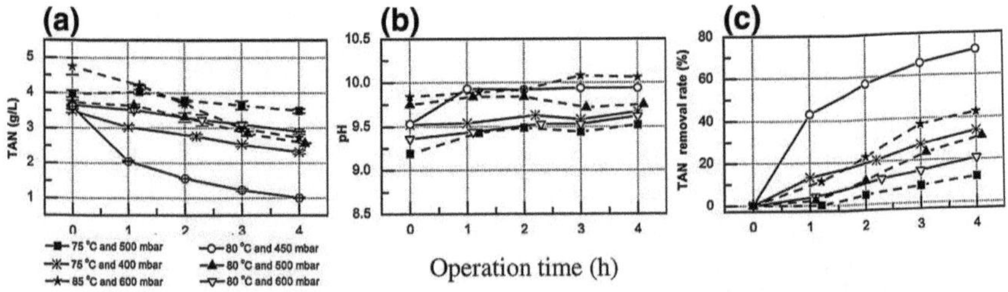

Fig. 13 Ammonia stripping experiment **a** TAN concentration, **b** pH of liquid sample and **c** accumulated TAN removal rate at different operation (Nie et al. 2015)

Technical stripping was performed to remove ammonia from the liquid fraction of the digester with production recycled to control ammonia concentration. Pre-treatment of ammonia stripping under different condition showed that at 450 mbar, 80 °C, the TAN was reduced by over 70 % at 4 h operation (Fig. 13). In this way, an organic loading rate of 5.3 gVS/L/d was achieved with an average free ammonia nitrogen (FAN) concentration of 0.77 g/L and a specific gas yield of 0.39 L/gVS (Nie et al. 2015), which was higher than for the raw chicken manure digestion (Niu et al. 2013b).

Compared with more rigorous wastewater discharge standards, removing ammonia from the digestate and recycle ammonia-depleted digestate back to the fermenter as in this study seems to be a good choice to cope with high ammonia loads (Table 8). However, strong foaming at the initial phase of some stripping tests was observed and would in practice need to be dealt with. For example, when TAN concentration in the liquid was 6.07 g/L, the total volume of reaction liquid and thick foam could add up to double of the filling level at 82.4 °C and ambient pressure. Although concentration of TAN decreased as the stripping went on, foaming could still be observed under some conditions. Other than the small and dense bubbles occurring at the initial phase, bubbles formed during the middle and later stage were big and busted more easily. Hence, foaming happening at later stages was acceptable (Myint et al. 2007).

The ammonia distribution and microbial growth rate in the digestion of manure are shown in Fig. 14a with the free ammonia distribution in liquid and gas under different pH and temperature (Fig. 14b). FA was determined as the main inhibitor leading to a suppression of methane formation in the process. FA is TAN-, temperature- and pH-depended with calculation according to the equilibrium equation:

$$\frac{\mathrm{NH_3}}{\mathrm{TAN}} = \left(1 + \frac{10^{-\mathrm{pH}}}{10^{-\left(0.09018 + \frac{2729.92}{\mathrm{T(k)}}\right)}}\right)^{-1}$$

The initial inhibition and inhibition threshold were given in the region as showed in Fig. 14 for the cattle manure digestion, pig manure digestion and chicken manure

Table 8 Comparison of this study with some works focusing on ammonia inhibition using CSTR at mesophilic temperature

Substrate	OLR kgVS/(m^3d)	T °C	pH	FAN	TAN	Holding time[a]	Remarks	Reference
		–		g/L	g/L			
Swine manure	–	37	8.1	0.75	5.9	18	Inhibited steady	Hansen et al. (1998)
SHW + OFMSW	3.7	34	7.9	0.34	4.1	24	Steady	Cuetos et al. (2008)
Sewage sludge	3	35	8	0.53	3.5	20	Inhibited steady	Duan et al. (2012)
Chicken manure	2.5	35	8.2	0.9	5	30	Inhibited steady; VFA accumulated	Niu et al. (2013b)
Chicken manure	6	40	7.8	0.58	5.59	31	Inhibited steady	Nie et al. (2015)
Chicken manure	6	40	8	0.86	6.96	40	Inhibited steady; VFA accumulated	Nie et al. (2015)

SHW solid slaughterhouse waste; *OFMSW* organic fraction of municipal solid waste
[a]The days during which the stable state was kept

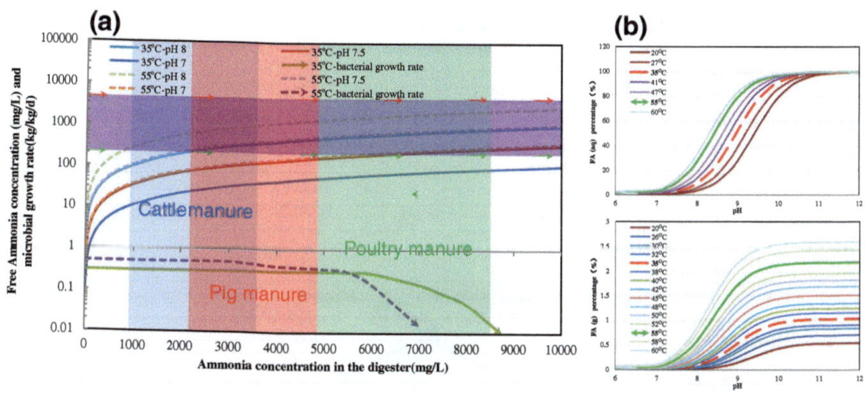

Fig. 14 Ammonia and free ammonia concentration in the digester of livestock manure (**a**), free ammonia distribution in liquid and gas (**b**)

digestion, which illustrated that the pig manure and chicken manure are more suitable for mesophilic digestion keeping the FA in a low range. Chicken manure owning to the highest TAN production between 5000 and 8500 mg/L which led to inhibition. Our experiments show that mesophilic digestion is more feasibility for the poultry manure treatment due to high ammonia concentrations.

4 Energy Conversion and Engineering Application

A combined heat and power (CHP) generator is typically a reciprocating gas engine. The CHP technology makes the bio-CH_4 utilization more efficiency ideal for anaerobic digestion facilities. Energy conversion was 40 % of electrical energy and 50 % of useable thermal energy. Full-scale application of livestock manure digestion has many benefits. Engineering application of chicken manure digestion is summarized in Fig. 15. A full-scale reactor fed with 10 tons per day can be produce 6000 m^3CH_4/d, which equals 1800 kWh/d of electric power. From the calculation and the results obtained in this study, several recommendations can be proposed. The recommendations can be distinguished into three parts:

(1) For engineering application, a high solid loading rate of 10 % of chicken manure can be stably achieved for both mesophilic and thermophilic conditions, when feeding ammonia-stripped CM keeping the TAN concentration below 4000 mg/L.
(2) The mesophilic process is more suitable for full-scale design of CM methane fermentation as it has a high tolerance to ammonia concentration. Keeping the process indicator of VFA below 5000 mg/L is important for process stability. Managers of such reactors should pay more attention to the functional resilience of microbial communities.
(3) Once the anaerobic process was inhibited, washing and dilution proved an effective way for the recovery process.

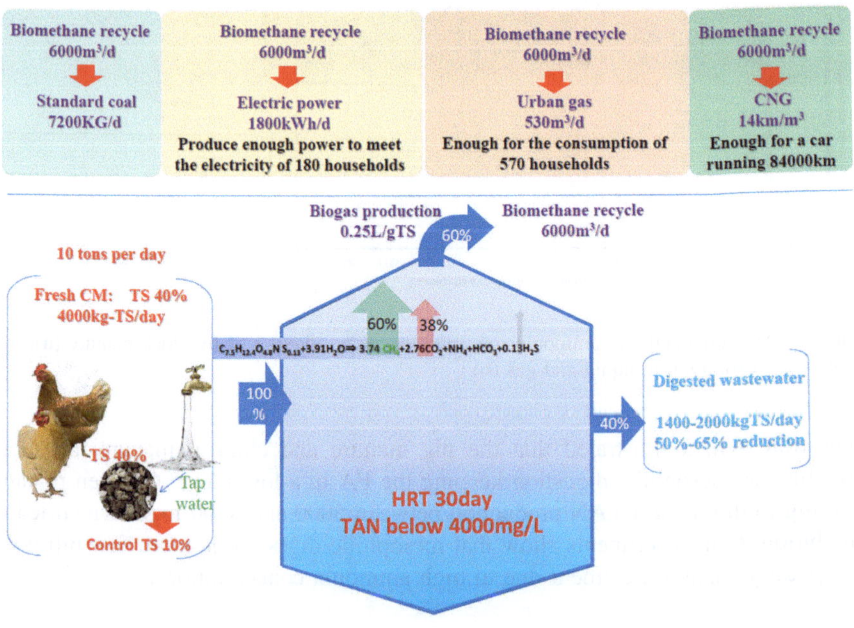

Fig. 15 Proposed case study based on the experimental calculation

5 Conclusions

Livestock manure waste will increase following the yearly livestock production increase. Engineering application of anaerobic digestion for various types of livestock manure wastes has long been established and has already made a significant contribution to improving the recovery of resources. Anaerobic digestion provides local sources of renewable energy in the form of bio-methane. Pre-treatment, co-digestion and two-stage anaerobic fermentation may potentially lead to higher energy recoveries. Ammonia is an inhibitor of the anaerobic process and should be controlled, keeping the TAN lower than 5000 mg/L and free ammonia lower than 500 mg/L to yield a steady process. Washing and dilution and keeping the biomass at a resilience level can make recovery of the process possible. Poultry manure had much higher ammonia levels than pig manure and cattle manure, which was more suitable for mesophilic digestion.

Livestock manure digestion is easily inhibited by ammonia due to high nitrogen contents in the substrate and high TS content. Compared with the low TS digestion, the high TS 10 % digestion with ammonia control in the safety region has many advantages with low cost and high energy recovery rate. Moreover, pre-treatment of chicken manure by ammonia stripping is an efficiency way to generate a stable process. Co-digestion to control ammonia concentration is another way to solve the problem.

References

Babaee A, Shayegan J, Roshani A (2013) Anaerobic slurry co-digestion of poultry manure and straw: effect of organic loading and temperature. Iran J Environ Heal Scien Engineer 11

Chen Y, Varel V, Hashimoto A (1980) Effect of temperature on methane fermentation kinetics of beef-cattle manure. Biotechnol Bioeng Symp 10:325–339

Coats ER, Gregg M, Crawford RL (2011) Effect of organic loading and retention time on dairy manure fermentation. Bioresour Technol 102(3):2572–2577

Cuetos MJ, Gomez X, Otero M, Moran A (2008) Anaerobic digestion of solid slaughterhouse waste (SHW) at laboratory scale: influence of co-digestion with the organic fraction of municipal solid waste (OFMSW). Biochem Eng J 40(1):99–106

Dorian JP, Franssen HT, Simbeck DR (2006) Global challenges in energy. Energy Policy 34 (15):1984–1991

Duan NN, Dong B, Wu B, Dai XH (2012) High-solid anaerobic digestion of sewage sludge under mesophilic conditions: feasibility study. Bioresour Technol 104:150–156

Engineers ASoA (2003) Manure production and characteristics. Engineering practices subcommittee of the ASAE. Agricultural Sanitation and Waste Management Committee, ASAE D384.1 FEB03

FAO (2013) Global livestock production and health atlas (GLiPHA)

Hansen KH, Angelidaki I, Ahring BK (1998) Anaerobic digestion of swine manure: inhibition by ammonia. Water Res 32(1):5–12

Jain M, Singh R, Tauro P (1981) Anaerobic digestion of cattle and sheep wastes. Agric Wastes 3 (1):65–73

Krishania M, Vijay VK, Chandra R (2013) Methane fermentation and kinetics of wheat straw pretreated substrates co-digested with cattle manure in batch assay. Energy 57:359–367

Li Y-Y, Gadow S, Niu Q (2015) Biomass energy using methane and hydrogen from waste materials. In: Topical themes in energy and resources, Springer, Japan, pp 131–157

Liu K, Tang YQ, Matsui T, Morimura S, Wu XL, Kida K (2009) Thermophilic anaerobic co-digestion of garbage, screened swine and dairy cattle manure. J Biosci Bioeng 107(1):54–60

Maghanaki MM, Ghobadian B, Najafi G, Galogah RJ (2013) Potential of biogas production in Iran. Renew Sustain Energy Rev 28:702–714

Myint M, Nirmalakhandan N, Speece RE (2007) Anaerobic fermentation of cattle manure: Modeling of hydrolysis and acidogenesis. Water Res 41(2):323–332

Nie H, Jacobi HF, Strach K, Xu CM, Zhou HJ, Liebetrau J (2015) Mono-fermentation of chicken manure: ammonia inhibition and recirculation of the digestate. Bioresour Technol 178:238–246

Nielsen H, Mladenovska Z, Westermann P, Ahring BK (2004) Comparison of two-stage thermophilic (68 C/55 C) anaerobic digestion with one-stage thermophilic (55 C) digestion of cattle manure. Biotechnol Bioeng 86(3):291–300

Niu Q (2014) Process performance and microbial community dynamics during methane fermentation of chicken manure, vol Doctor, Tohoku university, Sendai, pp 1–185

Niu Q, Qiao W, Qiang H, Hojo T, Li Y-Y (2013a) Mesophilic methane fermentation of chicken manure at a wide range of ammonia concentration: stability, inhibition and recovery. Bioresour Technol 137:358–367

Niu QG, Qiao W, Qiang H, Hojo T, Li YY (2013b) Mesophilic methane fermentation of chicken manure at a wide range of ammonia concentration: stability, inhibition and recovery. Bioresour Technol 137:358–367

Niu Q, Takemura Y, Kubota K, Li Y-Y (2015) Comparing mesophilic and thermophilic anaerobic digestion of chicken manure: microbial community dynamics and process resilience. Waste Manage 43:114–122

Robinson TP, Wint GRW, Conchedda G, Van Boeckel TP, Ercoli V, Palamara E, Cinardi G, D'Aietti L, Hay SI, Gilbert M (2014) Mapping the global distribution of livestock. Plos One 9(5)

Xie S, Lawlor PG, Frost JP, Hu Z, Zhan X (2011) Effect of pig manure to grass silage ratio on methane production in batch anaerobic co-digestion of concentrated pig manure and grass silage. Bioresour Technol 102(10):5728–5733

Anaerobic Digestion of Palm Oil Mill Residues for Energy Generation

Nastaein Qamaruz-Zaman, Nurashikin Yaacof and Hossein Faraji

Abstract Malaysia is one of the world's largest palm oil exporter (39 % of world palm oil production and 46 % of world exports). In the process of producing palm oil, a considerable amount of water is needed, leading to the generation of large volumes of wastewater also known as palm oil mill effluent (POME). Anaerobic digestion of palm oil mill effluents (POME) has started as early as the 1990s using the anaerobic lagoon system comprising a series of ponds in combination with aerobic and pre-treatment ponds to effectively meet the effluent discharge standards. This conventional open pond system requires long hydraulic retention times, large land area and at the same time release uncontrolled greenhouse gas and odour to the atmosphere. Of late, there has been an emergence of more advanced anaerobic digesters in palm oil mills replacing the conventional lagoon system. This chapter discusses anaerobic technology for POME moving from a purely effluent treatment focus using conventional lagoons to more advanced controlled systems for energy recovery purposes using closed tank digesters. The issues of palm oil mill residues for energy recovery will also be looked at taking into consideration POME co-digestion with other materials and possible environmental impacts.

Keywords Palm oil mill effluent · Anaerobic digestion · Energy recovery · Methane

N. Qamaruz-Zaman (✉) · N. Yaacof · H. Faraji
School of Civil Engineering, Universiti Sains Malaysia, 14300 Nibong Tebal, Penang, Malaysia
e-mail: cenastaein@usm.my

N. Qamaruz-Zaman
Solid Waste Management Cluster, Universiti Sains Malaysia, Engineering Campus, 14300 Nibong Tebal, Penang, Malaysia

O.P. Karthikeyan et al. (eds.), *Recycling of Solid Waste for Biofuels and Bio-chemicals*, Environmental Footprints and Eco-design of Products and Processes, DOI 10.1007/978-981-10-0150-5_7

1 Introduction

Palm oil mill effluent (POME) is wastewater that results from palm oil milling and typically originates from three main sources; sterilizer condensate, clarification wastewater and hydrocyclone wastewater. Anaerobic digestion presents a sustainable approach towards treating palm oil wastewater. This chapter begins with description of the palm oil milling and the generation of palm oil mill wastewater, to give readers an appreciation towards the characteristics of the effluent requiring treatment to meet the discharge standard. It is the objective of this chapter to present the evolvement of the anaerobic treatment from primarily to treat effluents using open lagoon pond systems, to more advanced anaerobic systems for energy recovery. This chapter also aims to highlight the support provided by the Clean Development Mechanism (CDM) in financing these modern digesters which saw the growth of advanced anaerobic digesters in the recent years. The efforts of the CDM and life cycle analysis encourages both energy recovery from palm oil mills whilst giving due considerations to the impact on the environment from such technologies.

2 Palm Oil Milling

Elaeis guineensis is a species of oil palm tree that is cultivated extensively for oil production given its highest yield of oil per unit area of all oil bearing plants. The plant can grow up to 50 m and may live for up to 200 years. That said, the plants are seldom allowed to grow for more than 30 m which makes harvesting less accessible, thus are usually cut and replanted. The use of palm oil is aplenty. The crude palm oil (CPO) is commonly used in a wide variety of food products such as cooking oil, shortenings and margarine. In addition to the crude palm oil, the palm kernel oil can also be extracted from the plant. The latter is a raw material in the production of non-food products which include toiletries candles, soaps, cosmetics, and detergents.

While Indonesia is the largest producer of palm oil in the world, Malaysia is the leading exporter of palm oil, accounting for 46 % of global exports. Together they make up over 88 % of palm oil exports. China, the European Union, India and Pakistan are the largest importers of palm oil (MPOB 2007).

The production of palm oil in brief, involves the reception of fresh fruit bunches from the plantations, sterilizing and threshing of the bunches to free the palm fruit, mashing the fruit and pressing out the crude palm oil. The crude oil is further treated to purify for storage and export. Figure 1 shows the palm oil production process flow diagram at a palm oil mill in Nibong Tebal, Malaysia.

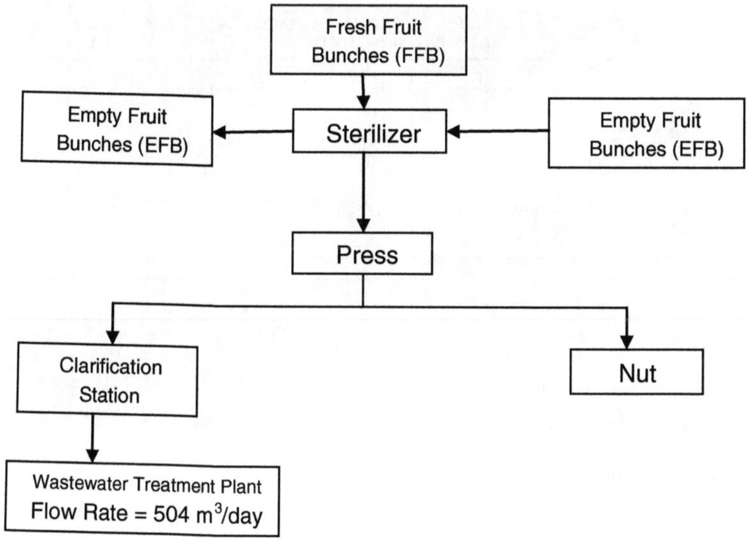

Fig. 1 Simplified production process flow diagram

2.1 *Fresh Fruit Bunches (FFB) Ramp*

Fresh fruit arrives from the field as bunches or loose fruit at the Fresh Fruit Bunches ramp as shown in Fig. 2. Lorries will unload Fresh Fruit Bunches (FFB) in this area where the bunches will be graded as:

- Hard bunches—Stalks which have fruits still attached on them after the sterilization and stripping process are called hard bunches, and have to be recycled back to sterilizers for further cooking. Hard bunches are detected by visible inspection.
- Ripe bunches.
- Over ripe bunches.
- Rotten bunches/Disease bunches.

2.2 *Sterilizer*

The second stage of palm oil processing is to facilitate the mechanical stripping of fruits by using sterilizers or pressure vessels. Steam will soften the fruit mesocarp for digestion and help the release of oil while conditioning the nuts to minimize kernel breakage. In addition, sterilization inactivates the lipases in the fruits, and prevents build-up of free fatty acids (FFA). This is because a low oil yield could result since the fat-splitting enzymes would hydrolyse much of the oil during the fruit pulping process (Liew et al. 2015).

Fig. 2 Fresh Fruit Bunch ramp

In the sterilizer, steam at 40 psig and a temperature of 145 °C is supplied in single-peak, double-peak or triple-peak cycles to cook the bunches. Steam consumption varies from 140 kg/ton FFB for a single-peak cycle to 224 kg/ton FFB for a triple—peak cycle. Figure 3 shows a horizontal cylindrical autoclave-type sterilizer operating at 60–90 min cooking time. The length of the sterilizer is dependent on the number of cages required for operation of the mill where 2.5–10 tons of FFB are usually accommodated by each cage (Sivasothy et al. 1986).

After sterilization, the softened FFB undergo threshing whereby these FFB are rolled and threshed in a revolving slated steel drum to separate the fruits from the bunch stalks (shown in Fig. 4a, b). The total oil loss absorbed on the stalks depends

Fig. 3 Horizontal cylindrical autoclaves

Fig. 4 Revolving slated steel drum for fruit recovery (**a**) and palm oil drum strippers (**b**)

on the sterilizing conditions where prolonged sterilization will increase oil loss in stalks. Irregular feeding of the stripper may also result in increase of oil loss in stalks. Considering these facts, stringent monitoring of operating conditions are practiced at the mills, thus, the bunch stalks rarely contain oil and are removed. The bunch stalks are often disposed by incineration to provide fuel for boilers, while the ash is reclaimed as potash fertilizer. Remaining unburnt bunch stalks are transported to nearby plantations for use as mulch for plantation fertilization.

Following sterilization, the fruits are placed in a steel vat known as a fruit digester (Fig. 5). Here, steam is injected again and fruits are mashed by sets of stirring mechanical arms to loosen the fibre from the nuts of the fruits. At this stage, any oil extracted from such mashing are sent to the purification process.

Fig. 5 Fruit digesters

2.3 *Press Station*

The Press Station as shown in Fig. 6 places the fibre-nut mash in a perforated press cage which is pressed to squeeze the oil and moisture from the fibre-nut mash. The extracted crude oil from such pressing flows into a crude oil tank into the Clarification Station for further purification. The press cake from the pressing will fall into a cake breaker conveyor and transported to the nut station for further processing.

Fig. 6 Press station

2.4 Clarification Station

The extracted crude oil from the press station contains a mixture of oil, water and solids from the bunch fibres. In order to give a clear stable product of acceptable appearance, the water and impurities must be removed from the crude oil which is done in the Clarification Station (Fig. 7).

The process of removing water and impurities is done step by step, which includes separation, purifying and drying to produce purified oil. First, hot water is added to the mixture containing the crude oil to dilute and reduce the viscosity of the mixture. A vibrating screen is used to remove some of the solids. Then, the oil mixture is heated to 85–90 °C and left to settle in the clarification tank between 1 and 3 h. Oil from the top of the clarification tank is skimmed off while the lower layer is sent to the centrifugal separator to recover the remaining oil. Oil extracted from both the top and lower layer is purified in the centrifuge and sent to the vacuum dryer for drying. The final crude palm oil is then cooled and stored in storage tanks.

2.5 Nut Station

The press cake from the pressing process is processed at the nut station to produce the palm kernel oil. Figure 8 shows a depericarper, a rotating steel drum used to separate the nuts from the fibre. The nuts from the depericarper are dried in the nut

Fig. 7 Clarification station

Fig. 8 Nut station

silo prior to feeding them to the mills to facilitate the cracking and separation of the palm kernel from the nut shell. The cracked nuts from the mills are subsequently fed into a winnower, a blowing machine where the lighter shell fragments and any remaining fibre are blown off by air-jets. Only the heavier palm kernels with parts of the nutshells still attached remain.

A hydrocyclone or a clay bath is used to further separate the mixture of palm kernel and nutshells. The clay bath principle works on the specific gravity of kernels of 1.07 and the shell of 1.17. Separation is achieved when the kernels float while the shells sink in a clay bath mixture of specific gravity of 1.12. A hydrocyclone on the other hand uses centrifugal force to separate the kernel from the shell using water. The kernels are then dried in hot air silos to reduce the moisture content to less than 7 %. About 0.4 Mt of kernels are produced with every Mt of CPO.

Palm kernel oil extraction results in the production of the Palm Kernel Cake (PKC). The palm kernel cake contains moderate nutrition with 16–18 % of crude protein and 13–20 % crude fiber making it popular as feeding material in ruminant diets. Its use in non-ruminant diets like poultry diets are limited due to the high fibre content of PKC (Sharmila et al. 2014).

3 Palm Oil Mill Effluent

3.1 Water Consumption in the Palm Oil Mill

Palm oil milling typically requires huge quantities of water for its process, for the operation of boilers and the hydrocyclone separator (DOE 1999; Chavalparit et al. 2006). The production of palm oil has grown over the years in Malaysia, from 4.1 million tonnes in 1985 to 18.9 million tonnes in 2011 (PalmOilWorld 2011). About 1.50 m^3 of water is extracted from freshwater resources for the processing of 1 tonne of FFB. Nearby freshwater resources, such as rivers are usually utilized as sourcing from natural water resources incurs very little pumping and treatment costs.

The feed water for the boiler is evaporated into steam with the steam quality dependent on the feed water temperature as well as the temperature and pressure of the steam. The boiler at the palm oil mill in Nibong Tebal, Malaysia (POMNT) is fed using water sourced from Jabatan Bekalan Air (JBA) at 375 m^3/day and treated using flocculation and clarification, to prevent corrosion of the boiler metal, scale formation, foaming, and priming.

Using a large quantity of water, the hydrocyclone separates wet kernels from the palm shells for further processing in the kernel silo. The hydrocyclone separators use the flow of water to separate two components of different densities by centrifugal force. The density of palm kernels is lower than that of the palm shells. At POMNT, the process water is sourced from the nearby Sungai Kerian at 7200 m^3/day which is treated using alum and soda ash.

In the end, about 50 % (0.75 m^3) of the water source for the palm oil milling eventually becomes palm oil mill effluent (POME) with the remaining 50 % ending up as used water. Due to the low contamination level, the used water rarely requires treatment and is usually discharged straight into the drains or rivers.

3.2 The Generation of Palm Oil Mill Effluent

Palm oil mill effluent is wastewater that results from palm oil milling and typically originates from three main sources; sterilizer condensate, clarification wastewater and hydrocyclone wastewater. The wastewater from the sterilizer condensate (or sterilizer waste) resulted from the FFB sterilization process in a horizontal sterilizer or horizontal cylindrical autoclaves (DOE 1999). About 36 % of the total POME constitutes sterilizer condensate (DOE 1999; Wu et al. 2010). The sterilizer at POMNT produces on average of 10.8 m^3 sterilizer condensate per day reaching a maximum of 14.4 m^3/day. The condensate must be removed quickly to prevent flooding of the bearings of the cage bogies and contamination of the sterilizer condensate with the hydrocarbon lubricant.

The clarification wastewater (or separator sludge) discharged from the process operation of clarification of the extracted crude palm oil is by far the major contributor to POME, producing about 60 % of the total effluent. The palm oil mill in Nibong Tebal produces an average of 12.6 m^3/day of wastewater from its clarifier with a maximum value of 25.9 m^3/day. The solids content in clarification wastewater is much higher compared to wastewater from the sterilizer or hydrocyclone, due to the presence of a higher proportion of both soluble and insoluble carbohydrate constituents in the wastewater (Ho et al. 1984). The smallest contributor to the POME is wastewater from the hydrocyclone wastewater, contributing only about 4 % of the effluent discharge (DOE 1999; Wu et al. 2010).

Palm oil mill effluent is a mixture of water, residue oil, and fine suspended components with very high organic matter content. The suspended solids in POME are mainly cellulose matter of vegetative origins like cell walls, organelles, short fibres, nitrogenous compounds (from proteins to amino acids), water-soluble carbohydrates (ranging from hemicelluloses to simple sugars including glucose, reducing sugars, and pectin), pentose (insoluble carbohydrate), free organic acids, lipids, minor organic and mineral constituents (Liew et al. 2015). Physically, fresh POME is thick, hot, acidic, odorous sand looking viscous brownish or grey slurry.

Although the palm oil mill effluent is more polluting than domestic sewage, about a hundred times over (Ma and Augustine Ong 1985; Khalid and Wan Mustafa 1992), the effluent is non-hazardous as the entire milling process is devoid of any chemicals (Ma and Augustine Ong 1985; Khalid and Wan Mustafa 1992; Igwe and Onyegbado 2007). That said, the COD and BOD values of POME are high enough to cause serious pollution and environmental problems for the rivers. As shown in Table 1, all parameters are excessive in the POME compared to

Table 1 Characteristics of POME compared to municipal sewage (Damayanti et al. 2011)

Parameter	POME values	Sewage
pH	5.6	7
COD	45	0.6
Total solids	43	0.8
Total dissolved solids	35	0.23
Suspended solids	8.2	0.56
NO_3^-	0.10	0.01
NO_2^-	0.30	0.06
NH_4^+	0.20	0.03
Total nitrogen	0.5	0.01

All parameter's units in g L^{-1} except pH

sewage, based on data compiled from Felda Bukit Besar Palm Mill, Johor, Malaysia and municipal wastewater, respectively. Many improvements have been made to address the issue of POME treatment with concerted efforts of the policy makers, government agencies, scientists as well as the plant owners. One rising effort is the utilization of POME for energy production through the implementation of engineered anaerobic digestion systems for treating POME.

4 Fundamentals of Anaerobic Digestion

In anaerobic treatment, large organic molecules enter a digester and are converted, mainly, into methane and carbon dioxide by the action of bacteria in the absence of oxygen. The process of anaerobic digestion proceeds in three main stages; (i) hydrolysis, (ii) acid formation and (iii) methanogenesis.

Hydrolysis involves the conversion of the complex waste (particulate and soluble polymers) into soluble products by extracellular enzymes secreted by hydrolytic bacteria. The once complex insoluble organic polymers become more easily available for use by acidogenic bacteria in the next stage. Proteins present in the waste are converted into amino acids, fats into long chain fatty acids and carbohydrates into simple sugars.

In the acid formation step (acidogensis/acetogenesis) step, the organic monomers of sugars and amino acids released earlier are degraded by fermentative bacteria to produce volatile fatty acids (VFA) namely propionic, butyric and valeric acids, together with acetate, hydrogen (H_2) carbon dioxide (CO_2). The degradation of amino acids also produces ammonia.

Methane is produced from the raw materials of the previous stage during methanogenesis, the last stage. This is done in two ways; one through hydrogenotrophic methanogenesis producing methane by utilizing H_2 and CO_2 by the hydrogen-consuming bacteria in a syntrophic co-culture with the OHPA bacteria. The other by methanogenic aceticlastic bacteria which grow on acetate as the substrate, releasing methane and carbon dioxide. Of these, acetic acid and the closely related acetate are the main precursors to methane production accounting to about 75 % of the methane production.

The most important advantages of anaerobic wastewater treatment processes are the high percentage of stabilization obtained and the low percentage of conversion of organic matter to biological cells. The small quantities of sludge growth minimize the problems of biological sludge disposal, as well as the requirements for the inorganic nutrients, nitrogen and phosphorus. The end product of the anaerobic process, methane gas is a highly valuable material that can be utilized for energy consumption at site or sold to consumers, provided a consistent and sufficient quality gas is produced.

5 Anaerobic Pond for the Purpose of Effluent Treatment

Since the 80's, biological treatment, consisting of anaerobic, facultative and aerobic pond systems (Wong 1980) has basically been the technique available for the treatment of POME in Malaysia. In 2011, only 13 % of 426 palm oil mills in Malaysia have installed biogas plants at their premises, with another 39 % slowly upgrading to biogas plants, while the remaining half still uses conventional open ponds (Chin et al. 2013). In fact, the use of sophisticated and controlled anaerobic reactors is still in its infancy and only a few mills have adopted the system. The pond system can achieve a reasonable degree of treatment using a relatively unsophisticated technology that is generally low in construction and operating costs. The pond technology is also easily maintained, thus the risks in treatment malfunction is greatly avoided. These pond methods are regarded as conventional POME treatment method whereby long retention times and large treatment areas are required.

Figure 9 shows the pond system at POMNT comprising of de-oiling tank, acidification ponds, anaerobic ponds and facultative or aerobic ponds. The size of the pond depends on the capacity of the palm oil mill as well as the area available for the ponds. POMNT has 9 ponds inclusive of 3 unit of anaerobic ponds, 3 unit aeration ponds and 1 unit each for the acid pond, cooling pond and polishing pond, respectively. In addition, the POME treatment needs 1 unit 500 tons de-oiling tank, 2 units concrete oil traps, and a belt press filter.

5.1 *Acidic Pond*

Wastewater from the milling process first flows into an oil trap which is built of concrete for the recovery of residual oil; equalization and cooling of the effluent. The effluent will flow into the Acidic Pond thereafter. Here, the Acidification Phase of the anaerobic process is initiated, where the acid forming anaerobic bacteria or acidogens convert the hydrolysed complex organics into free fatty acids.

5.2 *Cooling Pond*

The cooling pond cools the wastewater to a temperature suitable for the growth of bacteria for anaerobic treatment. The wastewater from the cooling pond then flows to the anaerobic pond by gravity flow. Approximately 2–3 days are allocated for de-oiling/cooling and initial sludge settling/acidification.

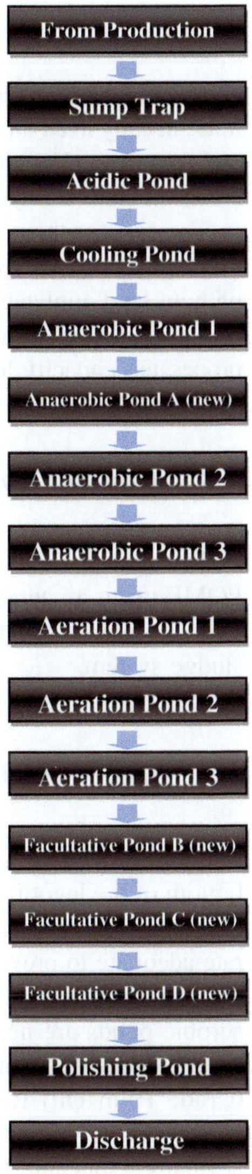

Fig. 9 Process flow for effluent treatment at a palm oil mill in Nibong Tebal, Malaysia

5.3 *Anaerobic Pond*

Anaerobic bacteria in the Anaerobic Pond decompose organic matter to carbon dioxide, methane gas, other gaseous end products, organic acid and cell tissue. The principle biological reactions are acid formation and methane fermentation. The high organic loads produce strict anaerobic conditions (no dissolved oxygen) throughout the pond.

$$CHONS + H_2O \rightarrow CO_2 + CH_4 + NH_3 + \text{other end products} + H_2S + \text{heat}$$

In terms of sizing, a palm oil mill with a processing capacity of 54 tons palm oil per hour, usually need an anaerobic pond that measures about 60 m in length, 30 m wide and 6 m deep (Yacob et al. 2006). Retention time of anaerobic ponds are the longest in the ponding system between 20 to about 200 days (Poh and Chong 2009).

The cost for the ponding system can be assumed to match that of open digesting tank. In 1986, the cost for initiating open digesting tank for POME treatment without land application for mills processing 30 tons FFB/h was RM 750,000. The cost was estimated to increase to RM 1,147,642 in the year 2006 for the same processing capacity and ponding system (Poh and Chong 2009).

5.4 Aeration Pond

In the Aeration Pond, a greater amount of dissolved oxygen is made available in the POME using an aerator. Through aeration, the aerobic microorganism in the pond will decompose most of the soluble organic matter. This produces the activated sludge system.

5.5 Secondary Facultative Pond

The facultative ponds act as stabilization ponds where heavy solids will settle to the bottom of the lagoon and lighter solid will float. It is important that the surface area of the facultative lagoon is large enough to provide an atmospheric oxygen transfer rate adequate to prevent anaerobic conditions on the lagoon surface. The depths of the facultative and aerobic ponds are more shallow, about 1.5 m. Facultative and aerobic ponds are necessary to further reduce BOD concentration in order to produce effluent that complies with the Environmental Quality (Prescribed Premises) (Crude Palm Oil) Regulations 1977, under the jurisdiction of Section 51 of the Environmental Quality Act, 1974. Table 2 presents the effluent discharge standards normally applicable to crude palm oil mills in Malaysia.

5.6 Belt Press Filter

The effluent from the prevailing ponds are then passed through a Belt Press Filter. Palm oil mill in Nibong Tebal (POMNT) uses the SD-3 Model from Green Mark Projects Sdn. Bhd. with a capacity of 30 m^3 of slurry with up to 2 % solids contents. The belt press filter can be operated either manually or automatically for a maximum continuous operating time of 16 h per day. The machine belt press filter is capable of handling 25–30 ton of POME.

Table 2 Effluent discharge standards for crude palm oil mills (Department of Environment 1999)

Parameter	Unit	Limits
Biochemical Oxygen Demand (BOD), 3-day, 30 °C	mg/L	100
Chemical Oxygen Demand (COD)	mg/L	*
Total solids	mg/L	*
Suspended solids	mg/L	400
Oil and grease	mg/L	50
Ammoniacal nitrogen	mg/L	150 (value of filtered sample)
Total nitrogen	mg/L	200 (value of filtered sample)
pH	–	5–9
Temperature	°C	45

*No discharge standard after 1984

The effluent from the belt press filter which is the filtrate will be sent to the polishing pond. Wastewater from the polishing pond will be discharged by gravity flow through a pipeline into the nearby watercourse. The effluent discharge into Sungai Kerian is controlled through a meter with control ball valve at a quantity of 6–12 m^3/h. The industrial wastewater production from the mill has been determined to be 504 m^3 per day.

The solids scraped from the belt press filter usually contain 62–85 % moisture with N (2–2.5 %), P_20_5 (0.5–1 %), K_2O (1–1.5 %), CaO (1.5–2 %) and MgO (0.5–1 %). Due to its nutritional values, the solids are utilized for compost production at the mill.

The palm oil mill at Nibong Tebal has since added 1 anaerobic and 3 facultative pond to the nine 9 effluent treatment ponds, making it a total of 13 effluent ponds. The upgrade was done in order to meet the Department of Environment Limit (Second Schedule) according to the Environmental Quality (Prescribed Premises) (Crude Palm Oil) Regulations, 1977, as shown in earlier in Table 2.

The preference by the POMNT to upgrade the ponding system with more treatment ponds, rather than adoption of a more effective technology is an indication that most mill owners employ anaerobic treatment for the sole purpose of effluent treatment. The choice of more sophisticated anaerobic reactors is often weighed against the high costs involved and scepticism on return in investment in energy production.

5.7 *Issues with Anaerobic Pond Treatment*

5.7.1 Ineffective Pond Design Consideration

Empirical data from actual field experience for hydraulic retention time (HRT), solids retention time (SRT), influent and effluent concentrations and sludge age are common parameters used to design anaerobic pond systems. The performance of the pond and its design is measurable by BOD, as the pollution parameter. Areal or surface BOD loading (λ_s) is the weight of BOD applied per unit area of pond per day and is given as (Wong 1980):

$$\lambda_s = \frac{10 L_i Q}{A}$$

where

A Area of pond in hectares
Q Flow rate of wastewater in m^3/day
L_1 Influent BOD in kg/m^3.

Volumetric loading rates are sometimes used particularly with anaerobic ponds. The weight of BOD applied per unit volume per day (λ_v) then (Wong 1980):

$$\lambda_v = \frac{L_i Q}{AD}$$

where

D pond depth
and AD retention time = t.

Then (Wong 1980)

$$\lambda_v = \frac{L_i}{t}$$

The effectiveness of the ponds in effluent treatment can be maximized by considering biological kinetics and incorporating bio-kinetic factors in the anaerobic pond design. There are many literature with results of research on advanced anaerobic treatment such as membrane anaerobic system (Fakhrul Razi and Noor 1999), high rate up-flow anaerobic sludge fixed film (Zinatizadeh et al. 2006), anaerobic hybrid digester and modified anaerobic baffled reactor (Faisal and Unno 2001), for the treatment of POME. That said, literature on anaerobic stabilization pond system with regards to its biological behaviour is still limited, but is highly needed as the ponding system is widely used in palm oil mills.

Bio-kinetic coefficients are useful tools to obtain information on the rate of microbial growth and consumption of substrate, which is essential to determine the volume of the reactor and understanding the process control well through system simulation. Meanwhile, bio-kinetic coefficients also play an important role to illustrate the development of microorganism and substrate balances, the prediction of effluent concentration, the development of process design factors and the effects of kinetic coefficients on the process of design, performance and stability (Tchobanoglous et al. 2003).

Another issue with the pond design are the formation of scum on the POME surface and the tendency for solids build-up at the bottom of the pond. Furthermore, the sludge and scum will clump together inside the pond and thus lowering the treatment efficiency. Pond maintenance becomes an important trait. The system requires regular desludging by either using submersible pumps or excavators to maintain the design efficiency.

5.7.2 Under-Production of Methane Gas

Anaerobic ponds at pam oil mills are conducted only in the mesophilic temperature range in Malaysia, despite the raw POME being discharged around 80–90 °C (Zinatizadeh et al. 2006) which actually makes thermophilic processes feasible. A mesophilic treatment requires a cooling pond or cooling tower prior to digestion to bring the temperature down to the mesophilic range. The cooling pond/tower may be eliminated if a thermophilic system is adopted instead.

Anaerobic systems operating at 55 °C have higher substrate degradation rate and biogas production rate, acting four times faster than operation in the mesophilic temperature of 35 °C. Thermophilic digesters are able to tolerate higher OLRs and operate at shorter HRT while producing more biogas (Ahn and Forster 2002; Kim et al. 2006; Yilmaz et al. 2008). If all POME in Malaysia would be treated at thermophilic temperature and recovered biogas was to be fully utilized for electricity energy generation, it would generate 2250 million kWh which contributes approximately 4 % of the 1999 national electricity demand (Yeoh 2004).

However, in thermophilic condition, the production of volatile fatty acid is higher compared to the mesophilic temperature range (Yu et al. 2002). There is a risk of accumulation of volatile fatty acid which could inhibit methanogenesis leading to biomass washout and ultimately reactor instability. Due to the high risk involved, many mill owners prefer digesters operating at mesophilic temperature as the process is more stable and easier to control.

5.7.3 Untapped Energy Resource

POME contains a high concentration of organic matter, COD concentration in the range of 45,000–65,000 mg/L, BOD of 18,000–48,000 mg/L and oil and grease greater than 2000 mg/L (Chin et al. 1996).The effluent is thick, oily, dark brown in colour and smells obnoxious (Oswal et al. 2002; Bhatia et al. 2007; Fang et al. 2011).

The methane concentration from the palm oil anaerobic ponds is fairly consistent at 54.4 % in the gaseous mixture. However, the amount of methane emitted from the anaerobic ponds is influenced by seasonal cropping of oil palm and the mill activities (Yacob et al. 2006). Yacob et al. (2006) found that one anaerobic pond can emit about 1043 kg of methane each day. For a mill with four anaerobic ponds, a total of 4172.4 kg of methane gas can be produced each day. If carbon credit price of €10 per ton of carbon is used (Menon 2007) and 300 working days, it is likely that RM 1,027,975 per year (€228,438.9) revenue can be obtained from the anaerobic ponds and captured as renewable energy.

6 Utilization of Anaerobic Digestion for Energy Production

6.1 Design of Energy-Intensive Anaerobic Digesters

Up-flow or down-flow filters, fluidized beds (Idris et al. 2003), up-flow anaerobic sludge blanket (UASB) systems (Chaisri et al. 2007) or up-flow floe digesters (Zinatizadeh et al. 2006) are modern high rate anaerobic digester technologies, however, their applications in palm oil mill effluent treatment are scarce. At best, these reactors are designed and performance only evaluated at bench-scale. Removal efficiency of COD over 90 % under very short hydraulic retention times and high loading rates have been reported using advanced anaerobic digesters, however, the results of a full-scale implementation may differ owing to the fact that working conditions are not as easily controlled or predicted.

Investors shy away from these advanced anaerobic digesters due to the excessive capital and operating costs compared to the conventional anaerobic digestion system. In fact, justification for such a need has no basis as currently, there is no environment legislation on bio-methane emission from POME in Malaysia.

6.2 Floating Cover System

An in situ and cost effective alternative means to recover bio-methane from anaerobic ponds is to retrofit the existing ponding/lagoon system with a closed digester system (Lam and Lee 2011). A closed digester system simply means covering the exposed pond with floating plastic membranes to help capture the produced biogas and retain the gas product within the floating plastic enclosure. The cover material are usually made of linear low density polyethylene liners that are resistant to bad weather, biological degradation, UV radiation and prevent pooling of rainwater.

The captured biogas can be used as a mixing mechanism in the anaerobic ponds. Effective mixing encourages contact between the POME with microorganisms (particularly methanogens) and thus promoting better organic material degradation and higher emission rates of biogas. The biogas accumulated from the anaerobic ponds goes through a biogas recycling system and then pumped back to the bottom of the ponds thus, creating a mixing effect in the anaerobic pond. In addition, the biogas can be cleaned to recover methane gas and combusted as a boiler fuel or to generate electricity.

Effluent from the ponds will still be discharged to facultative and aerobic ponds for further organic matters degradation. A belt press filter is still required to recover the solids which can be used as organic fertilizer in plantations.

A geo-membrane cover has been used at a palm oil mill in Honduras, over two mesophilic anaerobic lagoons with a volume of 7500 m^3 (Environmental Fabrics

Inc 2009). The palm oil mill processing capacity is 20 ton/h with wastewater production of 340 m^3 per day. The floating cover system was able to generate 10,000 m^3 day of biogas and with the production of 6 GW/year using two 633 kW generator sets, an annual saving of 1 Million USD is achievable.

6.3 *Single Stage Continuously Stirred Tank Reactors (CSTR)*

In the continuously stirred tank reactor (CSTR), reactor feeding should be continuous for maximum efficiency, for example once a day. When influent substrate concentration of 3–8 % total solids (TS) is added, an equal amount of effluent is withdrawn. A large reactor volume is needed to cater for the large addition of water volume. The operating temperature is maintained constantly at mesophilic or thermophilic temperature. The system can also be operated at high substrate concentration in the range of 16–22 % TS (Gunaseelan 1997), so called a semi-dry digestion. However, the use of semi-dry anaerobic digesters at palm oil mill has not been reported.

In the 1980s, Kek Seng (M) Berhad introduced the single stage CSTR in Malaysia (Chua and Gian 1986), where, at that time, the use of closed anaerobic reactors was only applied at two or three palm oil mills (Chua and Gian 1986). The biogas generated was captured and used as boiler fuel, estimated to be about 1407 tonne/year. A total of 29,547 tonne CO_2 of GHG emission was avoided per year. The biogas can also be used for the mill's electricity consumption while any excessive biogas needs to be flared off. For CSTR, aerobic/facultative ponds or extended aeration system sometimes are necessary still to meet the effluent discharge standard of the Department of Environment.

In Thailand, the anaerobic CSTR at Asian Palm Oil Co., Ltd. Krabi has been in operation since December 2001. The anaerobic digester measures 13.5 m in diameter and 15.3 m in height giving an effective volume of 2100 m^3 (Chotwattanasak and Puetpaiboon 2011). The reactor comprises of gear motor, torque tube, scrapper set and draft tube. Palm oil mill wastewater from the production process was pre-treated by oil recovery tank and primary ponds before feeding to the anaerobic digester by centrifugal pump at feeding rate 300 m^3/day from the tank bottom as up flow.

The plant is able to produce about 0.51 m^3 CH_4/kg COD·day from 1 m^3 (or 20 m^3 biogas) of palm oil mill effluent when operated at an average organic loading rate of 4.53 kg COD/m^3·day and hydraulic retention time of 7 days. About 92 and 64 % BOD_5 and COD removal efficiency, respectively, can be achieved in this anaerobic digester.

In September 2005, a 500 kW biogas engine (Shengli biogas combustion engine Model 500GF-RZ) was installed at Asian Palm Oil Mill for the continuous production of renewable electricity. After more than 17,500 h of operating time, a second biogas engine was operated at the same site. Biogas from the digester was scrubbed to reduce hydrogen sulphide (H_2S) using iron chip to prevent corrosion in any part of the engine. Hydrogen sulphide, nitrogen, particulates and water vapor are also removed from the gas stream. The 500 kW gas engine produced 20 m^3 of gas from 1 m^3 of wastewater or an electricity generation of 2.5 kWh/m^3 of biogas. The efficiency of the gas engine using biogas with a CH_4 content of 65 % achieved 35 % efficiency. The amount of power generation was about 2.2 million kWh/year which when sold to PEA creates a total average income of 4.9 Million Baht/year.

6.4 Expanded Granular Sludge Bed (EGSB)

EGSB is an advanced anaerobic digestion system, one of the few advanced anaerobic digesters in use at site to treat palm oil mill effluent. The EGSB at Kilang Kelapa Sawit (KKS) Labu, Negeri Sembilan, Malaysia comprises two reactors made of steel measuring 6 m diameter and 16 m in height (Wang et al. 2015). The reactor is separated into a three phase separator (top portion), reactor body (middle portion) and liquid distributor (end) which are operated in series using a set of valves and two dosing tanks. The influent is pumped through the bottom of the reactor while the treated effluent exits at the top. The suspended solids from the digestate is treated using dissolved air flotation (DAF) with some of the anaerobic sludge recycled back to the reactor.

The EGSB is operated at 35 °C with a hydraulic retention time (HRT) of 10 days. Its superficial hydraulic velocity was about 0.13 m/h, while superficial airflow velocity was about 1.76 m/h. When half of sludge was recycled via DAF, the anaerobic system removed 94.9 % of COD and from 71,179 to an average effluent COD of 3587 mg/L. About 28 m^3 biogas can be produced from 1 m^3 POME with 65–70 % CH_4, 25–36 % of CO_2, and 800–1500 ppm of H_2S.

The effluent is further purified using a combination of ultrafiltration (UF) and reverse osmosis. The two UF modules have a nominal molecular weight cut-off (MWCO) of 100,000 g/mol and the reverse osmosis (ESPA-2 RO membrane (Hydranautics, USA) have a 99.6 % NaCl rejection rate. A set of biogas purifier and a biogas gas engine generator set were used to transform biogas (methane) into electrical energy. The pilot plant performance assessment showed that an integrated anaerobic and aerobic biological treatment system of POME was possible to achieve a 'zero discharge' integration system. However, the economics of this approach needs to be further addressed and justified to warrant such application in the palm oil mills.

7 Issues of Energy Recovery from Anaerobic Digestion

Biogas emission from the anaerobic ponds consisting of methane, carbon dioxide and traces of hydrogen sulfide is the main emission during POME treatment using the ponding system (Subramaniam et al. 2008). The biogas if left un-harvested, is a greenhouse gas which harms the quality of the air. Instead, if the biogas is harvested and used as energy in the palm oil mill, the impact from the POME is removed and POME becomes a savings to the environment (Subramaniam et al. 2008). This is due to the reuse of the treated POME as fertilizers at the plantations. However, the harvesting of biogas at palm oil mills through controlled anaerobic digestion is still limited.

Vijaya et al. (2010) also recommended palm oil mills to capture biogas for conversion into renewable energy in order to significantly reduce climate change impact from palm oil mill operations. Despite that knowledge, industries are still reluctant to use advanced anaerobic reactors due to, among others, the high cost of investment and the lack of infrastructure associated with national grid connection.

The Clean Development Mechanism (CDM) helps efforts to reduce carbon emission to the environment by allowing developing countries such as Malaysia, Thailand, Indonesia and Africa (as exporters of palm oil) to attract foreign investments to invest in local renewable energy projects such as anaerobic digestion (Menon 2002). The Kyoto Protocol sets binding targets for 37 industrialized countries and the European community for reducing greenhouse gas (GHG) emissions. Under Article 12 of the Kyoto Protocol, the Clean Development Mechanism (CDM) allows a country with an emission-reduction or emission-limitation commitment under the Kyoto Protocol (Annex B Party) to implement an emission-reduction project in developing countries. Such projects can earn saleable certified emission reduction (CER) credits, each equivalent to one tonne of CO_2, which can be counted towards meeting Kyoto targets. The CDM through the Kyoto Protocol is the first global, environmental investment and credit scheme of its kind, providing a standardized emission offset instrument, CERs.

For anaerobic digestion of palm oil mill, the utilization of methane gas as a renewable energy from the anaerobic digestion can be used to obtain certified emission reduction (CER) credit (Tong and Jaafar 2006). In Malaysia, examples of CDM to recover methane from palm oil mill effluent includes the registration by Kim Loong Power Sdn. Bhd. (Project, 0867) and United Plantations Bhd. (Project, 1153). Payback period for investment on anaerobic bioreactors can be short if carbon credit prices remain high (Menon 2007).

That said, the benefits of the CDM in encouraging use of controlled amerce digestion at POME was short-lived. In 2012, the CDM program under the Kyoto Protocol expired leaving only the projects that have been accepted for CDM before the end of 2012 to be continued to mid-2015. The absence of this Emissions Trading Scheme means there is no longer financial support through sales of CERs to assist the development of new biogas plants in palm oil mills (Ji et al. 2013). Despite the known advantages of the more advanced anaerobic bioreactors against conventional ponding system, the route to energy recovery at POME via AD may take a while considering the financial and technical constraints.

8 Conclusions

The real-life applications of advanced reactors are still restricted to simple techniques such as the use of floating covers, one stage CSTRs and only recently the use of expanded granular sludge bed albeit only at pilot scale. The palm oil mill owners are now aware that anaerobic processes are not only are useful in treating palm oil mill effluent, but that the technology with modern approaches in the design and operation is useful for resource recovery specifically methane gas. Practising concept of knowledge transfer between mill operators and biogas technology providers can help to reduce any scepticism of the technology and ensuring a sustainable reactor. In addition, various means to finance biogas projects and other incentives are needed to promote and help the realisation of anaerobic reactors at both new and existing palm oil mills.

Acknowledgments The authors gratefully acknowledge the valuable information provided by Messrs. United Palm Oil Industries, Nibong Tebal, Pulau Pinang, Malaysia.

References

Ahn JH, Forster CF (2002) The effect of temperature variations on the performance of mesophilic and thermophilic anaerobic filters treating a simulated paper mill wastewater. Process Biochem 37(6):589–594

Bhatia S, Othman Z, Ahmad AL (2007) Pretreatment of palm oil mill effluent (POME) using *Moringaoleifera* seeds as natural coagulant. J Hazard Mater 145(1–2):120–126

Chaisri R, Boonsawang P, Prasertsan P, Chaiprapat S (2007) Effect of organic loading rate on methane and volatile fatty acids productions from anaerobic treatment of palm oil mill effluent in UASB and UFAF reactors. Songklanakarin J Sci Technol 29:311–323

Chavalparit O, Rulkens WH, Mol APJ, Khaodhair S (2006) Options for environmental sustainability of the crude palm oil industry in Thailand through enhancement of industrial ecosystems. Environ Dev Sustain 8:271–287

Chin KK, Lee SW, Mohammad HH (1996) A study of palm oil mill effluent treatment using a pond system. Water Sci Technol 34:119–123

Chin MJ, Poh PE, Tey BT, Chan ES, Chin KL (2013) Biogas from palm oil mill effluent (POME): Opportunities and challenges from Malaysia's perspective. Renewable and Sustainable Energy Reviews 26(2013):717–726

Chotwattanasak J, Puetpaiboon U (2011) Full scale anaerobic digester for treating palm oil mill wastewater. J Sustain Energy Environ 2(2011):133–136

Chua NS, HF Gian (1986) Biogas production and utilization—Keck Seng's Experience. In: Proceedings of national Workshop on Recent Developments in Palm Oil Milling Techniques and Pollution Control, Kuala Lumpur, 5–6 Aug 1986

Damayanti A, Ujang Z, Salim MR, Olsson G (2011) The effect of mixed liquor suspended solids (MLSS) on biofouling in a hybrid membrane bioreactor for the treatment of high concentration organic wastewater. Water Sci Technol 9(63.8):1701–1706

Department of Environment (1999) Industrial Processes & the Environment (Handbook No. 3)—Crude Palm Oil Industry. Ministry of Science, Technology and the Environment, Malaysia

Environmental Fabrics Inc (2009) Biodigestion Processes in Palm Oil Mills. Methane to Markets Partnership Meeting CINTERMEX Monterrey, Mexico 27–29 Jan 2009

Faisal M, Unno H (2001) Kinetic analysis of palm oil mill wastewater treatment by a modified anaerobic baffled reactor. Biochem Eng J 9:25–31

Fakhrul Razi A, Noor MJMM (1999) Treatment of palm oil mill effluent (POME) with the Membrane Anaerobic System (MAS). Water Sci Technol 39:159–163

Fang C, O-Thong S, Boe K, Angelidaki I (2011) Comparison of UASB and EGSB reactors performance, for treatment of raw and deoiled palm oil mill effluent

Gunaseelan VN (1997) Anaerobic digestion of biomass for methane production: a review. Biomass Bioenergy 13:83–114

Ho CC, Tan YK, Wang CW (1984) The distribution of chemical constituents between the soluble and the particulate fractions of palm oil mill effluent and its significance on its utilization/ treatment Agric. Wastes 11(1):61–71

Idris AB, Noor MJMM, Al-Mamun A (2003) Modelling of anaerobic fluidized bed bioreactor in the treatment of palm oilmill effluent. Asian J Microb, Biotechnol and Env Exp Sci 5:137–44

Igwe JC, Onyegbado CC (2007) A review of palm oil mill effluent (POME) water treatment. Glob J Environ Res 1(2):54–62

Ji CM, Eong PP, Ti TB, Seng CE, Ling CK (2013) Biogas from palm oil mill effluent (POME): opportunities and challenges from Malaysia's perspective. Renew Sustain Energy Rev 26 (2013):717–726

Khalid AR, Wan Mustafa WA (1992) External benefits of environmental regulation: resource recovery and the utilization of effluents. Environmentalist 12(4):277–285

Kim JK, Oh BR, Chun YN, Kim SW (2006) Effects of temperature and hydraulic retention time on anaerobic digestion of food waste. J Biosci Bioeng 102:328–332

Lam MK, Lee KT (2011) Research review paper Renewable and sustainable bioenergies production from palm oil mill effluent (POME): win–win strategies toward better environmental protection. Biotechnol Adv 29(2011):124–141

Liew WL, Kassim MA, Muda K, Loh SK, Affam AC (2015) Conventional methods and emerging wastewater polishing technologies for palm oil mill effluent treatment: a review. J Environ Manage 149:222–235

Ma AN, Augustine Ong SH (1985) Pollution control in palm oil mills in Malaysia. J Am Oil Chem Soc 62:261–265

Menon R (2002) Carbon credits and clean development mechanism. Palm Oil Eng Bull 65:11–14

Menon RN (2007) Dialogue session with the palm oil industry and stakeholders. Palm Oil Eng Bull 83:11–14

MPOB (2007) Malaysian Oil Palm Statistics 2007. Economics & Industry Development Division, MPOB

Oswal N, Sarma PM, Zinjarde SS, Pant A (2002) Palm oil mill effluent treatment by a tropical marine yeast. Bioresour Technol 85:35–37

PalmOilWorld (2011) Malaysian Palm Oil industry, PalmOilWorld.org. http://www.palmoilworld.org/about_malaysian-industry.html. Accessed 3 Sept 2015

Poh PE, Chong MF (2009) Review development of anaerobic digestion methods for palm oil mill effluent (POME) treatment. Bioresour Technol 100:1–9

Sharmila A, Alimon AR, Azhar K, Noor HM, Samsudin AA (2014) Improving nutritional values of Palm Kernel Cake (PKC) as poultry feeds: a review. Malays J Anim Sci 17(1):1–18

Sivasothy K, Shafil AF, Lim NBH (1986) Analysis of the steam consumption during sterilization. Workshop on recent developments in palm oil milling technology and pollution control, Kuala Lumpur, p 19, Malaysian Palm Oil Board, Kuala Lumpur

Subramaniam V, Ngan MA, May CY, Sulaiman NMN (2008) Environmental performance of the milling process of Malaysian palm oil using the life cycle assessment approach. Am J Environ Sci 4:310–315

Tchobanoglous G, Franklin LB, Stensel HD (2003) Wastewater engineering, treatment and reuse, 4th edn. McGraw Hill, New York, pp 96–97

Tong SL, Jaafar AB (2006) POME Biogas Capture, Upgrading and Utilization. Palm Oil Engineering Bulletin, 78:11–17

Vijaya S, Ma A, Choo Y (2010) Capturing biogas: a means to reduce greenhouse gas emissions for the production of crude palm oil. Am J Geosci 1:1–6. doi:10.3844/ajgsp.2010.1.6
Wang J, Mahmood Q, Qiu J-P, Li Y-S, Chang Y-S, Li X-D (2015) Anaerobic treatment of palm oil mill effluent in pilot-scale anaerobic EGSB reactor. Biomed Res Int 2015:398028
Wong KK (1980) Application of ponding systems in the treatment of palm oil mill and rubber mill effluents. Pertanika 3(2):133–141
Wu TY, Mohammad AW, Md Jahim J, Anuar N (2010) Pollution control technologies for the treatment of palm OIL mill effluent (POME) through end-of-pipe processes. J Environ Manag 91:1467–1490
Yacob S, Hassan MA, Shirai Y, Wakisaka M, Subash S (2006) Baseline study of methane emission from anaerobic ponds of palm oil mill effluent treatment. Sci Total Environ 366: 187–196
Yeoh BG (2004) A technical and economic analysis of heat and power generation from biomethanation of palm oil mill effluent. In: Electricity Supply Industry in Transition: Issues and Prospect for Asia Conference, Thailand. Citeseer, pp 14–16
Yilmaz T, Yuceer A, Basibuyuk M (2008) A comparison of the performance of mesophilic and thermophilic anaerobic filters treating paper mill wastewater. Bioresour Technol 99:156–163
Yu HQ, Fang HHP, Gu GW (2002) Comparative performance of mesophilic and thermophilic acidogenic up flow reactors. Process Biochem 38:447–454
Zinatizadeh AAL, Mohamed AR, Abdullah AZ, Mashitah MD, Hasnain Isa M, Najafpour GD (2006) Process modeling and analysis of palm oil mill effluent treatment in an up-flow anaerobic sludge fixed film bioreactor using response surface methodology (RSM). Water Res 40:3193–3208

Landfill Bioreactor Technology for Waste Management

Martina Di Addario and Bernardo Ruggeri

Abstract The content of this chapter focuses on the anaerobic BioReactor Landfill (BRL), an alternative to the traditional "dry tomb" landfill, with the aim of a more rapid degradation of the organic fraction of Municipal Solid Waste (MSW) through leachate recirculation. This technology is mainly based on the moisture increase of waste through liquid injection into the landfill body, as waste degradation is strongly dependent on humidity. The main advantage of bioreactor landfills is the rapid stabilization of the organic fraction, which can be reached in 5–10 years compared to 30–50 years under traditional operation. Biogas production can be therefore increased to higher volumes in less time, improving the efficiency of energy recovery. The main factors regarding MSW biodegradation in BRLs will be presented. Focusing on the different and interconnected phenomena, which take place in such a complex and heterogeneous system, it will be possible to gain an overview of the whole process. A number of studies have shown the positive effects of leachate recirculation on MSW degradation, either at laboratory-scale or on-site. The state of art of both, lab- and full-scale experimentations, will be presented, with a focus on an Italian case study.

Keywords Bioreactor landfill · Bioenergy · Anaerobic digestion · Leachate recirculation · Landfill gas · Municipal solid waste

M. Di Addario · B. Ruggeri (✉)
DISAT, Department of Applied Science and Technology, Politecnico di Torino, C.so Duca degli Abruzzi 24, 10129 Turin, Italy
e-mail: bernardo.ruggeri@polito.it

M. Di Addario
e-mail: martina.diaddario@polito.it

O.P. Karthikeyan et al. (eds.), *Recycling of Solid Waste for Biofuels and Bio-chemicals*, Environmental Footprints and Eco-design of Products and Processes, DOI 10.1007/978-981-10-0150-5_8

1 Introduction

Waste disposal represents a crucial aspect of current societies. Despite the greater awareness towards issues like recycle and reuse, we are still far from a real "recycling society". An efficient waste management plan should be oriented towards prevention, minimization, recycle, reuse and energy recovery. Then there is the landfill, the last necessary step that cannot be forgotten. The current idea of landfill is always related to environmental impacts to atmosphere, groundwater and land. New technologies for bioreactor landfills could reduce these risks and its reputation in popular imaginary imagination could be reconsidered.

Traditional landfills are thought of as a dry tomb where liquid inputs are limited to reduce environmental impact. In such conditions of low water content, the waste takes 30–50 years to decompose completely. Leachate and biogas, the two major pollutants produced by anaerobic decomposition of Municipal Solid Waste (MSW), are captured and removed from the landfill in order to avoid leakages into the surrounding environment. However, the time needed for waste stabilization increases the risks of long-term uncontrolled releases, which can be dangerous for environment and health.

In the last decades, the concept of BioReactor Landfill (BRL) has been introduced as an alternative to the traditional "dry tomb" landfill, with the aim of a more rapid degradation of the organic fraction of MSW. This new technology is mainly based on the increase of waste moisture through leachate recirculation. The waste degradation processes are strongly dependent on moisture content. The liquid injected in the landfill body stimulates microbial activity by promoting redistribution of substrates and nutrients and the diffusion of microorganisms between the micro-environments of the landfill (Sanphoti et al. 2006). The main advantage of bioreactor landfills is the rapid stabilization of organic fraction, which can be reached in 5–10 years compared to 30–50 years of the traditional one (Clement et al. 2010). Pommier et al. (2007) affirm that the water content of MSW greatly affects the methanization process in a landfill, by increasing both the specific microbial growth rate and the bioavailability of the solid substrate. Biogas production can be therefore increased with more volumes generated in less time, improving the energy recovery efficiency. Moreover, long-term environmental impact and post-closure care costs can be reduced (Warith 2002). On the contrary, the construction of bioreactor landfills requires additional operations in the design and management compared to the traditional landfill. The major initial investment and running costs related to the liquid recirculation system can be offset by a number of economic benefits arising from the management of the bioreactor landfill, including the sale of electrical energy from biogas and lower costs for treatment and disposal of leachate (Berge et al. 2009).

From the pioneering work of Pohland (1975), a number of studies have shown the positive effects of leachate recirculation on MSW degradation, either at laboratory-scale or on site. In the works carried out in bench reactors and in lisymeters, the conditions could be optimized in order to promote higher degradation rates, such as: waste shredding, pH control or nutrients addition (Francois et al. 2007; Mali Sandip et al. 2012; Huang et al. 2012). The experimentations on full landfill plant are more difficult to conduct because it is difficult to obtain a homogeneous distribution of liquids and to monitor its effects. Moreover, it is not always possible to generalize the results obtained in situ because each landfill has its own characteristics, both in terms of quality, quantity of waste and management options (Barina 2005).

The aim of this chapter is to provide an overview on waste biodegradation and biogas production through the anaerobic BRL technology. The main factors affecting MSW decomposition will be treated, in order to understand the different and interconnected processes taking place in such a complex system. More attention will be taken in the description of the biological reactions for landfill gas production and how they are affected by leachate recirculation. The state of the art in the lab-scale and full-scale experimentation will be presented, with a more detailed case study of a retrofit BRL in Northern Italy.

2 Main Factors Regarding Biogas Production in BRLs

A landfill is a complex system in which different and interconnected processes take place: biological processes, physico-chemical processes, hydrological and geo-technical behaviour are strictly related to each other. Biological processes play the main role, due to the bio-degradation of the organic matter and biogas production, but they cannot be considered apart from all the other processes.

Landfill biogas is a by-product of anaerobic digestion of the organic fraction of MSW placed within a BRL. Biogas is collected by vertical or horizontal drainpipes and is used to produce heat or energy. The microbes active in anaerobic digestion undertake a complex process involving many classes of bacteria and several intermediate steps of digestion. Four main degradation phases can be identified: hydrolysis, acidogenesis, acetogenesis and methanogenesis. Complex organic matter placed within BRLs must first be hydrolyzed into simpler organics, after which they are fermented to Volatile Fatty Acids (VFA) by acidogens. VFA are then converted into acetate and H_2 gas by the hydrogen-producing acetogens. Finally, the acetate and H_2 gas are converted into CH_4 and CO_2 via methanogens. Under stabilized methanogenic conditions, which is the stage of interest from a beneficial recovery perspective, landfill gas is composed of approximately 55–60 % methane and 40–45 % carbon dioxide with trace amounts of other gases (He et al. 2007). Humic substances such as humic acids and fulvic acids are also formed. These substances are resistant to further degradation. Nitrogen compounds, proteins and amino acids, are decomposed into ammonia, which is hydrolyzed and it is mainly present as ammonium. Denitrifying bacteria also take part to nitrogen degradation by reducing nitrites and nitrates into N_2 gas and N_2O. Sulfur compounds are reduced into H_2S, escaping with the biogas, and HS^-, dissolved in the leachate.

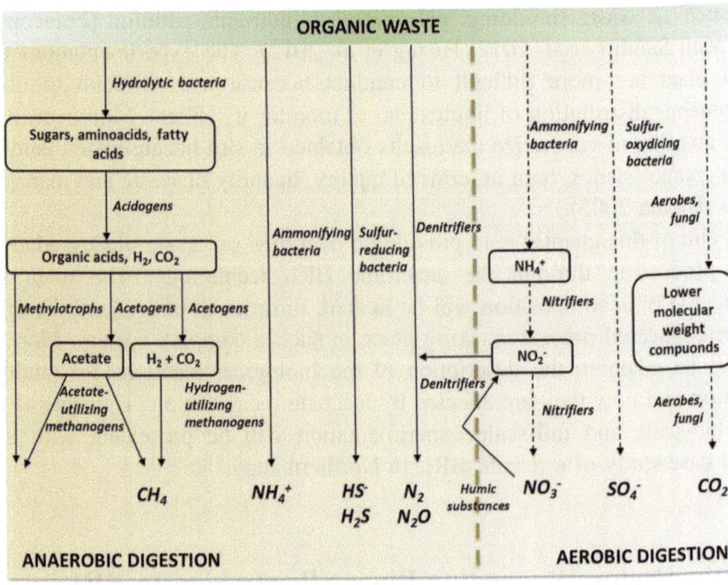

Fig. 1 Schematic diagram of solid waste decomposition with the role of microorganisms in BRLs. *Symbols continuous* line, anaerobic digestion; *dotted* line, aerobic digestion

In addition to anaerobic digestion, aerobic processes can also take place in a BRL. Landfill aeration is more and more chosen as an after-care option to rapidly stabilize refuses and minimize emission potential (Ritzkovski and Stegmann 2011). During aerobic digestion bacteria and fungi transform organic compounds in lower weight substances. CO_2, H_2O and humic substances are produced instead of methane. The higher energy yields of aerobic processes increase microbial growth and reproduction (Sang et al. 2012). Ammonium is converted into nitrites and then nitrates through nitrifying bacteria. Sulfur is oxidized into SO_4^-, thus reducing odorous impacts deriving by H_2S. Figure 1 illustrates the different paths of anaerobic and aerobic digestion, with the role of the different microorganisms involved.

Concerns of global warming related to methane have led to considerable interest in controlling landfill gas emissions into the atmosphere. Energy recovery is one of the most effective methods of controlling landfill gas emissions. It has been widely demonstrated that gas generation rate increases in a BRL compared to a traditional landfill: more LFG can be collected in less time, thus increasing also the energy recovery.

Biogas production is affected by different factors such as, MSW composition and age, density, moisture content, temperature, nutrients, pH and redox potential, inhibitors (VFA, oxygen or ammonium), operational and climatic conditions. Moreover, in a bioreactor landfill, additional factors linked to liquid distribution should be considered. In order to understand the BRL behavior it is important to gain an overview of the whole process, considering the main factors involved and their correlations.

2.1 *Landfill Heterogeneity*

The landfill body is characterized by highly heterogeneous material. Its physico-chemical properties change both spatially and temporally, so that it is difficult to assess a common degradation phase in the entire amount of waste. The main factors affecting the biochemical activity will be presented by describing their changes both in time and space.

2.1.1 MSW Composition

The MSW disposed in a landfill consists of a wide variety of refuse fractions. A merceological characterization classifies the different fractions by material, such as organic and food waste, gardening waste, wood, plastic and rubber, paper and cardboard, textiles and leather, glass and inerts, metals. Another classification, highlighting the waste biodegradability, divides the refuse into three main fractions: rapidly biodegradable (waste with high organic content, food and gardening waste); low biodegradable (materials more difficult to decompose, paper, wood, textiles); non biodegradable (all the remaining materials not taking part to biogas production). The concept of biodegradability is linked to the type of carbon present in the waste: not all organic carbon is "biogasifiable". For example, food waste contains 48 % (on wet basis) of organic carbon, but only 80 % of this can be transformed in biogas (Magnano 2010). The presence of lignocellulosic materials make the degradation harder. Definitively, biodegradability depends on the nature of carbon:

(i) Sugar carbon: easily biodegradable;
(ii) Hemicellulosic/cellulosic carbon: not easily biodegradable;
(iii) Lignocellulosic carbon: hardly biodegradable;
(iv) Fossil carbon: not biodegradable.

Nowadays, with the increase of separate collection, composition trends are changing: some of the fractions are currently recycled, composted or incinerated and only the untreated part is sent to the landfill. Each landfill has its own composition, depending on the waste management strategies followed. For example, most of the developing countries still present portions of food or organic waste higher than 50 % by weight (Sanphoti et al. 2006; Jiang et al. 2007; Mali Sandip et al. 2012). However, differences in waste composition can be found also in a single landfill. The growing phase of a landfill takes years: the bottom layers, where the rapidly biodegradable fraction has been already decomposed, coexist with the upper levels, where fresh waste has been just disposed. In fact, the four degradation phases, previously described, are not subsequent in the whole landfill body, but they can take place simultaneously in different zones.

2.1.2 Density

The typical landfill density is in the range between 500 and 1500 kg m^{-3}. It depends on the type of waste and its particle size: a waste well shredded reduces the voids inside the waste mass and increases compaction. Higher densities promote liquid distribution and the moisture increase among the refuse. Moreover, as landfill compaction increases, so does the available volume for waste disposal, thus reducing operational costs. On the other hand, low void volume and high compaction can be an obstacle for gas movements and biogas collection.

Density varies in the landfill body: spatially because the bottom is more compacted by the weight of the upper layers, and temporally because, as the biodegradation goes on, settling of the waste mass takes place.

In the BRL, additional settlement takes place due to the introduction of liquid into waste. Benson et al. (2007) compared the settling behaviour in a conventional and a bioreactor landfill in North America and stated that, after approximately 3 years waste in the bioreactor settled was 22–25 %, whereas waste in the conventional landfill settled with less than 5 %. Moreover, the rate of settlement in the recirculation landfill also varied with time, with an average rate of approximately 14 %/yr during the first 16 months, and approximately 6 %/yr thereafter. In contrast, waste in the conventional landfill settled at a relatively uniform rate of approximately 1.5 %/yr.

2.1.3 Temperature

Temperature influences both bacterial growth and chemical reactions in a landfill. In anaerobic conditions and with an adequate thermal isolation, temperatures can reach 35–50 °C. Mesophilic bacteria work in the range 30–35 °C, while thermophilic bacteria at higher temperatures of 45–50 °C. Most of the landfills operate in the mesophilic field. Due to the heterogeneity of the waste mass, temperature varies among different zones: the lower levels are less influenced by the external environment, due to the isolation by the above layers. If no proper isolation can be reached, the external environment can affect biodegradation with seasonal variations of biogas production, higher in the summer and lower in the winter.

Different climatic conditions lead to different management and operational strategies for BRLs. Zhao et al. (2008) investigated the effects of the temperature in a cold climate BRL. They affirm that gas production during the winter months was very low or did not occur; suggesting that, in addition to maintaining optimal moisture levels within the waste mass, temperature control must be a key design consideration in cold climates. In the case of tropical landfills (Sanphoti et al. 2006), where high temperatures and evaporation occurs, there are other drawbacks due to the need of supplemental water addition. During the dry season, leachate recirculation may be insufficient to maintain moisture contents and additional liquids are necessary to stabilize moisture levels as well as stimulate biological activity.

2.1.4 pH

Bacteria involved in anaerobic digestion have different sensitivity towards pH. Methanogen bacteria are the most affected by pH variations, while acetogens can operate in acidic environments where methanogenesis is inhibited. Anaerobic digestion is stable for pH in the range 6.5–7.5. pH varies during the different degradation phases. During the acidogenic phase high amount of VFAs cause a decrease of pH. Due to their capability of forming metal complexes, fatty acids increase the mobility of many metals (Lagerkvist and Cossu 2005). Under acid fermentation, leachate pH is about 4–6. In the following phases, as a consequence of the consumption of VFA, the system pH increases and it is buffered at about 7 by the dissolution of CO_2 in the leachate. Neutral pH during the methanogenic phase promotes the immobilisation of metals through the formation of metal hydroxides, sulphides and complexes with organic materials (Lagerkvist and Cossu 2005). There is no evidence of differences between the leachate pH measured in traditional and bioreactor landfills. There could be a sudden change in pH due to the recirculation of high organic waste water or sludge in addition to the leachate itself, but it depends on the amount of liquid recirculated. Liquid waste with an acidic pH ($pH < 4$) should not be introduced to a landfill unless the liquid can be neutralized prior to addition. Liquid waste with a basic pH may be introduced over a large area so that they are readily neutralized. If the buffer capacity of the waste body is not sufficient to establish a neutral environment, pre-treatment of the liquid is needed to adjust pH.

2.1.5 Moisture Content

Water plays a fundamental role in different biodegradation phenomena occurring in a landfill. Water molecules are consumed during the hydrolysis of the polymeric macromolecules. Microorganisms are composed by 80 % of water and their activity takes place in aqueous environments, where substrates and bacteria are present as solutes. Optimized Moisture Content (MC) can enhance MSW degradation, as it provides an aqueous environment containing the necessary nutrients and microbes and improves mass transfer in the landfill cells (Hao et al. 2008). MC mainly depends on the waste composition. It ranges from 60 % by weight for food residues and 1 % for non-biodegradable waste. Waste composition also affects the liquid distribution in the landfill body, as every fraction could be more or less hydrophobic. The initial moisture content could limit biological reactions, as it is not enough to guarantee nutrients and microorganisms distribution in all the micro-environments of the landfill. Depending on the initial water content, biological reactions may be strongly water limited. The threshold value, under which no biodegration can be observed, ranges from 0.15 to 0.50 kg water/kg dry waste. Generally, the initial moisture content values range between 0.25 and 0.65 kg water/kg dry waste (Pommier and Lefebvre 2009). Several authors suggested that moisture content close to field capacity (approximately 50 % w/w) may be optimal for landfilled waste, resulting in faster methanisation and three- to four-folds higher methane yields than at 30–40 % moisture content (Bayard et al. 2009).

In addition to its influence on biodegradation, the amount of moisture also influences liquid and gas transport: when water occupies a large fraction of the pores within solid waste, the resistance to water flow is lower, but the movement of gas through refuse is hampered. Moreover, MC influences also geo-technical stability, due to the higher weight of the waste mass and the higher local water pressure (Imhoff et al. 2007).

The optimization of moisture content should consider the liquid transfer among the refuse and the humidity needed for biologic reactions. Generally, at landfill-scale, only macro-porosities among the refuse are considered for the liquid transport. However, biological reactions occur at the lower level, the micro-porosities of organic particles. Therefore, also the necessary liquid, able to penetrate micro-pores, should be considered. The minimum quantity of liquid to recirculate should be estimated through the determination of the liquid already present in the waste and the additional quantity that the waste can retain. This liquid quantity, as the other properties described above, varies in space and time in the landfill body: it decreases with increasing density; it decreases with depth; it increases with decreasing particles size; it decreases with time due to waste degradation.

A distinction needs to be considered between Field Capacity (FC) and Water Holding Capacity (WHC). FC is the maximum moisture content reached by the waste, before the liquid enters the gravitational field. Generally, once the FC is reached, leachate production starts. WHC is not a moisture measurement, but it is the quantity of added liquid that the waste can retain not only in micropores (capillary water), but also in macropores. Therefore, the amount of liquid at maximum WHC can be higher than necessary to reach FC. An example of WHC estimation will be illustrated in the case study presented after.

2.2 Recirculation Modes

The choice of a proper recirculation mode represent a crucial aspect of BRLs design and it requires understanding of moisture movements within the waste matrix. An ideal injection system should: distribute the liquid uniformly throughout the waste, resulting in minimal increases in local pore water pressures, and be easy and

Table 1 List of the main recirculation systems for BRLs

Injection systems		Advantages	Disadvantages
Pre-capping surface methods	Pre-wetting Spray irrigation Drip irrigation Surface ponds	– Applied during waste disposal – Simple – Cheap – Efficient wetting	– Incompatible with closure – Odours
Post-capping subsurface methods	Vertical injection wells Horizontal trenches Permeable blankets	– Compatible with closure – Large volume of liquids – Liquids added under pressure – Better moisture distribution	– More difficult construction and maintenance – Costs

economical to install in new or retrofitted BRL (Imhoff et al. 2007). The choice of the proper injection method depends on different factors: the state of the landfill (as built or retrofit), the source of liquid, the available equipment, the landfill management choices, and also costs and current regulations. Predominantly used injection systems are listed in Table 1, sorted by the operational phase in which they can be installed, i.e. before or after the final capping.

Horizontal distribution lines buried in trenches filled with gravel are the most common method for leachate distribution (Benson et al. 2007; Knox et al. 2007; Vigneron et al. 2009; Oonk et al. 2013; Abichou et al. 2013). Vertical wells allow the operator to retrofit landfill areas that are already filled with waste and minimize the interference of their construction with routine landfill operations (Pradeep et al. 2014). Benson et al. (2007) reported that vertical injection lines and infiltration galleries are less effective than horizontal trenches. However, there is still a lack of regulation and a clear scientifically based conceptual design, which includes appropriate infrastructure to achieve the prefixed objectives (Knox et al. 2007).

Together with the proper injection system, in order to maximize waste stabilization, leachate recirculation rate and frequency must also be carefully selected. It is recommended that leachate should be introduced slowly, since high flow rates may deplete buffering capacity and remove methanogens (Šan and Onay 2001).

Recirculation regimes can be adapted to the different degradation phases with lower rates in the acidogenic phase, in order to avoid flushing of organic matter (Jiang et al. 2007). Once gas production is established, flow rates and frequency can be increased. Moreover, periodic liquid injections are preferred to waste saturation, because they assure liquid flow among the refuse.

In order to enhance waste decomposition and biogas production, not only higher water content is needed, but it is also important that the liquid could flow through the refuse and distribute nutrients and microorganisms (Bayard et al. 2009). Since MSW is a highly heterogeneous material, it is challenging to obtain a uniform liquid distribution avoiding preferential pathways and dry spots. Moreover, due to the higher weight of the landfill body, settlement and mechanical compaction can occur. Also operational problems such as flooding or clogging should be taken into account.

A number of studies evaluated optimal injection modes for different types of waste and recirculation rates range from 2.7 to 30 % of waste volume per day (Hao et al. 2008). Such a wide range confirms the difficulty in generalizing results obtained for different landfills.

2.3 *Ammonium Accumulation*

A crucial aspect of leachate recirculation is the accumulation of substances which cannot be consumed by microorganisms, mainly ammonium. NH_4^+ is produced through the degradation of organic nitrogen (Price et al. 2003) and it tends to accumulate since there are no removal mechanisms under strict anaerobic conditions. An ammonium content of 1500–3000 mg/L can inhibit methanogenesis (Francois et al. 2007).

NH_4^+ can be removed from leachate via ex situ treatments such as physico-chemical methods (air stripping, activated carbon adsorption, filtration, ion exchange, precipitation) and biological methods (aerobic and anaerobic treatment). The most common method is the biological treatment through nitrification followed by denitrification, due to its lower costs. However, these approaches are likely to produce NO_x and N_2O, which are significant pollutants and contribute to climate change.

Several laboratory studies demonstrated the feasibility of in situ ammonium removal. Onay and Pohland (1998) developed a system to simulate a BRL divided into three zones (anoxic, anaerobic and aerobic) for nitrification/denitrification treatment inside the landfill body. Their results showed a nitrogen removal up to 95 %. Tallec et al. (2007) combined ex situ leachate nitrification with in situ denitrification, i.e. the liquid recirculated is enriched with nitrates which can be reduced by anaerobic microorganisms among the waste. However, this operation causes higher emission of N_2O.

A novel method investigated in more recent studies is the anaerobic oxidation of ammonia-nitrogen through ANAMMOX (ANaerobic AMMonium OXidation). This is an autotrophic process which involves a complete conversion of ammonium to nitrogen gas without the addition of organic matter (Sri Shalini and Joseph 2012). In fact, ANAMMOX bacteria use ammonium as the electron donor and nitrite as the electron acceptor, i.e. a previous partial nitrification is needed to provide nitrites. It has been demonstrated that the intrusion of small quantities of O_2 into the landfill body can trigger in situ nitrification and promote the growth of ANAMMOX populations, thus contributing to nitrogen removal from the solid matrix (Valencia et al. 2011; Lubberding et al. 2012).

2.4 Waste Stabilization

BRLs have been suggested as a more sustainable alternative to traditional landfilling, but they should be designed from the beginning with the aim of reaching the final storage quality in the proper time (Valencia et al. 2009). It is important to define the main parameters assessing waste stabilization, that is when waste residues reach the same characteristics as those materials in their surrounding environment without the potential to produce pollution in the short, mid and long-term. As described above, the removal of leachable organic carbon can be optimized and accelerated by operating a landfill as a bioreactor. However, as bioreactor landfills mature, organic content of solid waste is enriched with compounds that are non-biodegradable by nature, such as xenobiotic substances and humic substances (Batarseh et al. 2010). In order to achieve an inert residual material, alternative solutions that might change the form in which heavy metals and salts are present in the residues and their ability to leach must be implemented.

At the closure stage of bioreactor landfills: leachate quantity will be a finite amount, with limited need for off-site transfer, treatment and/or disposal, LFG generation will be at its declining stage and long-term environmental risk will be minimized (Warith 2002). Cossu et al. (2003) proposed the PAF model, a combination of mechanical-biological Pretreatment, Aeration and Flushing. Batarseh et al. (2010) studied the effects of flushing combined with in situ aeration to produce stable solid waste cells within a short time periods after the bioreactor operation is no longer significantly reducing the organic content of leachate. Landfill aeration contributes towards an accelerated, controlled and sustainable conversion of conventional anaerobic landfills into a biological stabilized state associated with a minimized emission potential (Ritzkovski and Stegmann 2011; Sun et al. 2014). Moreover, existing equipment can be used, as air can be injected into the landfill with the same devices used for extracting gas or injecting leachate (Bilgili et al. 2007).

A number of technical criteria for assessing waste stabilization have been proposed in several studies. Morris et al. (2003) selected five parameters: pH and specific conductance, dissolved organic matter (BOD and BOD/COD ratio), and inorganic macro-components (ammonia). Ponsa et al. (2008) affirm that biochemical parameters such as volatile solids (VS), total and dissolved organic carbon (TOC, DOC) and chemical oxygen demand (COD), lack precision when determined on heterogeneous materials such as MSW. They chose stability indices related to aerobic (respiration index, RI) and anaerobic conditions (biochemical methane potential, BMP) in order to evaluate biological stabilization of the organic fraction of MSW. Cossu et al. (2012) outlined the main limits of RI: high cost of the respirometers; long testing time; low representability of the parameter in presence of toxic or inhibiting substances, which can alter the oxygen consumption, or in the presence of inert organics (such as plastic or paper) which can dilute the biodegradable matrix resulting in lower respiration indices. In order to overcome such limits, they proposed a standardization of BOD/COD ratio as a biological stability index for MSW.

3 Lab-Scale and Full-Scale Experimentations

From the 1970s until today, a number of studies have shown the positive effects of leachate recirculation on MSW degradation. In the works carried out in laboratory the operating conditions can be optimized in order to promote degradation and accelerate it. Landfill experimentations are more difficult to optimize because the non-homogeneous distribution of liquids and its effects are difficult to monitor. Moreover, it is not always possible to generalize the results obtained in situ because each landfill has its own characteristics, both in terms of quality and management. The results of different studies will be presented to outline the role of leachate recirculation in enhancing biogas production. More attention will be paid in the presentation of a case study, carried out at Politecnico di Torino to simulate a retrofit BRL in Northern Italy.

3.1 Experimentations on Lab-Scale Bioreactors and Lysimeters

The experimental devices used to simulate BRLs have to respect, as much as possible, the following criteria: anaerobic environment in a closed system, mesophilic temperature, leachate recovery and recirculation, biogas monitoring and recovery. In such devices the waste is previously shredded in order to increase the contact surface and its exposition to microbial activities. Due to the lower volumes, liquid distribution can be homogeneous, MC can be easily increased and FC can be reached in short times. Moreover, it is possible to test and compare different operational options, such as:

- pH adjustment with a basic medium to contrast VFAs inhibition (Warith 2002; Sanphoti et al. 2006; Francois et al. 2007; Elagroudy et al. 2009);
- nutrient addition, generally nitrogen, phosphorous and micronutrients (Warith 2002; Mali Sandip et al. 2012);
- sludge addition to enhance methanogenic population (Warith 2002; Sponza and Agdad 2004; Mali Sandip et al. 2012);
- gravel layers among the refuse to improve hydraulic conditions and contact surface (Mali Sandip et al. 2012);
- inoculum of aerobic cultures and/or aeration to avoid VFA accumulation (O'Keefe and Chynoweth 2000; Erses et al. 2008; Sun et al. 2014; Mali Sandip et al. 2012).

It has been widely demonstrated that at lab-scale methanogenesis is reached in short times and fresh waste can be stabilized in approximately 300–400 days. Table 2 reports data of methane production from different laboratory studies simulating leachate recirculation. They confirm a wide range of methane yields, depending on the different operational conditions adopted.

Table 2 Cumulative Methane Production data from laboratories studies

Authors	Characteristics	Cumulative methane production (L/kg dry waste)	Days of operation
Šan and Onay (2001)	76 % organic waste Solo leachate recirculation	78	270
Sanphoti et al. (2006)	29 % organic waste Water addition	55	180
Francois et al. (2007)	29 % organic waste 8 years old waste Sludge addition	43	400
Francois et al. (2007)	29 % organic waste Sludge addition	73	400
Erses et al. (2008)	45 % organic waste Solo leachate recirculation	158	630
Bayard et al. (2009)	40 % organic waste Solo leachate recirculation	185	170
Mali Sandip et al. (2012)	69 % organic waste Initial aeration Gravel mixing Sludge addition	36	270
Di Addario et al. (2014)	15 % organic waste Low biodegradable waste Solo leachate recircualtion	72	750

3.2 *Full-Scale Experiences*

The construction of bioreactor landfills requires additional operations in the design and management compared to traditional landfill: gas production must be managed intensively so that the increased gas is collected; liquid injection must avoid surface seeps; liquid must be compatible with refuse decomposition; liners and leachate collection system must be enhanced and constructed with conservative design; geo-technical stability must be ensured through proper slope and material selection (Barlaz and Reinhart 2004). However, the major initial investments and costs related to the liquid injection can be offset by a number of economic benefits arising from the management of the BRL, including the use of biogas for electric power and lower costs for treatment and disposal of leachate (Berge et al. 2009). Even if the positive effects of leachate recirculation have been widely demonstrated, there is still a lack of regulations and guidelines on BRL design and management. Most of the current regulations generally still encourage landfills to remain dry.

Reviewing different full-scale BRLs experiences, it is possible to obtain useful indications. Some examples of anaerobic BRLs are reported in Table 3, by comparing their main findings. Other useful information can be in the Yolo County

Table 3 Main findings from different full-scale BRLs experiences

Landfill	Location	Main findings	Methane production
Trail road landfill (Warith 2002)	Ontario, Canada	– 25 % air space recovery – decreased organic load	Not reported
CSWMC landfill cells (Morris et al. 2003)	Delaware, USA	– Accelerated decrease of contaminants – Increased rates of settlements	Accelerated
Landfill Q (Benson et al. 2007)	Northeast American region	– Same leachate quality of conventional landfill	Increased and accelerated
Landfill S (Benson et al. 2007)	Upper Midwest American region	– Increased rates of settlements	14 % more CH_4 than the conventional landfill
La Vergne (Vigneron et al. 2009)	Nantes, France	– Lower fugitive gas emissions	Increased
Loches (Vigneron et al. 2009)	Tours, France	– Acceleration of settlements in the first years	Increased
Landgraaf (Oonk et al. 2013)	Netherlands	– Anaerobic conversion completed after 4 years of recirculation – Preferential paths of liquid flow – Positive effects of post aeration	Increased and accelerated
Asan (Chung et al. 2015)	Korea	– Positive effects of ex-situ nitrification + in situ denitrification – Accelerated waste stabilization	Increased and accelerated

Project XL Bioreactor in California and the Florida Bioreactor Demonstration Project, which have been widely disseminated on their websites http://www.yolocounty.org/recycle/bioreactor.html and www.bioreactor.org.

3.3 *Case Study: Experimental Simulation and Prediction of Landfill Biogas Production from Low Biodegradable Waste Under Leachate Recirculation*

The aim of this study was the to evaluate the effects of leachate recirculation on Low Biodegradable Fraction (LBF) degradation, through an experimental and a modelling approach The waste used for the lab-scale experimentation was collected from the Cerro Tanaro landfill, located in Asti (Italy). In Cerro Tanaro, a full scale leachate recirculation system is under construction, financed by the EC with Life+ Program, in order to conduct full-plant tests. The project, called Bio.Lea.R. (Biogas Leachate Recovery), aims to the optimization of landfill management with a retrofit bioreactor landfill technology (http://biolear.eu/en/). Through the experimental simulation at laboratory-scale, we tested under optimized conditions how much the MSW already present in the landfill could benefit from the moisture increase. The MSW tested was 5 years old on average and it had already lost its RBF. A Trickle-Bed reactor was used: it is a three-phase reactor where the liquid moves downward in a closed loop, the gas, produced from waste decomposition, moves upward and it is collected at the top of reactor, while the solid bed consists in the mass of pre-treated municipal solid waste. Both liquid and gas were analysed in order to monitor the degradation trends. The experimental results were fitted with the Gompertz equation to estimate the Bio Methane Potential and kinetic parameters characterizing methane production.

3.3.1 Materials and Methods

Experimental Device

A schematic representation of the experimental apparatus is showed in Fig. 2. The reactor, placed in a thermostatic room at 31 °C, consists of a glass column of 35 cm height, 20 cm outer diameter and 18.2 cm inner diameter. At the top of the reactor there is a glass dome with a central inlet for the liquid. The liquid is distributed on the bed of waste through a plate with 16 holes of 19 mm diameter. The mass of

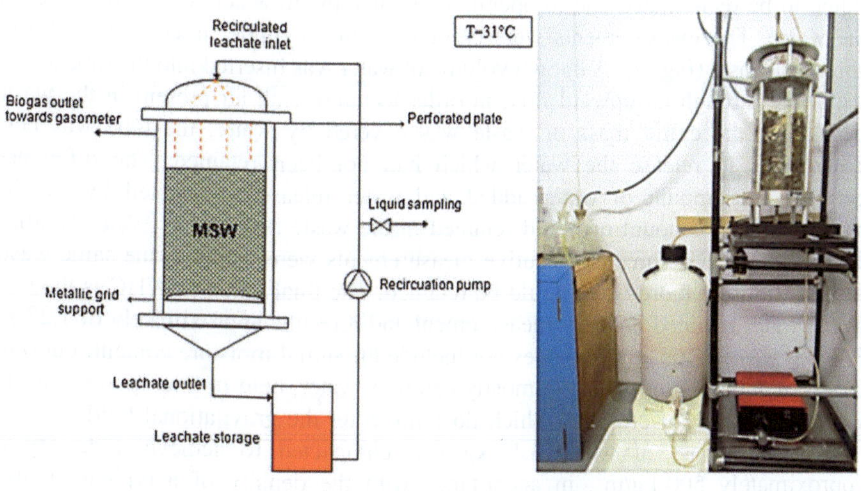

Fig. 2 Schematic representation and photo of the experimental equipment

waste 3.12 kg was placed on a metal grid supported by a spacer made of plastic material, to avoid that the waste could block the liquid outflow on the bottom; the bottom has a slight slope to avoid leachate stagnation. The leachate output is collected in a 25 L tank, sealed to maintain anaerobic conditions. A volumetric pump with a nominal capacity of 1 L/min is used to recirculate the liquid from the tank to the head of the reactor. The pump has an adjustable flow rate from 10 to 100 % of nominal capacity. The operation frequency of the pump is regulated by a timer. Along the tube for leachate recirculation liquid samples can be collected and analysed according to the analytical methodology described below. The biogas outlet is located on the top of the reactor and the cumulative biogas production is measured with the water replacement method.

Waste Preparation

The bioreactor was filled with 3.12 kg LBF, constituted by mixing together 5 samples taken from different areas and depths of the landfill. Cerro Tanaro landfill has been in service since 2004, thus it is assumed that the rapidly biodegradable fraction has been previously decomposed in the landfill, i.e. the waste placed in the reactor was constituted by 42 % w/w LBF and 58 % Non-Biodegradable Fraction (NBF). The waste was shredded with a kitchen blender to a size of 1–5 cm range, in order to increase the contact surface, accelerate the degradation by reducing mass transfer phenomena and facilitate compaction. The initial moisture content in the waste was measured by weighing different LBF samples before and after putting them in an oven at 105 ± 5 °C for 24 h. The average initial moisture content of the MSW used was 40.72 ± 7.48 % w/w.

Moreover, the Water Holding Capacity (WHC), as weight of retained water on the weight of refuse, was evaluated in order to assess the minimum quantity of liquid to be recirculated and to operate at the maximum water content possible for the waste. The measurements were carried out in a cylinder filled with 350 g of shredded waste (Fig. 2). A known volume of water was inserted into the tube to wet the waste through an upward flux, in order to remove all air present in the waste bed. Once the entire mass of waste was covered by water, the tube was bent downwards to release the water which had not been retained. The difference between the amount of water added and water released, evaluated by weight, represented the amount of liquid retained in the waste bed, i.e. its Water Holding Capacity (WHC). Three consecutive measurements were made on the same waste sample, until no more water could be retained. The final value of WHC is the sum of the water retained in every measurement and it results approximately of 1.27 kg H_2O/kg waste. This quantity does not include the initial moisture content, but only the extra added liquid which is mostly capillary water, held in micropores, and the water retained in macropores which does not enter the gravitational field.

The total mass of waste 3.12 kg was compacted to achieve a density of approximately 500 kg/m^3, in accordance with the density of a typical landfill, which is around 500–700 kg/m^3.

Leachate Injection Modes

At the beginning of the test the leachate storage tank was filled with 11 L of tap water. This quantity of liquid was recirculated in the reactor for the entire duration of the test, without further additions, in order to avoid to open the tank and to prevent possible air infiltrations into the system. Frequency and flow rate of recirculation were higher at the beginning of experimentation in order to obtain a quick saturation of the waste and thus reduce the time required to start the degradation. The liquid level in the tank decreased due to the quantity of liquid retained by the waste after recirculation. At approximately day 50th, no further reduction of liquid the level in the tank was noticed: the liquid entering the bioreactor was no more adsorbed by the solid mass and we assumed that WHC was reached. Nevertheless, it was necessary to maintain recirculation, at lower frequencies and flow rates. From day 57, it was noticed that the gas produced was not able to move upward and exit from the top of the reactor. This was due to the compaction of the refuse, due to the greater weight of the saturated MSW, which caused biogas entrapment within the solid mass and, during leachate recirculation, biogas bubbles were carried downward by the higher flow rate of the liquid. Owing to this problem, instead of liquid recirculation, another solution was adopted in order to inject liquid and collect biogas at the same time: the solid bed was flooded once a week so that, while the liquid level increased through the bed, the gas could flow upstream and be collected in the gasometer. The operation of flooding was accomplished by closing with a clamp the tube for the leachate bottom outlet and then switching on the recirculation pump. Once the waste height was flooded and all gas was collected, the bottom outlet was opened and the liquid could flow back into the storage tank.

Analytical Method

The liquid samples were collected on the recirculation tube, by a proper valve, and the following parameters were analyzed: pH with a CRISON pH meter (MicropH 2001 model); Red-Ox potential, ROP (mV), and ammonium, NH_4^+ (ppm), with a HANNA Multiparameter meter (HI 9829); Chemical Oxygen Demand, COD (mg O_2 L^{-1}), determined according to the Italian standard method IRSA-CNR 29/2003 vol. 2 n. 5130; Biochemical Oxygen Demand, BOD (mg O_2 L^{-1}), according to the Italian standard method IRSA-CNR 29/2003 vol. 2 n. 5120. The measured parameters provide the information needed to follow LBF decomposition: an acidic environment is unfavorable to methanogenesis; ROP showed if the environment is able to support methanogenesis; COD is an indirect measurement of the total oxidisable organic content, while BOD takes into account only the biodegradable one; the trend of ammonium concentration shows accumulation phenomena typical of bioreactor landfills, where ammonium cannot be consumed during decomposition. The biogas collected in the liquid gasometer, as described above, was sampled to analyze its composition by an off-line gas chromatographic analysis device (Varian, CP 4900) equipped with two columns: a molecular sieve type for H_2, CH_4,

CO, O_2 and N_2 (95 °C injection temperature, 200 kPa, Helium as the carrier), and a Poraplot U column for CO_2 determination (85 °C injection temperature, 200 kPa, Argon as the carrier).

Cumulative Methane Production Modelling

The kinetic Gompertz model equation, in Eq. (1) (Mali Sandip et al. 2012), was used to predict the maximum cumulative volume of methane in the experimental simulation:

$$BM = BMPexp\left\{-exp\left[\frac{R_m e}{BGP}(\lambda - t) + 1\right]\right\} \quad (1)$$

where BM is the cumulative methane production per mass unit of waste mass (NL kg^{-1}) at time t, BMP is the maximal methane production (NL kg^{-1}), R_m is the maximum methane production rate (NL kg^{-1} d^{-1}), e is the Euler constant (2718), λ is the lag phase (d) and t is the time (d).

The Gompertz model was used for curve fitting using a non-linear best-fit procedure of experimental data, in order to estimate the characteristic parameters of the methane production curve: BMP and the two kinetic constants R_m and λ. Equation (1) were used under the assumption that the system is spatially homogeneous. Even though the waste mass was not homogenous, this assumption can be acceptable due to the small volume of the reactor and the reduced particles size compared to the full-scale landfill. Using the Gompertz equation, it is possible to calculate half-transformation time, $t_{1/2}$, i.e. the time needed to produce 50 % of maximum biogas production. A landfill can be thought of as a non-renewable resource, such as an oil field or an ore deposit, and it is extremely important to estimate its $t_{1/2}$ in order to predict its behaviour over the long term.

The prediction of biogas production in the landfill is more complex than for the reactor, because there are many different parameters to take into account, such as waste physical properties, environmental and hydrological variable conditions, landfill management options, some unpredictable inhibitory phenomena and/or some synergistic ones. For example, the presence of sulfate could generate a different biological pathway of electron acceptor and influence methanogenesis adversely (Gurljala and Sulfita 1993; Ruggeri et al. 2015). The difficulties in collecting all the data needed increase model uncertainty, due to the variation of parameters both spatially and temporally and the inaccuracies in their measurement. Generally, the most used is the first-order generation model (Amini and Reinhart 2011), as the zero-order outcomes present relatively high errors while the second-order has a more complicated procedure which is not justified by the increase in accuracy.

The first order model for methane production in landfills is:

$$Q = ML_0 \, k \, e^{-k(t-t_l)} \quad (2)$$

where Q is the methane generation (m^3 $year^{-1}$), M is the disposed waste (ton), L_0 is the methane generation potential (m^3 ton^{-1}), t is the time after waste placement (year), t_l is the lag phase between the placement and the start of generation, k is the first-order rate constant ($year^{-1}$).

Usually, models like Eq. (2) are intended for single batches or single years, every batch has to be summed for all the years desired in order to build the whole gas generation curve, valid for the entire landfill. The first-order model has two main adjustable parameters: the methane generation potential L_0, and the methane generation rate constant k ($year^{-1}$). Both of them can be defined through lab-experimentation, pilot-scale cells or ranges present in the literature, in order to obtain a best fit to field data and minimize residual errors between the predicted and the experienced methane recovery. However, in the case of a retrofit bioreactor landfill with insufficient experimental data and lack of literature guidelines, it becomes challenging to predict the effect of leachate recirculation and moisture increase on CH_4 production without introducing a large degree of uncertainty.

3.3.2 Results and Discussion

Leachate Quality

Monitoring of pH showed that the initial environment is basic reaching neutral values around pH 7. It is important to point out that the pH development from the alkaline range towards neutral values, indicates good development conditions for mathanogen bacteria consortium activities.

ROP is an important parameter to monitor degradation conditions, because it indicates if the medium is able to favour reducing reactions necessary for methane production. Initially ROP showed positive values, indicating the presence of oxygen. From day 57, ROP decreased to negative values, the environment became reducing and methane production started. Ammonium concentrations showed a decreasing trend. The decreasing trend was due to the activity of aerobic nitrifying bacteria, which converted ammonium to nitrite and then nitrate in presence of oxygen. Once reducing reconditions have been established, no mechanisms exists for oxidising NH_4^+. Ammonium concentrations reached 5.5 ppm at the end of the test, without showing an accumulation phenomena with concentrations sufficiently low to not inhibit methanogenesis, i.e. 1500–3000 mg L^{-1}.

In Fig. 3, the trends of both COD and BOD during the test are reported. COD trends indicated the consumption of organic and inorganic substances able to be oxidized present in the leachate. At the beginning more intense degradation kinetics occurred, then, when the ROP established at quasi steady-state conditions, the slope decreased because organic substances left were more xenobiotic being more

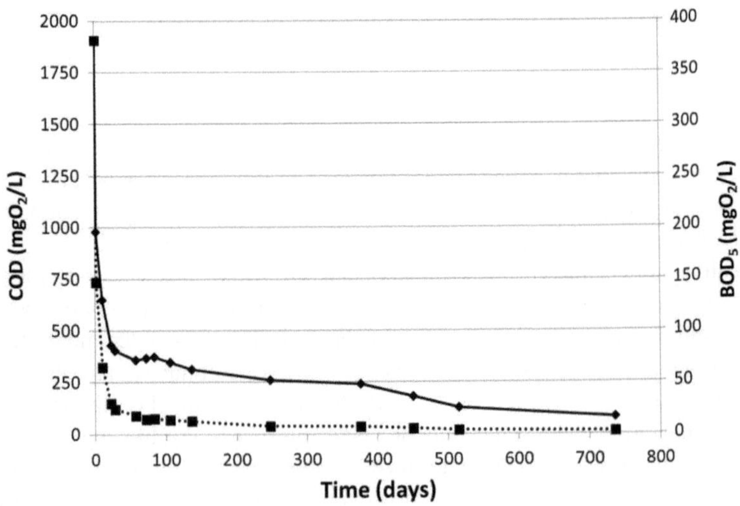

Fig. 3 Leachate COD and BOD5 variation with time

difficulties to be decomposed. The BOD_5/COD ratio was 0.2 at the beginning, confirming that the waste used is a low biodegradable waste with low organic content.

Biogas Production

Gas production was analysed both in quantity and quality, through measurements of volume and concentration, as previously reported. No biogas, neither CH_4 nor CO_2 was produced until day 15, as marked indication of biological activity, due to waste composition, free of rapidly biodegradable fraction. However, making a comparison with the typical times needed in a landfill to start the biodegradation process, an acceleration of kinetics can be noticed, probably due to the waste shredding and the higher quantity of water present, which is an indispensable vector for proteins and enzymes able to attack organic waste.

Biogas composition data are showed in Fig. 4. Methane production began on day 57. Even if landfill biogas has a typical concentration of around 50 % *v/v* CH_4, the reactor used for this simulation never reached such percentages. Methane concentration reached its maximum of 33 % *v/v* at day 133. These results agree with those of Huang et al. (2012) where methane concentration reaches 33 % *v/v*, but using "fresh" waste containing Readily Biodegradable Fraction (RBF) too, different to that used in the present study, which only constituted LBF. CO_2 concentrations were always lower than CH_4. This is due to partial carbon dioxide dissolution in the leachate at neutral pH values. Moreover, we propose that,

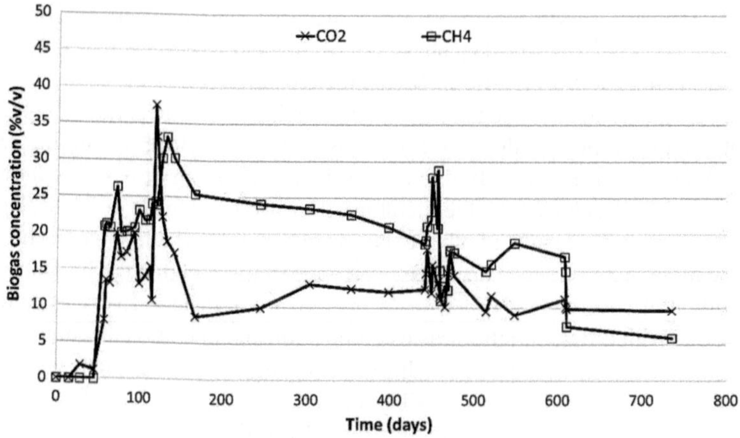

Fig. 4 Biogas composition with time

between the methanogen bacterial consortium, the activity of methanogens using H_2 and CO_2 prevailed over acetate-using methanogen bacteria (Deublein and Steinhauser 2008). Probably in this case, the contribution of acetate-using methanogens was low, as confirmed by very low H_2 concentration in the biogas.

Methane Production Modelling

The saturation Gompertz model of Eq. (1), was used to fit the experimental data of methane production obtained in the laboratory-scale bioreactor. It was possible to find the characteristic parameters of methane-specific production as reported in Fig. 5, where Gompertz simulation together with experimental data are shown. Bio

Fig. 5 Prediction of methane specific productivity with Gompertz equation and experimental data

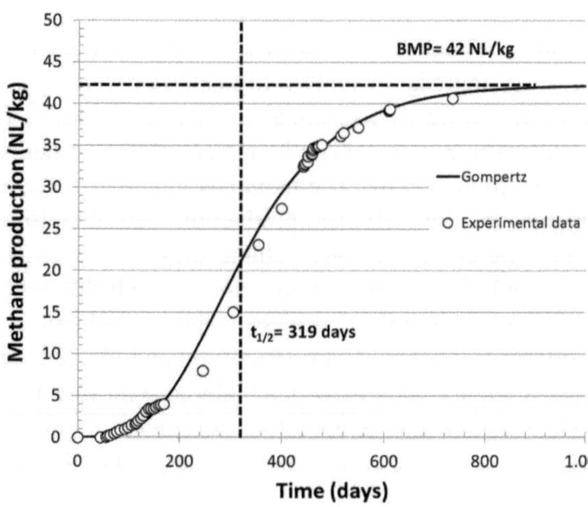

Methane Potential (BMP) 42.25 NL kg^{-1}, maximal daily methane production rate (R_m) 0.12 NL kg^{-1} d^{-1} and lag phase (λ) of 148 days were estimated by best-fit procedure using the Solver tool in Microsoft Office Excel.

The curve obtained can be used in order to estimate kinetic parameters useful to evaluate degradation trends; $t_{1/2}$ time was 319 days, BMP could be reached approximately after 3 years. By expressing the BMP per kg of dry mass and comparing it with values present in the literature, the estimation made of 72 NL kg^{-1} dry mass, agrees with the results of other experimental studies which range from 185 NL kg^{-1} dry mass to 36 NL kg^{-1} dry mass. The wide range found in the literature is due to the different choices made in term of operative parameters. Every experimental study works with different initial waste composition and operational conditions, such as nutrients or sludge addition, pH adjustment or high organic content. In this work we wanted to follow as much as possible the conditions of the Cerro Tanaro landfill, where leachate is recirculated, without external additions.

One of the main limits of the experimental simulation is that it depicts an optimal and ideal condition, such as waste shredding, uniform liquid distribution and high moisture content. The final moisture content of the MSW in the reactor was 59.8 % w/w. Such conditions can be hardly reached in full-scale landfills. However, this approach remains useful in order to obtain a quick response of the maximum beneficial effects of water content increase. Moreover, it permits to obtain some kinetic parameters that can be used in the full-scale landfill modelling.

Finally, the experimental simulation of the CerroTanaro landfill showed that leachate recirculation has benefits even in the case of MSW with low biodegradable organic content. Despite the difficulty in degrading this type of waste, due to the increase of moisture content, it was possible to reduce the times required to reach methanogenesis.

4 Summary and Conclusions

Traditional landfill technology is obsolete nowadays, due to the long times required for waste degradation and the high environmental risks. Bioreactor landfill is a valid alternative towards a sustainable disposal of MSW. Through leachate recirculation, waste moisture content can be increased, thus establishing a favorable environment for microbial activity. The main advantage of BRLs is the rapid stabilization of the organic fraction. Biogas production can therefore be increased in less time, improving energy-recovery efficiencies.

A landfill is a complex system in which different and interconnected processes take place: biological processes, physico-chemical processes, hydrological and geo-technical behavior are strictly related to each other. Moreover, it is characterized by highly heterogeneous material with physico-chemical properties changing both spatially and temporally. For these reasons an overview of all processes involved is needed to understand and predict the bioreactor landfill behaviour.

A crucial aspect of BRLs design is the choice of a proper recirculation mode, which requires understanding of moisture movements within the waste matrix. Several studies tested different recirculation equipment and different injection rates. The optimal recirculation rates range from 2.7 to 30 % of waste volume per day. Such a wide range confirms the difficulty in generalize the result obtained for landfill working at different operating conditions, each one with its own characteristics, both in terms of quality and management.

Experimentations both in on laboratory- and full-scale confirmed the advantages of BRLs in increasing bioenergy recovery from MSW, compared to the conventional landfills. The case study presented showed that leachate recirculation is able to enhance methane production, even in the case of a retrofit BRL with low degradable organic content.

Even if the positive effects of leachate recirculation have been widely demonstrated, there is still a lack of regulations and guidelines on BRL. Efforts are needed to develop a scientifically based conceptual design, which includes appropriate equipment and operational choices to achieve the prefixed objectives of energy recovery and waste stabilization. In that context, BRL modelling will be an important tool to simulate gas and leachate production and optimize the time and cost of operating bioreactor landfills.

References

Abichou T, Barlaz MA, Green R, Hater G (2013) Liquid balance monitoring inside conventional, Retrofit, and bio-reactor landfill cells. Waste Manage 33:2006–2014

Amini HR, Reinhart DR (2011) Regional prediction of long-term landfill gas to energy potential. Waste Manage 31:2020–2026

Barina G (2005) Leachate recirculation at large scale: investigation on design and waste degradation. In: International Workshop "Hydro-Physico-Mechanics of Landfills", LIRIGM, Grenoble 1 University, France, 21–22 Mar 2005

Barlaz MA, Reinhart DR (2004) Bioreactor landfills: progress continues. Waste Manage 24:859–860

Batarseh ES, Reinhart DR, Berge ND (2010) Sustainable disposal of municipal solid waste: post bioreactor landfill polishing. Waste Manage 30:2170–2176

Bayard R, Benbelkacem H, Zhang Y, Gourdon R (2009) Impact of leachate injection modes on landfill gas production. In: CISA Publisher (ed) Proceedings Sardinia 2009, twelfth international waste management an landfill symposium, S. Margherita di Pula, Cagliari, Italy, 5–9 Oct 2009

Berge ND, Reinhart DR, Batarseh ES (2009) An assessment of bioreactor landfill costs and benefits. Waste Manage 29:1558–1567

Benson CH, Barlaz MA, Lane DT, Rave JM (2007) Practice review of five bioreactor/recirculation landfills. Waste Manage 27:13–29

Bilgili MS, Demir A, Ozkaya B (2007) Influence of leachate recirculation on aerobic and anaerobic decomposition of solid wastes. J Hazard Mater 143:177–183

Chung K, Kim S, Baek S, Lee NH, Park S, Lee J, Lee H, Bae W (2015) Acceleration of aged-landfill stabilization by combining partial nitrification and leachate recirculation: a field-scale study. J Hazard Mater 285:436–444

Clement R, Descloitres M, Gunther T, Oxarango L, Morrra C, Laurent JP, Gourc JP (2010) Improvement of electrical resistivity tomography for leachate. Waste Manage 30:452–464

Cossu R, Raga R, Rossetti D (2003) The PAF model: an integrated approach for landfill sustainability. Waste Manage 23:37–44

Cossu R, Lai T, Sandon A (2012) Standardization of BOD_5/COD ratio as a biological stability index for MSW. Waste Manage 32: 1503–1508

Di Addario M, Ruggeri B, Chiampo F (2014) Enhanced biogas production of low biodegradable fraction of municipal solid waste via leachate recirculation: experimental simulation. In: 13th International conference on clean energy (ICCE-2014), Istanbul, Turkey, 8–12 June 2014

Deublein D, Steinhauser A (2008) Biogas from waste and renewable resources. WILEY-VCH Verlag GmbH & Co. KGaA, Weinheim

Elagroudy S, Abdel-Razik M, Abd El-Azeem M, Ghobrial F, Warith M (2009) Effects of biological degradation on settlement of bioreactor landfills. In: CISA Publisher (ed) Proceedings Sardinia 2009, twelfth international waste management an landfill symposium, S. Margherita di Pula, Cagliari, Italy, 5–9 Oct 2009

Erses AS, Onay TT, Yenigun O (2008) Comparison of aerobic and anaerobic degradation of municipalsolid waste in bioreactor landfills. Bioresour Technol 99:5418–5426

Francois V, Feuillade G, Matejka G, Lagier T, Skhiri N (2007) Leachate recirculation effects on waste degradation: study on columns. Waste Manage 27:1259–1272

Gurljala KR, Sulfita JM (1993) Environmenal factors influencing methanogenesis from refuse in landfill samples. EnvironSci Technol 27:1176–1181

Hao Y, Wu W, Wu S, Sun H, Chen Y (2008) Municipal solid waste decomposition under oversaturated condition in comparison with leachate recirculation. Process Biochem 43:108–112

He R, Liu XW, Zhang ZJ, Shen DS (2007) Characteristics of the bioreactor landfill system using an anaerobic-aerobic process for nitrogen removal. Bioresour Technol 98:2526–2532

Huang FS, Hung JM, Lu CJ (2012) Enhanced leachate recirculation and stabilization in a pilot landfill bioreactor in Taiwan. Waste Manage Res 38(8):849–858

Imhoff PT, Reinhart DR, Englund M, Guérin R, Gawande N, Han B, Jonnalagadda S, Townsend TG, Yazdani R (2007) Review of state of the art methods for measuring water in landfills. Waste Manage 27:729–745

IRSA-CNR (2003) Metodi analitici per le acque. Manuali e Linee Guida N. 29/2003 (in Italian)

Jiang J, Guodong Y, Zhou D, Yunfeng H, Zhonglin H, Xiangming F, Shengyong Z, Chaoping Z (2007) Pilot-scale experiment on anaerobic bioreactor landfills in China. Waste Manage 27:893–901

Knox K, Beaven RP, Rosevear A, Braithwaite P (2007) A technical review of leachate recirculation. In: CISA Publisher (ed) Proceedings Sardinia 2007, twelfth international waste management an landfill symposium, S. Margherita di Pula, Cagliari, Italy, 1–5 Oct 2007

Lagerkvist A, Cossu R (2005) Leachate recirculation: concepts and applications. In: CISA Publisher (ed) Proceedings Sardinia 2005, tenth international waste management and landfill symposium, S. Margherita di Pula, Cagliari, Italy, 3–7 Oct 2005

Lubberding HJ, Valencia R, Salazar RS, Lens PNL (2012) Release and conversion of ammonia in bioreactor landfill simulators. J Environ Manage 95:S144–S148

Magnano E (2010) Biogas da discarica. EPC Libri, Roma

Mali Sandip T, Khare Kanchan C, Biradar Ashok H (2012) Enhancement of methane production and bio-stabilization of municipal solid waste in anaerobic bioreactor landfill. Bioresour Technol 110:10–17

Morris JWF, Vasuki NC, Baker JA, Pendleton CH (2003) Findings from long-term monitoring studies at MSW landfill facilities with leachate recirculation. Waste Manage 23:653–666

O'Keefe DM, Chynoweth DP (2000) Infuence of phase separation, leachate recycle and aeration ontreatment of municipal solid waste in simulated landfill cells. Bioresour Technol 72:55–66

Onay TT, Pohland FG (1998) *In situ* Nitrogen management in controlled bioreactor landfills. Water Res. 32(5):1383–1392

Oonk H, van Zomeren A, Rees-White TC, Beaven RP, Hoekstra N, Luning L, Hannen M, Hermeks H, Woelders H (2013) Enhanced biodegradation at the Landgraaf bioreactor test-cell. Waste Manage 33(10):2048–2060

Pohland F (1975) Sanitary Landfill Stabilization with Leachate Recycle and Residual Treatment. Report for EPA Grant No. R-801397, USEPA National Environmental Research Center, Cincinnati, OH

Pommier S, Chenu D, Quintard M, Lefebvre X (2007) A logistic model for the prediction of the influence of water on the solid waste mathanization in landfills. BiothechnolBioeng 97:473–482

Pommier S, Lefebvre X (2009) Impact of moisture content on the biodegradation of heterogeneous solid waste: simulations by a new modelling framework. In: Third international workshop hydro-physico-mechanics of landfills, Braunschweig, Germany, 10–13 Mar 2009

Ponsa S, Gea T, Alerm L, Cerezo J, Sanchez A (2008) Comparison of aerobic and anaerobic stability indices througha MSW biological treatment process. Waste Manage 28:2735–2742

Pradeep J, Ko JH, Kumar D, Powell J, Kim H, Maldonado L, Townsend T, Reinhart DR (2014) Case study of landfill leachate recirculation using small-diameter vertical wells. Waste Manage 34:2312–2320

Price GA, Barlaz MA, Hater GR (2003) Nitrogen management in bioreactor landfills. Waste Manage 23:675–688

Ritzkovski M, Stegmann R (2011) Landfill aeration worldwide: concepts, indications and findings. In: CISA Publisher (ed) Proceedings Sardinia 2011, tenth international waste management and landfill symposium, S. Margherita di Pula, Cagliari, Italy, 3–7 Oct 2011

Ruggeri B, Tommasi T, Sanfilippo S (2015) $BioH_2$ & $BioCH_4$ through anaerobic digestion from research to full-scale applications. Springer, London

Šan I, Onay TT (2001) Impact of various leachate recirculation regimes on municipal solid waste degradation. J Hazard Mater B87:259–271

Sang NN, Soda S, Ishigaki T, Ike M (2012) Microorganisms in landfill bioreactor for accelerated stabilization of solid wastes. J Biosci Bioeng 114:243–250

Sanphoti N, Towprayoon S, Chaiprasert P, Nopharatana A (2006) The effects of leachate recirculation with supplemental water addition on methane production and waste decomposition in a simulated tropical landfill. J Environ Manage 81:27–35

Sponza DT, Agdad ON (2004) Impact of leachate recirculation and recirculation volume on stabilization of municipal solid wastesin simulated anaerobic bioreactors. Process Biochem 39:2157–2165

Sri Shalini S, Joseph K (2012) Nitrogen management in landfill leachate: application of SHARON, ANAMMOX and combined SHARON–ANAMMOX process. Waste Manag 32:2385–2400

Sun X, Sun Y, Zhao Y, Wang Y (2014) Leachate recirculation between alternating aged refuse bioreactors and its effects on refuse decomposition. Environ Technol 35:799–807

Tallec G, Bureau C, Peu P, Benoist JC, Courant P, Budka A (2007) Gaseous emission during recirculation of leachate enriched with nitrate in a full-scale bioreactor-landfill. In: CISA Publisher (ed) Proceedings Sardinia 2007, eleventh international waste management and landfill symposium, S. Margherita di Pula, Cagliari, Italy, 1–5 Oct 2007

Valencia R, van der Zon W, Woelders H, Lubberding HJ, Gijzen HJ (2009) Achieving "Final Storage Quality" of municipal solid waste in pilot scalebioreactor landfills. Waste Manage 29:78–85

Valencia R, van der Zon W, Woelders H, Lubberding HJ, Gijzen HJ (2011) Anammox: an option for ammonium removal in bioreactor landfills. Waste Manag 31:2287–2293

Vigneron V, Budka A, Jimenez E, Hermkes H, Rospars A, Jean B, Belbeze P (2009) Bioreactor landfill: a sustainable waste treatment process. In: CISA Publisher (ed) Proceedings Sardinia 2009, twelfth international waste management an landfill symposium, S. Margherita di Pula, Cagliari, Italy, 5–9 Oct 2009

Warith M (2002) Bioreactor landfills: experimental and field results. Waste Manage 22:7–17

Zhao X, Musleh R, Maher S, Khire MV, Voice TC, Hashsham SA (2008) Start-up performance of a full-scale bioreactor landfill cell under cold-climate conditions. Waste Manage 28:2623–2634

Biotransformation of Nitrogen in Landfills

Kurian Joseph, S. Suneethi and S. Sri Shalini

Abstract A new and promising trend in solid waste management is to operate landfills as bioreactors in which moisture addition (often leachate recirculation) is used to create a solid waste environment capable of actively degrading the readily biodegradable organic fraction of the waste and produce bioenergy. Although the organic strength of leachate is significantly reduced in bioreactor landfills, ammonia-nitrogen (NH_4^+-N) remains an issue, because there is no degradation pathway for ammonia-nitrogen in anaerobic systems. Ammonia-nitrogen removal methods often include complex sequences of physical, chemical, and/or biological processes, including chemical precipitation, nanofiltration, air stripping, and biological nitrification/denitrification via various reactor configurations. This chapter will present the facts about the ammonia nitrogen profile and nitrogen transformation pathways in bioreactor landfills including promising novel removal mechanisms (SHARON (Single reactor system for high activity ammonia removal over nitrite), ANAMMOX (Anaerobic Ammonium Oxidation), CANON (Complete Autotrophic Nitrogen removal Over Nitrite), OLAND (Oxygen Limited Autotrophic Nitrification and Denitrification) etc.) with the support of laboratory-scale and field-scale case studies. In a study on ex situ strategy for NH_4^+-N removal by ANAMMOX applied using AnMBR (Anaerobic Membrane Bioreactors), NH_4^+-N removal efficacy of 85.13 ± 9.67 % was achieved with an average nitrogen removal rate (NRR) of 5.54 ± 0.63 kg NH_4^+-N/m^3/day (d) at a nitrogen loading rate (NLR) of 6.51 ± 0.20 kg NH_4^+-N/m^3/d at a 1.5 d HRT. An in situ strategy for NH_4^+-N removal by SHARON in a bioreactor landfill gave 85 % NH_4^+-N removal efficacy with 98.5 % nitrite accumulation, while the ANAMMOX process gave 73 % of NH_4^+-N removal efficacy with a specific ANAMMOX activity of 0.96 mg NH_4^+-N/mg MLVSS (Volatile Suspended Solids)/d with an NLR of 1.2 kg N/d. Bioreactor

K. Joseph (✉) · S. Suneethi · S. Sri Shalini
Centre for Environmental Studies, Anna University, Chennai 600025, India
e-mail: kuttiani@gmail.com

S. Suneethi
e-mail: sundar.suneethi@gmail.com

S. Sri Shalini
e-mail: srishalini10@gmail.com

O.P. Karthikeyan et al. (eds.), *Recycling of Solid Waste for Biofuels and Bio-chemicals*, Environmental Footprints and Eco-design of Products and Processes, DOI 10.1007/978-981-10-0150-5_9

landfills integrated with a combined SHARON-ANAMMOX processes provides promising results for nitrogen management (Total nitrogen removal—84 % and ammonia-nitrogen removal efficacy—71 % at NLR of 1.2 kg N/m^3/d in 147 days). Further activities need to focus on full-scale demonstration of in situ ammonia-nitrogen removal in bioreactor landfills and assessing the effect of different environmental conditions affecting important operational parameters of the processes.

Keywords Bioreactor landfill · Leachate · Nitrogen removal · Leachate recirculation · Municipal solid waste · Nitrogen transformations

1 Landfills and Waste Decomposition

Landfilling is the primary source of disposal of municipal solid waste (MSW) in most parts of the world. MSW is composed of recyclables (paper, plastic, glass, metals, etc.), toxic substances (paints, pesticides, used batteries, medicines), compostable organic matter (fruit and vegetable peels, food waste), soiled biodegradable waste and household biomedical waste (blood stained cotton, sanitary napkins, disposable syringes) (Reddy and Galab 1998; Jha et al. 2003). It has been reported to contain about 4 % protein, which is a major source of soluble nitrogen (Barlaz et al. 1990; Tchobanoglous et al. 1993; El-Fadel et al. 1997). Landfilling is the simplest, cheapest and most cost-effective method of managing waste practised in most of the low to medium income developing nations (Berge et al. 2005; Burton and Watson-Craik 2002; Karthikeyan et al. 2007). Essentially the landfill design should incorporate the following components.

- A **liner system** at the base and sides of the landfill to prevent migration of leachate or gas to the surrounding environment.
- A **leachate collection and treatment system to** collect and extract leachate from within and from the base of the landfill for treatment to meet regulatory requirements.
- A **final cover** of the landfill to enhance surface drainage, preventing infiltration of water and supporting surface vegetation.
- A **surface water drainage system** to collect and remove all surface runoff from the landfill site.
- An **environmental monitoring system** to periodically collect and analyse air, surface water, soil and ground water samples around the landfill site.

Major environmental concerns of municipal landfills revolve around quantity and quality of leachate, gas generation, and decomposition processes occurring therein. Many researchers have investigated factors controlling the sequential phases of MSW stabilisation (Reinhart and Townsend 1998; Kim and Pohland 2003). Waste stabilization proceeds in five sequential and distinct phases. An

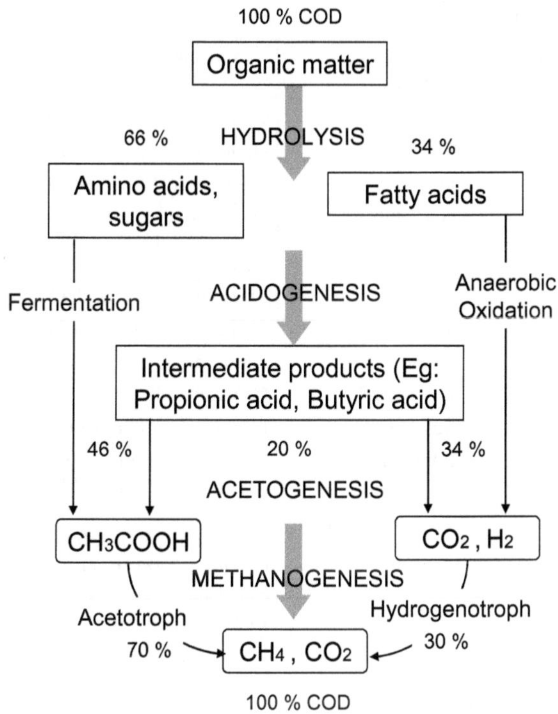

Fig. 1 Anaerobic fermentation pathways. Adapted from Renou et al. (2008)

acclimation period is observed until sufficient moisture develops and supports an active microbial community for biochemical decomposition. A transformation from an aerobic to anaerobic environment occurs, as evidenced by the depletion of oxygen within landfill media. Anaerobic decomposition (Fig. 1) is initiated with hydrolysis of solid waste, followed by the microbial conversion of biodegradable organic content resulting in the production of intermediate volatile organic acids (VOAs).

These intermediate acids are consumed by methanogenic bacteria generating methane and carbon dioxide. During the final state of landfill stabilization, nutrients and available substrates become limiting, and the biological activity shifts to relative dormancy. Environmental conditions which most significantly impact upon biodegradation during the methanogenic phase of anaerobic digestion in landfills include pH, temperature, nutrients, toxins, moisture content, particle size and oxidation-reduction potential etc., as summarized in Table 1.

Most studies have shown that the amount of moisture in the waste, which can vary widely within a single landfill, is a critical factor in the rate of decomposition (Barlaz et al. 1990), controlling the speed at which these phases can takes place (Valencia et al. 2009a, b). Increase in MSW moisture content creates a more favourable environment for biological decomposition of organic matter in the

Table 1 Factors influencing waste degradation in anaerobic digestion of landfills

Sl. No.	Influencing factors	Comments
1.	Moisture	Optimum: 60 % and above
2.	Oxygen	Optimum redox potential for methanogens −300 to −100 mV
3.	pH	Optimum pH for methanogenesis: 6–8
4.	Alkalinity	Optimum alkalinity for methanogenesis: 2000 mg/L Maximum organic acid concentration for methanogenesis: 3000 mg/L Maximum acetic acid/alkalinity ratio for methanogenesis: 0.8
5.	Temperature	Optimum temperature for methanogenesis; 34–38 °C
6.	Hydrogen	Partial hydrogen pressure for acetogenesis: $<10^{-6}$ atm
7.	Nutrients	Generally adequate
8.	Sulphate	Increase in sulphate decrease in methanogenesis
9.	Inhibitors	Cation concentration producing moderate inhibition (ppm) Ammonium (Total): 1500–3000 Sodium: 3500–5500 Potassium: 2500–4500 Calcium: 2500–4500 Magnesium: 1000–1500 Heavy metals: No significance influence Organic compounds: Inhibitory effect only in significant amount

Source Yuen et al. (1994)

landfill, and therefore accelerates the five phases of decomposition of waste (Townsend et al. 1995).

Conventional sanitary landfills, often called "dry tomb" landfills, are designed and operated to minimize contact between water and solid waste. The low moisture content in such landfills results in slow degradation rate reducing landfill life, furthering the burden of long post-closure care.

1.1 Bioreactor Landfill

The underlying principle of bioreactor landfills (Fig. 2) is that by optimizing operational control and environmental conditions within the waste (especially moisture content), more rapid and complete degradation of waste may be achieved (Berge et al. 2005).

A bioreactor landfill includes all of the protective containment systems of a conventional modern landfill, but in addition includes systems that accelerate the decomposition of the wastes and concomitant production of methane gas. These systems include pumps and piping to recirculate liquids, known as "leachate," that

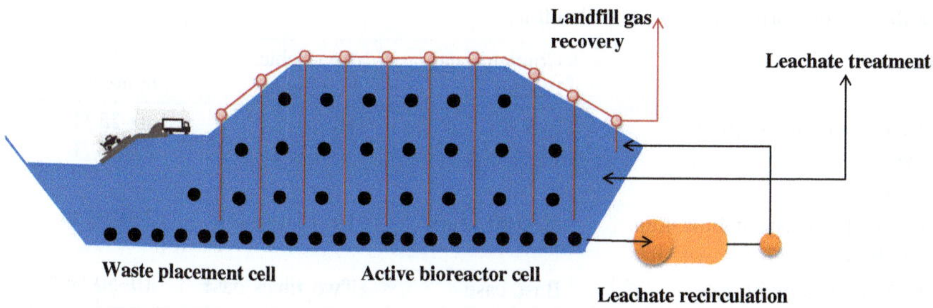

Fig. 2 Schematic view of a bioreactor landfill. Adapted from Walsh and O'Leary (2002)

drain to the bottom of the landfill back into the waste material. Leachate recirculation helps to keep the waste wet.

Leachate recirculation methods include spraying the leachate onto the working face, digging ponds or trenches into the landfill and filling them with leachate (some designs include filling the ponds or trenches with an aggregate material), and installing subsurface leach fields or injection wells. Especially, vertical and horizontal injection, and strategy of combining the horizontal and vertical leachate injection have been included (Townsend et al. 1995; Reinhart and Al-Yousfi 1996; Benson et al. 2007; Valencia et al. 2009a, b). Low pressure surface application uses a sprinkler to irrigate leachate at the tip face. Some systems use perforated pipes in a trench filled with drainage material. Deep vertical trenches, of approximately 6–8 m deep and 1.5 m wide filled with tyres and covered with geofabric have been used, both for the injection of liquid wastes and for recirculation.

Leachate recirculation also circulates nutrients and soluble organics, a further aid for decomposition and methane production. Minimizing the time period for maximum biodegradation reduce leachate and gas emissions after landfill closure, ease the requirement of leachate treatment, and reclamation of landfill. As leachate is recirculated, it is treated in situ, decreasing its organic strength and reducing the post closure care costs. It is reported that the regimes of leachate recirculation are dependent upon the different phases of waste stabilization in order to improve energy recovery efficacy (Jianguo et al. 2007). It was further recounted that it is appropriate to adopt a lower rate of 2.7 % for the acidogenic phase of waste and a higher rate of 5.3 % for methanogenic phase of waste. Sequential leachate recirculation between the methanogenic and acidogenic waste cells in the bioreactor could also promote energy recovery (Jianguo et al. 2007).

Research at several demonstration project landfill bioreactor sites has shown that the gas production rate typically is enhanced by a factor of 2–3 from what normally would be produced without liquids addition, leachate recirculation, and the other aspects of bioreactor operation (Reinhart and Townsend 1998). Table 2 compares conventional landfills with anaerobic and aerobic bioreactor landfills.

Benson et al. (2007) analysed five landfills in North America to provide a perspective of current practice and technical issues that differentiate bioreactor and recirculation landfills. Leachate generation rates, leachate depths and temperatures,

Table 2 Comparison of bioreactor landfills

Item	Conventional landfill	Anaerobic bioreactor	Aerobic bioreactor
Typical settlement after: 2 years 10 years	2–5 % 15 %	10–15 % 20–25 %	20–25 % 20–25 %
Anticipated waste-stabilization time frame	30–100 years	10–15 years	2–4 years
Methane generation rate	Base case	Two times base case	10–50 % base case
Average capital cost	Low	Medium	High
Average O&M cost	Low	Medium	High
Average closure/post closure cost	High	Medium	Low

Adapted from Benson et al. (2007)

and liner temperatures were similar for landfills operated in a bioreactor/recirculation or conventional mode. Gas production data indicate accelerated waste decomposition from leachate recirculation at one landfill. Ambiguities in gas production data precluded a definitive conclusion that leachate recirculation accelerated waste decomposition at the four other landfills. Analysis of leachate quality data showed that bioreactor and recirculation landfills generally produce stronger leachate than conventional landfills during the first two to three years of recirculation. Thereafter, leachate from conventional and bioreactor landfills is similar, at least in terms of conventional indicator variables (BOD, COD, pH). While the BOD and COD decreased, the pH remained around neutral and ammonia concentrations remained elevated. Settlement data collected from two of the landfills indicate that settlement occurs much faster in landfills operated as bioreactors or with leachate recirculation. The analysis also indicated that more detailed data collection over longer time periods is needed to draw definitive conclusions regarding the effects of bioreactor and recirculation operations.

1.2 *Carbon and Nitrogen Cycles in Landfills*

Organic compounds are usually present in fresh waste. The successive phases of biodegradation during waste deposition cause a reduction in COD and BOD, and the only substances left in the leachate are scarcely biodegradable compounds. COD and BOD may differ depending on the age of the landfill (Bodzek et al. 2006). From the studies carried out by Tengrui et al. (2007), old landfill leachate showed COD and BOD as 1650 and 75 mg/L, respectively. Erses et al. (2008), obtained initial young leachate COD concentrations of aerobic and anaerobic bioreactor landfills which were 17,900 and 38,000 mg/L, respectively. Whilst in Canziani et al. (2007), the COD and BOD concentrations were 6316 ± 5877 and 2950 ± 3537 mg/L, with a BOD/COD ratio of 0.43 ± 0.12. In the same study,

NH_4^+-N concentration was 1497 ± 1190 mg/L with a ratio of NH_4^+-N/COD in the range of 0.12–0.98. In Bohdziewicz et al. (2008b), the BOD/COD ratio of a young landfill leachate was >0.3. According to Valencia et al. (2011), experiments carried out in bioreactor landfill simulators demonstrated that more than 40 % of the total N was transferred into the liquid and gas phases during the incubation period of 380 days. Ammonium, an end product of protein degradation and important parameter to consider during landfill closure, tends to accumulate up to inhibitory levels in the leachate of landfills, especially in landfills with leachate recirculation.

The high strength nitrogen-rich wastewaters with low organic content such as partially treated landfill leachates from landfill bioreactors and old landfills are common wastewater sources obtained during acid or methanogenic phases which are characterized by anaerobic conditions. During anaerobic digestion, proteins are broken down into amino acids (hydrolysis) which during further degradation in the acidogenesis phase release ammonium. A similar process occurs within a sanitary landfill, where under acidogenic phase as depicted in Fig. 1, the NH_4^+-N concentration may gradually rise to over 1000 mg/L (Berge et al. 2005). In MSW lysimeter studies by Swati et al. (2007), an increase in NH_4^+-N concentration 69–76 % and Total Kjeldahl Nitrogen (TKN) by 70–74 %, was observed irrespective of the type of operation, owing to incapacity of natural biodegradation and physico-chemical attenuation processes in a landfill environment to attenuate their levels. Nitrogen exists in different oxidation states, based on the aerobic/anaerobic environment. In aerobic environments, bacteria oxidize NH_4^+-N, which formed due to death and decomposition of plants and animals, to NO_2^--N and NO_3^--N. NH_4^+-N concentrations of approximately 3000 mg/L can significantly inhibit methanogenesis (Burton and Watson-Craik 1998). The oxidation state of NH_4^+ and NH_3 is −3 while that of NO_2^--N and NO_3^--N is +3 and +5, respectively. The oxidation state of nitrogen in most organic compounds is −3 (Metcalf and Eddy 2002; Hammer and Hammer 2000).

The landfill environment is a complex heterogeneous system in which different types of microorganisms coexist. Dominance of microorganisms varies with prevailing conditions and organisms-substrate specificity during waste stabilization (He and Shen 2006). Normally, anaerobic/anoxic conditions are prevalent in a landfill enabling methanogenesis and, potentially, denitrification (Fu et al. 2009). Nitrogen in MSW is removed via ammonification and solubilization processes, resulting in accumulation as NH_4^+-N in the leachate (Burton and Watson-Craik 1998). NH_4^+-N is stable under anaerobic conditions. NH_4^+-N concentration is reduced during waste decomposition through leaching as there is a lack of degradation pathway in landfill environment, except for autotrophic ammonia removal such as anaerobic ammonium oxidation (ANAMMOX) (Robinson 1995; Burton and Watson-Craik 1998; Swati et al. 2007; Liang and Liu 2008). Leaching of NH_4^+-N from the landfilled waste into the environment continue over a long period.

Anaerobic fermentation of organic matter is related to the nitrogen profile in a landfill environment. In young landfills, the high concentration of organic matter is reduced by rapid anaerobic fermentation with end products of volatile fatty acids (VFAs). As the landfill ages, methanogenic bacteria in the waste convert the VFAs to CH_4 and CO_2. Organic material content is reduced with landfill age, with the result that an older leachate has a relatively low but non-biodegradable organic fraction (100–3460 mg/L of COD) compared to 13,000–50,000 mg/L of COD in young landfill leachate (Renou et al. 2008; Berge et al. 2006; Karthikeyan et al. 2007; Lau et al. 2001). The ratio of BOD/COD from 0.70 to 0.04 is also reduced rapidly with aging of landfills (Renou et al. 2008). In a landfill environment, high accumulations of NH_4^+-N occur owing to degradation of amino acids and proteins. Landfills could be operated as a bioreactor to treat organical-rich MSW by providing aeration and recirculating the generated leachate to enhance biodegradation and stabilization of the landfilled waste (Pohland 1980; Karthikeyan et al. 2007). But aeration and moisturization favours ammonification resulting in accumulation of NH_4^+-N at higher concentrations (>3000 mg/L) than that of conventional landfills, even after stabilization of the organic fraction (Connolly et al. 2004).

Results of studies on nitrogen management in bioreactor landfills by Price et al. (2003) suggest that landfills have significant capacity to convert nitrate to nitrogen gas that can be safely released to the atmosphere, thus providing a viable alternative for the long-term management of nitrogen in landfills. Although the consumption of organic carbon limited nitrate reduction rates, this could easily be managed in a full-scale landfill. The simplest way to enrich the leachate from a particular landfill cell in organic carbon would be to add fresh refuse to the top of the cell. If a landfill cell was no longer receiving fresh refuse, then leachate from another section of the landfill that contains a higher BOD, or a liquid waste with degradable organics could be added to provide sufficient carbon to drive denitrification.

2 Ammonia-Nitrogen in Leachate

Nitrogen cycling that may occur in landfills depicted in Fig. 3 which illustrates that processes such as ammonification, sorption, volatilization, nitrification, denitrification, ANAMMOX, and nitrate reduction may all occur in landfills.

However, in bioreactor landfills, moisture addition and/or recirculating leachate increases the rate of ammonification, resulting in accumulation of higher levels of NH_4^+-N, even after the organic fraction of the waste is degraded (Berge et al. 2005). It has been suggested that NH_4^+-N is one of the most significant long-term pollution problem in landfills, and it is likely that the presence of ammonia-nitrogen will determine when the landfill is biologically stable and when post closure monitoring may end. Thus an understanding of the fate of nitrogen in bioreactor landfills and possible mechanisms for NH_4^+-N removal is critical to both successful and economic operation.

Ammoniacal nitrogen removal strategies in bioreactor landfills would be to treat on site, pump to external wastewater facility and/or in situ treatment. Potential in situ removal mechanisms include conventional biological nitrogen removal (BNR) approach of nitrification—denitrification and volatilization.

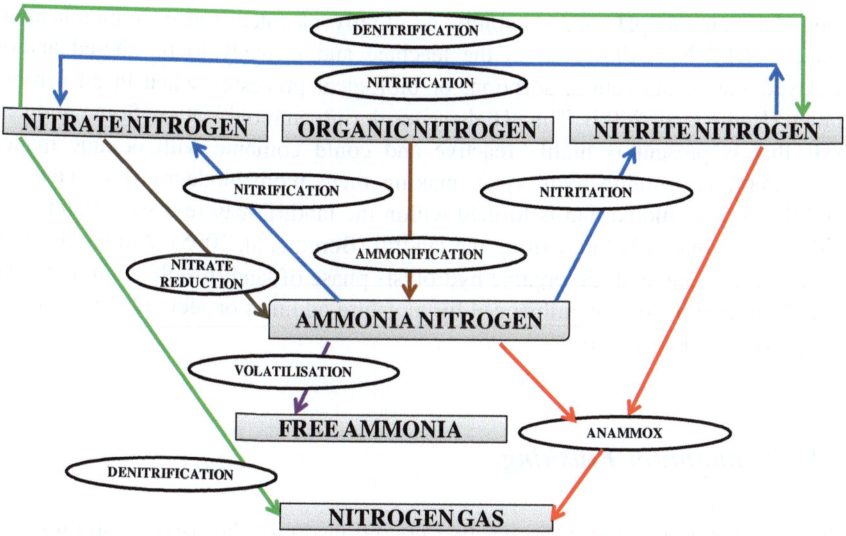

Fig. 3 Nitrogen cycles in landfill environment. *Source* Sri Shalini and Joseph (2012)

In practice when the NH_4^+-N concentration of the effluent ranges between 100 and 5000 mg/L, biological treatment by autotrophic nitrogen removal is preferred (Mulder 2003). Various biological and advanced wastewater treatment processes have been employed to treat landfill leachate, i.e. supernatant from anaerobically digested sludge and other nitrogenous wastewaters (Amokrane et al. 1997; Renou et al. 2008; He et al. 2007; Valencia et al. 2005; Ahn et al. 2004; Fux et al. 2002). Major biological treatment processes of both aerobic and anaerobic types comprise the activated sludge process (ASP), sequencing batch reactor (SBR), rotating biological contactor (RBC) etc., Traditionally NH_4^+-N rich wastewaters were treated by adopting combined autotrophic nitrification and heterotrophic denitrification methods (Berge et al. 2006; Han et al. 2001; Ruiz et al. 2006). Preferential selection for ammonia oxidisers is required to make the nitrification work, by denitrifying nitrite instead of nitrate.

2.1 Ammonification

Ammonification is the process in which the proteins present in the waste are used by heterotrophic bacteria, the major source of ammonia-nitrogen (Berge et al. 2005). Ammonification is a two-step process consisting of the enzymatic hydrolysis of proteins by aerobic and anaerobic microorganisms releasing amino acids and the subsequent deamination or fermentation (depending on aerobic vs. anaerobic conditions) of the acids to CO_2, NH_4^+-N and VFAs (Berge et al. 2005). Throughout deamination, amine groups are liberated to form ammonia or ammonium, depending on the pH, and alkalinity is slightly elevated. Once ammonification ensues, NH_4^+-N is dissolved in the leachate and is ready to be altered and/or removed via volatilization, sorption, or biological processes when in an aerobic milieu (Berge et al. 2005). The pH also rises during ammonification. Some remnant NH_3 that is present is highly reactive and could combine with organic matter (i.e. carboxyls, quinine hydroxyls), making them more biodegradable. Thus, in landfills, any ammonia that is formed within the landfill may re-dissolve and react with organic matter before exiting the landfill (Berge et al. 2005). Ammonification occurs in the course of the organic hydrolysis phase of landfill stabilization and the rate of ammonification is influenced by moisture addition or recirculating leachate (Berge et al. 2005, 2006).

2.2 *Ammonium Flushing*

Flushing operations under bio-reactive conditions could be used to optimize the removal of ammonia-nitrogen from landfilled waste. The 'Flushing bioreactor' approach encourages the ingress of water (hence leachate production) by re-injection of leachate or increased infiltration through cap and cover materials. Increases in moisture content in domestic waste have been shown to (a) increase CH_4 production, (b) enhance contaminant removal, and (c) ultimately reduce the period requiring landfill monitoring and control. Reduced concentrations of NH_4^+-N in leachate can be achieved by high moisture flux to wash out contaminants from the waste mass. The rate of reduction of NH_4^+-N is governed by moisture flux, therefore the greater the volumes of leachate produced, the lower the resulting leachate contaminant concentrations (Purcell et al. 1999).

The volume of water that is passed through the landfill, the nitrogen content of the waste, and the ammonia nitrogen concentration in the bulk liquid could drive the mass of ammonia-nitrogen that can be leached from the refuse (Berge et al. 2005). In order to reduce ammonia nitrogen concentrations either by washout or by dilution requires water addition. Flushing volumes between 5 and 7.5 m^3/tonne of waste were required to effectively decrease nitrogen concentrations in the landfill (Berge et al. 2005). The efficacy of flushing requires hydraulic conductivity of the refuse, making it difficult to introduce liquid in areas of lower permeability (Berge et al. 2005). Once there is significant decrease in hydraulic conductivity, the time required for leaching to occur increases, as does the ammonification process (Berge et al. 2005). The rate of NH_4^+-N removal is directly proportional to the rate of flushing. But this flushed water should be treated ex situ or re-used for recirculation, wherein the reintroduction of NH_4^+-N to the landfill is continuous and so is its solubilisation into the leachate (Berge et al. 2005).

2.3 Ammonium Sorption

The sorption of NH_4^+-N onto different organic and inorganic compounds has been reported in various studies, as summarised in Berge et al. (2005). The option of ammonium sorption onto the waste enables temporary storage of ammonium prior to nitrification and volatilization, and may also result in the slow dissolution of ammonium over time (Berge et al. 2005). Sorption is dependent on pH, temperature, ammonium concentration, and ionic strength of the bulk liquid. For ammonia to sorb to the waste particles, it must be in the form of ammonium (NH_4^+). At pH levels expected in a landfill, the dominant form of the ammonia species is the ammonium ion (Berge et al. 2005). Ammonium sorption is directly related to ionic strength of the bulk liquid, pH, temperature and ammonium concentration (Berge et al. 2005).

2.4 Ammonium Volatilization

In conventional landfills, ammonia makes up approximately 0.1–1.0 % (dry volume basis) of landfill gas exiting the landfill. But volatilization only occurs when free ammonia (NH_3) (FA) is present. At an alkaline pH (>10.5–11.5) the majority of the NH_4^+-N present in leachate is in the form of NH_3 gas (Berge et al. 2005). Ammonium volatilization increases with respect to increase in temperature (Berge et al. 2005, 2006). In case of aerobic landfills where there is air flow, it can agitate the leachate thereby enabling a removal pathway for dissolved NH_3 to volatilise and leave the landfill. Air injection also contributes to dilution effects and increases the partitioning of dissolved ammonia-nitrogen to the gaseous phase (Berge et al. 2005).

2.5 Nitrification and Denitrification

Nitrification is 2 step aerobic processes carried out by *Nitrosomonas* (Eq. 1) and *Nitrobacter* (Eq. 2), respectively. The denitrification process occurs under anoxic conditions carried out by denitrifying bacteria (Eqs. 3–6).

$$0.1667NH_4^+ + 0.25O_2 \rightarrow 0.1667NO_2^- + 0.333H^+ + 0.1667H_2O \quad (1)$$

$$0.50NO_2^- + 0.25O_2 \rightarrow 0.50NO_3^- \quad (2)$$

$$NO_3^- + 2e^- + 2H^+ \rightarrow NO_2^- + H_2O \quad (3)$$

$$NO_2^- + e^- + 2H^+ \rightarrow NO + H_2O \tag{4}$$

$$2NO + 2e^- + 2H^+ \rightarrow N_2O + H_2O \tag{5}$$

$$N_2O + 2e^- + 2H^+ \rightarrow N_2 + H_2O \tag{6}$$

The main implications of nitrification—denitrification are requirements of O_2, affected particularly by the age of the waste (old and new) and specific temperatures (Berge et al. 2005, 2006). As O_2 can be limiting, it is considered to be an operational parameter of concern during field-scale applications. O_2 is part of non-uniform applications with several different O_2 concentrations and some areas with no O_2 at all (Berge et al. 2005, 2006). Waste age is a critical factor, with younger waste exhibiting higher capacity for aerobic conditions at high temperature. The fresh waste also contains higher concentrations of organic carbon, displaying a higher O_2 demand (Berge et al. 2005, 2006). With respect to temperature, aerobic landfills promote thermophiles, >70 °C, while nitrifiers and denitrifiers are mesophiles (25–40 °C).

A landfill bioreactor in situ approach is considered feasible when waste stabilization (active phase) is completed, applying forced aeration at the bottom of landfill bioreactor and recirculation of leachate (Valencia et al. 2005). Periodical aeration was carried out at the top of landfilled waste by He and Shen (2006), whereas semi-aeration through natural ventilation was performed by Huo et al. (2008). The semi-aerobic landfill system was more favourable with simultaneous occurrence of nitrification and denitrification.

Further studies on this approach were carried out by Giannis et al. (2008), who achieved NH_4^+-N removal from 750 to 10 mg/L in 250 days. The process of nitrification contributed to NH_4^+-N conversion to NO_3^--N, but the concentrations of NO_3^--N (10–30 mg/L) was not removed by denitrification, suggesting that partial nitrification might be occurring. Even anaerobic landfill bioreactor studies focusing on in situ treatment performed by Jianguo et al. (2007) concluded that the NH_4^+-N of 11,000 mg/L was reduced to 5000 mg/L in 15 weeks of operation. About 23 % removal was observed in 28–30 weeks. Thus in situ practice disposed the leachate effectively, but removed NH_4^+-N to a certain extent and accelerated the stabilization of the waste. Other limitations were due to carbon requirement in denitrification and aeration needs. In situ nitrification and denitrification are feasible in bioreactor landfill conditions, with the possibility of denitrification occurring in older portions of landfill, even under low biodegradable C/N conditions (Berge et al. 2005).

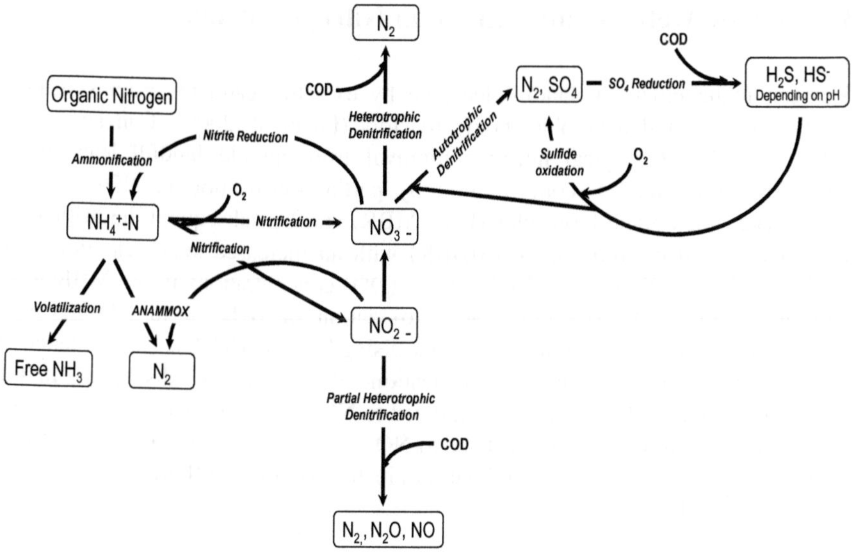

Fig. 4 Potential nitrogen transformation pathways in landfill environment. Adapted from Berge et al. (2005)

2.6 *Ammonia Nitrogen Profile in Bioreactor Landfill*

In a study of 50 German landfills, NH_4^+-N concentrations did not show a significant decrease even 30 years after landfill closure (Krumpelbeck and Ehrig 1999). It was reported by Ehrig (1998), that there was no observable change in NH_4^+-N concentrations from acidic and methanogenic phase of landfill and that, the mean NH_4^+-N concentration was reported to be 740 mg/L. The total NH_4^+-N concentration beneficial for microbial activity in anaerobic system was in the range of 50–200 mg/L. Presence of NH_4^+-N at about 300–3000 mg/L, inhibited methanogenic activity, and interfered in the anaerobic degradation pathway (Berge et al. 2005). The optimum range of temperature preferred for mesophiles and thermophiles were 29–38 and 49–57 °C, with pH 6.6–7.6 (O'Flaherty et al. 2006). Methanogens are sensitive to NH_4^+-N at high temperature and pH, as the inhibitory component is FA. Rise in pH result in increase in NH_3 concentration. NH_3 concentrations of 0.1–1.0 mg/L were reported to inhibit the biological oxidation of NO_2^--N to NO_3^--N (Anthonisen et al. 1976). The expected pathways of nitrogen transformations in a landfill condition, as depicted in Fig. 4, includes ammonification, sorption, volatilization, nitrification, denitrification, anaerobic ammonium oxidation (ANAMMOX), and NO_3^--N reduction. Environmental conditions such as moisture and temperature affect transformation and removal of nitrogen. Thus, within one landfill cell, there may be many nitrogen transformation processes occurring simultaneously or sequentially (Berge et al. 2005).

3 Ex Situ Methods for Ammonia Nitrogen Removal

Nitrogen management in landfill leachate by ex situ treatment systems usually involve aerobic and anaerobic/anoxic reactors (Fu et al. 2009). Canziani et al. (2007) achieved biological nitrogen removal from an old landfill leachate of 500–3000 mg/L NH_4^+-N concentration by partial nitrification to NO_2^--N in a pure oxygen membrane bioreactor (PO—MBR) and by subsequent denitrification in a moving bed biofilm reactor (MBBR) with an increased solid retention time (SRT) of 45 days. When dissolved oxygen (DO) concentrations in the MBR were kept in the range 0.2–0.5 mg/L, 90 % oxidation of NH_4^+-N to NO_2^--N was achieved with stable inhibition of nitrite oxidising bacteria (NOB). The sequence of operation was nitrification/partial nitrification—denitrification—discharge. In the nitrification stage, NH_4^+-N was first oxidized to NO_2^--N by *Nitrosomonas* bacteria, and the NO_2^--N produced was oxidized to NO_3^--N by *Nitrobacter* and finally to N_2 as indicated by Eqs. (7) and (8) and partial nitrification Eq. (9) (Paredes et al. 2007; He and Shen 2006).

$$NH_4^+ + 2O_2 + 2HCO_3^- \rightarrow NO_3^- + 2CO_2 + 3H_2O \quad (7)$$

$$4NO_3^- + 5C + 2H_2O \rightarrow 2N_2 + 4HCO_3^- + CO_2 \quad (8)$$

$$NH_4^+ + 0.75O_2 + HCO_3^- \rightarrow 0.5NO_2^- + 0.5NH_4^+ + CO_2 + 1.5H_2O \quad (9)$$

There was a high demand for O_2 during the nitrification process. When the degradable organic carbon concentration was high in the environment, heterotrophic microorganisms would outcompete nitrifiers for O_2 and nutrients. In contrast, most of the denitrifying bacteria exist in an environment in which organic compounds are present, and use organic matter as a carbon source and electron donors (He and Shen 2006). Denitrification was inhibited in the presence of O_2, and limited to anoxic environments. Therefore, treatment of NH_4^+-N required spatial separation of nitrifying and denitrifying units or temporal separation of each step by alternating the supply of aeration and no aeration in the same unit (He and Shen 2006). Furthermore, for high NH_4^+-N, low COD effluent, air-stripping pretreatment was usually needed and external carbon sources such as CH_3OH should be added to adjust COD/NH_4^+-N ratio (He and Shen 2006). This increased costs and enhanced difficulty of management which are obvious disadvantages (Valencia et al. 2005, 2009a, b, 2011).

In the study by Renou et al. (2008), application of MBR for treating landfill leachate of 1000 mg N/L resulted in >80 % of nitrogen being removed as TKN, while Bohdziewicz et al. (2008a) reported only 46 % NH_4^+-N removal treating leachate of influent NH_4^+-N concentration of 795 mg N/L using an Anaerobic MBR (AnMBR). Adoption of traditional autotrophic nitrification and heterotrophic denitrification methods leads to emissions of NO_x and N_2O that plays a significant role in global warming and climate change (Valencia et al. 2005). The ex situ methods are good at nitrogen removal but not suited for landfill bioreactor concepts, as leachate recirculation is pertinent for optimal performance.

3.1 *"Wuhrmann" Process*

The Wuhrmann process (Fig. 5) or post-denitrification, can achieve nitrification and carbonaceous oxidation before the leachate enters the anoxic zone for denitrification (Liu and Liptak 1999). Endogenous respiration could provide the required carbon source since all available extracellular carbon would have been removed. The efforts of Wuhrmann helped to develop other single sludge nitrification/denitrification systems; although this process was never used at full scale. Operational problems include high turbidity levels of the clarified effluent, ammonia release from cell lysis in the anoxic zone, and high nitrate levels due to low denitrification rates (USEPA 1993).

3.2 *"Ludzack-Ettinger" Process*

In the process based on Ludzack-Ettinger (Fig. 5), raw wastewater is directed first into an anoxic zone followed by an aerobic zone, which is called pre-denitrification (Liu and Liptak 1999). Since nitrification occurs after the anoxic zone, the return activated sludge (RAS) stream recycles nitrates. As such, this process typically operates with a high RAS return rate (75–150 % Q). The raw wastewater serves as

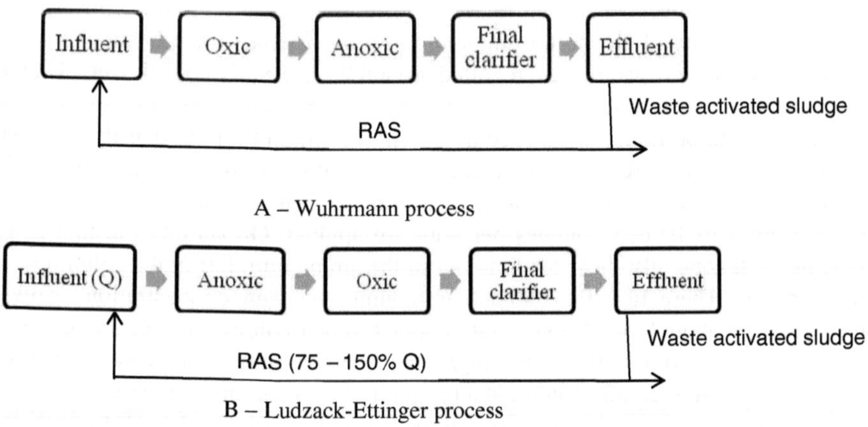

Fig. 5 Schematic diagrams of the ex situ nitrogen removal processes. Adapted from Liu and Liptak (1999)

a carbon source for denitrification and thus has a higher denitrification rate than the Wuhrmann process (Liu and Liptak 1999).

3.3 "4-Stage Bardenpho" Process

The Bardenpho process provides a TN removal capability that cannot be obtained in the Ludzack-Ettinger process (Liu and Liptak 1999). The four-stage Bardenpho process subjects the nitrate that was not recycled from the primary aerobic zone to anoxic conditions in a secondary anoxic zone. Although biological denitrification is occurring in both anoxic reactors, the carbon source is different; the carbon source of the primary anoxic zones is supplied by influent wastewater, whereas endogenous respiration is responsible for any denitrification in the secondary anoxic zone (Liu and Liptak 1999). A small secondary aerobic zone prior to secondary clarification strips away any N_2 gas entrained in the solids and nitrifies any ammonia released from cell lysis (Liu and Liptak 1999).

3.4 Ion Exchange Process

Ion exchange technology involves passing a liquid through a column or bed of a specific natural or synthetic zeolite resin and the exchange of one ion for another (USEPA 1993). The column is run until unacceptable breakthrough of the ion(s) of concern is achieved (reflecting the exhaustion of the exchange sites for this point of equilibrium). A highly concentrated regenerant is then passed through the column to displace the removed ions from the exchange sites. The regenerant can be processed further for recovery and reuse with a blow down of a concentrated waste or recyclable product, or passed to waste. Clinoptilolite, a natural occurring product, is the zeolite of choice for ammonium-nitrogen removal applications.

Total exchange capacities are reported around 2 milli equivalents/L (meq/L), but the capacity in wastewater applications for ammonium appears to range from 0.2 to 0.5 meq/L. Clinoptilolite has a specific gravity of 1.6 and a bulk density of about 0.75. In wastewater treatment applications, prior filtration must be used to avoid blinding the fine 20×50 (0.84×0.33 mm) mesh media. Hydraulic loading rates range from 5 to 10 bed volumes per hour are applied. On startup, the first 5–10 volumes will generally show the drag out of the ammonium left in the column after regeneration. Thereafter, the effluent ammonium-nitrogen concentrations slowly increase from about 0.5–1.0 mg/L until about 130 bed volumes is reached and then progressively deteriorate to about 5 mg/L at 170 bed volumes. The column may be removed from service after final effluent quality becomes unacceptable.

Regeneration is preceded by two to three bed volumes of backwash to flush out accumulated solids, and is accomplished by a high pH salt solution using flow rates of about 10 bed volumes per hour and 10–20 bed volumes of regenerant. Regeneration is followed by a rinse cycle of 2–3 bed volumes to minimize high pH as well as salt and ammonia concentrations in the product effluent on startup.

3.5 *Struvite Precipitation*

Struvite (magnesium ammonium phosphate (MAP)) or lime is usually employed as the precipitant for the removal of ammonia-nitrogen. Struvite or MAP precipitates in the presence of magnesium (Mg^{2+}), ammonium (NH_4^+), and phosphate (PO_4^{3-}) according to the Eq. 10, when the thermodynamic solubility product, Ks, is exceeded:

$$Mg^{2+} + NH_4^+ + PO_4^{3-} + 6H_2O \rightarrow MgNH_4PO_4 \cdot 6H_2O \downarrow \tag{10}$$

Mg^{2+}, NH_4^+ and PO_4^{-3} are required in equimolar quantities to form $MgNH_4PO_4{\cdot}6H_2O$. However, because of the common ion effect, other chemical species in solution will form Mg^{2+}, NH_4^+, and/or PO_4^{-3} compounds. This effect requires that additional $Mg^{2+}$and/or PO_4^{-3} is needed to precipitate the maximum mass of NH_4^+ (Turker and Celen 2010). The struvite precipitation process depends on two main factors: the molar ratio of Mg^{2+}: NH_4^+-N: PO_4^{-3}-P and the pH of the sample (Munch and Barr 2001). In a study by Karthikeyan and Joseph (2008), seven different molar ratios (Magnesium: Ammonium: Phosphate) were used to optimize the removal efficiency of ammonia-N. Around 58 % of ammonia removal was obtained in 1:1:1 molar ratio. Maximum removal efficiency of 76 and 78 % achieved at the molar ratio of 2:1:1.5 and 1.5:1:2, respectively. Leachate pH varied from 6.5 to 6.9 in supernatant samples obtained from different molar ratio.

4 In Situ Methods for Ammonia Nitrogen Removal

Control of in situ conditions in a landfill bioreactor is carried out by moisture addition (such as leachate recirculation) and/or aeration for promoting active degradation of readily biodegradable organic fraction of the waste. Besides that, aeration and moisturization favors ammonification, results in accumulation of ammonia at a higher concentration (5000 mg/L, Connolly et al. 2004) than that of conventional landfills, even after stabilization of the organic fraction. Onay and Pohland (1998) had expounded the potential of in situ attenuation of high residual leachate ammonia-nitrogen concentrations by nitrification and denitrification. This was performed by utilizing compost as the waste matrix, and by adopting the air inlet at the bottom of the reactor to simulate 3 components of landfill bioreactor systems, encompassing anoxic, anaerobic and aerobic zones. Since O_2 penetrates in the interstices of the landfilled waste, vertical aerobic/anoxic/anaerobic biological zones are formed naturally in landfill ecosystems. System performances of both combined and separate operations with leachate recycling had resulted in 95 % nitrogen conversion, whereas combined operation without recycling had conversion efficiency per cycle ranging from 30 to 52 % by nitrification and from 16 to 25 % by denitrification. Simultaneous nitrification and denitrification in aerobic landfill bioreactor cell occurs. While in another study, an efficacy of 99.5 % by in situ ammonia removal in a biofilter comprising of old waste by making use of anaerobic and aerobic sections was attained (He et al. 2007).

Thus an in situ approach is feasible when waste stabilization (active phase) is completed, by applying forced aeration at the bottom and recirculation (Valencia et al. 2005). Periodical aeration was carried out at the top of landfilled waste by He and Shen (2006), whereas semi-aeration through natural ventilation was performed by Huo et al. (2008). The semi-aerobic landfill system bodes well with simultaneously occurring nitrification and denitrification. Further studies on this approach carried out by Giannis et al. (2008) had achieved ammonia removal from 750 to 10 mg/L in 250 days. The process of nitrification had contributed to ammonia conversion to nitrate, but the levels of nitrate (10–30 mg/L) was not removed by denitrification, and suggested that other processes like partial nitrification might be occurring. Even anaerobic landfill bioreactor studies focusing on in situ treatment performed by Jianguo et al. (2007) had concluded that the ammoniacal nitrogen of 11,000 mg/L had reduced to 5000 mg/L in 15 weeks of operation. In 28–30 weeks, an average removal rate of 23 % was observed. Thus in situ practice disposes the leachate effectively, removes ammonia to a certain extent and accelerates the stabilization of the waste. The common limitations observed just like ex situ method are due to carbon requirement issues in denitrification and aeration needs (Karthikeyan and Joseph 2006). The précis of the ex situ and in situ modes of treatment of landfill leachate is presented in Table 3, where the performance of the particular system involved and its corresponding process efficiency has been reported.

The operation sequence in this hybrid process is given as ex situ nitrification—recirculation—in situ denitrification. Oxidation of ammonia in the external reactor to nitrate, nitrite followed by nitrogen gas occurs. As the landfill conditions normally favor denitrification owing to prevalence of anaerobic/anoxic condition, the absence of aerobic condition in the landfill is compensated for by external nitrification. The nitrified leachate is recycled back into landfill for in situ denitrification process (Fu et al. 2009). Ammonia accumulation is removed by this means but results in acidic pH of leachates, thereby reducing the buffering capacity of the leachate. External aeration brings in oxygen to the anoxic system, where denitrification occurs. Also the presence of O_2, NO_2^- and NO_3^- in leachate affects the methanogenesis of waste (He et al. 2007).

Table 3 Treatment technologies for nitrogen removal in landfill leachate

SI No	Mode of operations	Treatment technology	Concentration of nitrogen (as NH_4^+ or TKN)		Removal performance (%)	References
(I)	Ex situ		Influent (mg/L)	Effluent (mg/L)		
1.	Aerobic treatment (suspended growth processes)	Conventional activated sludge process	1400	175	87.5 NH_4^+	Renou et al. (2008)
2.		Sequencing batch biofilm reactors	1100	11	99 NH_4^+	Tengrui et al. (2007)
3.	Aerobic treatment (attached growth processes)	Trickling filters	115	12	90 NH_4^+	Renou et al. (2008)
4.	Anaerobic Treatment (Suspended growth processes)	Hybrid UASB & RO	196	137	30 NH_4^+ (UASB)	Bohdziewicz et al. (2008a)
			137	16	88.7 NH_4^+ (RO)	
5.	Anaerobic treatment (attached growth processes)	Anaerobic filter	1800	225	87.5 NH_4^+	Renou et al. (2008)
6.	Aerobic	Membrane bioreactor (MBR)	1000	200	>80 TKN	Renou et al. (2008)
7.	Anaerobic	Membrane bioreactor (MBR)	795	429	46 NH_4^+	Bohdziewicz et al. (2008a)
8.			10,000	1454	85 NH_4^+	Suneethi and Joseph (2013)
(II)	In situ					
9.	Aerobic treatment	Landfill bioreactor	500	10	98 NH_4^+	Giannis et al. (2008)
10.	Anaerobic treatment		2280	1755	23 NH_4^+	Jianguo et al. (2007)
11.	Anaerobic treatment	Landfill bioreactor	697	190	73 NH_4^+	Sri Shalini and Joseph (2013)

4.1 Bioaugmentation

The option of nitrogen removal in the side stream for landfill leachate is bioaugmentation. Bioaugmentation with endogenous nitrifiers is referred to as bioaugmentation batch-enhanced process (BABE). Bioaugmentation is accomplished by seeding activated sludge with an external source of nitrifying bacteria (i.e. external bioaugmentation) or making process improvements to increase the activity of or enrich the nitrifier population (i.e. internal bioaugmentation) (USEPA 2009).

In the BABE process only a small fraction of return sludge is used, with a distinct advantage of temperature, since the cultivated biomass is present in the nitrifying population. Usually 1 or 2 tank systems could be used to include denitrification. The BABE reactor does not need biomass retention. It can work with decreased load and SRT. BABE reactors are designed by simulation with easy testing of variables to monitor the nitrogen removal.

Emerging in situ bioaugmentation technologies used to enhance nitrifier growth and shown to be successful in bench, pilot, and/or full-scale trials are described briefly below (USEPA 2008):

- The Bio-Augmentation Regeneration/Reaeration (BAR) process was developed in the U.S. and is identical to the Regeneration-DeNitrification (R-DN) process developed independently in the Czech Republic. It mainly works by recycling ammonia-laden filtrate or centrate from dewatering of aerobically digested sludge to the head of the aeration tank. The sidestream is fully nitrified, seeding the aeration tank with additional nitrifying bacteria which allows for reduced SRT. There are numerous full-scale applications in the Czech Republic, USA and Canada. The Aeration Tank 3 (AT3) is similar to the BAR process except that it sends a smaller fraction of the return activated sludge (RAS) to the aeration tank in order to stop the nitrification process at the nitrite stage.
- The BABE process uses a sequencing batch reactor (SBR) to grow nitrifiers by feeding it RAS and reject water from the sludge dewatering process. After treatment, concentrated nitrifiers are recycled to the head of the aeration tank.
- The Mainstream Autotrophic Recycle Enhanced N-removal (MAUREEN) Process was developed for two-sludge treatment configuration at the Blue Plains Advanced Wastewater TreatmentPlant in Washington, DC. The process involves sidestream treatment of waste activated sludge (WAS) from the second stage to preferentially select aerobic ammonium oxidising bacteria (AOB) for bioaugmentation to the first sludge stage.

4.2 Nitritation

The process of producing NO_2^- from NH_4^+-N though activity of AOB under aerobic condition is nitritation. It is the preliminary step in a nitrification system.

Excess NO_2^- is then utilized by NOB for conversion to NO_3^- to complete the nitrification process.

The single reactor system for high activity ammonia removal over nitrite (SHARON) process, developed in the 1990s at Delft University of Technology, is a nitritation/partial nitrification system to oxidize half the influent ammonia-nitrogen to nitrite (Ganigue et al. 2009). As this nitritation process is limited to nitrite rather than nitrate in a conventional process, 25 % of aeration energy is saved (Mulder et al. 2006), 30 % sludge is reduced and overall 20 % less CO_2 is emitted. The stoichiometry of the process is given by the Eq. 11

$$NH_4^+ + 0.75O_2 + HCO_3^- \rightarrow 0.5NO_2^- + 0.5NH_4^+ + CO_2 + 1.5H_2O \quad (11)$$

This process can be applied to leachate because ammonia (50 %) gets oxidised to nitrite by AOB, which being an acidifying process can be neutralised by the bicarbonates present in the leachate.

The biochemistry behind the AOBs in the SHARON process involves (i) oxidation of ammonia-nitrogen to hydroxylamine (NH_2OH) by membrane bound enzyme ammonia mono-oxygenase (amo), where oxygen and dinitrogen tetroxide are the electron acceptors for this enzyme, (ii) oxidation of hydroxylamine to nitrite by the hydroxylamine oxidoreductase (HAO). In the SHARON process, the actual competitor is nitrite oxidising bacteria which prevent nitrite accumulation. Both the AOB and NOB are inhibited by free ammonia (FA) and/or free nitrous acid (FNA) (Ganigue et al. 2009). But, the NOB is more sensitive than AOB to FA.

The important operational parameters of the SHARON process are pH, temperature, hydraulic retention time (HRT)/sludge retention time (SRT), dissolved oxygen (DO) and bicarbonate to ammonium ratio (Mulder et al. 2006; Zhang et al. 2008). pH around 8 contains more NH_3 and less HNO_2, which clearly promotes AOB but suppresses NOB. Higher temperatures of 30–45 °C are optimal for the growth of AOBs. Optimum HRT is 1–1.54. Low DO concentration (0.5–1.5 mg/L) is more restrictive for the growth of NOB than AOB, which will result in nitrite accumulation. The oxidation of ammonium to nitrite requires bicarbonate for every mole of ammonium conversion.

4.3 *ANAMMOX (Deammonification)*

The anaerobic ammonium oxidation (ANAMMOX) process discovered by Mulder and co-workers in 1995 was seen as a new and promising alternative to conventional treatment systems to remove ammonia nitrogen (Reginatto et al. 2005). In comparison with conventional processes, this autotrophic process involves a complete conversion of NH_4^+-N to N_2 gas without the addition of organic matter (Dongen et al. 2001). Hence, it consumes at least 50 % less oxygen and saves up to 90 % of operational costs related to sludge disposal due to the lower growth rate of the microorganisms responsible for ANAMMOX (than the ones responsible for the nitrification/denitrification processes) besides reducing CO_2 emission (Reginatto et al. 2005). The ANAMMOX process with cell synthesis is given in the Eq. (12)

$$\begin{aligned} & NH_4^+ + 1.32NO_2^- + 0.132HCO_3^- + 0.512H^+ \\ & \rightarrow 1.02N_2 + 0.26NO_3 + 0.132CH_2O_{0.5}N_{0.15} + 2.19H_2O \end{aligned} \quad (12)$$

The ANAMMOX reaction is carried out by members of deeply branched Planctomycetes, such as '*Candidatus Brocadia anammoxidans*, '*Candidatus Kuenenia stuttgartiensis*', '*Candidatus Scalindua*', '*Candidatus Anammoxoglobus*' and '*Candidatus Jettenia*' which make use of NH_4^+-N as the electron donor (energy source) and NO_2^--N as the electron acceptor. These autotrophs utilize dissolved CO_2 or HCO_3^- from for cell biosynthesis (Berge et al. 2005), with slow growth rates of the bacteria and low biomass yield (0.13 g dry weight/g NH_4^+-N oxidized) (Chamchoi and Nitisoravut 2007; Trigo et al. 2006; Third et al. 2005; Li et al. 2009). Kinetic parameters were reported by Strous et al. (1998), for an enriched culture of *Candidatus Brocadia anammoxidans*. The substrate consumption rate was 45 nmol $NH4^+$-N/mg protein/min with a 10.3/d duplication period. Doubling times as low as 5.3 and 8.9 d were achieved by Park et al. (2010), whilst 11 and 14 d were reported by Strous et al. (1998, 2006), Li et al. (2009) from ANAMMOX-enriched reactors.

High NH_4^+-N removal rates could be obtained using the ANAMMOX process in two ways: two reactors in series, with a partial nitrification reactor as a first step, and a separate unit for the anaerobic oxidation of NH_4^+-N as a second step. With this configuration, the two biological processes can be controlled separately (Paredes et al. 2007). The second option was to use biofilm systems where classical nitrification is developed by the ammonium oxidizers in the outer aerobic layers, and anaerobic oxidation takes place in the deeper zones of the biofilm (Paredes et al. 2007). Application of the ANAMMOX process or coupling partial nitrification with ANAMMOX seems promising. It could result in 60 % savings in O_2 generation, 100 % savings of external carbon source addition, less sludge production and CO_2 emission, with a total reduction in treatment cost by 90 % (Karthikeyan and Joseph 2006). Application of the Complete Autotrophic Nitrogen removal Over Nitrite (CANON) process to achieve high removal of NH_4^+-N from wastewater could be achieved in a single O_2-limited treatment step (Third et al. 2001) for stable NH_4^+-N removal even during varying feed compositions and periods of NH_4^+-N limitations (Third et al. 2001; Padin et al. 2009).

Application of coupling partial nitrification with the ANAMMOX process was adopted by Liang and Liu (2008) for treating landfill leachate (NH_4^+-N of 1500–2500 mg/L). An integrated partial nitritation—ANAMMOX reactor—underground soil infiltration system was applied. ANAMMOX operation was performed in an upflow fixed bed biofilm reactor achieving 67 % NH_4^+-N and 77 % NO_2^--N removal within 97 days. The effluent of the partial nitritation process yielded a suitable influent for the ANAMMOX process, by yielding 50 % partial conversion of NH_4^+-N to NO_2^--N (ratio 1:1.3) favouring anaerobic ammonium oxidation. Nearly 60 % of NH_4^+-N removal was achieved at the end of ANAMMOX process, and

97 % removal was obtained at the end of the combined treatment train. From the initial COD of 1100–2500 mg/L 89 % removal of COD was achieved with the final effluent containing 30–250 mg/L, compared to almost 32 % COD removal by ANAMMOX process. The main limitation of the process could be ascribed to the low yield (0.14 g VSS/g NH_4^+-N) and slow growth rate of ANAMMOX bacteria (0.003 h^{-1}; 0.072 d^{-1} at 32 °C) resulting in slow removal of NO_3^--N (requiring half the time in aerobic nitrification) (Strous et al. 1998; Trigo et al. 2006; Third et al. 2005).

The investigation of aquatic humic substances (AHS) degradation by the ANAMMOX process was conducted by Liang et al. (2009) where the initial partial nitritation reactor was run for 166 days continuously using raw leachate, with NH_4^+-N of 1430–2720 mg/L and COD of 1170–2600 mg/L. Upon removal of VFA and acquiring the proper mixture of NO_2^--N to NH_4^+-N ratio, this effluent with NH_4^+-N of 506–885 mg/L and COD 303–954 mg/L was further treated in an ANAMMOX reactor. The pretreatment in partial nitritation enabled removal of biodegradable organics from the raw leachate, resulting in higher content of AHS in the feed to the ANAMMOX reactor (228 mg/L), reducing their content in the effluent from the ANAMMOX reactor to 91 mg/L. Dissolved Organic Carbon (DOC) content of the effluent was also reduced from 288 to 136 mg/L in the ANAMMOX reactor.

Direct application of the ANAMMOX process was adopted by Xu et al. (2007) to treat NH_4^+-N rich leachate using a Sequencing batch biofilm reactors (SBBR). The system was started up on 58 d and stabilized in 33 d, with DO of 1.2–1.4 mg/L, with alternate periods of aeration and anoxic condition. The leachate was used by spiking it with NH_4Cl to about 450 mg/L prior to feeding as influent to SBBR. The organic load was in the rage of 1876 ± 547 mg/L of COD and 1048 ± 436 mg/L of BOD_5. NLR was optimized to 300 mg/L/d, with pH around 7.3–7.8 without addition of alkali or acid. It was proposed that the repeated alteration between aeration and anoxic period neutralised the acidity generated in the aeration phase through the alkalinity produced in the anoxic phase. The ratio of NH_4^+-N/NO_2^--N was in the range of 1.058–1.074 in the aeration phase, and 0.558–0.776 in the anoxic phase, as compared to the theoretical value of 0.758 in the ANAMMOX reaction (Strous et al. 1998). It was proven that anoxic condition favoured ANAMMOX activity when weighed against oxic condition.

Guo and Qi (2006) treated aged landfill leachate in an UASB-ANAMMOX bioreactor (HRT 24 h) and achieved about 80 % total nitrogen removal efficiency from influent containing 900 mg TN/L and 88 % NH_4^+-N removal from an influent of 350 mg NH_4^+-N/L. During the study period for >200 days, average COD removal was 24 % from an influent of 1000 mg/L. Alkalinity concentrations of both the influent and effluent during the steady phase of ANAMMOX activity were 1 g/L and pH of influent and effluent were 8.3. This study indicated that alkalinity and pH could also be used to monitor ANAMMOX activity. The ratio of NO_2^--N/NH_4^+-N was in the range of 0.96–1.49, as compared to the stoichiometric value of 1.24 (Strous et al. 1998).

Table 4 Ammonia-nitrogen removal from leachate in different in situ and ex situ nitrification and denitrification studies

Sl. No	Reactor details	Nitrogen loading rate	Ammonia-nitrogen removal	References
In situ *nitrification studies*				
1.	A1—Landfill with leachate recirculation—aeration—0.084 L/(min kg) A2—Landfill without recirculation—aeration—0.086 L/(min kg)—both A1 and A2 operated for 250 d	Highest ammonia concentration—A1—1700 mg/L A2—1800 mg/L	A1—120 mg/L at 250 d A2—200 mg/L at 250 d	Bilgili et al. (2007)
2.	AR1—Landfill with aeration—2.77 L/min operated for 984 d AR2—Landfill with aeration—6.44 L/min operated for 300 d	Ammonia in waste—1000 mg/L ammonia in microcosm studies—500 mg/L	At 22 °C—30 % ammonia removal by in-situ; 35 °C—60 %—in-situ removal; 45 °C—50 %—in-situ removal	Berge et al. (2007)
In situ *denitrification studies*				
3.	AN1—Landfill with leachate recirculation AN2—Landfill without recirculation both AN1 and AN2 for 500 d	Highest ammonia concentration—AN1—2100 mg/L AN2—1950 mg/L	AN1—1000 mg/L AN2—1200 mg/L	Bilgili et al. (2007)
4.	Ra—Landfill with nitrite addition Rb—Landfill With nitrate addition	4000 mg/L (NO_2-N/NO_3-N)	Below 200 mg/L in 150 d	Fu et al. (2009)
In situ *nitrification and denitrification studies*				
5.	Three reactor system (nitrification, denitrification and methanogenic reactors) aeration—93 L of O_2/d	Separate reactors: NH_4-N—1.5 g in nitrification NO_3-N—6 g in denitrification combined with internal recycle: NH_4-N—0.509 g in nitrification NO_3-N—1.05 g in denitrification	Total: 95 % N recovery 99 % to NO_3-N 91–93 % N recovery 30–52 % to NO_3N—10 d 16–25 % to N_2—10 d	Onay and Pohland (1998)

(continued)

Table 4 (continued)

Sl. No	Reactor details	Nitrogen loading rate	Ammonia-nitrogen removal	References
6.	Landfill with in situ denitrification, methanogenesis and nitrification (aeration-826.3–8.5 mg O_2/mg N/d)	TN—4190.9 mg/kg of dry refuse extractable NH_4-N—285.7 mg/kg of dry refuse	Below 8 mg/L of NH_4-N	Long et al. (2008)
7.	Landfill with in situ denitrification, in situ methanogenesis and in situ nitrification	TN—12,330 mg/kg dry refuse highest NH_4-N 1930 mg N/kg	72 % removal in 357 d	Long et al. (2009)
8.	Landfill with leachate recirculation—in situ nitrification and denitrification	TN (inorganic)—1.0 kg N/d highest NH_4-N—652 mg/L	46 % removal in 315 days	Sri Shalini and Joseph (2013)
In situ and ex situ *nitrification and denitrification studies*				
9.	Landfill with in situ nitrification and denitrification and ex situ methanogenesis in UASB R1—aeration—0.5 L/min (8–10 h 19–21 h) R2—Without aeration	TN—4.388 ± 0.76 mg/dry g waste R1—Highest NH_4-N—800 mg/L, TN—900 mg/L R2—Highest NH_4-N—1200 mg/L, TN—1500 mg/L	R1—NH_4-N—186 mg/L in 105 d, TN—289 mg/L in 105 d R2—Above 1000 mg/L of NH_4-N and TN	He and Shen (2006)
10.	Landfill with in situ denitrification, ex situ methanogenesis in UASB and ex situ nitrification in ALSB aeration—0.0002 L/min m^3	UASB—0.015 g/L d ALSB—0.14 g/L d combined reactors: highest ammonia—1037 mg/L	UASB—66 % ALSB—100 % combined reactors with ammonia concentration below 200 mg/L	He et al. (2007)
11.	Ex situ nitrification (activated sludge reactor), landfill with in situ denitrification and in situ methanogenesis	0.18 kg NH_4-N/m^3 d	90 % removal in 132 d	Liang et al. (2008)

Table 5 Comparison of conventional and novel ammonia nitrogen removal processes

Sl No	System	Nitrification—denitrification	OLAND	CANON	SHARON	ANAMMOX
1.	Number of Reactor	Two	One	One	One	One
2.	Feed	Ammonium rich wastewater				
3.	Discharge	N_2, NO_2^-, NO_3^-	NH_4^+, N_2	N_2, NO_3^-	NH_4^+, NO_2^-	N_2, NO_3^-
4.	Conditions	Oxic, anoxic	Oxic, anoxic	Oxygen limited	Oxic	Anoxic
5.	Oxygen requirements	High	Low	Low	Low	None
6.	pH control	Yes	–	None	None	None
7.	Biomass retention	None	Yes	Yes	Yes	Yes
8.	COD requirement	Yes	None	None	None	None
9.	Alkalinity consumption (g $CaCO_3$/g N)	7.07/−3.57	3.6	3.68	–	–
10.	Sludge production	High	Low	Low	Low	Low
11.	Reactor capacity (kg N/m^3/day)	0.05–4	1	1–3	1	6–12
12.	NH_4^+ loading (kg N/m^3/day)	2–8	0.1	2–3	0.5–1.5	10–20
13.	N-removal efficiency	95 %	85 %	90 %	90 %	90 %
14.	Process complexity	Separate oxic anoxic system, methanol dosing	Aeration needs to be tuned to ammonia loading	Aeration needs to be tuned to ammonia loading	–	Preceding partial nitrification needed
15.	Application status	Established	Laboratory studies	Laboratory studies	Full scale plants	Full scale plants
16.	Investment costs	Medium	Medium	Medium	Medium	Low
17.	Operational costs	High	Unknown	Low	Low	Very low

Sources Ahn (2006), Schmidt et al. (2003) and Jetten et al. (2002)

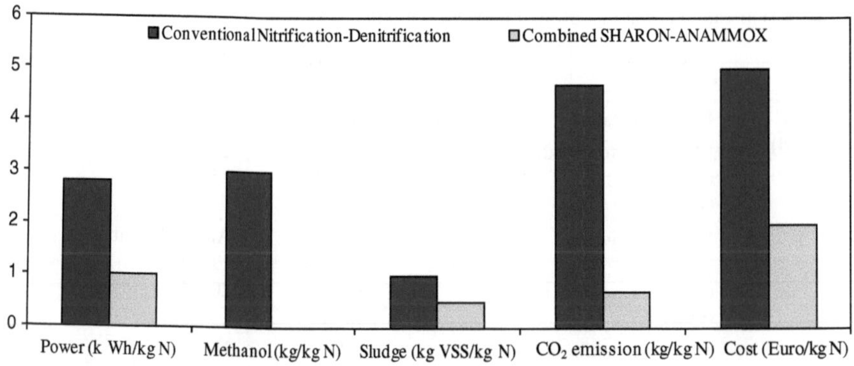

Fig. 6 Comparison between conventional nitrification-denitrification and combined SHARON-ANAMMOX processes. Adapted from Henze et al. (2008)

5 Case Studies on Ex Situ and In Situ Ammonia Nitrogen Management for Landfill Bioreactors

Conventional ex situ and in situ nitrification and denitrification studies for ammonia-nitrogen removal in landfill leachate are summarized in Table 4. To overcome the challenges of conventional processes, innovations such as completely autotrophic nitrogen removal over nitrite (CANON), oxygen-limited autotrophic nitrification and denitrification (OLAND), SHARON and ANAMMOX processes have been developed. A comparison between conventional and innovative processes of ammonia removal presented in Table 5 indicates that SHARON and ANAMMOX have several advantages over conventional processes.

Henze et al. (2008) compared conventional nitrification and denitrification processes with that of the combined SHARON-ANAMMOX processes in terms of consumption of power, methanol, sludge production, CO_2 emission and operational costs. This is depicted in Fig. 6. It demonstrated the savings in overall operational costs of the SHARON-ANAMMOX processes against that of conventional processes. These studies paved way for the application of in situ SHARON and ANAMMOX processes for bioreactor landfills.

Liang and Liu (2007) performed the SHARON process in bench-scale FBBRs treating leachate achieving ammonia: nitrite ratios of 1.0–1.4 and partial nitritation efficiency of 94 % at ammonia loading rates of 0.2–1.0 kg N/d. Further studies by Liang and Liu (2008) showed that increasing ammonium load in leachate accomplished good NH_4-N/NO_2-N ratio. The SHARON process can be used as pre-treatment for the ANAMMOX process with the influent NH_4-N/NO_2-N ratio around 1:1.32.

Spagni et al. (2008) studied nitrogen removal via nitrite in a SBR containing sanitary landfill leachate. Nitrification and nitrogen removal rate were higher than 98 and 95 %, respectively. Wei et al. (2010) studied the SHARON process in a UASB-SBR system. The removal efficiencies of ammonia-nitrogen were 99.7 %

Table 6 Operational features of the bioreactor landfills

Phases	Operation	Control bioreactor	SHARON bioreactor	ANAMMOX bioreactor
Phase I (0–42 d)	Feed/seed ratio	100/0	80/20	60/40
	Mined waste (Feed) (v/v) (w/w)	43 L 45.4 kg	34.4 L 36.3 kg	25.8 L 27.3 kg
	Enriched biomass (Seed) (v/v) (w/w)	Nil Nil	AOB biomass 8.6 L 8.6 kg	AnAOB biomass 17.2 L 20.6 kg
	Tap water addition	10 L	Nil	Nil
	Leachate recirculation	Weekly	Weekly	Weekly
	Leachate Sampling	Weekly	Weekly	Weekly
Phase II (43–188 d)	Ammonium source addition (1000 ppm) (NH_4Cl −127.8 g)	Yes	Yes	Yes
Phase III (189–315 d)		–	Start- up of SHARON bioreactor	Combining SHARON-ANAMMOX process

Adapted from Sri Shalini and Joseph (2013)

and TN was 99.2 %. Vilar et al. (2010) conducted the SHARON process in a continuous stirred tank reactor (CSTR) fed with anaerobically pre-treated leachate at a nitrogen load of 1.1 kg N/m^3/d at 36 °C. Valencia et al. (2011) showed the intrusion of small quantities of oxygen caused in situ nitrification and promoted the growth of ANAMMOX bacteria in the bioreactor which contributed to removal of nitrogen (40 %) from solid matrix in 380 days.

Sri Shalini and Joseph (2013), demonstrated the feasibility for the application of in situ SHARON and ANAMMOX processes in bioreactor landfills for ammonia-nitrogen removal. The operational features of the bioreactor landfills are given in Table 6. The SHARON and ANAMMOX bioreactor landfill was inoculated with AOB and anaerobic ammonium oxidising bacteria (AnAOB) biomass enriched from mined municipal solid waste from a laboratory-scale SHARON and ANAMMOX reactors which was successfully operated in batch mode. The details of the AOB and AnAOB biomass enrichment from municipal solid waste is detailed in Sri Shalini et al. (2015).

The variations in ammonia nitrogen, nitrite- and nitrate-nitrogen in the different bioreactor landfills is depicted in Fig. 7. Bioreactor landfills with the SHARON process having DO <1.0 mg/L gave 98.5 % nitrite accumulation, 85 % of ammonia-nitrogen removal efficiency with AOB population of a most probable number (MPN) of 5.1 × 10^6/mL. The start-up of the ANAMMOX process in bioreactor landfill obtained ammonia to nitrite ratio of 1:0.6–1:1.8. It contributed for the maximum ammonia-nitrogen removal efficiency of 73 % with specific

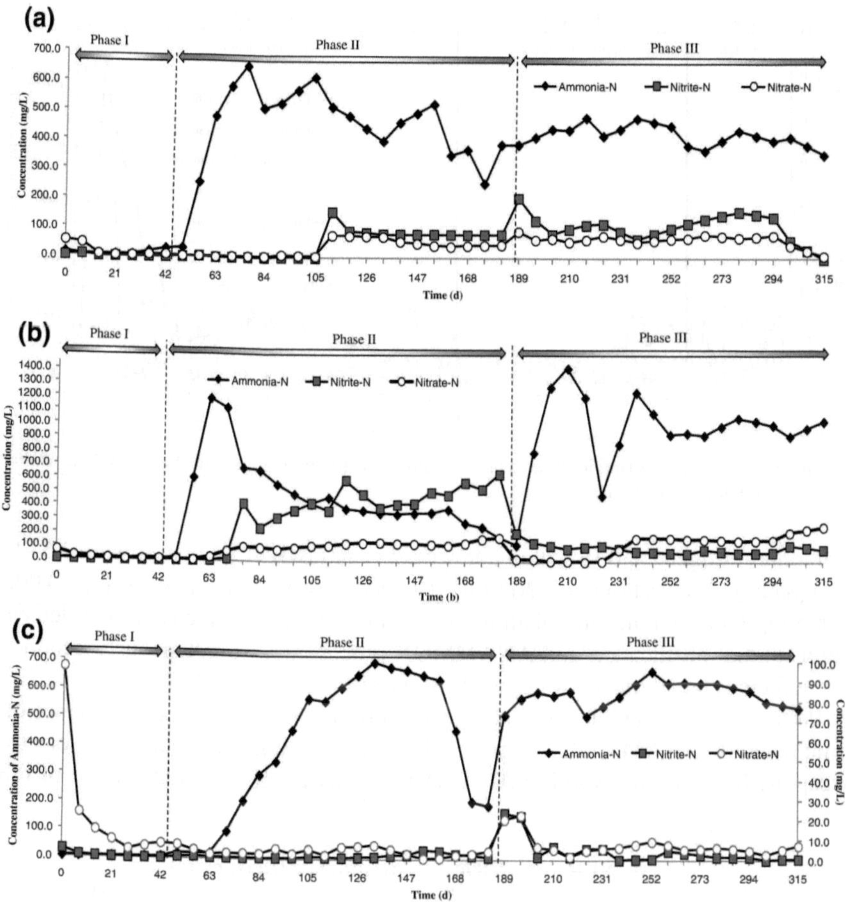

Fig. 7 Nitrogen transformations in bioreactor landfills. **a** Control bioreactor. **b** SHARON bioreactor. **c** ANAMMOX bioreactor. *Source* Sri Shalini and Joseph (2013)

ANAMMOX activity of 0.96 mg Amm-N/mg MLVSS/d. Bioreactor landfill gave evolution of 95 % of N_2 gas as the end product confirmed the ANAMMOX process.

Bioreactor landfill operated as combined SHARON-ANAMMOX bioreactor landfills had various advantages. For combining the SHARON-ANAMMOX process in bioreactor landfill, the leachate from the bioreactor landfill operated with SHARON process was recirculated into the bioreactor landfill operated with ANAMMOX process (Table 6). Operational parameter results showed optimum pH (7.0–8.0) and temperature (30–33 °C) for both the SHARON and ANAMMOX processes prevailed in the bioreactor landfill. Conductivity (15.3–17.7 mS/cm) and salinity (8.8–11.5 psu) concentrations were maintained in the reactor.

The nitrogen transformations in the combined SHARON-ANAMMOX processes are given in Fig. 7c (189–315 days). Higher removal rate of

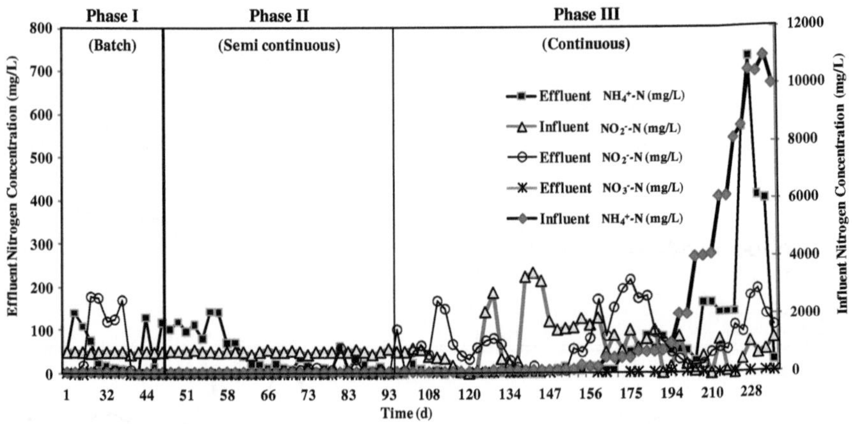

Fig. 8 Nitrogen transformations at varying nitrogen loading rates in the AnMBR. *Source* Suneethi and Joseph (2011b)

ammonia-nitrogen (>30 % within a few weeks) and BOD/COD ratios were 0.5–0.7. Adequate concentrations of NH_2OH (0.0004–0.001 mg/L) and N_2H_4 (0.002–0.005 mg/L) accumulated. Alkalinity concentrations were increasing and decreasing as partial nitrification and ANAMMOX process was simultaneously occurring but sufficient bicarbonates were available in bioreactor landfill (286–914 mg/L).

Partial nitritation efficiency (PNE) reduced from 53 to 6 % showed evidence of a simultaneous ANAMMOX process taking place (utilising nitrite as electron acceptor) based on the specific ANAMMOX activity (SAA) in the range of 0.3–

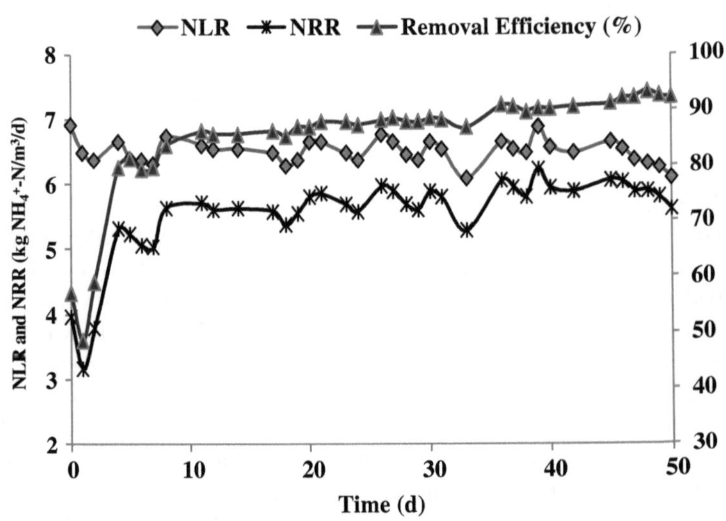

Fig. 9 Nitrogen removal performance in AnMBR treating landfill leachate. *Source* Suneethi and Joseph (2013)

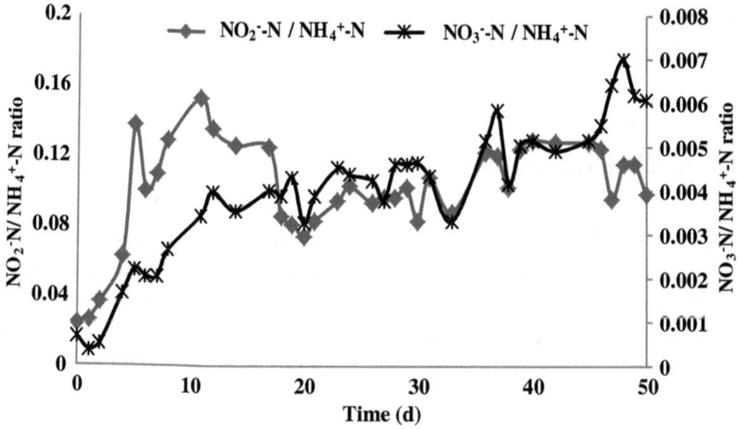

Fig. 10 Molar ratio during landfill leachate treatment in AnMBR. *Source* Suneethi and Joseph (2013)

0.7 mg Amm-N/mg MLVSS/d. Higher biomass build up was developed in the bioreactor landfill (1000–200 mg/L). FA (<61.3 mg/L) and FNA (<0.0075 mg/L) concentrations showed only NOBs were inhibited; AOB and ANAMMOX populations were not inhibited. In situ combined SHARON-ANAMMOX processes in landfill bioreactors gave total nitrogen removal of 84 % and ammonia-nitrogen removal efficiency of 71 % at NLR of 1.2 kg $N/m^3/d$ in 147 days. The nitrogen could be removed effectively in SHARON and ANAMMOX processes in bioreactor landfill which was successfully started and stably operated for longer operational days (Sri Shalini and Joseph 2013).

In the study on ex situ strategy for NH_4^+-N removal by Suneethi and Joseph (2011b, 2013), ANAMMOX process was applied using AnMBR to remove NH_4^+-N from landfill leachate with low COD. The profile of NH_4^+-N, NO_2^--N and NO_3^--N in the AnMBR during the startup period is depicted in Fig. 8. NO_3^--N generated was not significant (mean—1.64 mg/L) during the study period. The average effluent NH_4^+-N, NO_2^--N, NO_3^--N were 64.22 ± 114.95, 54.15 ± 62.83 and 1.64 ± 1.79 mg/L, respectively.

With the actual landfill leachate, the AnMBR was operated at NLR 6.51 ± 0.20 kg NH_4^+-$N/m^3/d$ at 1.5 d HRT and attained a NH_4^+-N removal efficacy of 85.13 ± 9.67 % as indicated in Fig. 9. The stoichiometric ratio (NH_4^+-N removed: NO_2^--N converted: NO_3^--N produced) indicating the ANAMMOX process of 1:1.32:0.26 (Strous et al. 1998) was verified during the landfill leachate study in the AnMBR. The ratio acquired was 1:0.10:0.003 as depicted in Fig. 10. This was lower than the ratio obtained during the simulated leachate study in the AnMBR (1:0.84:0.02) (Suneethi and Joseph 2011b). Gradual acclimation of ANAMMOX biomass was initiated from the anaerobic seed obtained from biosolids digester (Suneethi and Joseph 2011a) from simulated to actual leachate in the AnMBR. The nitrogen profile changes from N_2H_4, NH_2OH, NH_3 and HNO_2

concentrations were indicative of AOB and ANAMMOX activity with poor/low NOB activity enabling sustained NH_4^+-N removal.

6 Research Directions

From the detailed discussion on nitrogen biotransformation in bioreactor landfills, it could be understood that the incursion of small quantities of O_2 is beneficial for the biodegradation process of MSW and can also trigger in situ nitrification. This would promote the growth of ANAMMOX bacteria. The presence of ANAMMOX bacteria in Bioreactor Landfills simulators could to a certain extent contribute to the removal of N from the solid matrix. Hence it would be interesting to plot the metabolic pathways of Nitrogen transformations both in situ and ex situ when treating different aged leachate and various stages of stabilised waste. The toxicity and effects of nitrite oxidation inhibitors such as heavy metals that are commonly found in the landfill leachate is rarely researched. The studies on combined application of biomolecular, biokinetics and mass balance approach to track down the Nitrogen profile from the micro to the macro level of the physico-chemical environment and the biosystem would be beneficial to develop bioreactor landfill management strategies. Further activities need to focus on field-scale demonstration of in situ ammonia-nitrogen removal in landfill bioreactors and assessing the effect of different environmental conditions affecting the important operational parameters of the processes and determination of SHARON and ANAMMOX pathway in the landfill bioreactors.

7 Summary

This chapter presented the bioreactor landfill concept, ammonia-nitrogen profile in landfills, and different ex situ and in situ removal methods for nitrogen management in landfills. The various nitrogen profile in the form of NH_4^+-N, NO_2^--N and NO_3^--N in the solid waste environment and its further changes have been discussed. The NH_4^+-N removal processes including conventional and advanced technologies and its variations with the support of laboratory-scale and field-scale case studies were also discussed. Bioreactor landfill integrated with SHARON and ANAMMOX processes provides promising results for nitrogen management (Total nitrogen removal—84 % and ammonia-nitrogen removal efficacy—71 % at NLR of 1.2 kg N/m^3/d in 147 days). It resulted in significant saving of energy, reduction in investment costs, space requirements and operating costs. Full-scale demonstrations for combined SHARON-ANAMMOX in bioreactor landfills are needed.

References

Ahn YH, Hwang IS, Min KS (2004) Anammox and partial denitritation in anaerobic nitrogen removal from piggery waste. Water Sci Technol 49(5):145–153

Ahn YH (2006) Sustainable nitrogen elimination biotechnologies: a review. Process Biochem 41:1709–1721

Amokrane A, Comel C, Veron J (1997) Landfill leachate pretreatment by coagulation—flocculation. Water Res 31:2775–2782

Anthonisen AC, Loehr RC, Prakasam TBS, Srinath EG (1976) Inhibition of nitrification by ammonia and nitrous acid. J Water Pollut Control Fed 48:835–852

Barlaz MA, Ham RK, Schaefer DM (1990) Methane production from municipal refuse: a review of enhancement techniques and microbial dynamics. Crit Rev Environ Control 19:557–584

Benson CH, Barlaz MA, Lane DT, Rawe JM (2007) Practice review of five bioreactor/recirculation landfills. Waste Manag 27:13–29

Berge ND, Reinhart DR, Dietz J, Townsend TG (2005) The fate of nitrogen in bioreactor landfills. Crit Rev Environ Sci Technol 35:365–399

Berge ND, Reinhart DR, Dietz J, Townsend TG (2006) In situ ammonia removal in bioreactor landfill leachate. Waste Manag 26:334–343

Berge ND, Reinhart DR, Fasce PE, Batarseh ES (2007) Strategy for complete nitrogen removal in bioreactor landfills. J Environ Eng 133:1117–1125

Bilgili SM, Demir A, Ozkaya B (2007) Influence of leachate recirculation on aerobic and anaerobic decomposition of solid wastes. J Hazard Mater 143:177–183

Bodzek M, Moysa MEL, Zamrowska M (2006) Removal of organic compounds from municipal landfill leachate in a membrane bioreactor. Sci Technol 198:16–23

Bohdziewicz J, Neczaj E, Kwarciak A (2008a) Landfill leachate treatment by means of anaerobic membrane bioreactor. Desalination 221:559–565

Bohdziewicz J, Neczaj E, Kwarciak A (2008b) The application of hybrid system UASB reactor-RO in landfill leachate treatment. Desalination 221:128–134

Burton SAQ, Watson-Craik IA (1998) Ammonia and nitrogen fluxes in landfill sites: applicability to sustainable landfilling. Waste Manag Res 16(1):41–53

Burton SAQ, Watson-Craik IA (2002), Ammonia fluxes in landfills. R&D Technical Report P1-306/TR, ISBN: 1-84432-067-7

Canziani R, Emondi V, Garavaglia M, Malpei F, Pasinetti E, Buttiglieri G (2007) Effect of oxygen concentration on biological nitrification and microbial kinetics in a cross-flow membrane bioreactor (MBR) and moving-bed biofilm reactor (MBBR) treating old landfill leachate. J Memb Sci 286:202–212

Chamchoi N, Nitisoravut S (2007) ANAMMOX enrichment from different conventional sludges. Chemosphere 66:2225–2232

Connolly R, Zhao Y, Sun G, Allen S (2004) Removal of ammoniacal—nitrogen from an artificial landfill leachate in down flow reed beds. Process Biochem 39:1971–1976

Dongen U, Jetten MSM, van Loosdrecht MCM (2001) The SHARON-ANAMMOX process for treatment of ammonium rich wastewater. Water Sci Technol 44:153–160

El Fadel M, Findikakis AN, Leckie JO (1997) Environmental impacts of solid waste landfilling. J Environ Manag 50:1–25

Ehrig HJ (1998) Water and element balances of landfills. Springer, Berlin

Erses AS, Onay TT, Yenigun O (2008) Comparison of aerobic and anaerobic degradation of municipal solid waste in bioreactor landfills. Bioresour Technol 99:5418–5426

Ganigue R, Gabarro J, Sanchez-Melsio A, Ruscalleda M, Lopez H, Vila X, Colprim J, Balaguer MD (2009) Long-term operation of a partial pilot plant treating leachate with extremely high ammonium concentration prior to an anammox process. Bioresour Technol 100:5624–5632

Fu Z, Yang F, An Y, Xue Y (2009) Characteristics of nitrite and nitrate in situ denitrification in landfill bioreactors. Bioresour Technol 100:3015–3021

Fux C, Boehler M, Huber P, Brunner I, Siegrist H (2002) Biological treatment of ammonium-rich wastewater by partial nitritation and subsequent anaerobic ammonium oxidation (anammox) in a pilot plant. J Biotechnol 99:295–306

Giannis A, Makripodis G, Simantiraki F, Somara M, Gidarakos E (2008) Monitoring operational and leachate characteristics of an aerobic simulated landfill bioreactor. Waste Manag 28:1346–1354

Guo ZH, Qi ZS (2006) Treating leachate mixture with anaerobic ammonium oxidation technology. J Cent South Univ Technol 6:663–667

Hammer MJ, Hammer Jr MJ (2000) Water and wastewater technol, Fourth Edition. Prentice—Hall Inc., New Jersey

Han DW, Yun HJ, Kim DJ (2001) Autotrophic nitrification and denitrification characteristics of an upflow biological aerated filter. J Chem Technol Biotechnol 76(11):1112–1116

He R, Shen D (2006) Nitrogen removal in bioreactor landfill system with intermittent aeration at the top of the landfilled waste. J Hazard Mat 136:784–790

He R, Liu X, Zhang Z, Shen D (2007) Characteristics of the bioreactor landfill system using an anaerobic—aerobic process for nitrogen removal. Bioresour Technol 98:2526–2532

Henze M, Loosdrecht MCMV, Ekama G, Bridjanovic D (2008) Biological wastewater treatment, principles, modelling and design. IWA Publishing, pp 139–151

Huo SL, Xi B, Yu HC, Fan SL, Jing S, Liu HL (2008) A laboratory simulation of in situ leachate treatment in semi-aerobic bioreactor landfill. Water SA 34(1):133–140

Jetten MSK, Schmid M, Schmidt I, Wubben M, Dongen UV, Abma W, Sliekers O, Revsbech NP, Beaumont HJE, Ottosen L, Volcke E, Laanbroek HJ, Campos-Gomez JL, Cole J, van Loosdrecht M, Mulder JW, Fuerst J, Richardson D, van de Pas K, Mendez-Pampin R, Third K, Cirpus I, van Spanning R, Bollmann A, Nielsen LP, Op den Camp H, Schultz C, Gundersen J, Vanrolleghem P, Strous M, Wagner M, Kuenen JG (2002) Improved nitrogen removal by application of new nitrogen-cycle bacteria. Rev Environ Sci Biotechnol 1:51–53

Jha MK, Sondhi OAK, Pansare M (2003) Solid waste management—a case study. Indian J Environ Prot 23(10):1153–1160

Jianguo J, Guodong Y, Zhou D, Yunfeng H, Zhonglin H, Xiangming F, Shengyong Z, Chaoping Z (2007) Pilot—scale experiment on anaerobic bioreactor landfills in China. Waste Manag 27:893–901

Karthikeyan OP and Joseph K (2006) Anammox a novel process for nitrogen management in bioreactor landfills—a review. National seminar on integrated municipal solid waste management, Periyar Maniammai College of Technology for Women, Thanjavur, Tamilnadu, India

Karthikeyan OP, Joseph Kurian (2008) Chemical precipitation of NH_4-N as struvite from landfill leachate effect of molar ratio upon recovery. J Solid waste Technol Manag 34(1):20–26

Karthikeyan OP, Swati M, Nagendran R, Joseph K (2007) Performance of bioreactor landfill with waste mined from a dumpsite. Environ Monit Assess 135(1–3):141–151

Kim J, Pohland F (2003) Process enhancement in anaerobic bioreactor landfills. Water Sci Technol 48:29–36

Krumpelbeck I, Ehrig HJ (1999) Long-term behaviour of municipal solid waste landfills in Germany. In: Proceedings 7th international waste management and landfill symposium, Cagliari, Italy

Lau IWC, Wang P, Fang HHP (2001) Organic removal of anaerobically treated leachate by Fenton coagulation. J Environ Eng Sci 127:666–669

Liang SH, Bei X, Hai Y, Shi-lei F, Jing S, Hong L (2008) In situ simultaneous organics and nitrogen removal from recycled landfill leachate using an anaerobic-aerobic process. Bioresour Technol 99:6456–6463

Liang Z, Liu JX (2007) Control factors of partial nitritation for landfill leachate treatment. J Environ Sci (China) 19:523–529

Liang Z, Liu J (2008) Landfill Leachate Treatment with a novel process: anaerobic ammonium oxidation (Anammox) combined with soil infiltration system. J Hazard Mat 151:202–212

Liang Z, Liu JX, Li J (2009) Decomposition and mineralization of aquatic humic substances (AHS) in treating landfill leachate using the Anammox process. Chemosphere 74:1315–1320

Li X, Du B, Fu H, Wang R, Shi J, Wang Y, Jetten MSM, Quan Z (2009) The bacterial diversity in an anaerobic ammonium-oxidizing (anammox) reactor community. Syst Appl Microbiol 32:278–289

Liu DHF, Liptak BG (1999) Environmental engineer's handbook, CRCnet base 1999. CRC Press LLC, International Standard Book Number 0-8493-2157-3 and International Standard Series Number 1523-3197

Long Y, Lao H, Hu L, Shen D (2008) Effects of in situ nitrogen removal on degradation/stabilization of MSW in bioreactor landfill. Bioresour Technol 99:2787–2794

Long Y, Lf Hu, Shen DS (2009) Nitrogen transformation in the hybrid bioreactor landfill. Bioresour Technol 100(9):2527–2533

Metcalf and Eddy (2002) Wastewater engineering: treatment and reuse, 4th edn. Mc-Graw Hill Science Engineering, New York

Munch E, Barr K (2001) Controlled struvite crystallization for removing phosphorus from anaerobic digester side streams. Water Res 35(1):151–159

Mulder A (2003) The quest for sustainable nitrogen removal technologies. Water Sci Technol 48 (1):67–75

Mulder JW, Duin JOJ, Goverde J, Poiesz WG, van Veldhuizen HM, van Kempen R, Roeleveld P (2006) Full-scale experience with the SHARON process through the eyes of the operators. Water Environ Found 5256–5270

O'Flaherty V, Collins G, Mahony T (2006) The microbiology and biochemistry of anaerobic bioreactors with relevance to domestic sewage treatment. Rev Environ Sci Biotechnol 5:39–55

Onay TT, Pohland FG (1998) In-situ nitrogen management in controlled bioreactor landfills. Water Res 32(5):1383–1392

Padin JRV, Pozob MJ, Jarpac M, Figueroaa M, Francoa A, Mosquera-Corral A, Camposa JL, Mendeza R (2009) Treatment of anaerobic sludge digester effluents by the CANON process in an air pulsing SBR. J Hazard Mat 166:336–341

Paredes D, Kuschk P, Mbwette TSA, Stange F, Muller RA, Koser H (2007) New aspects of microbial nitrogen transformations in the context of wastewater treatment—a review. Eng Life Sci 7(1):13–25

Park H, Rosenthal A, Jezek R, Ramalingam K, Filios J, Chandran K (2010) Impact of inocula and growth mode on the molecular microbial ecology of anaerobic ammonia oxidation (anammox) bioreactor communities. Water Res 44:5005–5013

Pohland FG (1980) Leachate recycle as a landfill management option. J Environ Eng 107:1057–1071

Price GA, Barlaz MA, Hater GR (2003) Nitrogen management in bioreactor landfills. Waste Manag 23:675–688

Purcell BE, Butler AP, Sollars CJ, Buss SE (1999) Leachate ammonia flushing from landfill simulators. Water Environ J 13:107–111

Reddy S, Galab S (1998) An integrated economic and environmental assessment of solid waste management in India. The Case of Hyderabad, India

Reginatto V, Teixera RM, Pereira F, Schmidell W, Furigo A, Menes R, Etchebehere C, Soares HM (2005) Anaerobic ammonium oxidation in bioreactor treating slaughterhouse wastewater. Braz J Chem Eng 22:593–600

Reinhart DR, Al-Yousfi BA (1996) The Impact of Leachate Recirculation of MSW Landfill Leachate and Gas Characteristics. Waste Manag Res 14:337

Reinhart DB, Townsend TG (1998) Landfill bioreactor design & operation. CRC Press LLC., Boca Raton. ISBN 1-56670-259-3

Renou S, Givaudan JG, Poulain S, Dirassouyan F, Moulin P (2008) Landfill leachate treatment: review and opportunity. J Hazard Mat 150(3):468–493

Robinson HD (1995) The technical aspects of controlled waste management—a review of the composition of leachates from domestic wastes in landfill sites. Report for the UK Department of the Environment Waste Sci. Res., Aspinwall & Company, Ltd., London, UK

Ruiz G, Jeison D, Rubilar O, Ciudad G, Chamy R (2006) Nitrification—denitrification via nitrite accumulation for nitrogen removal from wastewaters. Bioresour Technol 97:330–335

Schmidt I, Sliekers O, Schmid M, Bock E, Fuerst J, Kuenen JG, Jetten MSM, Strous M (2003) New concepts of microbial treatment processes for the nitrogen removal in wastewater. FEMS Microbiol Rev 27:481–492

Sri Shalini S, Joseph K (2012) Nitrogen management in landfill leachate: application of SHARON, ANAMMOX and Combined SHARON-ANAMMOX process—a review. Waste Manag 32:2385–2400

Sri Shalini S, Joseph K (2013) Start-up of the SHARON and ANAMMOX process in landfill bioreactors using aerobic and anaerobic ammonium oxidising biomass. Bioresour Technol 149:474–485

Sri Shalini S, Joseph K, Wong JWC (2015) Aerobic and anaerobic ammonium oxidising bacterial enrichment from municipal solid waste. Int J Environ Sci Technol 12:1107–1122

Spagni A, Marsili-Libelli S, Lavagnolo MC (2008) Optimisation of sanitary landfill leachate treatment in a sequencing batch reactor. Water Sci Technol 58:337–343

Strous M, Heijnen JJ, Kuenen JG, Jetten MSM (1998) The sequencing batch reactor as a powerful tool for the study of slowly growing anaerobic ammonium-oxidizing microorganisms. Appl Microbiol Biotechnol 50:596–598

Strous M, Pelletier E, Mangenot S, Rattei T, Lehner A, Taylor MW, Horn M, Daims H, Bartol-Mavel D, Wincker P, Barbe V, Fonknechten N, Vallenet D, Segurens B, Schenowitz-Truong C, Medigue C, Collingro A, Snel B, Dutilh BE, Op den Camp HJM, van der Drift C, Cirpus I, van de Pas-Schoonen KT, Harhangi HR, van Niftrik L, Schmid M, Keltjens J, van de Vossenberg J, Kartal B, Meier H, Frishman D, Huynen MA, Mewes HW, Weissenbach J, Jetten MSM, Wagner M, Le Paslier D (2006) Deciphering the evolution and metabolism of an anammox bacterium from a community genome. Nature 440(7085):790–794

Suneethi S, Joseph K (2011a) Batch culture enrichment of ANAMMOX populations from anaerobic and aerobic seed cultures. Bioresour Technol 102:585–591

Suneethi S, Joseph K (2011b) ANAMMOX process startup and stabilization with an anaerobic seed in anaerobic membrane Bioreactor (AnMBR). Bioresour Technol 102:8860–8867

Suneethi S, Joseph K (2013) Autotrophic ammonia removal from landfill leachate in anaerobic membrane bioreactor (AnMBR). Environ Technol 34(24):3161–3167

Swati M, Karthikeyan OP, Joseph K, Nagendran R (2007) Landfill bioreactor: a biotechnological solution for waste management. J Sci Ind Res 66:670–674

Tchobanoglous F, Theisen H, Vigil SA (1993) Integrated solid waste management: engineering principles and management issues. McGraw-Hill, New York

Tengrui L, Al-Harbawi AF, Bo LM, Jun Z, Long XY (2007) Characteristics of nitrogen removal from old landfill leachate by sequencing batch biofilm reactor. Am J Appl Sci 4(4):211–214

Third KA, Sliekers AO, Kuenen JG, Jetten MSM (2001) The CANON system (completely autotrophic nitrogen-removal over nitrite) under ammonium limitation: interaction and competition between three groups of bacteria. Syst Appl Microbiol 24(4):588–596

Third KA, Paxman J, Schmid M, Strous M, Jetten MSM, Cord-Ruwisch R (2005) Enrichment of ANAMMOX from activated sludge and its application in the CANON process. Microb Ecol 459:236–244

Townsend TG, Miller WL, Earle JFK (1995) Leachate recycle infiltration ponds. J Environ Eng ASCE 121:465–471

Trigo C, Campos JL, Garridio JM, Mendez R (2006) Startup of the ANAMMOX process in a membrane bioreactor. J Biotech 126:475–487

Turker M, Celen I (2010) Chemical equilibrium model of struvite precipitation from anaerobic digester effluents. Turk J Eng Environ Sci 34:39–48

USEPA (1993) Nitrogen control manual, EPA/625/R-93/010, office of research and development. U.S. Environmental Protection Agency, Washington, DC

USEPA (United States Environmental Protection Agency) (2008) Emerging technologies for wastewater treatment and in-plant wet weather management. EPA 832-R-06-006

USEPA (United States Environmental Protection Agency) (2009) Of, state, and technology review. "Nutrient Control Design Manual State of Technology Review Report" no. EPA/600/R-09/012 January (2009)

Valencia R, van der Zon W, Woelders H, Lubberding HJ, Gijzen HJ (2005) Evolution and fate of nitrogen compounds in pilot scale bioreactor landfills. In: Proceedings Sardinia, tenth international waste management and landfill symposium, Cagliari, Italy

Valencia R, van der Zon W, Woelders H, Lubberding HJ, Gijzen HJ (2009a) Achieving "final storage quality" of municipal solid waste in pilot scale bioreactor landfills. Waste Manag 29 (1):78–85

Valencia R, Zon WVD, Woelders H, Lubberding HJ, Gijzen HJ (2009b) The effect of hydraulic conditions on waste stabilisation in bioreactor landfill simulators. Bioresour Technol 100 (5):1754–1761

Valencia R, van der Zon W, Woelders H, Lubberding HJ, Gijzen HJ (2011) Anammox: an option for ammonium removal in bioreactor landfills. Waste Manag 31(11):2287–2293

Vilar A, Eiroa M, Kennes C, Veiga MC (2010) The SHARON process in the treatment of landfill leachate. Water Sci Technol 61(1):47–52

Walsh P, O'Leary P (2002) Landfill bioreactor design and operation. Waste Age 72–76

Wei SH, Qing Y, GuoRi D, HongXun H, ShuJun Z, Ying YY, Yong PZ (2010) Achieving the nitrite pathway using FA inhibition and process control in UASB-SBR system removing nitrogen from landfill leachate. Sci China Chem 53:1210–1216

Xu ZY, Hui Z, Guangming Z, Yong X, Jiuhua D (2007) Mechanism studies on nitrogen removal when treating ammonium-rich leachate by sequencing batch biofilm reactor. Front Environ Sci Eng China 1(1):43–48

Yuen STS, Styles JR, McMahon TA (1994) Process-based landfills achieved by leachate recirculation—a critical review and summary. Centre for Environmental Applied Hydrology Report, University of Melbourne, November 1994

Zhang L, Zheng P, Tang C, Jin R (2008) Anaerobic ammonium oxidation for treatment of ammonium-rich wastewaters. J Zhej Univer SCI B 9:416–426

Biofuel Production Technology and Engineering

Hugo Iván Velázquez-Sánchez, Alexis Saldivar-García and Ricardo Aguilar-López

Abstract Biofuels are leading a group of alternative energy sources due the fact they can make use of organic waste as feedstock and be more environmentally friendly than fossil ones. One of the most attractive ones is the use of butanol as a gasoline enhancer or substitute, as both compounds share significant physico-chemical properties such as energy content. Current research results show that it's possible to use agro-industrial waste as feedstock thanks to the discovery of new species and saccharification technologies. In this work a basic outline of the state-of-the-art overview of biofuel technologies, their properties and current challenges is presented. The potential of the use of saccharification processes into biofuel producing ones as a way to take advantage of the wide array of agro-industrial waste currently generated as feedstock is discussed, and finally a brief introduction to the ABE (acetone-butanol-ethanol) fermentation system is given, as it is a pathway for butanol production by biological means by the bacterium *Clostridium*. Although wide array of sugars can be used, some of the current challenges and strategies to address the problems inherent to the biological system, such as low productivities and inhibitory effects caused by solvents accumulation into the reactor is discussed. Finally, this chapter will close with a brief analysis of the scope of these strategies within the context of bioprocess engineering, showcasing the efforts made in this context to adapt new fermentation regimes to increase the system's butanol productivity based on modelling and simulation techniques.

Keywords Agro-industrial waste · Lignocellulose · Saccharification · Biofuels · Butanol · *Clostridium*

H.I. Velázquez-Sánchez · A. Saldivar-García · R. Aguilar-López (✉)
Centro de Investigación y de Estudios Avanzados del Instituto Politécnico Nacional,
Av. Instituto Politécnico Nacional No. 2508, San Pedro Zacatenco,
07340 Mexico City, Mexico
e-mail: raguilar@cinvestav.mx

O.P. Karthikeyan et al. (eds.), *Recycling of Solid Waste for Biofuels and Bio-chemicals*, Environmental Footprints and Eco-design of Products and Processes, DOI 10.1007/978-981-10-0150-5_10

1 Introduction

It is well known that humanity is currently experiencing a major energy crisis because the main source of it comes from a non-renewable resource such as oil. The reality of the oil sector indicates that reserves are at a critical level, therefore now research effort is focused on the study and development of processes aimed to consolidate alternative energy production and distribution technologies (Demirbas 2009).

One of the key objectives set out in the approach to the use of alternative energy technologies is to reduce the environmental impact caused by both the exploitation and use of fossil fuel energy, especially considering the great environmental harm caused by extraction, refining and use of oil derivatives, which represents a threat to both terrestrial and marine ecosystems due to leakage of aromatic hydrocarbons (Brown 2003). Although organisms of bacterial or plant origin can remediate contaminated sites, is also known that many of these organisms may produce intermediate compounds that could exhibit greater toxic effect on the biota of affected sites than the original contaminants. Additionally, the use of petroleum fuels in conventional internal combustion engines also generates toxic compounds, resulting from the partial oxidation of these and their impurities, that appreciably contribute to increase the emission of greenhouse effect gases as CO_2, CH_4, NOx and SOx (Escobar et al. 2008).

Therefore, when looking for strategies to develop and exploit alternative energy sources, greater emphasis on generating strategies to provide added value to the operation should be given, as this aims to increase the success rate of the process for scale-up to commercial levels without suffering undue pressure from an economic perspective (Dufey 2006).

Within the overall progress made in research and development of alternative energy technologies, a vast set of technologies exists aimed to harness sunlight, air currents, geothermal activity or nuclear energy. However, although there is a wide range of possibilities, most of the technologies currently developed for the exploitation of these sources have very low conversion yields and high infrastructure costs associated with their exploitation. In addition, it must be considered that these types of alternative energy sources are often limited to generate electricity, but cannot easily cover the growing problem of fuel supply for the transport sector (Melaina et al. 2013). Attempts to incorporate the above mentioned technologies not only requires implementation, for example, of a solar panel system, but also the modification of the engine and its mechanical system to make it compatible with this new technology, and that would arise the need to replace almost all of existing vehicles today by those that could function with the new power supply, which implies further economic and social issues (Ogden et al. 2004).

It is noteworthy that none of the technologies presented in the previous paragraph meet the criteria mentioned above of adding additional benefits outside reducing pollution derived from their use, especially considering that another concerning issue today is the current rate of waste generation and the environmental pollution caused by the various human activities, and the use of renewable energies like wind and solar ones do not provide direct answers to said problems.

Biofuels, which can be defined as all compounds of organic nature derived from living beings and their metabolism, lead a group of alternative energy sources aimed to provide solution to the issues raised above, as they seem more suitable to address specific issues that other technologies cannot overcome, such the use of organic waste as raw material (Maddipati et al. 2011). Additionally, biofuels processes have extensive theoretical knowledge that supports them, as fermentation technology is almost as ancient as humanity (Naik et al. 2010). However, current research and development of biofuel technologies cannot achieve reasonable production levels to make them attractive from the economic point of view due both low conversion yields and availability of feedstock materials.

One of the most studied biofuels in the last decade is butanol, as opposed to the currently developed processes for the production of ethanol and biodiesel, which could offer better answers regarding fuel mileage yield, lower gaseous emissions and better physico-chemical properties while also being able to make use of agro-industrial waste as feedstock requiring minimal or null pre-treatment to obtain it (Lee et al. 2008a, b).

Therefore in this chapter there will be a general showcase of the state-of-the-art biofuel technology, which leads to the current advances and challenges of said technologies regarding the use of agro-industrial waste as feedstock, followed by a brief discussion about the down-side of the use of ethanol as fuel and the advantages offered by butanol for the automotive sector. Then butanol production via bioprocesses and the advances made in this field of bioengineering to make use of biomass-derived residues as substrate will be discussed. Finally, efforts made in our work group to find novel fermentation techniques that could allow to improve the productivity of the butanol producing systems from a kinetic modelling and simulation perspective will be introduced.

2 Biofuels

Biofuels can be classified according to the aggregation state in which they are useful as fuels. Within gaseous biofuels, the best known representative is the so-called biogas, which is a mixture consisting primarily of methane, carbon dioxide and other trace elements. This type of fuel is also found relatively easily in the environment and can come from natural sources, such as swamps or lakes, or from anthropogenic sources such as livestock. The main virtue of this type of compounds is their high calorific value, whereby large volumes of gas are not required to achieve acceptable production of energy. However the most important issues to overcome before its use as a reliable energy source is the large amount of infrastructure needed to efficiently collect such compounds and the difficulty and risk for its transport and storage, since pressure and temperature of the vessels and ducts employed for such task need a strict control to prevent explosions.

Solid biofuels, made up mostly of plant biomass as wood, straw or coal are abundant in nature. Yet, uncontrolled combustion does result in the generation of ashes with high contents of heavy metals or gaseous emissions containing sulphur or nitrogen oxides that not only contribute to increasing the greenhouse effect, but strong acids in the presence of atmospheric water vapour can be formed which precipitate as acid rain (Vamvuka et al. 2003). In addition, the process in which they can be obtained, can further affect the ecological balance by requiring the destruction of large-scale forest ecosystems.

Finally, for liquid biofuels, a wide array of compounds exist but, unlike solid or gaseous biofuels, liquid ones are not easily found in the environment. These are generally obtained mostly through fermentation processes performed by microorganisms of bacterial or fungal origin, or generated by algae with high lipid content (Gomez et al. 2008). Within this scenario, it cannot be anticipated that liquid biofuels would share most of the problems of exploitation, management or distribution of solid and gaseous ones, since most of the internal combustion systems are based on the use of liquid fuels; however these compounds still suffer from low conversion yields and, in some cases, high recovery costs of finished products. In contrast to the classic fossil fuel's production processes, such renewable biofuels tend to be restricted to recoverable or profitable compounds of interest, which affects the economic viability of their production (Pfromm et al. 2010).

Another current issue with biofuel technology is feedstock availability and price, as traditional biotechnological processes require the use of monomeric or dimeric sugars as substrate for cell growth (McNew and Griffith 2005). Nonetheless, it is known that within the diversity of waste generated due world trade and industrial activities, there is a vast variety of hydrocarbonated components that can be utilized as raw material for producing biofuels (Sharma et al. 2013).

3 Agro-Industrial Waste

In this context, agro-industrial waste can be defined as those residuals of organic nature generated or derived from the use, collection and processing of plant biomass. The agro-industrial wastes consist of a variety of components that can come from various sources. There are those generated by the exploitation of forest resources, such as tree bark, wood chips, sawdust, remains of chipboard, etc. that are not used efficiently and are not integrated into finished products or from agronomic nature which include stubble remains of fruits and vegetables, straw, seed shells, waste from processes for obtaining syrups or juices, pulp, etc. (Singh et al. 2012). These compounds mainly contain sugars arranged into molecules of high molecular weight that exhibit significant amounts of radical branching, cyclization and physico-chemical properties that confer biochemical stability, rendering them unsuitable for traditional fermentation (Lo et al. 2008).

Therefore, based on the composition and structure of the various agro-industrial residuals, a classification can be made in advance to propose an adequate mechanism for their decomposition into fermentable sugars and subsequent transformation into fuels.

Waste from logging and conversion of forest resources consist mainly of what is called lignocellulose, which is composed of cellulose fibres wrapped in an amorphous matrix of hemicellulose chains and skeletons of lignin (Martinez et al. 2009). It has a high mechanical and chemical resistance due to cyclic aromatic groups present in the matrix of lignin, rendering them more or less resistance against attack by microorganisms and preventing direct use as raw materials for liquid biofuels despite high availability (Yuea et al. 2014).

Cellulose is the most abundant organic polymer in nature. It is made up of long chains of glucose, usually linked by covalent bonds β-1,4. Glucose polymers forming the cellulose have a linear structure and tend to bind together by hydrogen bridge links, generating what is known as crystalline region, where cellulose fibres are organized into compact structures which is the reason behind their high chemical stability and resistance to attack by biological agents (Klemn et al. 2005). Additionally, there is another region within cellulose chains termed amorphous region, conformed by glucose polymers that do not arrange so tightly due a reduced formation of hydrogen bonds between the chains; thus amorphous regions are more susceptible to chemical and biological degradation (Aro et al. 2005).

In contrast, hemicellulose is a heterogeneous polymer, consisting of monomers bearing no specific type of sugar or with a single type of bond between them, making it more likely to have no crystal structures (Aspinall 1959). This polymer is comprised mostly of pentose sugars such as xylose or arabinose, hexoses such as mannose, glucose or galactose and uronic acids (Gírio et al. 2010). These chains tend to link to the cellulose through hydrogen bonds or by interaction with other polymers such as lignin and pectin. Hemicelluloses have a relatively short chain length with respect to cellulose, however this kind of polymer can represent from 15 to 35 % of the total dry weight of the plant material (Scheller and Ulvskov 2010).

With the statements in the preceding paragraphs it is easy to understand the importance of generate technologies that allow the use of the large amount of organic matter present in agro-industrial waste, not only to solve problems related to environmental pollution but also to take advantage more efficiently of natural resources and guarantee the supply of raw material for renewable liquid biofuel production to ensure economic and technical feasibility.

4 Saccharification Methods

Obtaining sugars from the structural constituents of plant biomass is not a novel research topic, as since the composition of the polymers described in the previous section are known, sugar extraction attempts for commercial purposes, particularly within the area of food supply, have been extensively studied (Harris 1949). The biggest obstacle present is inherent to the physico-chemical nature of agro-industrial waste which has hampered incorporating them and the use of chemical or physical agents that jeopardize both the product integrity, quality and safety for both human and bacterial feedstock (Sun and Cheng 2002).

The first techniques of degradation of plant biomass were based on the process developed during World War II by Giordani in 1939 (Kobayashi et al. 1962). These techniques involved the use of size reduction operations, such as grinding or milling, the resulting chips is given a chemical treatment with dilute sulphuric acid to remove or eliminate the hemicellulose present in the plant tissue, then the biomass is subjected to a treatment with a concentrated acid solution containing up to 60 % H_2SO_4 and allowed to dry to be then re-treated with the liquor obtained from the first stage of hydrolysis in a container at high temperature. This process has the advantage of having very good recovery yields of sugars, which provides concentrated sugar solutions for fermentation, however high consumption of acid solutions and the presence of inhibitory compounds for bacterial growth (such as furfural) require additional treatment for the use as feedstock (Qureshi et al. 2007).

Another hydrolysis techniques also involves the use of acid but exclusively in diluted form. Here the biomass is crushed and then treated with sulphur dioxide gas, the chips are then heated to 180 °C for 2–3 min and then compressed by expansion valves to generate a pulp. This pulp is subsequently washed and the soluble components recovered into the supernatant is subjected to a new round of contact with dilute sulphuric acid. This technique allows to obtain fermentable sugar solutions with up to 8 % w/w and without high concentrations of inhibitory compounds that could affect the bacterial growth for carrying out fermentation (Saha et al. 2005).

Finally, another chemical treatment performed to obtain fermentable sugars from plant biomass is the use of alkaline solutions, which may contain as active compound sodium or ammonium hydroxide. This treatment is, however, only used during pre-treatment stages, as they help to generate porosity into the cellulosic material and thereby allows for increased contact surfaces between the solid phase and the liquid medium containing hydrolysis agents, such as enzyme preparations (Chen et al. 2013a, b).

The techniques of biological degradation of lignocellulosic material is a more recent research topic, which has emerged due to the large amount of knowledge generated in the areas of bioengineering and molecular biology that allows a more detailed study of the organisms that have the capacity of feeding on such complex substrates or agro-industrial waste (Demirbas 2009). These technologies promise higher performance of polymer conversion to fermentable sugars due to use of highly specific enzymes involved in the hydrolysis. An additional advantage is the requirement of less aggressive operating conditions such as pressure, temperature and pH compared to chemical methods (Sandgren and Hiberg 2005). Despite this, the cost of producing the enzymes required to perform the operation is relatively high and the organisms, that produce proteins with such properties naturally, have complex nutritional requirements and require facilities that allow for solid state fermentation to optimize production of these enzymes (Pandey 2003).

5 Current Challenges of Biofuel Production

One of the processes currently established at large-scale is ethanol production as an additive or substitute for gasoline in internal combustion engines, which, despite being one of the oldest biotechnology techniques and with extensive knowledge derived from that vast experience, still requires additional efforts to become a solid long-term candidate to replace fossil fuels (Lee et al. 2008a, b). First of all it must be mentioned that the type of raw material (sugar cane and corn, respectively) used for production in Brazil and in the United States, which are the largest producers of ethanol for use as biofuel, comes from resources originally destined for both animal and human consumption, (Pimentel et al. 2007). Added to this, ethanol has technical limitations of use, storage and integration into current internal combustion engine technology, such as being more hygroscopic with respect to oil-based fuels. Therefore it cannot be transported efficiently over long distances through pipes without accumulating additional moisture, thus reducing its final titter, and contributing to internal oxidation of the pipes (Antoni and Zverlov 2007).

With the above, different biofuels have been looked for in nature to attempt to lessen the issues previously stated, and one of the most convincing candidates is butanol.

6 Butanol as Biofuel

Butanol is a short chain alcohol, which has been produced by biological means for a long time. It has the advantage of a higher energy content per litre of fuel than ethanol, less volatility and slightly lower octane number than gasoline, which can improve its yield over the ethanol and reduce engine gaseous emissions. Also due to higher density, a higher mass of fuel is injected into the engine, which, considering its similar calorific power versus gasoline, helps to attain comparable energy content per litre. It is therefore assumed that butanol would be a better fuel extender or substitute for current regular gasoline than ethanol (Table 1).

Traditionally, the methodology for obtaining butanol by fermentation is based on the degradation of various sugars (particularly glucose or sucrose) carried out by Gram-positive bacteria of the genus *Clostridium*, via a metabolic pathway called

Table 1 Comparison of different physico-chemical properties between regular gasoline, ethanol and n-butanol (Modified from Mužíková et al. 2014)

Fuel	Density (kg/m^3)	Reid vapour pressure (kPa)	Energy content per litre (MJ/L)	Research octane number	Solubility in water @ 20 °C (w/w)
Regular gasoline (US/UK)	740	45–90	32.6	95	0.01
n-butanol	810	2.3	29.2	94	20.1
Ethanol	794	17	23.5	106	Completely miscible

ABE (acetone-butanol-ethanol) (Qureshi et al. 2008). This biological process was originally exploited for the production of acetone as a solvent for the chemical and military industry, being of such importance that most of the explosives made and used during World War II were based on this technology. Back then, butanol was considered an undesirable product, however, the former Soviet Union began its use as fuel, partly due to the harsh weather conditions. By the mid-60s the process of obtaining ABE products at industrial-scale allowed to reach reaction volumes up to 300 m^3. Nonetheless, the boom in the petrochemical industry in the 70s provided solvents and chemicals at very low cost compared to the biological process, resulting in fermentation plants being dismantled (Zverlov et al. 2006).

Currently, there is a renewed interest in the study of the ABE system in order to adapt such technology to suit the energy demands of this time (Lee et al. 2008a, b). Still, biological system present native restrictions that have prevented consolidation as a mature technology such as low production yields, inhibitory effects of accumulated solvents with culture age and mechanisms of metabolic regulation present in Gram-positive bacilli, such as sporulation (Zheng et al. 2009).

7 *Clostridium* Cultures Using Agro-Industrial Waste as Substrate

One of the biggest problems of traditional fermentation processes, which are generally based on yeast cultures for the production of alcohol, is that the organisms do not have the ability to degrade and consume pentoses constitutively. This is a desirable attribute in the context of use of agro-industrial waste as feedstock for the generation of biofuels, because of the large amount of hemicellulose present in the structure of residues coming from both the wood and paper industry (Martín et al. 2007). Studies in the past decade have been conducted to find new species, use molecular techniques or metabolic engineering in order to generate strains with the ability to growth on a wider range of carbon sources. Generated strains existing today exhibit moderate yields of alcohol production which do not surpass titters over 20 g L^{-1}, preventing consideration as strong candidates for biofuel production (Atsumi et al. 2008; Peralta-Yahya et al. 2012).

In contrast, many strains of *Clostridia* display natural capabilities to metabolise a wider range of carbon sources including pentoses, being most attractive the species that can feed on xylose and mannose, which are the most abundant monosaccharides in w/w fraction generated from saccharification of lignocellulosic residues (Qureshi et al. 2007).

In addition, it is known that some wild-type *Clostridium* species have the ability to grow solely on lignocellulosic carbon sources, such as *Clostridium thermocellum* or *Clostridium celullolyticum* (Bayer et al. 1983; Keis et al. 2001). These species are widely recognized within the processes of degradation of plant biomass because they possess the ability to produce extracellular cellulolytic enzymes in large quantities organized into complex structures called cellulosomes (Lamed et al. 1983a, b). Within a cellulosome, a wide array of endoglucanases, cellobiohydrolases, xylanases and other degradative enzymes exist that allow the microorganisms to efficiently degrade almost all of the cellulosic or hemicellulosic substrate found in vegetal biomass (Nordon et al. 2009).

Table 2 summarizes current advances in ABE fermentation using agro-industrial waste as feedstock and the highest producing strains. It is noteworthy to emphasize that there are technologies that can reach practical yields that match the theoretical value of 0.45 g/g of ABEs from glucose, which is indicative that the research in this field is stepping in the right direction.

Table 2 Summary of the results of ABE production under batch fermentation (Modified from Liu et al. 2013)

Feedstock	Strain	ABE concentration (g/L)	ABE yield (g/g)	ABE Productivity (g/L * h)	References
Barley straw	*C. beijerinckii*	26.64	0.43	0.39	Qureshi et al. (2010)
Wheat straw	*C. beijerinckii*	21.42	0.41	0.31	Qureshi et al. (2008)
Corn fiber	*C. beijerinckii*	9.3	0.39	0.10	Qureshi et al. (2008)
Corn stover	*C. beijerinckii*	26.27	0.44	0.31	Qureshi et al. (2010)
Rice straw	*C. saccharoperbutylacetonicum*	13	0.28	0.15	Soni et al. (1982)
Bagasse	*C. saccharoperbutylacetonicum*	18.1	0.33	0.3	Soni et al. (1982)
Switch grass	*C. beijerinckii*	14.61	0.39	0.17	Qureshi et al. (2010)

(continued)

Table 2 (continued)

Feedstock	Strain	ABE concentration (g/L)	ABE yield (g/g)	ABE Productivity (g/L * h)	References
Domestic organic waste	*C. acetobutylicum*	9.3	0.38	0.08	Claassen et al. (2000)
Sago	*C. saccharobutylicum*	16.38	0.33	0.59	Liew et al. (2005)
Defibrated sweet potato slurry	*C. acetobutylicum*	5.87	0.29	0.12	Badr et al. (2001)
Cassava	Co-culture of *B. subtillis* and *C. butylicum*	9.71	0.21	0.135	Tran et al. (2010)
Crystaline cellulose	Co-culture of *C. thermocellum* and *C. saccharoperbutylacetonicum*	10.3	0.25	0.02	Nakayama et al. (2011)
Deshelled corn cobs	Co-culture of *C. cellulovorans* and *C. beijerinckii*	11.8	0.17	0.14	Wen et al. (2014)

8 Current Challenges for the ABE Fermentation Processes

Although, as described above, it can be assumed that the problem of availability of substrate for butanol production as a biofuel could be guaranteed, it should be mentioned that generally, most of the obstacles that impede the use of clostridial strains in ABE producing bioprocesses involve the biochemical limitations to bacterial growth by both substrate and solvents concentrations in the broth. Some authors (Lee et al. 2008a, b; Qureshi et al. 2010) report that ABE producing bacteria will not grow in solutions containing over 175 g/L of glucose or 14–21 g/L of solvents, as excess carbon sources disrupt incorporation by the microorganisms and the high solvent titters partially solubilize the cell's wall causing a depolarization of their membranes, which impedes stabilization after or during cell division.

Additionally, the production of butanol using agro-industrial waste as feedstock, needs to consider pre-treatment and saccharification techniques, as both acid and alkaline methods have a negative impact on growth of the culture because clostridial species regulate their metabolism. They sense hydrogen potential gradients between their cytoplasm and their surroundings, resulting in acid media causing what is it known as acid crash and alkaline ones cause inhibition due to poor proton exchange (Dürre 2007). Finally it should be noted that *Clostridium* species with cellulolytic capabilities usually do not have the enzymes necessary to perform ABE fermentation, so these strains cannot be used by themselves in the production processes of second generation biofuels (López-Contreras 2001).

Research advances made up to date utilise both molecular biology and engineering techniques to solve some of the issues of renewable butanol for biofuel production from agro-industry wastes. Genetic manipulation target overexpression of the key transcription factor that regulates sporulation (*spoA*) coupled to the downregulation of the expression of hydrogenases encoded by the gene cluster known as *hupCBA*. This approach resulted in increased solvent production in batch culture and reduction of inhibitory effects caused by the accumulation of solvents in the medium (Alsaker et al. 2004). Additionally there are reports of metabolic engineering via antisense RNA, targeting the transcript of the gene *ctfB*, to redirect the carbon flux from acetone forming pathways to butanol producing ones, has had moderate success (Tummala et al. 2003).

Recent studies using an engineering approach have explored the feasibility of implementing strategies for unconventional culture techniques in which it is proposed to feed a first reactor directly with agricultural shredded waste and inoculate it with bacteria of the genus *Clostridium thermocellum* to obtain fermentable sugars that could feed to a second reactor inoculated with *Clostridium acetobutyllicum* or *Clostridium beijerinckii* to obtain ABE products (Qureshi et al. 2007). These studies have shown that success can be further enhanced by implementing a co-culture of both a cellulolytic and a butanol-producing strain in the same reactor (Nakayama et al. 2011).

Despite all advances made in the ABE technology, these have been unable to achieve optimal yields based on capabilities of the *Clostridium* strains. Particularly, as yields are typically based on empirical knowledge, process efficiency cannot be assessed with regards to process engineering, such as analysis, design, optimization and process control.

Formulation of mathematical models could overcome such challenges if they could describe and, ideally, also predict the behaviour of the biological system under study over a wide range of operational conditions. This would allow to determine more appropriate procedures for the exploitation of the bacterial metabolism more precisely. In this light, it is necessary to analyse the ABE pathway to determine best strategies to successfully achieve the intended goal.

9 ABE Metabolic Pathway

The ABE metabolic pathways is composed of 19 main reactions (Fig. 1), in which acetate, butyrate, ethanol, acetone, lactate and hydrogen are co-products of the fermentation. The metabolism shows two distinctive phases, an acidogenic phase, in which acetate and butyrate are the main products during exponential growth phase of the culture, and a solventogenic phase, in which butanol, ethanol and acetone are the main products relating to stationary phase (Bahl et al. 1982).

Hexoses are metabolised into pyruvate via the Embden-Meyerhof-Parnas pathway, whereas pentoses are incorporated into the metabolism by the action of the UDP-glucose uridyl transferase (Durán-Padilla et al. 2014). Pyruvate is one of the key intermediates of the *Clostridium* metabolism, and under certain conditions *Clostridium* species are capable of transforming pyruvate to lactate through

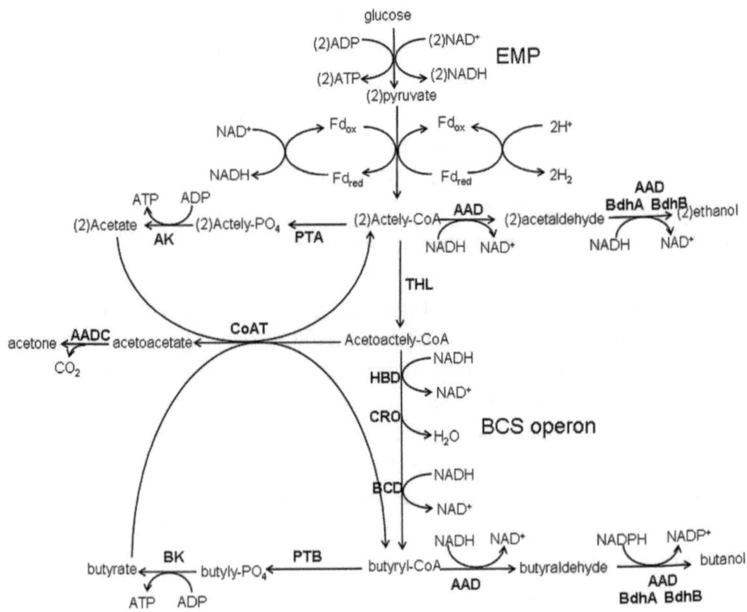

Fig. 1 ABE metabolic pathway

pyruvate dehydrogenase. The main reaction, however, is conversion of pyruvate to acetyl-CoA through pyruvate-ferredoxin oxidoreductase (Uyeda and Rabinowitz 1971) with the concurrent reduction of ferredoxin, a reason why lactate formation is not considered in most metabolic models.

NAD(P)H-Ferredoxin oxidoreductases are key enzymes for electron transport in the *Clostridium* genus (Gheshlaghi et al. 2009). In acid-producing cultures, ferredoxin-reductase shows high activity. Two reasons have been proposed, (i) necessity for regenerating NAD^- consumed in the Glyceraldehide-3-P-dehydrogenase reaction, or (ii) as acidogenic cultures are in exponential growth, high enzymatic activity is required for meeting energy requirements. The ferredoxin oxidoreductases regulatory mechanism appears to be related to NAD- and NADH concentration, which turns out to be an efficient regulatory system that prevents accumulation of NADH in the acidogenic phase, and increasing $NAD(P)^-$ reductases activity during the solventogenic phase in turn, for the higher NAD(P)H demand needed for ethanol and butanol production in that stage (Gheshlaghi et al. 2009).

Hydrogen formation is catalysed by the hydrogenase, which uses reduced ferredoxin as electron donor (Gheshlaghi et al. 2009). As well as ferredoxin reductases, the hydrogenase shows distinctive behaviour during acidogenesis and solventogenesis, as reduced ferredoxin competes with $NAD(P)^-$ reductases for it, meaning it also competes for reducing power with butanol and ethanol dehydrogenases during solventogenesis.

The ABE pathway has three important nodes in Acetyl-CoA, Acetoacetyl-CoA and Butyryl-CoA. Acetyl-CoA is a rigid node from which the carbon flux is distributed to five different metabolites. Two branches come out directly from Acetyl-CoA which end up in acetate and ethanol, from Acetoacetyl-CoA acetone is produced while butanol and butyrate are produced from butyryl-CoA. Acetate and butyrate are energy producing reactions in which 1 molecule of ATP is produced for each Acetyl/Butyryl-CoA consumed. Both compounds are re-assimilated during solventogenesis, this reaction is catalysed in two different ways, acetate and butyrate can either be converted to acetyl/butyryl-CoA through a reversible reactions or through the action of the CoA-transferase (Millat et al. 2014), which consumes one molecule of Acetoacetyl-CoA and one of acetate/butyrate to produce one molecule of acetoacetate and one of acetyl/butyryl-CoA. This mechanism sets up an efficient solvent producing strategy, given that it would be energetically unfavourable to re-assimilate both acetate and butyrate through the reversible reactions, the availability of a non-ATP consuming reaction lets the organism face the energy deficit while still being capable to produce ethanol and butanol in the stationary or solventogenic phase.

This theory is supported by findings made by Desai et al. (1999) and Lehmann et al. (2012) indicating that butyrate consumption during solventogenesis is catalysed through the reversible reaction, as opposed to acetate consumption which is related to acetoacetate production. Also Wang et al. (2013) and Desai et al. (1999) reported that acetate is produced even after the organism shifted to solventogenesis, which would be an indication that the production of acetate is necessary for energy generation.

There are several hypothesis trying to explain the factors that control the switch between acid generating metabolism and solvent generating metabolism. Some authors (Hüsemann and Papoutsakis 1988; Terracciano and Kashket 1986) suggest that the concentration of un-dissociated butyric acid is the controlling factor. However, Chen and Blaschek (1999) proposed that the switch is the result from extracellular and intracellular signals, like culture's pH or intracellular Acetyl-P and Butyryl-P concentrations respectively. Zhao and Tomas (2005) concluded that it is Butyryl-P and not Acetyl-P that controls the metabolism switch, while Wietzke and Bahl (2012) suggest that energy and redox balance are the key intracellular signals to start solvent production.

10 ABE Fermentation Kinetic Modelling

According to Mayank et al. (2012) and the authors knowledge there are 7 relevant kinetic models for ABE fermentation reported in literature (Shinto et al. 2007; Li et al. 2011a, b; Haus et al. 2011; Napoli et al. 2011; Thorn et al. 2013; Millat et al. 2014; Velázquez-Sánchez et al. 2014), moreover Metabolic Flux Analysis has been used to successfully describe the behaviour of the ABE pathway (Desai et al. 1999; Papoutsakis and Rice 1984). However, for the sake of simplicity and reproducibility, all reported kinetic models have been validated through experiments conducted using glucose as the only carbon source. Therefore, to ensure a fair comparison between results of this and the following sections of the chapter, all selected analyses are based on studies made under that condition, which implies that the fermentation stage of the analysed operational regimes are made after an upstream operation of lignocellulosic digestion.

Five of the six proposed models are structured metabolic models, which intend to recreate the bacterial metabolism at both an enzymatic and genetic level, but their inherent complexity makes it difficult to successfully apply them to the bioengineering field, such as in process design and optimization. Therefore the necessity for a model aiming to study the process at a much simpler level is evident. The model should hold sufficient predicting capacity to provide insight into the Clostridial metabolism and also help in the design of industrial-scale ABE fermentation.

Velázquez-Sánchez et al. (2014) set up the precedent of an unstructured phenomenological model that could describe the process with sufficient accuracy and in this revision an improved version of the model is used to study different working regimes and butanol production strategies. The kinetic model has a set of seven differential equations describing the behaviour of each product of interest. There are 6 production rate equations, one for each of the following compounds: Butanol (But), Acetone (Ace), Ethanol (Et), Acetate (Act), Butyrate (Sb) and Biomass (X). The biomass production model (Eq. 1) is a Haldane-Luong equation, taking into account the inhibitory effect of butanol, but not the one caused by the carbon source. Butyrate kinetics (Eq. 13) include a product formation rate (Eq. 5) considering glucose as its main substrate, and one consumption rate described by Eq. 2 for Butanol production.

$$\mu_x = mumaxx * \left(\frac{Sg}{ksg + Sg}\right) * \left(1 - \left(\frac{But}{kbut}\right)^m\right) \tag{1}$$

$$\mu_{but} = mumaxbut * \left(\frac{Sg}{ksgb + Sg}\right) * \left(\frac{Sb}{ksb + Sb}\right) \tag{2}$$

$$\mu_{act} = mumaxact * \left(\frac{Sg}{ksgact + Act}\right) \tag{3}$$

$$\mu_{et} = mumaxet * \left(\frac{Sg}{ksget + Sg}\right) \tag{4}$$

$$\mu_{sb} = mumaxsb * \left(\frac{Sg}{ksbsg + Sg}\right) \tag{5}$$

$$\mu_{ace} = mumaxace * \left(\frac{Sb}{kba + Sb}\right) * \left(\frac{Act}{kaa + Act}\right) \tag{6}$$

Butyrate consumption (Eq. 13) takes into account the reversible pathway that transforms butyrate into butyryl-P and, subsequently, into Butyryl-CoA, as well as the reaction catalysed by CoA-transferase. For acetate consumption, given the results reported by Desai et al. (1999) and Lehmann et al. (2012), only the reaction catalysed by the CoA-transferase is considered.

$$\frac{dSg}{dt} = (D * (Sga - Sg)) - X * \left(\left(\frac{\mu_x}{Yxsg}\right) + \left(\frac{\mu_{but} * (Ybx)}{Ybutsg}\right) + \left(\frac{\mu_{act} * (Yactx)}{Yactsg}\right) + \left(\frac{\mu_{et} * (Yetx)}{Yetsg}\right) + \left(\frac{\mu_{sb} * (Ysbx)}{Ysbsg}\right)\right) \tag{7}$$

$$\frac{dX}{dt} = (-D * X) + (\mu_x - kd) * X \tag{8}$$

$$\frac{dAce}{dt} = (-D * Ace) + (\mu_{ace} * (X * Yacex)) \tag{9}$$

$$\frac{dBut}{dt} = (-D * But) + (\mu_{but} * (X * Ybx)) \tag{10}$$

$$\frac{dAct}{dt} = (-D * Act) + X * \left((\mu_{act} * (Yactx)) - \left(\frac{\mu_{ace} * (Yacex)}{Yaceact}\right)\right) \tag{11}$$

$$\frac{dEt}{dt} = (-D * Etoh) + (\mu_{et} * (X * Yetx)) \tag{12}$$

$$\frac{dSb}{dt} = (D * (Sba - Sb)) + X * \left((\mu_{sb} * (Ysbx)) - \left(\frac{\mu_{but} * (Ybx)}{Ybutsb}\right) - \left(\frac{\mu_{ace} * (Yacex)}{Yacesb}\right)\right) \tag{13}$$

Having declared the mass balance equations, it is necessary to analyse the behaviour of the biological system and its response under multiple operating conditions to ensure stability and reproducibility of the results, with a view to scaling and control system strategies, using this kind of novel modelling technique.

11 Butanol Production Strategies and Analysis

Little has been reported in the literature regarding the effect of the operating regime in establishing ABE fermentation using *Clostridium* strains and the possible causes and effects associated with butanol production due the use of one or another production strategy, as there are generally are limited studies available to evaluate a particular condition against classic batch methods, so many of the novel proposed processes are still open for analysis.

Therefore, the following four different production strategies are being analysed:

1. Adding acetate/butyrate at the beginning of a batch fermentation and in the inlet of a Continuous Stirred Tank Reactor (CSTR).
2. CSTR in one, two and three stages.
3. Fed-Batch culture.
4. CSTR followed by a fed-batch stage.

11.1 Batch Fermentation

It has been proved that adding butyrate at the beginning of a batch fermentation improves butanol yields and productivity (Chen and Blaschek 1999; Lee et al. 2008a, b; Chang 2010; Wang et al. 2013), due to a larger carbon pool shifts carbon flux toward butanol and because the presence of butyrate causes feed-back inhibition of its forming pathway (Lee et al. 2008a, b; Wang et al. 2013).

Figure 2 showcases the possible behaviour of the system feeding either butyrate or acetate at the beginning of fermentation. Simulations performed with 4 g L^{-1} of butyrate added at the beginning of fermentation shows that final butanol concentration after 300 h is 12.7 g L^{-1}, an improvement of 33.7 % from the 9.6 g L^{-1} of butanol achieved without butyrate. Analysing the product formation rate it becomes evident that although butanol maximum productivity does not exceed that of the one obtained by the culture without butyrate, butanol production begins earlier in the fermentation and continues over a longer period, which causes the final titter to increase. Acetone experiences the opposite effect, final concentration drops by 15 % caused by a reduction in its formation rate during solventogenesis. There is no

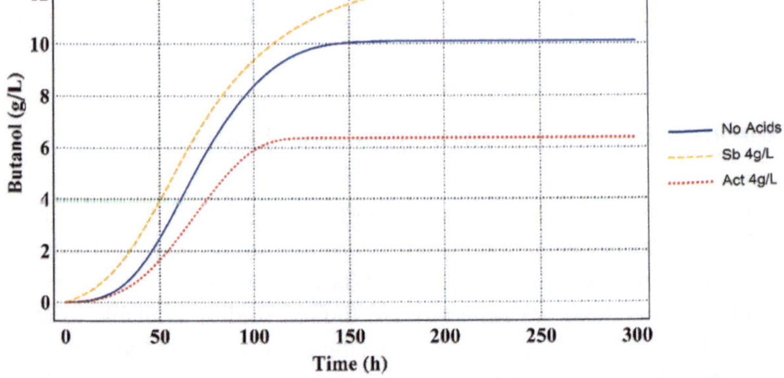

Fig. 2 Comparison of butanol production between classical batch fermentation vs. systems supplemented with butyrate or acetate at t_0 made by simulation using the author's kinetic model

agreement on the effect of adding acetate into a batch ABE fermentation. Chen and Blaschek (1999) report that the presence of acetate indeed increases both acetone and butanol final titters, however Holt et al. (1984) report that even though acetone concentration increases, butanol productivity experiences as much as a two fold decrease. Simulations made with 4 g L^{-1} of initial acetate agree with Holt et al. (1984) findings, as butanol final concentration after 300 h reduces as much as 33.2 %, while acetone concentration increases 30 %. Product formation rates show how butanol productivity not only drastically decreases, but its production starts later during fermentation and stops as much as 30 h earlier than in cultures without acetate. This agrees with findings indicating that butyrate re-assimilation is catalysed by the reversible reaction and acetate re-assimilation is catalysed by the CoA-transferase pathway.

11.2 Continuous Stirred Tank Reactor (CSTR)

It is clear that the addition of butyrate to batch fermentation causes a substantial increase in butanol titters. It is, however, also apparent that a batch culture is not the optimal strategy in industrial fermentation because time taken between the end of one culture and the beginning of a new one (sterilization and cleaning) would decrease productivity.

Therefore, butanol production in a CSTR is being taken as the optimal strategy for ABE fermentation, because it can keep production for longer periods, reducing time spent on up and downstream processes (Li et al. 2011a, b).

Further description of continuous culture will be given later in the chapter, while in this section the first working regime studied was using inlet feeding at a dilution rate of 0.04 h^{-1} containing 100 g L^{-1} of glucose and 4 g L^{-1} of butyrate. As previous results indicate that the addition of acetate diminishes butanol formation, this regime was not further studied. Simulations results show that butanol productivity increases 8 % with a final concentration of 3.01 g L^{-1} in the effluent. In terms of the production strategy, the observed increase justifies the addition of butyrate, however further economic studies are need to check if indeed the expense made on butyrate is feasible, because butyrate can be considered an expensive supply.

Simulations were based on a CSTR intended to study the dynamics of the fermentation over wide working regimes, more specifically from dilution rates from 0 to 0.1 h^{-1} and glucose concentrations in the feed from 0 to 200 g L^{-1}. The results show that solvents maximum production rate was between 0.02 to 0.08 h^{-1}, with a maximum butanol production in 0.05 h^{-1}, irrespective of the glucose concentration in the feed, while acid maximum production rates were found into dilution rates higher than 0.05 h^{-1}. This phenomena is typical for ABE fermentation, given that acidogenic and solventogenic phases are related to the growth phase of the culture. It is most likely that in CSTR operated with higher dilution rates, cells were not allowed to remain in the reactor for sufficient time to enter solventogenesis, while the opposite is observed at lower dilution rates (Li et al. 2011a, b). To shorten the favourable working regions, the cell's catalytic activity was evaluated also, showing that at dilution rates between 0.04 and 0.07 h^{-1} and glucose concentrations between 80 and 120 g L^{-1}, maximum butanol production per unit mass of cells in the reactor was achieved.

11.2.1 CSTR with Two and Three Stages

To further study the strategies under development with CSTRs, further analyses were made considering both two- and three-staged processes. One of the propositions made regarding a two-staged process indicates that high dilution rates in the first reactor are needed to get high acids concentrations which then, taking advantage of *Clostridium's* biphasic metabolism, would be converted to solvents at a posterior stage with low dilution rates (Bahl et al. 1982; Lai and Traxler 1994; Mutschlechner et al. 2000). The results indicate the absence of this phenomenon, as when the conditions set on the first reactor favour acids or solvents production, the second reactor displays an identical internal dynamic, increasing the concentrations of the components produced in the first stage. This behaviour is consistent with experimental evidence, indicating that indeed the conditions set on the first stage become the controlling conditions of the whole process, independently of pH and dilution rate of the posterior stages (Godin and Engasser 1990; Setlhaku et al. 2012).

Propositions made by De Gooijer et al. (1996) indicate that staged processes become favourable when product inhibition is strong, moreover three-staged CSTRs can achieve the closest performance to a CSTR followed by a plug flow reactor. Simulations on staged CSTRs were run in two different conditions, one in which the dilution rates increase in every stage and one with the opposite conditions, being the dilution rates 0.04, 0.06 and 0.08 h^{-1} for the former and 0.04, 0.03 and 0.015 h^{-1} for the latter, both using 100 g L^{-1} of glucose in the feeding of the first stage. Butanol concentration increased in both conditions, as for the system with increasing dilution rates butanol increased 80.56 % from the first stage to the second and from the second to the third 29.78 %, being the final butanol concentrations 2.9, 5.2 and 6.7 g L^{-1} in each stage respectively. This results indicates that there is little improvement with the addition of the third stage, indeed the overall productivity up to the third stage remains the same as if the process only had two stages, being 0.124 g L^{-1} h^{-1} in both scenarios. The system with decreasing dilution rates showed a greater improvement in butanol concentration with an increase of 120.7 % from the first to the second stage and 42.2 % from the second to the third stage, being the final butanol concentrations 6.4 and 9.1 g L^{-1} in the second and third stage respectively. Although the increase in butanol concentrations might be impressive, there is a downside in decreasing the dilution rates because Hydraulic Retention Times (HRTs) of the whole process increase, which result in decreased productivity; for the conditions tested the overall productivity up to the second and third stage were 0.11 and 0.073 g L^{-1} h^{-1}, respectively.

Further simulations show that the productivity when a third stage is added only improves if the total HRT of the whole process is lower than 56 h, nonetheless the conversion yield from glucose to butanol of the processes with a third stage improves by as much as 60.8 % in systems with HRT's lower than 80 h. Further research is needed to identify which of the conditions has an economic advantage, given that, for example, the conditions with low productivity but higher butanol concentrations could simplify the posterior purification processes up to the point where the production cost is lower than that of a system with higher productivity.

11.3 Fed-Batch Fermentation

Fed-Batch fermentation is a technique used to reduce substrate inhibition, a batch fermentation is run until either a desired substrate concentration or biomass concentration is achieved, then a solution with concentrated substrate is fed at a constant, linear, exponential or intermittent rate. In ABE fermentation, fed batch culture is recommended only with in-line product recovery (Qureshi et al. 1992; Ezeji et al. 2004), due to the strong inhibition effects of butanol, a simple fed-batch would not achieve sufficient productivity to offer any advantages over CSTR operation.

In order to study the behaviour of a fed-batch culture, a theoretical scenario was set up with a 30,000 L bioreactor and constant flow regimes between 34 and 800 L h^{-1} with a glucose feed concentration of 150 g L^{-1}. Maximum productivity was achieved between 200 and 400 L h^{-1}, but it is clear that productivity values reachable through simple fed batch operation are lower, 33 % less, than those of a CSTR. There is, however, an advantage in dealing with a single culture. Fed batch can reach butanol concentrations of 8 g L^{-1} in a single stage, 63 % more than that of a single stage CSTR.

For further analysis, a novel bioreactor configuration, in which the effluent of a CSTR was used as the feed to a fed batch reactor was tested, and the results of both this stage and the one discussed prior to are presented in Figs. 3 and 4. To the

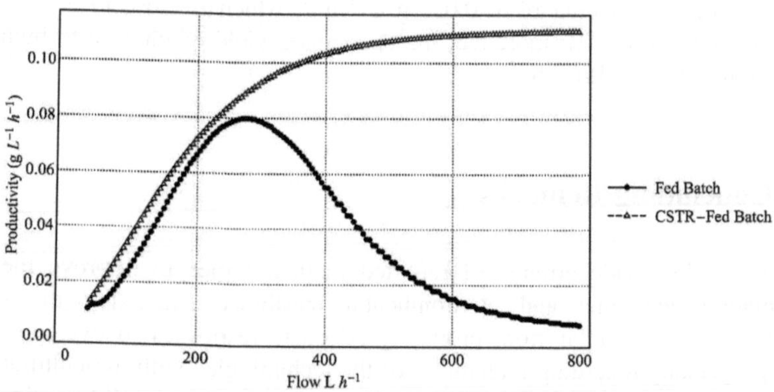

Fig. 3 Theoretical butanol productivity versus feeding flow rate of both a single fed batch reactor and a hybrid system composed of a CSTR and a fed batch fermenters obtained by simulation using the author's kinetic model

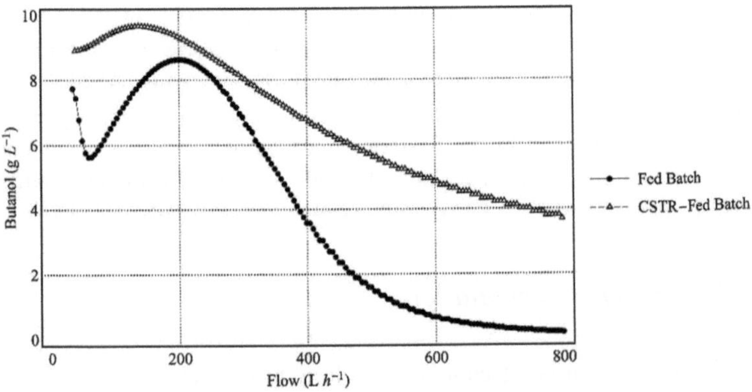

Fig. 4 Final theoretical butanol titter comparison between single fed batch and an hybrid arrange composed of a single stage CSTR and a fed batch reactor obtained via simulations made using the author's kinetic model

authors' knowledge, the work performed by Setlhaku et al. (2012) is the only report of this kind of configuration for ABE fermentation. The study indicates that this system is capable of reaching higher solvent volumetric titters and specific productivity, as well as higher butanol concentration before any product removal technique is applied. The simulations made using the enhanced model proposed here indicate that even though the productivity surpasses that of a single fed batch, the same does not apply for the CSTR. Productivity in this system increases asymptotically with flow, reaching more than 0.12 g L^{-1} h^{-1} in the CSTR at flows higher than 400 L h^{-1}. This sets up two problems, first the difficulty of finding a working region that offers an advantage of some kind, and second, the problems arising from feeding such high fluxes, as it depends on the capacity of the first stage to grow sufficiently before feeding the fed batch, which can lead to scale up troubles. This bioreactor configuration is advantageous in other ways, for instance it reaches 9.6 g L^{-1} of butanol at 0.055 g L^{-1} h^{-1}, which is close to 30 % slower than a three-stage CSTR, moreover the conversion yield of glucose to butanol is 30 % higher (0.24 g But gS^{-1}).

12 Concluding Remarks

As evidenced, a wide arrange of reported methodologies to improve the ABE fermentation economic and environmental feasibility exist, but the current state-of-the-art is still far from providing adequate responses to overcome challenges, especially regarding integration of the technologies with agricultural waste management and treatment. Nonetheless, Clostridial species are solid candidates for consideration as versatile biological systems for transformation of organic residues into energy compounds and also into commodity chemicals such as CO_2, solvents or even hydrogen. Further study in this area is encouraged and justified.

Further efforts made into the study of the ABE producing processes should incorporate engineering tools like design, optimization and control of bio-systems that could lead to the scale-up of this technology beyond semi-pilot stage. Factors to be considered are: analysis of feedstock availability and costs and the current efforts made into pre-treatment and co-culture techniques to re-evaluate their economic and technical feasibility at large-scale.

References

Alsaker KV, Spitzer TR, Papoutsakis ET (2004) Transcriptional analysis of spo0A overexpression in *Clostridium acetobutylicum* and its effect on the cell's response to butanol stress. J. Bacteriol 186:1959–1971

Antoni D, Zverlov V (2007) Biofuels from microbes. Appl Microbiol Biotechnol 77:23–35

Aro N, Pakula T, Penttila M (2005) Transcriptional regulation of plant cell wall degradation by filamentous fungi. FEMS Microbiol Rev 29(4):719–739. ISSN:0168-6445

Aspinall GO (1959) Structural chemistry of the hemicelluloses. Adv Carbohydr Chem 14:429–468

Atsumi S, Cann AF, Connor MR (2008) Metabolic engineering of Escherichia coli for 1-butanol production. Metab Eng 10:305–311

Badr HR, Toledo R, Hamdy MK (2001) Continuous acetone ethanol butanol fermentation by immobilized cells of *Clostridium acetobutylicum*. Biomass Bioenergy 20:119–132

Bahl H, Andersch W. Gottschalk G (1982) Continuous production of acetone and butanol by *Clostridium acetobutylicum* in a two-stage phosphate limited chemostat. Appl Microbiol Biotechnol 201–205

Bayer EA, Kenig R, Lamed R (1983) Adherence of *Clostridium thermocellum* to cellulose. J Bacteriol 2:818–827

Brown R (2003) Biorenewable resources. Iowa State Press, Blackwell Publishing Co., Iowa

Chang W (2010) Acetone-butanol-ethanol fermentation by engineered *Clostridium beijerinckii* and *Clostridium tyrobutyricum*. Food science and technology PhD Thesis, Ohio State University, USA

Chen C, Blaschek H (1999) Effect of acetate on molecular and physiological aspects of *Clostridium beijerinckii* NCIMB 8052 solvent production and strain degeneration. Appl Environ Microbiol 65(2):499–505

Chen Y, Stevens MA, Zhu Y, Holmes J, Xu H (2013a) Understanding of alkaline pretreatment parameters for corn stover enzymatic saccharification. Biotechnol Biofuels. 6(8)

Chen Y, Zhou T, Liu D, Li An, Xu S, Liu Q, Li B, Ying H (2013b) Production of butanol from glucose and xylose with immobilized cells of *Clostridium acetobutylicum*. Biotechnol Bioprocess Eng 18:234–241

Claassen PAM, Budde MAW, López-Contreras AM (2000) Acetone, butanol and ethanol production from domestic organic waste by solventogenic clostridia. J Mol Microb Biotech 2 (1):39–44

Demirbas A (2009) Progress and recent trends in biodiesel fuels. Energ Convers Manag 50:14–34

Desai RP, Harris LM, Welker NE, Papoutsakis ET (1999) Metabolic flux analysis elucidates the importance of the acid-formation pathways in regulating solvent production by *Clostridium acetobutylicum*. Metab Eng 1(3):206–213

De Gooijer CD, Bakker WAM, Beeftink HH, Tramper J (1996) Bioreactors in series: an overview of design procedures and practical applications. Enzym and Microb Technol 18(3):202–219

Dufey A (2006) Biofuels production, trade and sustainable development: emerging issues. Environmental economics programme/Sustainable markets group, Paper No 2, International Institute for Environment and Development (IIED)
Durán-Padilla VR, Dávila-Vazquez G, Chávez-Vela NA, Tinoco-Valencia JR, Jáuregui-Rincón J (2014) Iron effect on the fermentative metabolism of *Clostridium acetobutylicum* ATCC 824 using cheese whey as substrate. Biofuel Res J 4:129–133
Dürre P (2007) Biobutanol: an attractive biofuel. Biotechnol J 2:1525–1534
Escobar J, Lora S, Venturini O, Yanez E, Castillo E, Almazan O (2008) Biofuels: environment, technology and food security. Renew Sustain Energ Rev 13:1275–1287
Ezeji TC, Qureshi N, Blanschek HP (2004) Acetone butanol ethanol (ABE) production from concentrated substrate: reduction in substrate inhibition by fed-batch technique and product inhibition by gas stripping. Appl Microbiol Biotechnol 63:653–658
Gheshlaghi R, Scharer JM, Moo-Young M, Chou CP (2009) Metabolic pathways of clostridia for producing butanol. Biotechnol Adv 27(6):764–781
Gírio FM, Fonseca C, Carvalheiro F, Duarte LC, Marques S, Bogel-Łukasik R (2010) Hemicelluloses for fuel ethanol: a review. Bioresour Technol 101(13):4775–4800. ISSN:09608524
Godin G, Engasser M (1990) Two-stage continuous fermentation of *Clostridium acetobutylicum* : effects of pH and dilutio rate. Appl Microbiol Biotechnol 269–273
Gomez LD, Clare GS, McQueen-Mason J (2008) Sustainable liquid biofuels from biomass: the writing's on the walls. New Phytol 178:473–485
Harris EE (1949) Wood saccharification, advances in carbohydrate chemistry, vol IV. Academic Press Inc, New York
Haus S, Jabbari S, Millat T, Janssen H, Fischer R-J, Bahl H, King JR, Wolkenhauer O (2011) A systems biology approach to investigate the effect of pH-induced gene regulation on solvent production by *Clostridium acetobutylicum* in continuous culture. BMC Syst Biol 5:5–10
Holt RA, Stephens GM, Morris JG (1984) Production of solvents by *Clostridium acetobutylicum* cultures maintained at neutral pH. Appl Environ Microbiol 48(6):1166–1170
Hüsemann MH, Papoutsakis ET (1988) Solventogenesis in *Clostridium acetobutylicum* fermentations related to carboxylic acid and proton concentrations. Biotechnol Bioeng 32(7):843–852
Keis S, Shaheen R, Jones D (2001) Emended descriptions of *Clostridium acetobutylicum* and *Clostridium beijerinckii*, and descriptions of *Clostridium saccharoperbutylacetonicum* sp. nov. and *Clostridium saccharobutylicum* sp. nov. Int J Syst Evol Microbiol 51:2095–2103
Klemn D, Heublein B, Fink H-P, Bohn A (2005) Cellulose: fascinating biopolymer and sustainable raw material. Angew Chem Int Ed 44:3358–3393
Kobayashi T, Mitachi K, Kikuchi H (1962) Main-hydrolysis of White Birch Chips Wetted with Sulfuric Acid Solution of Medium Strength Using a Through Drying Operation. Agr Biol Chem 26(6):371–377
Lai M-C, Traxler RW (1994) A coupled two-stage continuous fermentation for solvent production by *Clostridium acetobutylicum*. Enzym Microbial Technology 16(12):1021–1025
Lamed R, Setter E, Kenig R, Bayer EA (1983a) The cellulosome: a discrete cell surface organelle of *Clostridium thermocellum* which exhibits separate antigenic, cellulose-binding and various cellulolytic activities. Biotechnol Bioeng 13:63–181. ISSN:0006-3592
Lamed R, Setter E, Bayer EA (1983b) Characterization of a cellulose-binding, cellulase-containing complex in *Clostridium thermocellum*. J Bacteriol 2:828–836
Lee SY, Park JH, Jang SH, Nielsen LK, Kim J, Jung KS (2008a) Fermentative butanol production by clostridia. Biotechnol Bioeng 101–2:209–228
Lee S, Cho MO, Park CH, Chung Y, Kim JH (2008b) Continuous Butanol production using suspended and immobilized *Clostridium beijerinckii* NCIMB 8052 with supplementary butyrate. Energy Fuels 13:3459–3464
Lehmann D, Hönicke D, Ehrenreich A, Schmidt M, Weuster-Botz D, Bahl H, Lütke-Eversloh T (2012) Modifying the product pattern of *Clostridium acetobutylicum*: physiological effects of disrupting the acetate and acetone formation pathways. Appl Microbiol Biotechnol 94(3): 743–754

Li R-D, Li Y-Y, Lu L-Y, Ren C, Li Y-X, Liu L (2011a) An improved kinetic model for the acetone-butanol-ethanol pathway of *Clostridium acetobutylicum* and model-based perturbation analysis. BMC Syst Biol 5(Suppl 1):S12

Li S-Y, Srivastava R, Suib SL, Li Y, Parnas RS (2011b) Performance of batch, fed-batch, and continuous A-B-E fermentation with pH-control. Bioresour Technol 102(5):4241–4250

Liew ST, Arbakariya A, Rosfarizan M, Raha AR (2005) Production of solvent (acetone-butanol-ethanol) in continuous fermentation by *Clostridium saccharobutylicum* DSM13864 using gelatinised sago starch as a carbon source. Malays J Microbiol 2(2):42–45

Liu H, Wang G, Zhang J (2013) In: Fang Z (ed), The promising fuel-biobutanol, liquid, gaseous and solid biofuels—conversion techniques, ISBN:978-953-51-1050-7, InTech, doi:10.5772/52535. Available from: http://www.intechopen.com/books/liquid-gaseous-and-solid-biofuels-conversion-techniques/the-promising-fuel-biobutanol

Lo YC, Bai MD, Chen WM, Chang JS (2008) Cellulosic hydrogen production with a sequencing bacterial hydrolysis and dark fermentation strategy. Bioresour Technol 99(17):8299–8303

López-Contreras AM (2001) *Clostridium beijerinckii* cells expressing Neocallimastix patriciarum glycoside hydrolases show enhanced lichenan utilization and solvent production. Appl Environ Microbiol 67:5127–5133

Maddipati P, Atiyeh HK, Bellmer DD, Huhnke RL (2011) Ethanol production from syngas by *Clostridium* strain P11 using corn steep liquor as a nutrient replacement to yeast extract. Bioresour Technol 102(11):6494–6501

Martín C, Marcet M, Almazán O, Jönsson LJ (2007) Adaptation of a recombinant xylose-utilizing saccharomyces cerevisiae strain to a sugarcane bagasse hydrolysate with high content of fermentation inhibitors. Bioresour Technol 98(9):1767–1773. ISSN:0960-8524

Martınez AT, Ruiz-Dueñas FJ, Martínez MJ, Del Río JC, Gutiérrez A (2009) Enzymatic delignification of plant cell wall: from nature to mill. Curr Opin Biotechnol 20(3):348–357. ISSN:0958-1669

Mayank R, Ranjan A, Moholkar VS (2012) Mathematical models of ABE fermentation: review and analysis. Crit Rev Biotechnol 1–29

McNew K, Griffith D (2005) Measuring the impact of ethanol plants on localgrain prices. Rev Agric Econ 27:164–180

Melaina MW, Heath G, Sandor D, Steward D, Vimmerstedt L, Warner E, Webster KW (2013) Alternative fuel infrastructure expansion: costs, resources, production capacity, and retail availability for low-carbon scenarios. Transportation energy futures series. 101 Prepared for the U.S. Department of Energy by National Renewable Energy Laboratory, Golden, CO. DOE/GO-102013-3710

Millat T, Voigt C, Janssen H, Cooksley CM, Winzer K, Minton NP, Wolkenhauer O (2014) Coenzyme A-transferase-independent butyrate re-assimilation in *Clostridium acetobutylicum*-evidence from a mathematical model. Appl Microbiol Biotechnol 98:9059–9072

Mutschlechner O, Swoboda H, Gapes JR (2000) Continuous two-stage ABE-fermentation using *Clostridium beijerinckii* NRRL B592 operating with a growth rate in the first stage vessel close to its maximal value. J Mol Microbiol Biotechnol 2:101–105

Mužíková Z, Šimáček P, Pospíšil M, Šebor G (2014) Density, viscosity and water phase stability of 1-butanol and gasoline blends. J Fuels 2014, ID 459287

Naik SN, Goud VV, Rout PK, Dalai AK (2010) Production of first and second generation biofuels: a comprehensive review. Renew Sustain Energy Rev 14:578–597

Nakayama S, Kiyoshi K, Kadokura T, Nakazato A (2011) Butanol production from crystalline cellulose by cocultured *Clostridium thermocellum* and *Clostridium saccharoperbutylacetonicum* N1-4. Appl Env Microbiol 77(18):6470–6475

Napoli F, Olivieri G, Russo M, Marzocchella A, Salatino P (2011) Continuous lactose fermentation by *Clostridium acetobutylicum*—assessment of acidogenesis kinetics. Bioresour Technol 102–2:1608–1614

Nordon RE, Craig SJ, Foong FC (2009) Molecular engineering of the cellulosome complex for affinity and bioenergy applications. Biotechnol Lett 31(4):465–476. ISSN:1573-6776

Ogden J, Williams RH, Larson ED (2004) Societal lifecycle costs of cars with alternative fuels/engines. Energy Policy 32(1):7–27

Pandey A (2003) Solid-state fermentation. Biochem Eng J 13(2–3):81–84. ISSN:1369-703X

Papoutsakis ET, Rice U (1984) Equations and calculations for fermentations of butyric acid bacteria. Biotechnol Bioeng 26(2):174–187

Peralta-Yahya PP, Zhang F, del Cardayre SB, Keasling JD (2012) Microbial engineering for the production of advanced biofuels. Nature 488:320–328

Pfromm PH, Amanor-Boadu V, Nelson R, Vadlani P, Madl R (2010) Bio-butanol versus bio-ethanol: a technical and economic assessment for corn and switchgrass fermented by yeast or *Clostridium acetobutylicum*. Biomass Bioenergy 34:515–524

Pimentel D, Patzek TW, Gerald C (2007) Ethanol production: energy, economic, and environmental losses. Rev Environ Contam Toxicol 189:25–41

Qureshi N, Maddox IS, Friedl A (1992) Application of continuous substrate feeding to the ABE fermentation: relief of product inhibition using extraction, perstraction, stripping, and pervaporation. Biotechnol Prog 8:382–390

Qureshi N, Saha BCJ, Cotta MA (2007) Butanol production from wheat stray hydrolysate using *Clostridium beijerinckii*. Bioprocess Biosyst Eng 30:419–427

Qureshi N, Ezeji TC, Ebener J, Dien BS, Cotta MA, Blaschek HP (2008) Butanol production by *Clostridium beijerinckii*. Part I. Use of acid and enzyme hydrolyzed corn fiber. Bioresour Technol 99:5915–5922

Qureshi N, Saha BC, Dien B, Hector RE, Cotta MA (2010) Production of butanol (a biofuel) from agricultural residues: Part I. Use of barley straw hydrolysate. Biomass Bioenergy 34:559–565

Saha BC, Iten LB, Cotta MA, Wu V (2005) Dilute acid pretreatment, enzymatic saccharification and fermentation of wheat straw to ethanol. Process Biochem 40:3693–3700

Sandgren M, Hiberg JS (2005) Structural and biochemical studies of gh family 12 cellulases: improved thermal stability and ligand complexes. Prog Biophys Mol Biol 89(3):246–291. ISSN:0079-6107

Scheller HV, Ulvskov P (2010) Hemicelluloses. Ann Rev Plant Biol 61:263–89. ISSN:1543-5008

Setlhaku M, Brunberg S, Villa A, Wichmann R (2012) Improvement in the bioreactor specific productivity by coupling continuous reactor with repeated fed-batch reactor for acetone—butanol—ethanol production. J Biotechnol 161(2):147–152. doi:10.1016/j.jbiotec.2012.04.004

Sharma B, Ingalls RG, Jones CL, Khanchi A (2013) Biomass supply chaindesign and analysis: basis, overview, modeling, challenges, and future. Renew Sustain Energy Rev 24:608–627

Shinto H, Tashiro Y, Yamashita N, Kobayashi G, Sekiguchi T, Hanai T, Kuriya Y, Okamoto M, Sonomoto K (2007) Kinetic modeling and sensitivity analysis of acetone-butanol-ethanol production. J Biotechnol 131(1):45–56

Singh R, Kapoor V, Kumar V (2012) Utilization of agro-industrial wastes for the simultaneous production of amylase and xylanase by thermophilic actinomycetes. Braz J Microbiol 1545–1552

Soni BK, Das K, Ghose TK (1982) Bioconversion of agro-wastes into acetone butanol. Biotechnol Lett 4(1):19–22

Sun Y, Cheng JY (2002) Hydrolysis of lignocellulosic materials for ethanol production: a review. Bioresour Technol 83(1):1–11

Terracciano JS, Kashket ER (1986) Intracellular conditions required for initiation of solvent production by *Clostridium acetobutylicum*. Appl Environ Microbiol 52(1):86–91

Thorn GJ, King JR, Jabbari S (2013) pH-induced gene regulation of solvent production by *Clostridium acetobutylicum* in continuous culture: parameter estimation and sporulation modelling. Math Biosci 241(2):149–166

Tran HTM, Cheirsilp B, Hodgson B, Umsakul K (2010) Potential use of *Bacillus subtilis* in a co-culture with *Clostridium butylicum* for acetone-butanol-ethanol production from cassava starch. Biochem Eng 48:260–267

Tummala SB, Welker NE, Papoutsaks ET (2003) Design of antisense RNA constructs for downregulation of the acetone formation pathway of *Clostridium acetobutylicum*. J Bacteriol 185:1923–1934

Uyeda K, Rabinowitz JC (1971) Pyruvate-ferredoxin oxidoreductase. IV. Studies on the reaction mechanism. J Bacteriol Chem 246:3120–3125

Vamvuka D, Kastanaki E, Grammelis P, Kakaras E (2003) Pyrolysis characteristics and kinetics of biomass residuals mixtures with poor coal. Fuel 82:1949

Velázquez-Sánchez HI, Montes-Horcasitas MC, Aguilar-López R (2014) Development of a phenomenological kinetic model for butanol productiong using *Clostridium Beijeinckii*. Revista Mexicana de Ingeniería Química 13(1):103–112

Wang Y, Li X, Blaschek HP (2013) Effects of supplementary butyrate on butanol production and the metabolic switch in *Clostridium beijerinckii* NCIMB 8052: genome-wide transcriptional analysis with RNA-Seq. Biotechnol Biofuels 6(1):138

Wen Z, Wu M, Lin Y, Yang L, Lin J, Cen P (2014) Artificial symbiosis for acetone-butanol-ethanol (ABE) fermentation from alkali extracted deshelled corn cobs by co-culture of Clostridium beijerinckii and Clostridium cellulovorans. Microb Cell Fact 13: 92–103

Wietzke M, Bahl H (2012) The redox-sensing protein Rex, a transcriptional regulator of solventogenesis in *Clostridium acetobutylicum*. Appl Microbiol Biotechnol 96(3):749–761

Yuea D, Youa F, Snyder SW (2014) Biomass-to-bioenergy and biofuel supply chain optimization: overview, key issues and challenges. Comput Chem Eng 66:36–56

Zhao Y, Tomas C (2005) Intracellular butyryl phosphate and acetyl phosphate concentrations in *Clostridium acetobutylicum* and their implications for solvent formation. Appl Environ Microbiol 71(1):530–537

Zheng Y, Li L, Xian M, Ma Y, Yang J, Xu X, He D (2009) Problems with the microbial production of butanol. J Ind Microbiol Biotechnol 36:1127–1138

Zverlov V, Berezina O, Velikodvorskaya GA, Schwarz WH (2006) Bacterial acetone and butanol production by industrial fermentation in the Soviet Union: use of hydrolyzed agricultural waste for biorefinery. Appl Microbiol Biotechnol 71:587–597

[illegible] (2008) [illegible]

Fast Pyrolysis of Agricultural Wastes for Bio-fuel and Bio-char

Suchithra Thangalazhy-Gopakumar and Sushil Adhikari

Abstract Fast pyrolysis is a thermo-chemical conversion method to produce liquid fuel from biomass. This process involves the rapid thermal decomposition of organic compounds in the absence of oxygen. The vapors formed are rapidly condensed to yield a liquid product called bio-oil. Since the major product is liquid, it is easy to store, and handle. Moreover, bio-oil can be readily transported to facilities where it can be most effectively used. Unlike other conversion methods for bio-fuel production, fast pyrolysis utilizes different types of feedstocks, therefore this process can be considered as a tool for solid waste management. Different types of feedstocks ranging from agriculture and forest residues to MSW, plastic wastes and animal manures have been utilized for pyrolysis studies around the world. Liquid fuel production using fast pyrolysis has received much attention in recent years. Bio-oil, the liquid product of fast pyrolysis, can be considered as an intermediate for fuel and chemical production. Bio-char, the solid product of pyrolysis, has multiple applications like bio-char from other methods. The chapter reviews different types of reactors used for fast pyrolysis, bio-oil properties, challenges and opportunities, and the current status of fast pyrolysis applications. In addition, the chapter discusses applications of bio-char and energy and economics of fast pyrolysis.

Keywords Biomass · Bio-char · Bio-oil · Fast pyrolysis · Reactors · Solid wastes

S. Thangalazhy-Gopakumar (✉)
Department of Chemical and Environmental Engineering, Faculty of Engineering, University of Nottingham Malaysia Campus, Jalan Broga, 43500 Semenyih, Selangor Darul Ehsan, Malaysia
e-mail: Suchithra.Thangalazhy@nottingham.edu.my

S. Adhikari
Department of Biosystems Engineering, Auburn University, Auburn, AL 36849, USA

O.P. Karthikeyan et al. (eds.), *Recycling of Solid Waste for Biofuels and Bio-chemicals*, Environmental Footprints and Eco-design of Products and Processes, DOI 10.1007/978-981-10-0150-5_11

1 Introduction

Energy is a vital ingredient for socio-economic development. Fossil fuel resources are finite, alternative sources are needed as supplements due to the fast past modernization. Long-term sustainability can be guaranteed if sources are renewable. Wind, solar, geothermal, and hydro have direct potential for renewable energy, however, biomass shows the highest capacity to supplement fossil fuels for both carbon-based fuels and chemicals.

Food production and consumption will inevitably produce organic residues/waste. These organic residues/wastes were formerly deposited in wastelands, where decomposition produces greenhouse gases such as CO_2 or methane. The organic part of these residues, termed biomass here, can be effectively used for energy applications. Therefore biomass can be considered as a sustainable source of fixed carbon for chemicals and fuels. There are different processes for the production of energy from biomass, which can be broadly classified as biochemical, thermo-chemical, catalytic and synthetic biology routes. Among these options, thermo-chemical methods have an advantage as they are somewhat feedstock agnostic. Bio-ethanol and biodiesel productions are the two major bioenergy routes, which depend on cellulosic biomass and lipid-rich biomass, respectively. Thermo-chemical routes, such as combustion, gasification and pyrolysis utilize all types of residues/feedstocks. Therefore, a successful implementation of thermo-chemical routes followed by catalytic upgrading would be a sustainable supplement for the increasing demands of fossil fuels.

This chapter discusses recent developments in the field of fast pyrolysis, one of the thermo-chemical biomass conversion methodologies. Bio-oil, the main product of fast pyrolysis of biomass, can be considered a substitute/additive for crude oil for fuels and chemicals, but there are challenges in achieving these aims. Since fast pyrolysis is feedstock agnostic and requires low capital investment, research into the chemical makeup arising from different feedstocks, bio-oil properties and potential applications are summarized along with upgrading of the bio-oils for the production of fuel and chemical feedstocks. Furthermore, major reactor configurations are discussed, as is progress in upscaling the technology from laboratory-scale to industrial-scales.

2 Fast Pyrolysis

Pyrolysis, thermal degradation of organic matter at high temperature in the absence of oxygen, involves changes in chemical composition and the physical phase. Pyrolysis has been practiced for centuries for the production of charcoal. Since the oil crisis in the mid-1970s, considerable effort has been directed towards the development of liquid fuels from ligno-cellulosic biomass (Czernik and Bridgwater 2004). Pyrolysis can also be utilized for the production of liquid fuels by changing process parameters like temperature, heating rate and residence time of vapors: called fast pyrolysis. In fast pyrolysis, the biomass is decomposed thermally at a medium temperature (about 500 °C) in the absence of oxygen. Fast pyrolysis allows high heat transfer rate to the biomass feedstock and a short residence time (about 2 s) in the reaction zone. The major products of fast pyrolysis are vapors, aerosols, char and gases. Vapors and aerosols are condensed quickly to form a liquid, which is known as bio-oil. The yields and properties of bio-oil depend on the feedstock used, process conditions, and the product collection efficiency. There are some essential criteria for producing liquid fuels from fast pyrolysis such as (i) high heat transfer rate for biomass particle reaction with relatively fine biomass feedstock of less than 3 mm, (ii) short residence time of about 2 s to avoid secondary reactions, (iii) control of reaction temperature at around 500 °C, (iv) avoiding cracking of vapors by char and ash removal, and (v) rapid cooling of vapors to increase bio-oil yields (Meier et al. 2013; Bridgwater et al. 1999; Bridgwater 1999; Bridgwater and Peacocke 2000).

Fast pyrolysis is the most effective liquefaction method as it can yield around 80 wt% of dry biomass as bio-oil. Pyrolysis is a feedstock agnostic process where any type of organic wastes can be converted into liquid fuel. Most of the thermo-chemical biomass conversion technologies have challenges with high ash content organic biomass such as animal manure, sludge, municipal solid wastes. High temperature processes such as gasification and combustion (which work above the ash fusion temperature) have major issues with slag formation due to ash content. In contrast, in slow pyrolysis and hydrothermal carbonization, most of the ash content of the biomass will remain with the major product bio-char. Therefore, energy application of bio-char (gasification, combustion) would produce the same issue as slag formation. Fast pyrolysis is a medium temperature process and a minor portion of ash will be captured in the major product: bio-oil. Though fouling (due to ash content) can be an issue during bio-oil processing, the frequency of occurrence of this issue will be low.

The fast pyrolysis process can be divided into four sections: biomass pre-treatment, pyrolysis, char recovery and liquid collection (Bridgwater and Peacocke 2000). The process flow diagram for fast pyrolysis is given in Fig. 1. Harvesting of biomass is normally not considered as a part of fast pyrolysis systems since it is common for all biomass-using processes. However, one of the advantages for fast pyrolysis is the utilization of mobile units/pilot plants. Mobile units can be utilized for seasonal feedstocks and pilot plants can be utilized for decentralized feedstocks. Since the major product is liquid, subsequent transport to existing refining facilities is easy. Therefore harvesting and handling cost of feedstock can be reduced (Badger and Fransham 2006). However, an economic study on mobile plants in 2011 showed a low probability for a positive Net Present Value (NPV) compared to stationary plants. The economics of mobile plants could be improved, if feedstock costs were reduced and crude oil price increased (Palma et al. 2011).

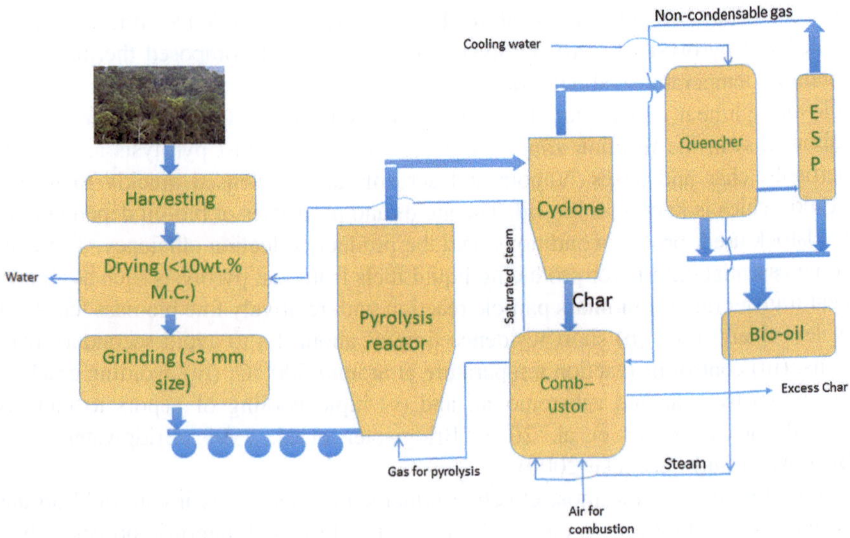

Fig. 1 Schematic of the fast pyrolysis process

In order to achieve heat and mass transfer restrictions of fast pyrolysis, biomass has to be dried to reduce the moisture content to below 10 wt% and ground to a particle size of less than 3 mm. In a self-sustained design of a fast pyrolysis plant, drying can be achieved by utilizing the heat from the bio-char recovery and bio-oil collection sections. The steam produced during the condensation of pyrolytic vapors could be passed through the combustion chamber (where char will be burned to recover heat) and the heat of this saturated steam could be used for drying. The heart of the fast pyrolysis process is the pyrolysis reactor. Reactors of different designs have been tested for fast pyrolysis in order to achieve high heating rate and heat transfer rate for biomass particles. A detailed description of reactor configurations is presented in Sect. 4. A char recovery system is installed downstream of the pyrolysis reactor. The char recovery system normally consists of a cyclone separator to remove the char and ash from the pyrolytic vapors. The recovered char will be sent to a combustion chamber. In the combustion chamber, the heat produced by burning of char will be transferred to the biomass pre-treatment section via saturated steam, while pyrolytic vapors from the cyclone will be sent to a quencher/condenser followed by an ESP (electrostatic precipitator) to recover all condensable compounds in liquid form which is called bio-oil. Bio-oil can be stored in a corrosive resistant vessel and transferred for further processing. The non-condensable gas from condenser and ESP will be recycled as pyrolyzing gas. One part of heat produced from char combustion can be transferred to the pyrolysis reactor through non-condensable gas. The steam produced from condenser could be used as saturated steam for drying by passing through the combustion chamber.

3 Biomass

According to ANSI/ASABE, any organic waste produced from living/dead matter other than fossil fuel can be considered as biomass (ANSI/ASABES593 2006). In this regard, there are many organic residues potentially available for bioenergy applications which can be categorized as forestland residues, farmland residues, and city residues (Loque et al. 2011). Forestland residues include tree residues and wood residues. Tree residues are solid wastes generated during the growth of forest such as leaves, needles, and branches while wood residues are generated during the wood harvest and processing such as branches, bark and saw dust (Loque et al. 2011). Farm land wastes are of two types such as straw residues and manures. Straw residues include all agriculture residues formed during the processing of seed, sugar or other food products. Manures, another type of farmland residues, are generated in large amounts from livestock production. In addition, the organic parts of urban land residue or municipal solid wastes (MSW) such as paper, cardboards, landscaping materials can be utilized as biomass (Loque et al. 2011).

According to International Energy Outlook 2013, the world energy consumption will increase from 524 quadrillion Btu in 2010 to 630 quadrillion Btu in 2020 and is estimated to further increase to about 820 quadrillion Btu in 2040, representing a 30-year increase of 56 % (EIA 2013). Crude oil is the major worldwide energy supply with 32 %, followed by coal 29 %, natural gas 21 %, nuclear 5 %, and renewable resources 13 %. Biomass contributed with 10 % to the total energy supply in 2011 (WBA 2014). Asia and Africa major consumers of bioenergy, however, most of the biomass consumption in these countries is as a heat source. Utilization of biomass for heating contributes significantly to air pollution in these countries. The utilization of biomass for biofuel production has increased in recent years, most of which occurs in developed countries. Brazil, however, ranks number one in the production of biofuel. A study on global potential of biomass for energy projected surplus agricultural land as major biomass source which would provide 76.7 % of total biomass energy (1300EJ) in 2050 (Ladanai and Vinterbäck 2009). According to the World Bioenergy Association (WBA), major biomass from forests are: fuel wood, charcoal, forest residues, black liquor, wood industry residues, and recovered wood; from agriculture: animal by-products, agricultural by-products, and energy crops; and from wastes such as MSW and landfill. The WBA has published the yield of major agricultural crops from different continents, as shown in Table 1.

Since fast pyrolysis is feedstock agnostic, different types of biomass have been utilized for fast pyrolysis. For example, animal manures when utilized as fertilizer results in nutrient leaching and water pollution. Another way of handling these wastes is anaerobic fermentation, which produces greenhouse gas emissions, if not handled efficiently. In this regard, fast pyrolysis is a potential candidate to utilize the organic portion of manure. Handling high ash-content biomass is a challenge for other thermo-chemical conversions, whereas in fast pyrolysis, most of the ash will remain in the byproduct: bio-char. However, most of the work done so far in fast

Table 1 Average yields of crops for year 2011 (tons/ha)

	World	Africa	Americas	Asia	Europe	Oceania
Cassava	12.7	10.7	13	19.7	–	10.9
Maize	5.16	1.92	6.84	4.81	6.66	6.81
Rapeseed	1.86	1.64	1.88	1.54	2.53	1.14
Sorghum	1.38	0.95	3.38	1.16	3.63	3.06
Soybeans	2.53	1.25	2.85	1.48	1.95	1.71
Sugar beet	54.9	52.8	55.1	49.2	56.2	–
Sugar cane	71.1	64.2	75.1	66.9	85.5	76.2

Source World Bioenergy Association (WBA 2014) (Reprinted with the permission from Dr. Bharadwaj V Kummamuru, World Bioenergy Association, Sweden (lead author))

pyrolysis was carried out on wood and agriculture wastes because of feedstock consistency and comparability between tests. More than one hundred different types of biomass have been tested ranging from agricultural wastes, energy crops, and forestry wastes and solid wastes, some examples are given in Table 5. The selection of biomass depends on their abundance in particular locations.

4 Reactor Configurations

A fast pyrolysis process normally starts with the preparation of feed, which is typically dried to have less than 10 % water content in order to minimize water content in the liquid oil product. The feed is then ground into small particles to reduce mass transfer resistance, then fast pyrolysis takes place, followed by the rapid and efficient separation of solid char from vapors and finally rapid quenching and collection of bio-oil (Fig. 1) (Bridgwater 2012a, b).

The fast pyrolysis process involves heat and mass transfer, phase transition, and chemical reactions in a few seconds or less. In this process, biomass will be suddenly exposed to optimum temperature with a reduced exposure to lower temperatures that favor char formation. In other words, the reactor configuration for fast pyrolysis process should be able to provide a high heating rate and heat transfer rates for biomass. Considerable research has focused on developing new reactor designs for a variety of feedstocks to control and improve the final liquids' quality and the collection systems (Bridgwater 2011, 2012a, b). Some reactor configurations can achieve liquid yields of 70–80 % based on dry biomass weight (Mohan et al. 2006). The choice of reactor mainly depends on the flexibility/ease in operation, feed size and the desired purity of the bio-oil.

Major reactor configurations are: bubbling fluidized bed, circulating fluidized bed, rotating cone, augur, and ablative reactors (Fig. 2). In addition to these reactors, other configurations were also investigated for biomass fast pyrolysis, which includes vacuum pyrolysis, fixed bed, entrained flow reactors, microwave

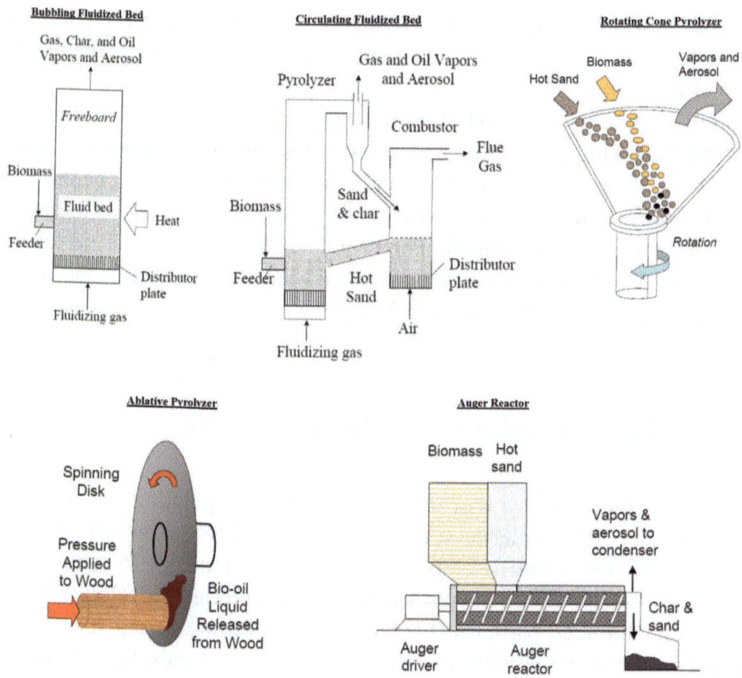

Fig. 2 Major reactor configurations for fast pyrolysis (reproduced with the permission from Prof. Robert C. Brown, Iowa State University)

pyrolyzer, plasma, and solar reactors. All of these reactors have advantages and drawbacks, depending on the types of raw biomass material and commercial scale of fast pyrolysis.

4.1 Bubbling Fluidized Bed Reactor

Fluidized bed is a well-developed technology, where the reactors are simple in construction and operation, which provide good and efficient temperature control, heating rate and heat transfer to biomass particles. Hence fluidized bed reactors are attractive and popular for biomass fast pyrolysis. Very small biomass particles of typically 2–3 mm is used in fluidized bed reactors to obtain good liquid yields. In this reactor configuration, the heat energy required is supplied by heating the walls of the reactor or by injecting hot inert gas. The distributor plate ensures that the fluidizing gas is well distributed. Normally, helium or nitrogen is used as the fluidizing gas and sand as the bed material. Shallow depth bed or a high rates of gas flow are usually both used to achieve short residence time for vapors (Scott et al. 1999). If the bio-char has a higher residence time, they act as a vapor cracking catalyst in the process and cause secondary cracking of vapors. Therefore char must be rapidly and effectively separated. This is normally done by passing product vapors and gases to a separation system, which includes one or more cyclones (Mohan et al. 2006).

In catalytic pyrolysis, instead of sand, catalysts such as ZSM-5 or silica are used as bed material. The gas, vapors and aerosol formed are transferred to a series of condensers to obtain the bio-oil (Agblevor et al. 2010; Shi 2012). The fluidized bed reactor provides high yields and good quality bio-oil and does not allow char to accumulate as it is rapidly eluted. The liquid yield from fluid-bed pyrolyzers are typically 70–75 wt% from wood on a dry-feed basis. The yield of the by-product char is typically 15 wt% (Bridgwater 2012a, b).

4.2 Circulating Fluidized Bed Reactor (CFB)

The circulating fluidized bed reactor is quite similar in operation to that of fluidized bed reactors but the residence time for char in CFBs is almost same as that of vapors. A second vessel is employed as a char combustor to reheat the solids which are circulated back to the reactor by passing through a cyclone separator for separating the entrained particles (Lappas et al. 2002). In this reactor, the hot sand circulating between the combustor and the pyrolyzer provide heat. However, circulating fluidized bed reactors are as not efficient as bubbling fluidized bed reactors in terms of temperature control and heat transfer. CFB reactors are suitable for larger throughput (Bridgwater 2012a, b). The char is more attired, therefore higher char contents appear in the condensed bio-oil. There is no char available for export since all char is burned in the combustor and used as heat supply. CFBs also do not have high heat transfer rates because they are dependent on gas flow rates. The biomass ash builds up in the circulating solids, which can act as a cracking catalyst, causing the loss of volatiles and improving the properties of the bio-oil products (Mohan et al. 2006; Bridgwater 2012a, b). CFB technology is often developed in larger application with enhanced flexibility for igniting multi-fuels and it has efficiency up to 95 % (Bridgwater 2004).

4.3 Rotating Cone Pyrolyzer

University of Twente invented a new reactor design for fast pyrolysis, a rotating cone pyrolyzer developed by the Biomass Technology Group (Biomass Technology Group) in the Netherlands (Wagenaar et al. 1994, 2008). The rotating cone pyrolyzer is an innovative reactor configuration achieving high heating rates and a short vapor residence time (Wagenaar et al. 1993). The working principle of a rotating cone pyrolyzer is quite similar to circulating fluidized bed reactor in terms of the raw materials inlet direction and processing materials where centrifugal forces are used to transport sand and biomass. Biomass and sand are fed to the bottom of a rotating cone and the biomass is pyrolyzed by hot sand while transported upward through a spiral motion along the hot sidewall of the cone. The pressure of the outgoing materials is slightly above atmospheric levels. The inner diameter of the cone has a maximum of 0.650 m. The rotational speed of the cone is 900 rpm. The reactor volume may range from 2 to 200 dm^3. The first industrial use of this reactor was for the pyrolysis of biomass (Wagenaar et al. 1994). Using sand in the process has the advantage of avoiding fouling of the cone wall and enhancing heat transfer. After leaving the impeller, the particles flow outwards from the cone and experience high heat transfer rate from the heated surface. This pyrolyzer is very compact in design, requires feeding with very small particles and can be used for high throughput (Bridgwater 2012a, b). The char produced is not a by-product since it is burned entirely for heat supply purpose. Liquid yields of 60–70 wt% on dry feed are typically obtained (Bridgwater 2012a, b).

4.4 Ablative Pyrolyzer

Ablative pyrolysis applies a mode of reaction similar to melting butter in a hot pan. The high heat from the wall causes wood to 'melt' on contact under high pressure. As the wood is pressed mechanically, the residual oil is evaporated to yield pyrolysis vapor, which can be collected by rapid cooling. The residual oil on the hot surface provides lubrication and helps to enhance evaporation rates of biomass. Liquid yields of 60–65 wt% on dry feed basis were typically obtained (Bridgwater 2011). Large particles including logs can be pyrolyzed without being ground. The char is deposited on the hot surface due to element cracking. A study comparing product yields for fluid bed- and ablative reactors (Peacocke et al. 1996) gave 59.4 wt% organics at 515 °C and 1.19 s residence time for a fluid bed reactor, and 62.1 wt% organics at 502 °C and 1.1 s residence time for the ablative reactor.

The main benefits of this reactor are that carrier gas is not required and larger biomass particles can be used (Scott et al. 1999). The rate of reaction in the reactor is strongly depended on the applied pressure, the reactor surface temperature, and the relative velocity between biomass and the hot surface. However, post-treatment of bio-oil is required for char removal.

4.5 Auger Reactor

The auger reactor is a compact design for continuous fast pyrolysis which does not require a carrier gas. Augers are moved with biomass feed through a heated cylindrical tube, which causes pyrolysis, producing bio-char, bio-oil and gas. Sand or steel shot are used as heat carriers. The mode of heat transfer in this reactor is mainly conduction. The rotating screw conveyors combine the biomass and the heat carrier effectively and pyrolysis reaction takes place. The design reduces energy cost and the vapour residence time can be manipulated by the heated zone length (Mohan et al. 2006; Thangalazhy-Gopakumar et al. 2010; Bridgwater 2012a, b). Energy costs for the operation are very low compared to other reactor designs (Adjaye and Bakhshi 1995a, b, c). However, this it is only suitable for small scale production of bio-oil.

4.6 Vacuum Pyrolysis Reactor

Though vacuum pyrolysis provides low heat transfer rates, the vapor residence time is comparable with fast pyrolysis and therefore utilized for bio-oil production. Vacuum pyrolysis is the thermal decomposition of organics under reduced pressure. Vapors are rapidly removed from the reactor by vacuum, and recovered as bio-oils by condensation. Compared to other fast pyrolysis systems, lower liquid and increased char yields were typically obtained (Bouchera et al. 2000; Garcia-Perez et al. 2002). The advantages of the process are that larger particles can be used and less char contaminates the liquid product. However, the process is very complex, has less efficient heat and mass transfer rates and is costly as it requires large-scale equipment and higher capacity vacuum (Mohan et al. 2006; Bridgwater 2012a, b). The pressure in the vacuum reactor is maintained at 15 kPa by using a vacuum pump to remove the vapor formed due to the heating of biomass.

4.7 Fixed Bed

Fixed bed reactors are reactors in which the feedstock, carrier system, catalyst, filter media and other substances remain stationary. It has some advantages, such retention of high amount of carbon. The main advantage is the simplicity of equipment design and low fabrication cost for the equipment. Inert condition of the environment is maintained by the presence of pure nitrogen flow. A cold trap is used to collect condensable vapors to condense them into bio-oil. In a fixed bed reactor, the flow rate of nitrogen needs to be controlled in order to obtain optimum yields. This reactor configuration is widely used for laboratory studies.

4.8 Microwave Pyrolyzer

Some laboratory studies have been conducted with microwave reactors for fast pyrolysis. Conventional thermal heating, heat transfers from the surface to the center of the material by conduction, convection and radiation. Conventional heating is inefficient and slow, as it depends on the convection current and thermal conductivity of the feedstock. Conventional heating is characterized by high-energy consumption therefore microwave reactors were developed to reduce the energy consumption of the process. Microwave heating is achieved by dielectric heating and transferring heat from electromagnetic energy to thermal energy which is a kind of energy conversion rather than conventional heating (Motasemi and Afzal 2013). However, microwaves consume high amount of electric power and the operating costs are very high.

According to the IEA (International Energy Agency) Bioenergy Task 34 for Pyrolysis (http://www.pyne.co.uk/), bubbling fluidized bed reactors are the most favorable configuration in terms of technical strength and market attractiveness, followed by circulating fluidized bed and auger reactors respectively. Rotating cone and ablative pyrolyzers are very difficult to scale up and therefore less attractive than other three reactors. In terms of bio-oil yield, both fluidized bed and ablative reactors perform better than rotating cone and auger reactors.

5 Bio-oil

Fast pyrolysis liquid, which is known as bio-oil, has a higher heating value (HHV) of about 17 MJ/kg when produced with about 25 wt% water that cannot readily be separated. Though known as bio-oil, this liquid will not mix with hydrocarbon liquids. Bio-oil is a multi-component dark brown organic liquid. The liquid is formed from rapid and simultaneous thermal depolymerization, fragmentation and quenching of cellulose, hemicellulose and lignin of biomass (Czernik and Bridgwater 2004; Mohan et al. 2006). Though bio-oil is homogeneous in appearance, it contains compounds of different sizes and structures, difference in their functional groups, polarity and density. Therefore, bio-oil should be mixed thoroughly before analysis.

5.1 Analysis

The composition or properties of bio-oil depends on the composition of the biomass (percentage of cellulose, hemicellulose, and lignin). In addition, the presence of alkali metals (in the form of ash) and other extractives in biomass. Due to the complexity, chemical characterization of bio-oil cannot be done with one or two simple analyses (Kanaujia et al. 2014; Staš et al. 2014).

For detailed characterization, bio-oil should be fractionated as water-soluble and -insoluble or pentane-soluble and -insoluble, and further fractionated to methanol-, ether- or toluene-soluble and so on (Sipila et al. 1998; Oasmaa et al. 2003; Garcia-Perez et al. 2007; Salehi et al. 2011). Each of these fractions will be analyzed for the compounds/functional groups present. Gas Chromatography (GC) equipped with Mass Spectroscopy (GC-MS) or Flame Ionization Detector (GC-FID) are the most common analytical instruments used for most of the chemical composition studies of bio-oil (Branca et al. 2003; Bhattacharya et al. 2009; Thangalazhy-Gopakumar et al. 2010; Kantarelis et al. 2013). However, GC can detect only volatile compounds. Some of the non-volatile compounds (especially anhydro-sugar compounds) can be detected by utilizing polar columns such as 1701. Most GC analyses quantify compounds such as phenol derivatives i.e. alkyl phenols, guaiacols, syngols, benzenediols, oxygenated aromatics, furans, pyrrols, higher aldehydes and ketones, and some anhydro-sugar compounds, all these compounds accounting for a maximum 40 wt% of the bio-oil (Staš et al. 2014). Another 25–30 wt% is the water content of the bio-oil. Water content in bio-oil is normally analyzed using the Carl-Fischer titration method. Remaining fraction of bio-oil remains unknown in most of the studies. High Performance Liquid Chromatography (HPLC) has been utilized to analyze organic acids, aldehydes, and sugars in bio-oil (Salehi et al. 2011; Choi et al. 2014). Another chromatographic techniques such as gel permeation-, ion-, and column chromatography are also utilized for bio-oil analysis. Fourier Transform Infrared Spectroscopy (FTIR) is another technology widely used for characterization of different functional groups present in bio-oil. Some of the common absorption bands identified in bio-oil are C=O stretching (aldehydes, ketones, acids), C=C stretching (aromatics), C–O stretching (alcohols, ethers) O–H stretching (phenols) (Pütün 2002; Onay and Kockar 2006; Sensoz et al. 2006). ^{1}H NMR (Nuclear Magnetic Resonance) studies have also conducted to identify the different H atoms present in bio-oils (Pütün 2002; Onay and Kockar 2006; Sensoz et al. 2006). Chemical formulae for bio-oil are developed by analyzing the elemental composition (Choi et al. 2014). FTIR spectra for bio-oils from wheat straw or HWSP (*Triticum aestivum*), timothy grass or HTGP (*Phleum pratense*) and pinewood or HPWP (*Pinus banksiana*) are given in Fig. 3.

In addition to the above analytical techniques, some methods used in the food industry such as total water, soluble sugars, and total phenol contents have been evaluated for bio-oil analysis. Water soluble total carbohydrates in bio-oil was evaluated using the phenol–sulfuric acid assay developed by the Association of Analytical Communities, Inc. (AOAC), Method 988.12 (44.1.30) (Albalasmeh et al. 2013; Rover et al. 2013). In this method, bio-oil is treated with phenol and concentrated sulfuric acid, and the treated sample taken for UV analysis ($\sim$490 nm). The results were comparable with sugar analysis of the bio-oil by HPLC. However, presence of water-soluble furan compounds influence the results. Similarly, the total phenol content in bio-oil was evaluated by the Folin–Ciocalteu (FC) colorimetry method (Rover and Brown 2013). In this method, gallic acid is used as standard and the results are reported as gallic acid equivalent. The FC reagent will react with phenols to form chromogens that can be detected spectrophotometrically (Ikawa et al. 2003). However, this method is non-specific and can be affected by the presence of other compounds.

Bio-oil properties depend on its chemical composition. Therefore, there are several properties that need to be considered for bio-oil.

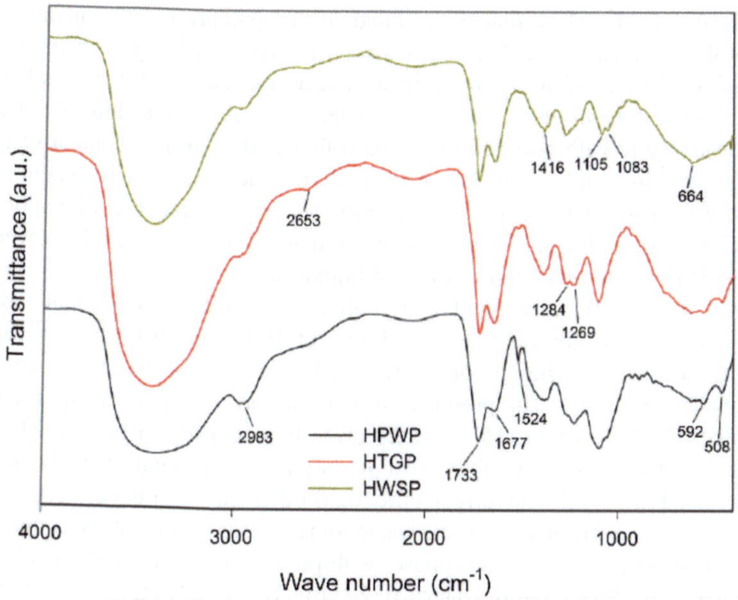

Fig. 3 FTIR spectra for bio-oil (Nanda et al. 2014) (Reprinted with the permission from Prof. Ajay K. Dalai, University of Saskatchewan, Canada)

5.2 *Properties*

Density: The density of the liquid is very high at around 1.2 g/mL as compared to light fuel oil at around 0.85 g/mL. This is due to the presence of some high molecular weight compounds such as syringols and sugar compounds. The feedstock used and the residence time of vapors also influence the density of bio-oil.

pH: The pH of the lignocellulosic bio-oil normally is about 2–3 which is due to the presence of the organic acid, such as acetic and formic acids. Therefore, the material of storage vessels must be made of acid-proof material such as stainless steel. The pH of bio-oil from sludge and animal manures are basic due to the presence of nitrogen compounds.

Higher Heating Value (HHV): The HHV of bio-oil is normally in the range of 16–19 MJ/kg, which is less than half of the values of conventional fuel. However, the energy density of bio-oil is 10 times higher than biomass. Therefore, it is easily transported. Bio-oil has about 42 % of the energy content of fuel oil on a weight basis, but 61 % on a volumetric basis (Abdullah et al. 2007; Bridgwater 2012a, b).

Solid Content: Some char, which is relatively fine, might be suspended in the liquid product, acting as a vapor cracking catalyst. Therefore char separation is needed to prevent vapor cracking which affect the yield of the liquid product. It is, however, difficult to achieve complete removal. Currently, two types of filtration methods are used cyclone separator and hot vapor filtration, to collect char from vapors before condensation as bio-oil.

Oxygen content: More than three hundred compounds are present in bio-oil; most of them are oxygenated. Therefore oxygen content of bio-oil is usually in the range of 35–40 %, and this is the primary reason for the vast differences between bio-oils and petroleum fuels. These oxygenated compounds make bio-oil polar, and thus non-miscible with non-polar petroleum fuels. High oxygen content results in low heating values and also makes the bio-oil unstable. The oxygen content can be reduced by increasing the process severity leading to cracking of vapors and removal of oxygen due to deoxygenation and dehydration reactions. However, this will also lead to the reduction of organic liquid yield.

Water content: Water content in bio-oil is the result from original moisture content of biomass (normally at a maximum of 10 %) and from the dehydration reactions occurring during pyrolysis (typically about 12 % based on dry feed) (Asadullah et al. 2007). The presence of water in bio-oil has advantages and disadvantages. Water content (15–30 wt%) gives homogeneity to bio-oil. The solubilizing effect of water for other polar hydrophilic compounds usually helps to ensure miscibility with the oligometric lignin-derived components (Bridgwater et al. 2002). However, increasing the water content leads to separation of the bio-oil into two distinct phases: aqueous phase (with polar compounds) and organic phase (non-polar compounds (Bridgwater 2012a, b). The disadvantages are lowering heating values, increasing ignition delay and decreasing the rate of combustion. Water content on the other hand is advantageous since it improves bio-oil flow characteristics and also leads to a more uniform temperature profile in a cylinder of a diesel engine (Czernik and Bridgwater 2004).

Volatility distribution: Bio-oil cannot be completely vaporized once it is condensed as a liquid and during vaporization, compounds in bio-oil will rapidly react and produce char/solid. Bio-oil contains substantial amounts of oxygenated compounds such as sugar and oligomeric phenolics, therefore, slow heating of oil during distillation results in polymerization of some reactive components. Consequently the oil starts boiling below 100 °C and stops at 250–280 °C (Czernik and Bridgwater 2004).

Viscosity and Aging: Viscosity of bio-oil varies over a wide range (35–1000cP at 40 °C), depending on feedstock, the process conditions and aging. An increase in temperature causes a decrease in viscosity thus the bio-oil can be pumped easily by heating the pipeline. The viscosity can also be reduced by the addition of polar solvents such as methanol or acetone (Diebold and Czernik 1997; Bridgwater 2012a, b). If the bio-oil is subjected to high temperature for a long period of time, viscosity will increase. This is a result of a chemical reaction between the various compounds that will result in the formation of larger molecules (Diebold 2000).

Corrosiveness: Corrosiveness of bio-oil is caused by the presence of organic acids, which result from the pH of bio-oil (2–3).. Corrosiveness is elevated at high temperature and also with the increase in water content of bio-oil. Storage tanks for bio-oil can be made up of polyolefins as they are more resistant to corrosion than steel (Czernik and Bridgwater 2004).

Combustion behavior: Bio-oil is combustible but not flammable. Due to the abundance of non-volatile components, bio-oil requires significant energy for ignition. But once ignited, it burns with a stable self-sustaining flame (Czernik and Bridgwater 2004). Cetane number is a parameter that determine the ignition quality. A high cetane number reduces ignition delay before combustion. However, the cetane number of bio-oil is difficult to measure which is very low (close to 5) due to the absence of straight hydrocarbons (Ikura et al. 2003).

Flash Point and Pour Point: The flash point is the lowest temperature where enough fluid can evaporate to form a combustible gas. Fuels with higher flash point are less flammable or hazardous. However, this property does not contribute to the performance of oil in engines. The pour point of a fuel is an indication of the lowest temperature, at which fuel starts flowing. The pour point of bio-oil is varies from −12 to −36 °C which depends on the biomass feedstock (Oasmaa and Peacocke 2001). Therefore, the low pour point indicates the low viscosity bio-oil is obtained.

5.3 *Standards and Norms*

ASTM D7544 -12 is the standard specification covers grades of pyrolysis liquid biofuel produced from biomass. These grades can divide into Grade G and Grade D. Grade G is for the use in industrial burners and Grade D is for use in commercial or industrial burners requiring lower solids and ash content. Table 2 summarizes requirements for pyrolysis liquid biofuels. However, these grades are not intended for use in residential heaters, small commercial boilers, engines or marine applications (Oasmaa et al. 2015).

Recently, CAS number 1207435-39-9 has been issued for fast pyrolysis bio-oil. In order to transport bio-oil, it is classified as Class-8 (corrosive) substances, however no UN number has been assigned for transportation (Oasmaa et al. 2015).

Table 2 ASTM standard for bio-oil (ASTMD7544 2012)

Property	Test Method	Grade G	Grade D
Gross heat of combustion, MJ/kg, min	D240	15	15
Water content, % mass, max	E203	30	30
Pyrolysis solids content, % mass, max	D7579	2.5	2.5
Kinematic viscosity at 40 °C mm^2/s, max	D455[A]	125	125
Density at 20 °C kg/dm^3	D4052	1.1–1.3	1.1–1.3
Sulphur content, % mass, max	D4294	0.05	0.05
Ash content, % mass, max	D482	0.25	0.15
pH	E70	Report	Report
Flash point, °C min	D93, Procedure B	45	45
Pour point, °C max	D97	-9	-9

A—without filtering

5.4 Applications

Bio-oil has several advantages and disadvantages in terms of applications. Bio-oil is a renewable carbon-based liquid that is derived from agricultural wastes, animal manure, forest residue, sludge, or any type of volatile feedstocks. Bio-oil can be readily stored, transported and used for production of value-added bio-products, such as chemicals. The significant disadvantage of bio-oil as a fuel is the acidity, high viscosity and high oxygen content. High acidity tends to corrode most steel and aluminium storage tanks, fuel lines and engine parts. Due to the high viscosity and high flash (40–100 °C), the heavy pyrolysis oil is not auto-igniting. Though bio-oil heating value is low as compared to conventional liquid fuels, it has higher energy density for higher efficiency energy production. Besides, bio-oil will be possessed with alkali metals (ash content), which cause the incomplete solid separation. These metals might cause the catalyst poisoning, slag formation, deposition of solids in combustion, erosion and corrosion (Bridgwater 2012a, b). The potential applications for bio-oil are as follows:

Furnaces and Boilers: Bio-oil has been tested in furnaces and boilers, for heat and power generation although at a presently unacceptable energetic and financial cost. Bio-oil has similar characteristics to light fuel, although significant differences in energy content, viscosity, pH, emission levels and ignition are observed. One of the difficulties with bio-oil is that different bio-oils show different combustion characteristic. The Red Arrow Products pyrolysis plant in Wisconsin uses bio-oil to generate heat (Freel et al. 1996).

Diesel Engines: The main concerns for operating diesel engines on bio-oils are the difficult in ignition due to high water content and low heating value, corrosiveness and thermally unstable compounds which cause coking. The most identified problems are fuel pumping, high CO emissions, wear and corrosion of lining, wax formation, poor atomization, and injector coking (Chiaramonti et al. 2007). Bio-oil could be efficiently used in pilot-ignited medium-speed diesel engines. Bio-oil emulsified with diesel showed that a stainless steel or similar corrosive resistant material should be used as the material of construction for injector and fuel pump (Chiaramonti et al. 2003a, b). The test at VTT Energy (84 kWe engine) showed that the bio-oil could be efficiently used in pilot-ignited medium-speed diesel engines (Solantausta et al. 1994).

Gasifiers: Gasification of bio-oil can be done to obtain maximum amount of the high-value components such as synthesis gas (syngas). In the Fischer-Tropsch process, the further processing of syngas produced by gasification of bio-oil with pure oxygen may become economically and technically feasible. The production of syngas from the pyrolysis of bio-oil was studied (Panigrahi et al. 2003; Wang et al. 2007). The bio-oil conversion to gas was 83 wt% at 800 °C. The compositions of product gases as the following ranges: syngas:16–36 mol%; CH_4:19–27 mol%; and C_2H_4:21–31 mol%. The heating values of the gas product were 1300–1700 Btu/SCF (standard cubic foot) (Panigrahi et al. 2003).

Chemicals: The chemicals commercially produced from bio-oil are food products such as food flavors and liquid smokes. Some of the fine chemicals extracted from bio-oil are acetic acid, levoglucosan, hydroxyaceladehyde, and levoglucosenone. In addition some other commercial chemicals have been produced from bio-oil such as phenol resins, wood adhesives, slow fertilizers, calcium or magnesium acetate (de-icers), BioLime (to remove sulfur dioxide from flue gas), MNRP (commercial resin), and plasticizers and emulsions (Bridgwater 2012a, b). The aqueous phase of bio-oil has been studied for the production of hydrogen through reforming. Though some aqueous phase reforming (under high pressure) has been conducted, most studies focus on steam reforming (Pan et al. 2012; Trane et al. 2012). Many catalysts have been tested in which Ni, Ru and Rh achieved high performance (Trane et al. 2012). However, carbon deposition on catalyst is still a challenge with bio-oil. Including water-gas shift reactions, bio-oil reforming yield an average of one mole of hydrogen from one mole of carbon in bio-oil.

In addition to the above applications, bio-oil has the potential for other end-use markets. The sticky, resin-like quality of bio-oil can substitute for some petroleum products in asphalt emulsions and so it can be utilized as an asphalt binder. Another potential application of bio-oil is in coal dust suppression. The environmental and physical requirements of bio-oil such as biodegradability, water immiscibility and the strength of its polymerization reactions have to be analyzed in order to coat the coal piles (Farag et al. 2002).

Transportation fuels: Bio-oil can be upgraded as transportation fuel, which is still in research.

5.5 *Upgrading*

In a nutshell, bio-oil has the potential to be available in large amounts, environmental friendly and competitively priced. Upgrading bio-oil in a cost-effective fashion is essential. In order to improve the fuel quality, bio-oil has to be completely deoxygenated. Hydro-treating and catalytic cracking are the two main processes used to upgrade bio-oil. Recent research and development of technologies that attempt to address this challenge will give bio-oil great promise as a domestically grown, green feedstock for production of fuels and chemicals using the current refining and transportation infrastructure. Bio-oil obtained through pyrolysis can be upgraded through hydrogenation, hydro-deoxygenation, catalytic cracking, molecular distillation, esterification, emulsification, catalytic pyrolysis and the co-pyrolysis of biomass with different material.

Hydro-deoxygenation (HDO) is a method for upgrading bio-oil by removing the oxygen under high pressure of hydrogen with a catalyst such Pt, NiMO or CoMO. Oxygen in the feed is converted into H_2O (Furimsky 2000). The H_2 consumption and severity of the operation required for achieving high HDO conversion depends on the content and type of the oxygen compounds in bio-oil. In order to convert bio-oil to commercial fuels, the first stage is performed under 573 K to remove oxygen compounds by readily undergo polymerization, this stage is also known as stabilization stage. The phenols which are formed by methoxyphenols, biphenols and ethers in the first stage are removed at about 623 K (Furimsky 2000). The compounds, such as ketones and carboxylic acids have a low HDO reactivity. Therefore, a higher H_2 pressure and a higher temperature are necessary for their conversion to oxygen free products. Most of the HDO studies are done with model compounds. Reaction mechanism and kinetic models have been developed for HDO of many oxygenated compounds (Furimsky 2000; Elliott 2007; Mortensen et al. 2011; Graça et al. 2013). General reaction can be expresses as Eq. 1 (Adhikari et al., 2013).

$$C_xH_yO_z + \left(x + z - \frac{y}{2}\right)H_2 \rightarrow xCH_2 + zH_2O \tag{1}$$

Wildschut et al. (2009) tested novel-metal catalysts (Ru/C, Ru/TiO_2, Ru/Al_2O_3, Pt/C, and Pd/C) and conventional bimetal catalysts (sulfided NiMo/Al_2O_3 and CoMo/Al_2O_3) for HDO of bio-oil. Their results showed an inverse relationship between increase in severity of HDO and oxygen content of bio-oil. Ru/C was identified as the most promising catalyst. Hydro-treatment of bio-oil with Ru/C as a catalyst at 350 °C and 200 bar, for 4 h upgraded the HHV of bio-oil from 20 MJ/kg to 43 MJ/kg, pH from 2.5 to 5.8. The upgraded bio-oil had two layers, where top oil had 11 wt% and bottom oil had 5 wt% oxygen content. The results of their study were very promising (Wildschut et al. 2009).

Catalytic cracking is the thermal conversion of bio-oil with the presence of cracking catalysts at a specific temperature higher than 350 °C and atmospheric pressure. Major products from this process are aromatic hydrocarbons. Main catalysts used for cracking are FFC catalysts (HY zeolite) and HZM-5. Catalyst coking is a major challenge in this process. The catalyst can be regenerated by combusting the coke at higher temperature (650–700 °C) (Graça et al. 2013). Comparisons of different acid catalysts such as HZM-5, HY, silicate-alumina have been undertaken (Sharma and Bakhshi 1993; Adjaye and Bakhshi 1995a, b, c; Williams and Horne 1995). Model compounds such as alcohols, acids, furans have been studied for reaction mechanism. Reactivities of each group of compounds towards cracking are different. Moreover, the synergistic effects will be different from individual reactivities (Adjaye and Bakhshi 1995a, b, c; Graça et al. 2013).

Molecular distillation was particularly suitable for bio-oil separation to fractionate as light, medium and heavy fractions. The light fraction mostly contains water and is strongly acidic, with poor stability but good fluidity. The middle fraction is more viscous than the light fraction due to lower water content. The heavy fraction, without volatile substances, was similar to a black solid in appearance and had a relatively higher heating value compared to the raw bio-oil (Wang et al. 2009). By using this method, acidity and water content are removed as light fraction from high heating value compounds. However, the yield is highest in the light fraction, which is the undesirable fraction as it has low or even zero heating value.

Catalytic esterification is considered as a promising method for increasing the pH of bio-oil. Esterification of bio-oil with methanol in the presence of ion-exchange resins as catalyst has showed a significant decrease in acid number and an increase in heating value (Wang et al. 2010). Li et al. showed that the oxygen containing groups were stabilized due to catalytic esterification (Li et al. 2011). In addition to catalytic esterification, the esterification in presence of supercritical CO_2 gave promising results (Cui et al. 2010).

Emulsification: To promote bio-oil as a combustion fuel. Acidity and viscosity problems can be tolerated, if stable bio-oil in diesel emulsions could be produced (Ikura et al. 2003). Emulsification is suggested to be carried out just after the production of bio-oil through fast pyrolysis. As the diesel percentage increases, the stability of bio-oil increases (Chiaramonti et al. 2003a, b). Crossley et al. have developed a recoverable noble metal catalyst (Pd) on nanotubes which can be used to make stable emulsified bio-oil for downstream processes (Crossley et al. 2010).

Catalytic pyrolysis: Catalysts can be introduced during the pyrolysis so that the cracking occurs simultaneous with pyrolysis. Catalytic pyrolysis is conducted in two ways. One is called *in situ* catalytic pyrolysis in which pyrolytic vapors pass through a catalyst bed where cracking occurs before condensation. The other is in-bed catalytic pyrolysis where the biomass will be mixed with the catalyst resulting in more uniform cracking of pyrolytic vapors (Thangalazhy-Gopakumar et al. 2011; Tan et al. 2013). The Zeolite ZSM-5 catalyst has been widely studied for catalytic pyrolysis. Other molecular sieve catalysts such as Al-MCM-41, Al-MSU-F, β-zeolite, SBA-15, silica-alumina and alumina have also been utilized (Pattiya et al. 2008; Carlson et al. 2010; Zhao et al. 2010). 30–40 wt% carbon from biomass can be recovered as aromatic hydrocarbon from biomass through this process (Carlson et al. 2010; Zhao et al. 2010; Thangalazhy-Gopakumar et al. 2011).

Co-pyrolysis: Recently co-pyrolysis is getting more attention as a method for recycling plastics and as well as improving bio-oil properties (Abnisa and Wan Daud 2014). Different types of materials were used for co-pyrolysis with biomass, such as polystyrene, high density polyethylene, bisphemol A, waste tire, polypropylene and plastic film. When biomass and tires are co-pyrolyzed, the C, O and H can be balanced due to the high C content and low O content in tires. In addition, it is worth highlighting that some physico-chemical properties of the tire pyrolysis liquids are fairly comparable to those found in commercial automotive diesel fuels (Martínez et al. 2014). Through co-pyrolysis with plastic, the oxygen content of the bio-oil can be reduced, resulting in higher hydrogen and carbon content and with a higher heating value (Sajdak and Muzyka 2014). The co-pyrolysis with plastic will increase the yield of the liquid product, which is the most desirable product (Bhattacharya et al. 2009).

6 Bio-char

Bio-char is the carbon-rich solid product from pyrolysis of biomass, which comprises mainly stable aromatic forms of organic carbon (Kim et al. 2012). Bio-char is mostly derived from the lignin part in the case of lingo-cellulosic biomass.

The balance between char, gas and oil products in pyrolysis depends largely upon the rate of heating where slow pyrolysis produces the most amount of char. The characteristics of the char depend on the extent of the pyrolysis (peak temperature), particle size of the feed as well as residence time in the reactor. The vapor residence time for volatiles and gases is also an important factor as longer residence time result in secondary reactions, most notably the reaction of tar on char surfaces and also the charring of tar (Mohan et al. 2006).

Bio-char is widely applied for many environmental applications in addition to energy applications. The high HHV of bio-char makes it a good source to generate energy (Thangalazhy-Gopakumar et al. 2010). Bio-char can be used as a carbon sink, storing large amounts of CO_2 in a stable and fixed form for centuries thus reducing greenhouse gases in the atmosphere.

Bio-char can improve water quality, reduce soil emissions of greenhouse gasses, reduce nutrient leaching, reduce soil acidity as well as reduce irrigation and fertilizer requirements. The complex structure of bio-char helps to attract more bacteria and fungi which are needed by plants to absorb more nutrients from the soil (Ahmad et al. 2014). In addition, bio-char has the ability to attract and retain water, due to the porous structure and high surface area, retaining also nutrients, phosphorus and agro-chemicals. Biochar therefore reduces leaching of fertilizers into surface or groundwater. An approximate 18 % increase in the water holding capacity of soil containing bio-char has been reported (Glaser et al. 2002). Another application of bio-char is as an adsorbent to remove heavy metals from water (Fonts et al. 2012). Recently some studies of fast pyrolysis char have reported the adsorption of metals and tetracycline from aqueous solutions (Liu et al. 2012; Cho et al. 2013). Table 3 provides some examples of metal removal studies of fast pyrolysis bio-char.

7 Current Status

Major industrial examples of conversion of biomass to bio-oil are DynaMotive Energy Systems Corp, Ensyn Technology and BTG Biomass Technology Group BV (BTG), KiOR (Table 4).

Ensyn Techonlogy produce bio-oil in a fast pyrolysis plant from wood biomass. The core technology is Rapid Thermal Processing (RTP), a fast pyrolysis technology for conversion of cellulosic wood feedstock. RTP is analogous to fluid catalytic cracking (FCC), which is a key process in gasoline production. Ensyn is producing RTP liquids with around 75 % yield. Ensyn has been operating for 25 years. Ensyn has a strategic relationship with UOP, a Honeywell company, to develop biorefineries. Current production capacity at the Ontario plant is being increased to 3 million gallon/year. Envergent Technologies LLC (www.envergenttech.com) utilize Honeywell's hydro-processing technology to upgrade RTP bio-oil to high-quality drop-in fuels.

Table 3 Fast pyrolysis bio-char as adsorbent for heavy metals

Feedstock	Adsorbate	Adsorption capacity (mg/g)	Reference
Pinewood	Pb (II)	4.13	Mohan et al. (2007)
	Cd (II)	n/a	
	As (III)	1.2	
Oak wood	Pb (II)	2.62	
	Cd (II)	0.37	
	As (III)	5.85	
Pine bark	Pb (II)	3	
	Cd (II)	0.34	
	As (III)	12.15	
Oak bark	Pb (II)	13.1	
	Cd (II)	5.4	
	As (III)	7.4	
P. terebinthus L.	Cr (VI)	3.97	Deveci and Kar (2013)
Almond shell	Ni (II)	22.22	Kılıç et al. (2013)
	Co (II)	26.95	
Municipal sewage sludge	Cd (II)	42.8	Chen et al. (2015)

Table 4 Bio-oil industries and their projected applications

Company's name	Bio-oil applications (current/R&D)	References
Ensyn technologies	– Food chemicals – Resins	Ensyn corporation. Retrieved from: http://www.ensyn.com/
BTG-BTL	– Heat and power (boilers, gas turbines, diesel engines) – Biobased chemicals	BTG Retrieved from: http://www.btg-btl.com/en
Anellotech	– Bio-aromatics/bio-BTX	Anellotech Inc Retrieved from: http://anellotech.com/

BTG-BTL (a daughter company of BTG) has specialized in the process of conversion of biomass into useful fuels and energy. The reactor used by BTG for fast pyrolysis is a modified Rotating Cyclone Reactor. Char and sand are separated from vapors through cyclones, passed a fluidized bed combustor. Hot sand is recycled back to the reactor. Empyro (a joint project of BTG with Tree Power) announced that the construction of its pyrolysis oil production plant has started at the AkzoNobel site in Hengelo (The Netherlands). The production capacity will then be gradually increased to its maximum of over 20 million liters of pyrolysis oil per year.

DynaMotive Energy Systems Corporation is a Canadian-based renewable energy company, which specializes in fast pyrolysis. Dynamotives' patented fast pyrolysis process involves rapid heating of a biomass feedstock in the absence of oxygen. The yield of the Dynamotive`s pyrolysis is typically 60–75 wt% for bio-oil and 15–20 wt% for bio-char. In order to produce high quality bio-oil, Dynamotive has a pyrolysis upgrading process, which is a two stages process, involving hydro-reforming and hydro-treating. In the hydro-reforming stage, hydrogen is added to the pyrolysis oil in a reactor in the presence of an industrial catalyst. Water, methanol and acetic acid are removed from the pyrolysis oil in this stage. The second stage is hydro-treating, where more hydrogen is added additionally removing remaining oxygen. Current status of Dynamotive is not available.

KiOR Inc. utilize a circulating fluid bed reactor for catalytic pyrolysis of biomass. The condensed bio-oil undergoes hydrotreating. The product can be used as a gasoline/diesel blend. Their technology Biomass Fluid Catalyst Cracking (BFCC) was planned for a commercial scale up to 40 million gallons/year. However, currently KiOR is out of business.

KIT Bioliq, Germany (http://www.bioliq.de/) in cooperation with Lurgi GmbH utilize the Lurgi/Auger reactor for fast pyrolysis. Their product called bioliqSyncrude is atomized in a gasifier for syngas production. PYTEC GmbH Thermochemische Anlagen (http://www.pytecsite.de/) is another company in Germany. PYTEC utilizes an ablative reactor for fast pyrolysis. Their pilot plant (design capacity 6 Mg/d dry mass) include a diesel-based CHP plant for power production (approximately 300 kW/h output) (Meier et al. 2007). Renewable Oil International is another well-known company producing bio-oil using a vacuum pyrolysis reactor (http://www.renewableoil.com). Anellotech Inc is another booming industry; their core process known as Thermo Catalytic Biomass Conversion (Bio-TCAT™) is used to produce chemicals from biomass (http://anellotech.com/). Pyrovc, utilizing vacuum pyrolysis for bio-oil production, has now developed a new vacuum reactors, which encounters heat transfer limitations.

Since the major industries currently practicing fast pyrolysis are located in North America and Europe, global bio-oil production is currently based on wood. In Malaysia and Indonesia, fast pyrolysis studies are based on solid wastes from the palm industry. Currently, Ensyn is constructing their plants in Malaysia for bio-oil production from empty fruit bunch palm biomass, whereas, BTG-BTL has already constructed a plant in Johor Bahru, Malaysia, which is dormant now. Other location for potential industrial applications for fast pyrolysis are Brazil, Cuba and Argentina in South America.

In addition to these major industries many research institutions and universities have built pilot plants, which have feedstock capacities from 10–1000 kg/h. Major institutions/universities in this list are Aston University, UK; University of Western Ontario, Canada; Virginia Tech, USA; University of Science and Technology of China, China; Mississippi State University, USA; USDA-ARS-ERRC USA; RTI international, USA; VTT Ltd., Finland; Iowa State University, USA; NREL USA; CANMET, Canada (Wagenaar et al. 1994). Some examples of different types of feedstock used in laboratory studies of fast pyrolysis around the world are given in Table 5.

Table 5 Laboratory studies of fast pyrolysis around the world

Feedstock	Reactor	Country	Bio-oil yield (wt%)	Bio-char yield (wt%)	Bio-oil HHV (MJ/kg)	Reference
Agriculture waste						
Rice straw	Fluidized bed	Thailand	54.1	~27	28.4	Pattiya and Suttibak (2012)
Rice husk			57.1	~31	25.4	
Sugarcane	Fixed bed	Bangladesh	56.0	33.0	23.5	Islam et al. (2010)
Empty fruit bunch	Fluidized bed	Malaysia	55.1	23.9	36.06	Abdullah et al. (2010)
Cassava rhizome	Fluidized bed	Thailand	64.93	19.97	15.8	Pattiya (2011)
Cassava stalk			61.71	20.91	12.7	
Palm kernel shell	Fluidized bed	Republic of Korea	48.7	~25	17.9	Kim et al. (2010)
Black cumin seed cake	Fixed bed	Turkey	48.21	~26	38.48	Sen et al. (2011)
Corncob	Fluidized bed	China	56.8	23.2	18.8	Zhang et al. (2009)
Jatorpha curcas de-oiled seed cake	Fluidized bed	India	48	35.1	n/a	Biradar et al. (2014)
Forest residue						
Pine wood	Auger reactor	USA	50	26	19.10	Thangalazhy-Gopakumar et al. (2010)
Eucalyptus wood	Fluidized bed	China	~58	~18	16.62	Chang et al. (2013)
Pinewood waste (*Pinus insignis*)	Spouted bed reactor	Spain	75	~17	14.6 (LHV)	Amutio et al. (2012)
Hybrid Poplar Wood	Fluidized bed	USA	63.3	12.7	24.48	Agblevor et al. (2010)
Solid wastes						
Furniture Saw dust	Fluidized bed	Republic of Korea	58.1	~30	n/a	Heo et al. (2010a, b)

(continued)

Table 5 (continued)

Feedstock	Reactor	Country	Bio-oil yield (wt%)	Bio-char yield (wt%)	Bio-oil HHV (MJ/kg)	Reference
Chicken litter	Fluidized bed	USA	23.39	33.85	27.98	Kim et al. (2009)
Turkey litter			26.32	24.49	26.24	
Rubber from tires	Fluidized bed	Germany	27.8	36.7	n/a	Kaminsky et al. (2009)
Rubber from hand gloves			72.2	1.1	n/a	
Algae						
C. vulgaris	Fluidized bed	USA	53	31	24.57	Wang et al. (2013)
C. vulgaris	Fixed bed	USA	52.7	25.7	18.6	Thangalazhy-Gopakumar et al. (2012)
Energy plants						
Switchgrass	Fluidized bed	USA	~43.0	~26	19.6	He et al. (2009)
Switchgrass	Fluidized bed	USA	60		18.445	Mullen et al. (2008)
Alfalfa bud and flower			45		25.852	
Alfalfa stem			53		26.542	
Miscanthus	Fluidized bed	Republic of Korea	69.2	~25	22.5	Heo et al. (2010a, b)
Catalytic pyrolysis						
Wheat spent grains	Fluidized bed (alumina catalyst)	UK	~53.0	~17	22.8	Sanna et al. (2011)
Brewers spent grains			~48.0	~15	24.07	
Corncob	Fluidized bed (HZM-5)	China	39.3	20.1	34.6	Zhang et al. (2009)
Hybrid poplar wood	Fluidized bed (HZM-5)	USA	33	12.6	30.5	Agblevor et al. (2010)

(continued)

Table 5 (continued)

Feedstock	Reactor	Country	Bio-oil yield (wt%)	Bio-char yield (wt%)	Bio-oil HHV (MJ/kg)	Reference
Co-pyrolysis						
50:50::pine wood/polystyrene	Auger	USA	64.9	12.1	28.42	Bhattacharya et al. (2009)
50:50::pine wood/HDPE			338.9	25.9	27.68	
50:50::pine wood/polypropylene			46	32.2	36.94	
50:50::pine cone: PE	Fixed bed	Romania	63.9	19.6	46.33	Brebu et al. (2010)
50:50::pine cone: PP			64.1	19.1	45.58	
50:50::pine cone: PS			69.7	21.5	46.43	
30:40:20:10::pine cone: PE: PP:PS			76	12.7	41.33	
66.6: 33.3::Swine Solid: plastic mulch film	Fixed Bed	USA	45.3	15.7	59	Ro et al. (2014)
80:20::walnut shell: tar sands	Fixed Bed	Turkey	31.84	30.3	27.45	Kar (2011)

8 Energy and Economics

Energy requirement for pyrolysis is mainly for drying, grinding, pyrolysis and condensation. Energy will be recovered in terms of the heating content of products, mainly bio-oil and bio-char. Therefore energy efficiency can be written as Eq. 2

$$\eta_{energy} = \frac{E_{Bio-oil} + E_{Bio-char}}{E_{Biomass} + E_{process}} \tag{2}$$

where $E_{Bio-oil}$, $E_{Bio-char}$, and $E_{Biomass}$ are the energy contents of bio-oil, bio-char and biomass respectively and $E_{Process}$ is the external energy required for the complete pyrolysis process from biomass pre-treatments to product collection (Jahirul et al. 2012). Energy recovery studies on three hardwoods by Stals et al. showed that 34–39 % energy can be recovered through a fast pyrolysis process (Stals et al. 2010). Daugaard and Brown found that the average enthalpy for fast pyrolysis for biomass having 8–10 wt% moisture content is about 1.5 MJ/kg (Daugaard and Brown 2003). Life cycle analysis by Steele et al. proved near carbon neutrality for bio-oil production. According to their study, for the production of bio-oil, based on 2000 dry tons/day input feed, 0.52 MJ/MJ energy was required for the complete process (from cradle to grave) whereas residual fuel oil required 1.2 MJ/MJ. The overall energy use for bio-oil was negative. A 70 % reduction in potential global warming emission was calculated for bio-oil as compared to residual fuel oil and it was 0.0323 kg CO_2 eq per MJ bio-oil production (Steele et al. 2012).

Reinger et al. conducted a detailed economic analysis to evaluate the capital and operating costs of a biomass fast pyrolysis plant with a capacity of 550 dry tonnes/day of wood chips (Ringer et al. 2006). In the design basis, there are five major processing areas; feed handling and drying, pyrolysis, char combustion, product recovery, and steam generation. The total capital investment of the project was estimated to be $48.29 million and a total operating cost of around $9.6 million. Feedstock accounted for the largest variable in operating cost. About 61 % of total capital investment ($29.7 million) was equipment and installation costs. The selling value for bio-oil/t was projected to be $7.62/GJ, LHV (Ringer et al. 2006).

The capital cost for a 2000 dry metric tons/day of a hybrid poplar wood chips in another design was $303 million accounting for upgrading of bio-oil to gasoline and diesel fuels (Jones et al. 2009). Similar work was done for 2000 dry tonne per day corn stover fast pyrolysis. Two scenario were considered for upgrading bio-oil to gasoline and diesel range fuels: hydrogen production from bio-oil for upgrading and hydrogen purchase for upgrading. The capital cost for the hydrogen production scenario was $287 million and $200 million for the hydrogen purchase scenario. Therefore the first scenario had higher fuel prices than second scenario (Wright et al. 2010).

9 Conclusions

Fast pyrolysis for the production of liquid fuels is a relatively new biomass conversion method which is emerging commercially. Major attributes of fast pyrolysis are high heat and mass transfer rates and low residence time in the vapor state. A wide range of organic wastes have been investigated as feedstock for fast pyrolysis. The basic principle behind fast pyrolysis feedstock selection is decentralized densification and centralized upgrading. Therefore research on different types of other feedstocks still continue around the world. The selection of this feedstock basically depends on regional abundance.

Many reactor configurations have been tested and developed in order to achieve fast pyrolysis quality attributes during the process. Though each configuration has its own benefits and challenges, none of the reactors is superior to another one. Norms and standards for bio-oil have been established recently. However, bio-oil cannot be used as it is. Even the utilization as a feedstock for energy or chemicals is still underway. Based on the proposed applications of bio-oil, many upgrading techniques are being researched. In addition to upgrading studies, some demonstration of raw bio-oil for CHP application with modified engines are being developed around the world. In a concluding remark, though fast pyrolysis bio-oil has a bright future, it has a long path to general implementation.

Acknowledgments Authors would like to acknowledge the students who worked for the fast pyrolysis section of the project 'Proj-In-FETBE-015' funded by Centre of Excellence for Research, Value Innovation and Entrepreneurship (CERVIE) at UCSI University, Kuala Lumpur, Malaysia. The students Chin Chun Man, Ng Han Ject, Chang Jang Sen, Kho Lik Yeh, Punit Purshotam, Rameshwaran, Kavindran are acknowledged for their contributions in the literature.

References

Abdullah, Gerhauser, Bridgwater (2007) Bio-Oil from fast pyrolysis of oil palm empty fruit bunches. J Phy Sci 18(1):57–74

Abdullah, Gerhauser, Sulaiman (2010) Fast pyrolysis of empty fruit bunches. Fuel 89(8): 2166–2169

Abnisa, Wan Daud (2014) A review on co-pyrolysis of biomass: an optional technique to obtain a high-grade pyrolysis oil. Energy Convers Manag 87:71–85

Adhikari, Thangalazhy-Gopakumar, Taylor (2013) Gasification and pyrolysis for fuel and power production In: Christopher L (ed) Integrated forest biorefineries, Challenges and Opportunities Royal Society of Chemistry, Cambridge, UK

Adjaye, Bakhshi (1995a) Catalytic conversion of a biomass-derived oil to fuels and chemicals I: model compound studies and reaction pathways. Biomass Bioenergy 8(3):131–149

Adjaye, Bakhshi (1995b) Production of hydrocarbons by catalytic upgrading of a fast pyrolysis bio-oil. part I: conversion over various catalysts. Fuel Process Technol 45:161–183

Adjaye, Bakhshi (1995c) Production of hydrocarbons by catalytic upgrading of a fast pyrolysis bio-oil. part II: comparative catalyst performance and reaction pathways. Fuel Process Technol 45(3):185–202

Agblevor, Beis, Mante et al (2010) Fractional catalytic pyrolysis of hybrid poplar wood. Ind Eng Chem Res 49(8):3533–3538

Ahmad, Rajapaksha, Lim et al (2014) Biochar as a sorbent for contaminant management in soil and water: a review. Chemosphere 99:19–33

Albalasmeh, Berhe, Ghezzehei (2013) A new method for rapid determination of carbohydrate and total carbon concentrations using UV spectrophotometry. Carbohydr Polym 97(2):253–261

Amutio, Lopez, Artetxe et al (2012) Influence of temperature on biomass pyrolysis in a conical spouted bed reactor. Resour Conserv Recycl 59(0):23–31

ANSI, ASABES593 (2006) Terminology and definitions for biomass production, harvesting and collection, storage, processing conversion and utilization. St. Joseph, MI, USA, American Society of Agricultural and Biological Engineers

Asadullah, Rahman, Ali et al (2007) Production of bio-oil from fixed bed pyrolysis of bagasse. Fuel 86(16):2514–2520

ASTMD7544 (2012) Standard specification for pyrolysis liquid biofuel. West Conshohocken, PA, USA, ASTM International

Badger, Fransham (2006) Use of mobile fast pyrolysis plants to densify biomass and reduce biomass handling costs—a preliminary assessment. Biomass Bioenergy 30:321–325

Bhattacharya, Steele, Hassan et al (2009) Wood/plastic copyrolysis in an auger reactor: chemical and physical analysis of the products. Fuel 88(7):1251–1260

Biradar, Subramanian, Dastidar (2014) Production and fuel quality upgradation of pyrolytic bio-oil from Jatropha Curcas de-oiled seed cake. Fuel 119(0):81–89

Bouchera, Chaalab, Roy (2000) Bio-oils obtained by vacuum pyrolysis of softwood bark as a liquid fuel for gas turbines. Part I: properties of bio-oil and its blends with methanol and a pyrolytic aqueous phase. Biomass Bioenergy 19:337–350

Branca, Giudicianni, Blasi (2003) GC/MS Characterization of liquids generated from low-temperature pyrolysis of wood. Ind Eng Chem Res 42(14):3190–3202

Brebu, Ucar, Vasile et al (2010) Co-pyrolysis of pine cone with synthetic polymers. Fuel 89(8):1911–1918

Bridgwater A, Meier D & Radlein D (1999) An Overview of fast pyrolysis of biomass. Org Geochem 30:1479 –1493

Bridgwater AV (1999) Principles and practice of biomass fast pyrolysis processes for liquids. J Anal Appl Pyrol, 51:3–22

Bridgwater (2004) Biomass fast pyrolysis. Thermal Sci 8(2):21–49

Bridgwater (2011) Review of fast pyrolysis of biomass and product upgrading. Biomass Bioenergy 1–27

Bridgwater (2012a) Review of fast pyrolysis of biomass and product upgrading. Biomass Bioenergy 38:68–94

Bridgwater (2012b) Review of fast pyrolysis of biomass and product upgrading. Biomass Bioenergy 38:68–94

Bridgwater, Peacocke (2000) Fast pyrolysis processes for biomass. Renew Sustain Energy Rev 4:1–73

Bridgwater, Toft, Brammer (2002) A techno-economic comparison of power production by biomass fast pyrolysis with gasification and combustion. Renew Sustain Energy Rev 6:181–248

Carlson, Jae, Lin et al (2010) Catalytic fast pyrolysis of glucose with HZSM-5: the combined homogeneous and heterogeneous reactions. J Catal 270:110–124

Chang, Zhao, Zheng et al (2013) Effect of hydrothermal pretreatment on properties of bio-oil produced from fast pyrolysis of eucalyptus wood in a fluidized bed reactor. Bioresour Technol 138(0):321–328

Chen, Zhou, Han et al (2015) Adsorption of cadmium by biochar derived from municipal sewage sludge: impact factors and adsorption mechanism. Chemosphere 134:286–293

Chiaramonti, Bonini, Fratini et al (2003) Development of emulsions from biomass pyrolysis liquid and diesel and their use in engines—part 1: emulsion production. Biomass Bioenergy 25:85–99

Chiaramonti, Bonini, Fratini et al (2003) Development of emulsions from biomass pyrolysis liquid and diesel and their use in engines—part 2: tests in diesel engines. Biomass Bioenergy 25: 101–111

Chiaramonti, Oasmaa, Solantausta (2007) Power generation using fast pyrolysis liquids from biomass. Renew Sustain Energy Rev 11(6):1056–1086

Cho, Baek, Jeon et al (2013) Removal characteristics of copper by marine macro-algae-derived chars. Chem Eng J 217:205–211

Choi, Johnston, Brown et al (2014) Detailed characterization of red oak-derived pyrolysis oil: integrated use of GC, HPLC, IC, GPC and Karl-Fischer. J Anal Appl Pyrol 110(0):147–154

Crossley, Faria, Shen et al (2010) Solid nanoparticles that catalyze biofuel upgrade reactions at the water/oil interface. Science 327(5961):68–72

Cui, Wang, Wei et al (2010) Upgrading bio-oil by esterification under supercritical CO2 conditions. J Fuel Chem Technol 38(6):673–678

Czernik, Bridgwater (2004) Overview of applications of biomass fast pyrolysis oil. Enegry Fuels 18(2):590–598

Daugaard, Brown (2003) Enthalpy for pyrolysis for several types of biomass. Energy Fuels 17(4):934–939

Deveci, Kar (2013) Adsorption of hexavalent chromium from aqueous solutions by bio-chars obtained during biomass pyrolysis. J Ind Eng Chem 19(1):190–196

Diebold (2000) A review of the chemical and physical mechanisms of the storage stability of fast pyrolysis bio-oils golden. Colorado, National Renewable Energy Laboratory

Diebold, Czernik (1997) Additives to lower and stabilize the viscosity of pyrolysis oils during storage. Enegry Fuels 11:1081–1091

EIA (2013) International energy outlook 2013. E. I. Administration. Washington, DC, USA, U.S. Department of Energy, U. S

Elliott (2007) Historical developments in hydroprocessing bio-oils. Enegry Fuels 21:1792–1815

Farag, LaClair, Barrett (2002) Technical, environmental and economic feasibility of bio-oil in New Hampshire's North Country. Durham, NH, New Hampshire Industrial Research Center (NHIRC) 1–95

Fonts, Gea, Azuara et al (2012) Sewage sludge pyrolysis for liquid production: a review. Renew Sustain Energy Rev 16:2781–2805

Freel, Graham, Huffman (1996) Commercial aspects of rapid thermal processing (RTM[TM]). In: Thermal biomass processing: bio-oil production & utilisation CPL Press, Newbery, UK

Furimsky (2000) Catalytic hydrodeoxygenation. Appl Catal A 199:147–190

Garcia-Perez, Chaala, Roy (2002) Vacuum pyrolysis of sugarcane bagasse. J Anal Appl Pyrol 65(2):111–136

Garcia-Perez, Chaala, Pakdel et al (2007) Characterization of bio-oils in chemical families. Biomass Bioenergy 31(4):222–242

Glaser, Lehmann, Zech (2002) Ameliorating physical and chemical properties of highly weathered soils in the tropics with charcoal—a review. Biol Fertil Soils 35(4):219–230

Graça, Lopes, Cerqueira et al (2013) Bio-oils upgrading for second generation biofuels. Ind Eng Chem Res 52(1):275–287

He, Ye, English et al (2009) Influence of pyrolysis condition on switchgrass bio-oil yield and physicochemical properties. Bioresour Technol 100(21):5305–5311

Heo, Park, Park et al (2010) Bio-oil production from fast pyrolysis of waste furniture sawdust in a fluidized bed. Bioresour Technol 101:S91–S96

Heo, Park, Yim et al (2010) Influence of operation variables on fast pyrolysis of miscanthus sinensis var. Purpurascens. Bioresour Technol 101(10):3672–3677

Ikawa, Schaper, Dollard et al (2003) Utilization of Folin–Ciocalteu phenol reagent for the detection of certain nitrogen compounds. J Agric Food Chem 51(7):1811–1815

Ikura, Stanciulescu, Hogan (2003) Emulsification of pyrolysis derived bio-oil in diesel fuel. Biomass Bioenergy 24(3):221–232

Islam, Parveen, Haniu (2010) Properties of sugarcane waste-derived bio-oils obtained by fixed-bed fire-tube heating pyrolysis. Bioresour Technol 101(11):4162–4168

Jahirul, Rasul, Chowdhury et al (2012) Biofuels production through biomass pyrolysis—a technological review. Energies 5(12):4952–5001

Jones, Valkenburg, Walton et al (2009) Production of gasoline and diesel from biomass via fast pyrolysis, hydrotreating and hydrocracking: a design case. DE-AC05-76RL01830 PNNL-18284 Rev 1

Kaminsky, Mennerich, Zhang (2009) Feedstock recycling of synthetic and natural rubber by pyrolysis in a fluidized bed. J Anal Appl Pyrol 85:334–337

Kanaujia, Sharma, Garg et al (2014) Review of analytical strategies in the production and upgrading of bio-oils derived from lignocellulosic biomass. J Anal Appl Pyrol 105:55–74

Kantarelis, Yang, Blasiak (2013) Production of liquid feedstock from biomass via steam pyrolysis in a fluidized bed reactor. Energy Fuels 27(8):4748–4759

Kar (2011) Co-pyrolysis of walnut shell and tar sand in a fixed-bed reactor. Bioresour Technol 102(20):9800–9805

Kılıç, Kırbıyık, Çepelioğullar et al (2013) Adsorption of heavy metal ions from aqueous solutions by bio-char, a by-product of pyrolysis. Appl Surf Sci 283:856–862

Kim, Agblevor, Lim (2009) Fast pyrolysis of chicken litter and turkey litter in a fluidized bed reactor. J Ind Eng Chem 15(2):247–252

Kim, Jung, Kim (2010) Fast pyrolysis of palm kernel shells: Influence of operation parameters on the bio-oil yield and the yield of phenol and phenolic compounds. Bioresour Technol 101(23):9294–9300

Kim, Kim, Cho et al (2012) Influence of pyrolysis temperature on physicochemical properties of biochar obtained from the fast pyrolysis of pitch pine (pinus rigida). Bioresour Technol 118:158–162

Ladanai, Vinterbäck (2009) Global potential of sustainable biomass for energy SLU, Swedish University of Agricultural Sciences, Department of Energy and Technology

Lappas, Samolada, Iatridis et al (2002) Biomass pyrolysis in a circulating fluid bed reactor for the production of fuels and chemicals. Fuel 81:2087–2095

Li, Gunawan, Lievens et al (2011) Simultaneous catalytic esterification of carboxylic acids and acetalisation of aldehydes in a fast pyrolysis bio-oil from mallee biomass. Fuel 90(7): 2530–2537

Liu, Liu, Jiang et al (2012) Modification of Bio-char Derived from Fast Pyrolysis of Biomass and Its Application in Removal of Tetracycline from Aqueous Solution. Bioresour Technol 121:235–240

Loque, Eudes, Yang (2011) Chapter 2 biomass availability and sustainability for biofuels. In: Simmons BA (ed) Chemical and biochemical catalysis for next generation biofuels The Royal Society of Chemistry. UK, Cambridge, pp 5–32

Martínez, Veses, Mastral et al (2014) Co-pyrolysis of biomass with waste tyres: upgrading of liquid bio-fuel. Fuel Process Technol 119:263–271

Meier, Schöll, Klaubert et al (2007) Practical results from PYTEC's biomass-to-oil (BTO) pocess with ablaive pyrolyser and diesel CHP plant. In: Bridgwater AV (ed) Success and visions for bioenergy: thermal processing of biomass for bioenergy, biofuels and bioproducts. CPL Scientific Publishing Service Ltd

Meier, van de Beld, Bridgwater et al (2013) State-of-the-art of fast pyrolysis in IEA bioenergy member countries. Renew Sustain Energy Rev 20:619–641

Mohan, Pittman, Steele (2006) Pyrolysis of wood/biomass for bio-oil: a critical review. Energy Fuels 20(3):848–889

Mohan, Pittman Jr, Bricka et al (2007) Sorption of arsenic, cadmium, and lead by chars produced from fast pyrolysis of wood and bark during bio-oil production. J Colloid Interface Sci 310(1):57–73

Mortensen, Grunwaldt, Jensen et al (2011) A review of catalytic upgrading of bio-oil to engine fuels. Appl Catal A 407(1–2):1–19

Motasemi, Afzal (2013) A review on the microwave-assisted pyyrolysis technique. Renew Sustain Energy Rev 28:317–330

Mullen, Boateng (2008) Chemical composition of bio-oils produced by fast pyrolysis of two energy crops. Energy Fuels 22(3)
Nanda, Mohanty, Kozinski et al (2014) Physico-chemical properties of bio-oils from pyrolysis of lignocellulosic biomass with high and slow heating rate. Energy Environ Res 4(3):21–32
Oasmaa, Peacocke (2001) A guide to physical property characterisation of biomass—derived fast pyrolysis liquids. VTT Publications 450, Technical Research Centre of Finland
Oasmaa, Kuoppala, Solantausta (2003) Fast pyrolysis of forestry residue. 2. physicochemical composition of product liquid. Enegry Fuels 17:433–443
Oasmaa, van de Beld, Saari et al (2015) Norms, standards, and legislation for fast pyrolysis bio-oils from lignocellulosic biomass. Energy Fuels 29(4):2471–2484
Onay, Kockar (2006) Pyrolysis of rapeseed in a free fall reactor for production of bio-oil. Fuel 85:1921–1928
Palma, Richardson, Roberson et al (2011) Economic feasibility of a mobile fast pyrolysis system for sustainable bio-crude oil production. Int Food Agribusiness Manage Rev 14(3):1–16
Pan, Chen, Liu et al (2012) Aqueous-phase reforming of the low-boiling fraction of rice husk pyrolyzed bio-oil in the presence of platinum catalyst for hydrogen production. Bioresour Technol 125:335–339
Panigrahi, Dalai, Chaudhari et al (2003) Synthesis gas production from steam gasification of biomass-derived oil. Energy Fuels 17(3):637–642
Pattiya (2011) Bio-oil production via fast pyrolysis of biomass residues from cassava plants in a fluidised-bed reactor. Bioresour Technol 102(2):1959–1967
Pattiya, Suttibak (2012) Influence of a glass wool hot vapour filter on yields and properties of bio-oil derived from rapid pyrolysis of paddy residues. Bioresour Technol 116:107–113
Pattiya, Titiloye, Bridgwater (2008) Fast pyrolysis of cassava rhizome in the presence of catalysts. J Anal Appl Pyrol 81:72–79
Peacocke, Dick, Hague et al (1996) Comparison of ablative and fluid bed fast pyrolysis products: yields and analyses. Biomass Energy Environ 1632–1637
Pütün (2002) Biomass to bio-oil via fast pyrolysis of cotton straw and stalk. Energy Sources 24(3):275–285
Ringer, Putsche, Scahill (2006) Large-scale pyrolysis oil production: a technology assessment and economic analysis. National Renewable Energy Laboratory (NREL), United States
Ro, Hunt, Jackson et al (2014) Co-pyrolysis of swine manure with agricultural plastic waste: laboratory-scale study. Waste Manag 34(8):1520–1528
Rover, Brown (2013) Quantification of total phenols in bio-oil using the Folin-Ciocalteu method. J Anal Appl Pyrol 104:366–371
Rover, Johnston, Lamsal et al (2013) Total water-soluble sugars quantification in bio-oil using the phenol–sulfuric acid assay. J Anal Appl Pyrol 104:194–201
Sajdak, Muzyka (2014) Use of plastic waste as a fuel in the co-pyrolysis of biomass. Part I: the effect of the addition of plastic waste on the process and products. J Anal Appl Pyrol 107: 267–275
Salehi, Abedi, Harding (2011) Bio-oil from sawdust: effect of operating parameters on the yield and quality of pyrolysis products. Energy Fuels 25(9):4145–4154
Sanna, Li, Linforth et al (2011) Bio-oil and bio-char from low temperature pyrolysis of spent grains using activated alumina. Bioresour Technol 102(22):10695–10703
Scott, Majerski, Piskorz et al (1999) A second look at fast pyrolysis of biomass—the RTI process. J Anal Appl Pyrol 51:23–37
Şen, Kar (2011) Pyrolysis of black cumin seed cake in a fixed-bed reactor. Biomass Bioenergy 35(10):4297–4304
Sensoz, Demiral, Gercel (2006) Olive bagasse (Olea europea L.) pyrolysis. Bioresour Technol 97(3):429–436
Sharma, Bakhshi (1993) Catalytic upgrading of pyrolysis oil. Enegry Fuels 7(2):306–314
Shi J (2012) Catalytic fast pyrolysis od biomass in a bubbling fluidized bed reactor with gallium promoted ZSM-5 catalyst. Department of Chemical Engineering, University of Massachusetts, Master of Science in Chemical Engineering

Sipila, Kuoppala, Fagernas et al (1998) Characterization of biomass-based flash pyrolysis oils. Biomass Bioenergy 14(2):103–113

Solantausta, Nylund, Gust (1994) Use of pyrolysis oil in a test diesel engine to study the feasibility of a diesel power plant concept. Biomass Bioenergy 7(1–6):297–306

Stals, Carleer, Reggers et al (2010) Flash pyrolysis of heavy metal contaminated hardwoods from phytoremediation: characterisation of biomass, pyrolysis oil and char/ash fraction. J Anal Appl Pyrol 89(1):22–29

Staš, Kubička, Chudoba et al (2014) Overview of analytical methods used for chemical characterization of pyrolysis bio-oil. Energy Fuels 28(1):385–402

Steele, Puettmann, Penmetsa et al (2012) Life-cycle assessment of pyrolysis bio-oil production. Forest Prod J 62(4):326–334

Tan, Zhang, Sun et al (2013) Recent progress of catalytic pyrolysis of biomass by HZSM-5. Chin J Catal 34(4):641–650

Thangalazhy-Gopakumar, Adhikari, Ravindran et al (2010) Physiochemical properties of bio-oil produced at various temperatures from pine wood using an auger reactor. Bioresour Technol 101(21):8389–8395

Thangalazhy-Gopakumar, Adhikari, Gupta et al (2011) Production of hydrocarbon fuels from biomass using catalytic pyrolysis under helium and hydrogen environments. Bioresour Technol 102(12):6742–6749

Thangalazhy-Gopakumar, Adhikari, Chattanathan et al (2012) Catalytic pyrolysis of green algae for hydrocarbon production using H+ZSM-5 catalyst. Bioresour Technol 118:150–157

Trane, Dahl, Skjøth-Rasmussen et al (2012) Catalytic steam reforming of bio-oil. Int J Hydrogen Energy 37(8):6447–6472

Wagenaar, Kuipers, Prins et al (1993) The rotating cone flash pyrolysis reactor. In: Bridgwater AV (ed) Advances in thermochemical biomass conversion. Blackie Academic & Professional, London. pp 1122–1133

Wagenaar, Prins, van Swaaij (1994) Pyrolysis of biomass in the rotating cone reactor: modelling and experimental justification. Chem Eng Sci 49(24, Part 2):5109–5126

Wagenaar, Venderbosch, Carrasco et al (2008) Rotating cone bio-oil production and applications. In: Progress in thermochemical biomass conversion. Blackwell Science Ltd pp 1268–1280

Wang, Dong, Yuan et al (2007) Characteristics of bio-oil-syngas and its utilization in Fischer–Tropsch synthesis. Energy Fuels 21(4):2421–2432

Wang, Gu, Liu et al (2009) Separation of bio-oil by molecular distillation. Fuel Process Technol 90(5):738–745

Wang, Chang, Fan (2010) Catalytic esterification of bio-oil by ion exchange resins. J Fuel Chem Technol 38(5):560–564

Wang, Brown, Homsy et al (2013) Fast pyrolysis of microalgae remnants in a fluidized bed reactor for bio-oil and biochar production. Bioresour Technol 127:494–499

WBA (2014) WBA global bioenergy statistics 2014, World Bioenergy Association

Wildschut, Mahfud, Venderbosch et al (2009) Hydrotreatment of fast pyrolysis oil using heterogeneous noble-metal catalysts. Ind Eng Chem Res 48:10324–10334

Williams, Horne (1995) The influence of catalyst type on the composition of upgraded biomass pyrolysis oils. J Anal Appl Pyrol 31:39–61

Wright, Daugaard, Satrio et al (2010) Techno-economic analysis of biomass fast pyrolysis to transportation fuels. Fuel 89:S2–S10

Zhang, Xiao, Huang et al (2009) Comparison of non-catalytic and catalytic fast pyrolysis of corncob in a fluidized bed reactor. Bioresour Technol 100(3):1428–1434

Zhao, Deng, Liao et al (2010) Aromatics production via catalytic pyrolysis of pyrolytic lignins from bio-oil. Energy Fuels 24(10):5735–5740

The Energy and Value-Added Products from Pyrolysis of Waste Plastics

Rashid Miandad, Mohammad Rehan, Abdul-Sattar Nizami, Mohammad Abou El-Fetouh Barakat and Iqbal Mohammad Ismail

Abstract Plastic usage in daily life has increased from 5 to 100 million tons per year since the 1950s due to their light-weight, non-corrosive nature, durability and cheap price. Plastic products consist mainly of polyethylene (PE), polystyrene (PS), polypropylene (PP) and polyvinyl chloride (PVC) type plastics. The disposal of plastic waste causes environmental and operational burden to landfills. Conventional mechanical recycling methods such as sorting, grinding, washing and extrusion can recycle only 15–20 % of all plastic waste. The use of open or uncontrolled incineration or combustion of plastic waste has resulted in air and waterborne pollutants. Recently, pyrolysis technology with catalytic reforming is being used to convert plastic waste into liquid oil and char as energy and value-added products. Pyrolysis is one of the tertiary recycling techniques in which plastic polymers are broken down into smaller organic molecules (monomers) in the absence of oxygen at elevated temperatures (>400 °C). Use of catalysts such as aluminum oxides, natural and synthetic zeolites, fly ash, calcium hydroxide, and red mud can improve the yield and quality of liquid oil. The pyrolysis yield depends on a number of parameters such as temperature, heating rate, moisture contents, retention time, type of plastic and particle size. A yield of up to 80 % of liquid oil by weight can be achieved from plastic waste. The produced liquid oil has similar characteristics to conventional diesel; density (0.8 kg/m^3), viscosity (up to 2.96 mm^2/s), cloud point (−18 °C), flash point (30.5 °C) and energy content (41.58 MJ/kg). Char produced from pyrolysis can be activated at standard conditions to be used in wastewater treatment, heavy metals removal, and smoke and odor removal. The produced gases from pyrolysis are

R. Miandad · M.A. El-Fetouh Barakat
Department of Environmental Sciences, Faculty of Meteorology, Environment and Arid Land Agriculture, King Abdulaziz University, Jeddah, Saudi Arabia

R. Miandad · M. Rehan · A.-S. Nizami (✉) · I.M. Ismail
Centre of Excellence in Environmental Studies (CEES), King Abdulaziz University, P.O. Box 80203, Jeddah 21589, Saudi Arabia
e-mail: nizami_pk@yahoo.com; anizami@kau.edu.sa

M.A. El-Fetouh Barakat
Central Metallurgical R&D Institute, Helwan 11421, Cairo, Egypt

O.P. Karthikeyan et al. (eds.), *Recycling of Solid Waste for Biofuels and Bio-chemicals*, Environmental Footprints and Eco-design of Products and Processes, DOI 10.1007/978-981-10-0150-5_12

hydrogen (H_2), carbon monoxide (CO) and carbon dioxide (CO_2) and can be used as energy carriers. This chapter reviews the challenges and, perspectives of pyrolysis technology for production of energy and value-added products from waste plastics.

Keywords Plastic waste · Liquid oil · Energy · Pyrolysis · Char · Hydrogen (H_2)

1 Introduction

1.1 World Energy Demands

The current world population of 7.2 billion is projected to increase by 1 billion in 2025 with an annual growth rate of 1 % (WHO 2014). The developing world is the place, where most of this growth will occur due to rapid growth in urbanization and population. Consequently, the energy demand has increased significantly in developing countries, especially in Asia (Ouda et al. 2013, 2015), that will further increase by 46–58 % with an annual rate of 3.7 % till 2025 (FAO 2010). The present global energy consumption is 524 quadrillion thermal British units (Btu), that will increase upto 820 quadrillion Btu by 2040 (US-EIA 2014). Global electricity demand will increase by 80 % from 2012 to 2040; in which non-OECD countries such as China (33 %), India (15 %), Southeast Asia (9 %) and Middle East (6 %) show highest electricity demand (IEA 2014). About 80 % of the present global energy requirements are fulfilled by thermal power plants including coal, natural gas, oil and nuclear plants, while only 13 % is produced from renewable sources (WEC 2013; Demirbas et al. 2016a).

Energy recovery or waste-to-energy (WTE) from non-food biomass sources such as agricultural and forestary residues and domestic and industrial waste is an eco-friendly approach for renewable energy production (Nizami et al. 2015a, b, c and 2016; Ouda et al. 2016; Rehan et al. 2016; Sadaf et al. 2015; Tahir et al. 2015). Fuel-wood is one of the major sources of bioenergy that produces about 87 % of global bioenergy. About 2 billion people in the world still depend on fuel-wood and charcoal for their daily energy production (UNWWD 2014). Among renewable energy, wind and solar energy are dominant choices, as they produce the least greenhouse gas (GHG) emissions and do not require water for energy production (Evans et al. 2009; IEA 2014). Nevertheless, around 1.3 billion people across the globe are still living without electricity and most of them (95 %) are habitants of sub-Saharan Africa and developing Asia (IEA 2014).

1.2 Energy Consumption in the Kingdom of Saudi Arabia (KSA)

In 2013, KSA was the world's 12th largest primary energy consumer country with a total energy consumption of 9 quadrillions British thermal units (Btu) (US-EIA

2014). The annual electricity demand of KSA has increased at an average rate of 5.8 % from 2006 to 2010 (MEP 2010). Oil fulfils 60 % of this energy demand, while natural gas covers the remaining 40 % energy requirements (KACARE 2012). The KSA's government has planned to double its energy generating capacity from 55 gigawatts (GW) to 120 GW by 2032. For this, the government has launched a special program; King Abdullah City of Atomic and Renewable Energy (KACARE) (Decree 2010). The vision of this initiative is to generate about half of the Saudi electricity capacity from different renewable sources, including solar, nuclear, wind, geothermal and WTE by 2032 (US-EIA 2014; Nizami et al. 2015a).

1.3 Plastic Consumption in the World and KSA

Plastics are composed of petroleum-derivative hydrocarbons, antioxidants, colorants and stabilizers (Hamid et al. 2000; Williams and Williams 1999; Perugini et al. 2005; Siddiqui and Redhwi 2009). The current global plastic consumption is round 154 million tons (Patni et al. 2013). Figure 1 shows the plastic consumption in different parts of the world. Thermoplastic and thermoset are commonly used plastic types. Thermoplastic softens when heating and hardens when cooled while, thermoset become hardened by curing and thus cannot be re-moulded. Thermoplastics

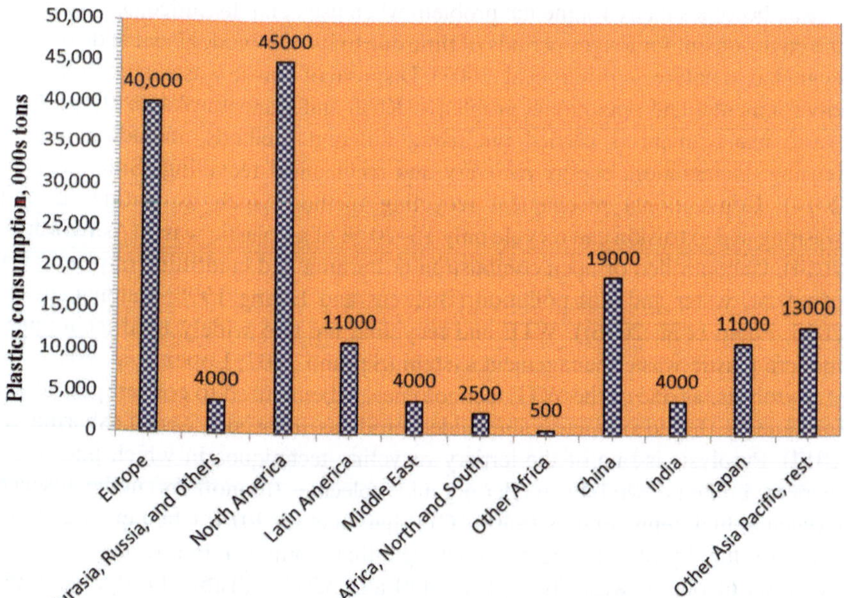

Fig. 1 Plastic consumption in different parts of the world (Patni et al. 2013)

are the most used plastic (around 80 %) in Western Europe (Jude et al. 2009). The plastic products are mostly manufactured from low density polyethylene (LDPE), high density polyethylene (HDPE), polypropylene (PP), polystyrene (PS), polyvinyl chloride (PVC) and polyethylene-terephthalate (PET) (Jude et al. 2009).

Plastic consumption has increased significantly worldwide over the last few decades due to their durability, light weight and low prices (Serrano et al. 2002; Aguado et al. 2008). As a consequence, global plastic waste generation was 280 million tons in 2011 that is expected to increase at an annual rate of 4 % till 2016 (Sriningsih et al. 2014). In the European Union (EU), plastic production remained stable (55–60 million tons per year) since 2000 (Beneroso et al. 2015; Williams 2005). According to Shah et al. (2010), plastic waste production in Western Europe was the same from 1993 to 2003 with a little increase (annual 64–93 kg per capita). In KSA, aroung 15 million tons of MSW is generated every year at an average rate of 1.4 kg per capita per day (Ouda et al. 2015). Plastic waste is the second largest waste stream with 17.4 % of total generated MSW (Siddiqui and Redhwi 2009; Nizami et al. 2015a). Most of the MSW produced in KSA ends up in landfills or dumpsites. Metal and cardboard are the waste materials, whose recycling is being carried out by up to 10–15 % of total produced MSW (Nizami et al. 2015a).

1.4 Potential of Waste Plastic to Energy

Plastic become an environmental problem when used and discarded, as it remains in the environment for longer periods of time due to its non-biodegradable or very slow degradation nature (Achilias et al. 2007). Disposal of plastic waste in landfills causes environmental and operational problems (Khan and Kaneesamkandi 2013). Plastic waste management is carried out using different methods, including reducing, reusing, incineration, energy recovery and mechanical recycling (Sriningsih et al. 2014). Conventional mechanical recycling methods such as sorting, grinding, washing and extrusion can recycle only 15–20 % of all plastic waste (Ashworth et al. 2014). Uncontrolled or open combustion or burning and landfilling of plastic waste cause air, water and soil pollution (Buekens and Huang 1998; Rahmanian et al. 2015; Eqani et al. 2016)). WTE and recycling are two widely used techniques in modern plastic waste management system (Serrano 2007; Lopez et al. 2012).

Pyrolysis, as one of the WTE technologies, is being used to convert plastic waste into energy (liquid oil) and value-added products (char and gases) (Sharma et al. 2014). Pyrolysis is one of the tertiary recycling techniques in which plastic poly mers are broken down into smaller organic molecules (monomers) in the absence of oxygen at high temperatures (>400 °C) (Sharma et al. 2014; Chen et al. 2014a, b). Most of the Pyrolysis studies concluded that optimum temperature range for pyrolysis of plastic waste is between 450 and 550 °C (Table 1). Pyrolysis of all types of plastic waste is possible except for PET and PVC, as they cause corrosion and pipeline obstruction problems (Lopez et al. 2012). The produced liquid oil has similar characteristics to conventional diesel (Syamsiro et al. 2014).

Table 1 Pyrolysis reactor types, conditions, feedstock used and comparative results

Reactor type	Experiment conditions			Feedstock	Catalyst used	Obtained results			References
	Temperature (°C)	Residence time (min)	Heating rate (°C/min)			Oil (%)	Char (%)	Syngas (%)	
Semi-batch	500	15–30	20	PP, PS, PET, PE, PVC	–	65.2	–	–	Lopez et al. (2011a)
Semi-batch	440	30	20	Industrial packing waste	ZSM-5	42.6	47.7	9.7	Lopez et al. (2011b)
Semi-batch	350	60	–	LDPE	NZ with Ni, Co, Mo	71.49	–	–	Sriningsih et al. (2014)
Semi-batch	450	–	–	PP	FCC	92.3	3.6	4.1	Abbas-Abadi et al. (2014)
Semi-batch	440	30		Mix Plastic	ZSM-5	80	18	2	Lopez et al. (2011c)
Packed bed	450	–	–	HDPE	HY-Zeolite	70	–	–	Syamsiro et al. (2013)
Packed bed	450	–	–	PS	HY-Zeolite	88	–	–	Syamsiro et al. (2013)
Fluidized bed	500	–	20	HDPE	HZSM-5	67.6	11.4	21	Hernández et al. (2007)
Bench scale	420	300	20	Municipal plastic	FCC	79.08	13.05	7.87	Lee (2009)
Two stage	450	–	–	HDPE	Natural zeolite	58	35	7	Syamsiro et al. (2014)
Fluidized bed	530	–	–	PE	–	92.3	0.1	7.6	Williams (2006)

The aim of this chapter is to review the challenges, and perspectives of pyrolysis technology for producing energy and value-added products from plastic waste.

2 Plastic Waste Disposal

2.1 *Recycling of Plastic Waste*

Plastic waste has increased with the increase of its consumption worldwide. Conventional techniques for the handling of plastic waste such as combustion or burning and landfilling has resulted in serious environmental problems. The US-EPA reported that in 2000, there was a 50 % increase in plastic waste dumping compared to early 1990s. Increase in plastic waste urged the scientists to look for alternative techniques for sustainable plastic waste management. Therefore, recycling of plastic waste is encouraged and practiced to overcome inherent plastic's problems. As a result, since the last few decades, recycling of plastic waste has increased at a very high rate. There are four different types of plastic recycling as follow;

- Primary recycling
  Primary recycling involves plastic waste recycling into a product that has similar characteristics as the original products. Mostly, PET plastic type recycling fits into this category.
- Secondary recycling
  Secondary recycling involves the recycling of plastic waste into a product that has different characteristics as compared to original products. PS, PE and PP are the main plastic types that fall into this category (Lerici et al. 2015).
- Tertiary recycling
  Tertiary recycling involves processes such as pyrolysis, in which plastic waste is converted into liquid oil at high temperatures. All types of plastic can recycle through tertiary recycling except PET and PVC. The PVC plastic produces hazardous chlorine gas and also corrodes the process apparatus (Lopez et al. 2012).
- Quaternary recycling
  Quaternary recycling involves the process of retrieving energy from the plastic waste through burning under controlled conditions. All types of plastic can be recycled via quaternary recycling.

Primary recycling is used when semi-cleaned plastic waste is available; although its use is very limited. While, secondary plastic recycling is used to recycle plastic waste into products that can be used as substitute for wood, concrete and metal. Mostly, secondary plastic recycling products are used as fences, benches, desks, chairs, etc. Tertiary recycling techniques such as pyrolysis convert plastic into liquid oil, while quaternary recycling burns plastic waste to produce energy. The latter was widely used due to high energy contents of plastic, but due to air pollution its use has limited (Ahmad 2015).

2.2 *Pyrolysis of Plastic Waste*

The pyrolysis process converts plastic waste through thermal decomposition to organic vapours and char and gases in the absence of oxygen. The produced organic vapours, are converted into oil as liquid fuel by a condensation process. It takes about 2–4 h to convert waste into useful energy products (Fonts et al. 2009); however there is a scope of time reduction and process optimization. The inorganic gradient remains unchanged and hence can be reused in plastic industries as a raw material for the manufacturing of new plastic products. Liquid oil is the main product from the pyrolysis process, while char and gases are by-products. About 80 % of plastic can be converted into liquid oil through pyrolysis (Lee 2009).

Dechlorination is used to stop the chlorine generation when pyrolysis of PVC plastic is carried out either at low temperature, as PVC decomposes at 250–320 °C (Lopez et al. 2011a, b) or can be removed via physical or chemical adsorption methods using different adsorbents (Lopez et al. 2012). The produced liquid oil has similar heating value (HHV) as compared to conventional diesel (Table 2), thus can be used as alternative fuel for various energy generating applications. According to Demirbas (2010), HHV value can be derived as per Eq. 1.

$$\mathrm{HHV(MJ/kg)} = 0.335\ \mathrm{C} + 1.423\ \mathrm{H} - 0.154\ \mathrm{O} - 0.145\ \mathrm{N} \tag{1}$$

where C is carbon content (wt%), H is hydrogen content (wt%), O is oxygen content (wt%), and N is nitrogen content (wt%).

Table 2 High heating values (HHV) of produced liquid oil from different feedstock

Feedstock	HHV (MJ/kg)	References
Tires	43.22	Wongkhorsub and Chindaprasert (2013)
Plastic	46.19	
Diesel	45.81	
PE Bag 1	41.45	Syamsiro et al. (2014)
HDPE waste	42.82	Syamsiro et al. (2014)
PE Bag 2	46.67	Syamsiro et al. (2014)
HDPE	45.86	Sharma et al. (2014)
HDPE	45.78	Kumar and Singh (2011)
Mixed plastic	44.40	Kim et al. (2010)
LDPE	38–39	Panda et al. 2010
HDPE	40.17	Kumar et al. (2010)
Mixed waste	40–40.5	Mani et al. (2011)

Pyrolysis is carried out in different types of reactors such as fixed bed (Wang et al. 2006), tube (Miskolczi et al. 2009), rotary kiln (Li et al. 2005) and two-stage reactors (Syamsiro et al. 2014) that have both advantages and disadvantages (Table 3).

2.2.1 Selection of Plastic Types for Pyrolysis

Pyrolysis of different feedstocks has been carried out by a number of researchers including, sewage sludge (Xiong et al. 2013), tires (Antoniou and Zabaniotou 2013), electronic waste (Yang et al. 2013) and plastic waste (Chen et al. 2014a, b). Plastic waste contains different types of plastic such as PS, PP, PE (e.g. LDPE and HDPE), PVC and PET. Pyrolysis has been successfully used to convert all types of plastic waste into liquid oil except PVC and PET (Table 4). Only few studies were carried out on pyrolysis of PVC due to production of hazardous chlorinated gas. Moreover, PE is also not suitable for thermal pyrolysis because of wax formation instead of liquid oil (Lee 2012). PS plastic decomposes at low temperature in comparison to PE and PP (Lee and Shin 2007).

2.2.2 Factors Affecting the Pyrolysis Process

There are different factors affecting the pyrolysis process such as temperature, retention time, feedstock composition, moisture content, heating rate and particle size of the feedstock. Temperature is the most influential and widely studied factor by scientists (Li et al. 1999; Yoshioka et al. 2004; Ji et al. 2006), as it affects thermal cracking and secondary reaction of the pyrolysis process. López et al. (2010) reported that solid residue (char) produced at different temperature remains the same, however the yield of liquid and gases changes. The liquid yield decreases and gas yield increases with temperature increase. Moreover, the increase in temperature affected the quality of the produced liquid oil.

At lower temperatures, long chain hydrocarbons are produced, while increase in temperature results in shorter carbon chains due to cracking of C–C bonds (Hernández et al. 2007; Lopez et al. 2009). Similarly, at higher temperature (around 600 °C), aromatic compounds are produced, while at lower temperature (around 460–500 °C) aromatic compounds with unsaturated hydrocarbons are produced (Lopez et al. 2011a). However, some scientists argue that aromatics are produced by secondary reactions due to increase in temperature (Kaminsk et al. 1999; Li et al. 1999, 2005).

Polyaromatic hydrocarbon (PAH) are also produced at high temperature that leads to increase in percentage composition of aromatic (Sánchez et al. 2009). At 600 °C, >C13 fraction decreases, while at the same temperature C10–C13 fraction increases which showing that higher temperature favours formation of short carbon chain compounds (Marcilla et al. 2009; Jung et al. 2010). However, at high temperature (500–600 °C), styrene production decreases. It shows that styrene is

Table 3 Advantages and disadvantages of pyrolysis reactors

Reactor Type	Advantages	Disadvantages	References
Fixed bed	• Used to identify the governing parameters that effect the pyrolysis products	• Low heating rate (HR) and low heat transfer coefficient • Temperature is not uniform when high mass is used as feedstock • Decomposition of feedstock at different temperature	Wang et al. (2006)
Rotary kiln	• Good for heating up the feedstock • Good mixing of waste (feedstock) during the pyrolysis process due to slow rotation of reactor • Residence time for feedstock is 1 h • Extensive pre-treatment for feedstock is not required • Maintenance of the reactor is simple	• Used for slow pyrolysis process at slow HR • Less information for heat transfer coefficient for heterogeneous MSW	Li et al. (2002, 2005), Chen et al. (2014a, b)
Fluidised-bed	• Heating rate is high • Feedstock blending is good • Used to find out the effect of temperature and residence time on the products of pyrolysis • Low thermal conductivity • High viscosity • Widely used for MPW	• Application at industrial scale for MSW is not common • Separation method for coke from bed-material is a challenge • Separation of bed material, recirculation and external heating and MSW pre-treatment increase cost of process which decreases its value economically	Al-Salem et al. (2010), Arena and Mastellone (2006), Mastral et al. (2002, 2003), Dai et al. (2001a, b), Chen et al. (2014a, b)
Tubular	• Consist of various tubes with fixed wall • Heated externally • Simple and safe • Coke and gas can be obtained from reactor Continuously • Suitable to use for both thermal and catalytic pyrolysis processes	• Required extensive pretreatment for MSW • Small channel for the passage of feedstock • Erosion of the reactor due to presence of sand and other solid contaminants present in the feedstock • Heat transfer co-efficient is not well defined	Aguado et al. (2002), Walendziewski (2002), Marculescu et al. (2007)

(continued)

Table 3 (continued)

Reactor Type	Advantages	Disadvantages	References
	• Thermoselect process, Compact power process and CNRS thermo-chemical convertor are the types of tubular reactor		
Multi stage	• Consist of two stages and three stages • Have the potential to control at different conditions • Provide solution to run pyrolysis technology independently • Produce HCl gas can be separated from volatile value products at different stages		Ohmukai et al. (2008), Zhao et al. (2011)
Plasma	• Convert waste into synthetic gas • High heat transfer which increase reactor temperature rapidly • Control process temperature, high process rate, low reaction volume • Produced syngas has optimum composition • Produced products are harmless to human health and environment	• Required high energy to run the process as temperature required is 1000 °C • Only applied for hazardous waste • Economically not favorable for MSW	Wang and Huang (2008), Huang and Tang (2007), George (1994)
Microwave	• Used for homogenous waste • Rapid increase of reactor temperature • Easy to control • Can be run at desire temperature • Lower and higher temperature for reaction vessels and for reaction mixture respectively	• Very fine feedstock particles are required to get rapid heating rate • To reduce secondary cracking solid-laden vapors has to be removed rapidly from reactor	Baghurst and Mingos (1992), Yin (2002)

Table 4 Suitability of plastic for pyrolysis

Resin	Suitability for pyrolysis	Temperature required	References
Polystyrene (PS)	Good and gives better fuel properties	Low temperature is required as compared to PP and PE	Lee and Shin (2007)
Polyethylene (PE)	Good, but mostly is converted into wax, if catalyst is not used	Required Temperature is high >500 °C	Miskolczi et al. (2009), Lee (2012)
Polypropylene (PP)	Good	Requires high temperature	Miskolczi et al. (2009)
PVC	Not suitable, Produce hazardous chlorinated gas, Dechlorination via low temperature (250–320 °C) or physical or chemical adsorption is required		Lopez et al. (2011a, 2012)
PET	Not suitable, and contains Heteroatom's		Thorat et al. (2013)

produced at low temperature, while at higher temperature it decomposes into other chemicals like toluene and ethyl-benzene (Lopez et al. 2009; Lee and Shin 2007). Demirbas (2004) reported that decrease in styrene yield at high temperature is due to secondary reactions. Pyrolysis of PP and PE require high temperature (>450 °C), as they produce high carbon chain fractions in the liquid due to presence of long carbon chain compounds (López et al. 2010).

Retention time also affects the pyrolysis process, however its impact is lower than temperature. According to Lopez et al. (2009), at shorter retention times, aromatic hydrocarbons are produced, especially when feedstock consist of PS plastic. About 60 % of aromatic hydrocarbons are produced from the pyrolysis of mixed plastic waste containing PS, PP and PE at 350 °C (Lee and Shin 2007). Moreover, the effect of retention time on the carbon chain fraction is less, as similar carbon chain compounds are produced at each temperature with different retention times. However at higher temperature, an increase in the >C13 fraction is observed. According to Lopez et al. (2009), this may be due to high retention time in reactors, as plastic and its derivatives decompose and generate high carbon chain compounds in comparison to light carbon chain compounds that are formed at low retention time.

López et al. (2010) studied catalytic pyrolysis of various feedstock including film, paper and char by using ZSM-5 zeolite catalyst. They reported that even at low temperature (440 °C), more than 90 % by weight decomposition is occurred when feedstock consist of 65 % PE. However, paper-rich samples produced less organic liquid (25.5 wt%) and gases (29.4 wt%), while glass-rich samples produced more char (12.9 wt%) in comparison to both paper and film-rich samples. High char production may be due to the reaction of the glass-rich samples with ZSM-5 zeolite catalyst.

The type of plastic also affects the quantity and quality of produced liquid oil from the pyrolysis process. According to Siddiqui and Redhwi (2009), PS plastic produces less viscous liquid oil due to shorter carbon chain compounds in comparison to PE and PP that produce long carbon chain compounds. Therefore, liquid oil produced from PE and PP has high viscosity (Syamsiro et al. 2014). During the winter season or areas where temperature is very low, high viscous oil affects the oil fluidity, resulting in poor atomization and poor engine performance (Kim et al. 2009; Oasmaa and Czernik 1999).

The effect of moisture contents in feedstock is limited. However its presence in paper, pulp and bamboo feedstocks leads to decrease in H_2 production and increase in tar concentration (Chen et al. 2014a, b). Luo et al. (2010) examined MSW pyrolysis with different feedstock and particle size range (0–5, 5–10 and 10–20 mm). They found that with smaller particle size, the gas fractions increase in comparison to liquid and char fractions. Moreover, there are some other factors that affect pyrolysis process such as flash and fast pyrolysis, which increase liquid oil yield production (Sharma et al. 2014). Furthermore, continuous stirring of feedstock inside the reactor decreases the overall process retention time (Abbas-Abadi et al. 2014), while insulation of the reactor saves energy (Chen et al. 2014a, b). Hernández et al. (2007) used sand to increase the temperature of the reactor in order to reduce the parasitic energy demand of the process.

2.3 *Effect of Catalyst on Pyrolysis*

Catalysts are used to improve the quality of produced liquid oil and overall performance of pyrolysis process. Different types of catalyst are used including, Red Mud (López et al. 2010), FCC (Lee 2009), Ca $(OH)_2$ (Sarker et al. 2011), and Fe_2O_3 (Sarkar et al. 2011). Saw dust was also used to improve the quality of the produced oil from the pyrolysis of rubber waste (Wang et al. 2014). Moreover, different types of natural and synthetic zeolite catalysts have also been used including, ZSM-5 zeolite (Lopez et al. 2010), HZSM-5 zeolite (Hernández et al. 2007) and natural zeolite (Syamsiro et al. 2014). Modifications in natural zeolite catalysts using different processes were also carried out such as zinc-modified zeolite via wet impregnation (Ciobanu et al. 2008), ZSM-5 modified with boron via impregnation (Zhou et al. 2014) and natural zeolite catalyst modified with Ni, Ni-Mo, Co and Co-Mo metals (Sriningsih et al. 2014).

Use of a catalyst has different effects on the pyrolysis process, however one of the main effects is to increase the quality of produced liquid oil (Wang and Wang 2011). The lighter fraction, especially gasoline is increased by the use of catalyst. According to Lopez et al. (2011a), catalysts also decrease energy consumption of the pyrolysis process. According to Miskolczi et al. (2009), the use of ZSM-5 catalyst decreased the impurities in liquid oil as well as nitrogen and sulphur contents.

The effect of catalyst varies with the type of plastic waste. By using ZSM-5 zeolite catalyst, it was found that plastic waste containing mostly PE type plastic achieve higher yields of liquid oil. The maximum liquid oil yield was obtained at 440 °C from PE plastic; however the same result was also achieved at 500 °C without using a catalyst. This means that the catalyst decreases the temperature requirement of the pyrolysis process (Lopez et al. 2011a).

In another study by Lee (2012), pyrolysis of PE plastic type without catalyst resulted in wax formation. The use of a catalyst with this wax produced liquid oil with a high H/C ratio and hydrocarbons. The textural properties of a catalyst plays a vital role in liquid oil's purifying characteristic. The ZSM-5 catalyst has a high micropore area and volume that means this catalyst has high internal porosity. While, red mud has low porosity and most of the macro and meso pores are located on the external surface of the catalyst. Moreover, red mud is less acidic than ZSM-5 catalyst and thus has less effect on product yield in comparison to the ZSM-5 catalyst (Lopez et el. 2011b).

Modification in zeolite catalyst has been under consideration for the last few decades. By making zeolite composites with zinc, adsorption capacity has increased due to narrower pores of zeolites. Modified zeolites catalysts have an average particle size of few microns, thus internal surface area can represent more than 99 % of the total surface area. Almost all active sites are located inside the pores increasing its adsorption capacity and thus enhancing catalyst purifying ability (Ciobanu et al. 2008).

3 Energy and Value-Added Products of Plastic Waste Pyrolysis

3.1 Liquid Oil as an Energy Source

Liquid oil is the main product of the pyrolysis process. According to Lopez et al. (2010) and Williams (2006), 1 kg of plastic can produce around 72–84 % by weight liquid oil. Theoretically, 1.047 MJ/kg energy is required to convert PE plastic into liquid oil by pyrolysis. The produced liquid oil has an energy value of around 43 MJ/kg that is much higher than the energy consumed by the process (Gao 2010). Pyrolysis of PE plastic converts feedstock into wax instead of liquid oil, however use of catalyst can convert PE into liquid oil (Lee 2009). According to Achilias et al. (2007), pyrolysis of PE and PP is difficult to carry out without catalyst. Moreover by using the ZSM-5 catalyst, a maximum yield from PE is obtained at low temperature (440 °C) in comparison to high temperature (500 °C) without using a catalyst (Lopez et al. 2011a). Use of catalysts not only reduce the process time, but also remove the impurities from produced liquid oil. According to Miskolczi et al. (2009), use of catalysts decrease sulphur content in produced liquid oil.

The produced liquid oil from different feedstock has different physical and chemical properties such as viscosity, density, cold flow properties and HHV. Viscosity and density are the main properties of liquid oil with regards to its operational use (Table 5). Liquid oil from plastic pyrolysis has viscosity of 1–2.96 mm^2/s. PS plastic liquid oil has a lower viscosity compared to PE and PP, as it produces

Table 5 Density and viscosity of pyrolysis liquid fuel

Feedstock	Density g/cm^3 (15 °C)	Viscosity mm^2/s (40 °C)	Reference
PE Bag 1	0.8544	1.739	Syamsiro et al. (2014)
HDPE waste	0.7991	2.319	
PE Bag 2	0.824	1.838	
LDPE	0.7787	1.89	Panda et al. (2010)
HDPE	0.790	2.1	Kumar et al. (2010)
Waste plastic	0.8355	2.52	Mani et al. (2011)

short carbon chain compounds (Williams and Williams 1999; Siddiqui and Redhwi 2009). While, PE and PP plastic types produce high viscous oils due to presence of long carbon chain compounds (Syamsiro et al. 2014; Panda and Singh 2013).

Density of the produced liquid oil also varies from 0.77 to 0.91 g/cm^3. However, the values for density are similar to the conventional diesel (0.815–0.870 g/cm^3) (Syamsiro et al. 2014). Therefore, pyrolytic liquid oil can be used as an energy source. Moreover, cold flow properties of produced liquid oil are below its freezing point. These properties are critical for the use of any petroleum product. Cold flow properties include cloud point (where crystals begin to appear) and pour point (below this temperature liquid does not flow) (Isioma et al. 2013; Gardy et al. 2014). Low cold flow properties show that the produced liquid oil has the potential to be used as an alternative fuel oil in those areas where temperature is extremely low. Moreover, liquid oil produced from pyrolysis of plastic waste has the same properties as conventional diesel (Table 6).

3.2 *Char as by-Product*

Char is an unburnt feedstock left in the pyrolysis reactor. It is produced in very low quantities, for example, 1.1–3 % char by weight is produced from 1 kg of feedstock (Lopez et al. 2011a; Williams 2006). Lopez et al. (2010) reported that increase in temperature decreases the quantity of produced char. Elemental and proximate analysis show that char contains volatile matter, fixed carbon, ash, etc. (Table 7).

Table 6 Comparison of pyrolysis liquid oil with conventional diesel

Parameters	Units	Pyrolytic liquid fuel	Conventional diesel	Reference
Viscosity	mm^2/s	1.89	1–4.11	Wongkhorsub and Chindaprasert (2013)
Density	g/cm^3	0.854	0.815–0.870	Syamsiro et al. (2014)
Kinematic viscosity	cSt	1.77	2.0–5.0	Syamsiro et al. (2014)
Pour point	°C	−18	Max 18	Syamsiro et al. (2014)
HHV	MJ/kg	41.58	46.67	Syamsiro et al. (2014)

Table 7 Elemental and proximate analysis of char (wt%)

Feedstock	Pyrolysis Temp (°C)	Moisture content	Volatile matter	Fixed carbon	Ash	C	H	N	Cl	HHV (MJ/kg)	Ref
PS	450	0.91	37.44	57.28	4.37	91.14	4.09	0.09	–	36.29	Syamsiro et al. (2013)
HDPE	450	3.09	19.14	57.99	19.78	65.88	2.01	1.50	–	23.04	Syamsiro et al. (2013)
Mix plastic	440	0.7	–	–	–	23.0	2.2		0.4	9.4	Lopez et al. 2012
Packaging waste	440	–	–	–	–	60.9	10.2	0.2	0.2	41.5	Lopez et al. (2010)
PE bag	450	1.12	35.29	14.13	49.47	–	–	–	–	19.80	Syamsiro et al. (2014)
HDPE	450	1.74	44.47	25.88	27.33	–	–	–	–	26.35	Syamsiro et al. (2014)

Temp Temperature; *Ref* Reference

According to Jamradloedluk and Lertsatitthanakorn (2014), char produced from HDPE plastic type has volatile matter (51.40 %), fixed carbon (46.03 %), moisture content (2.41 %), and a small amount of ash. Moreover, char produced from HDPE and PS has HHVs of 23.04 and 36.29 MJ/kg respectively (Syamsiro et al. 2013). High heating value of char shows its potential to be used as a source of energy. Jamradloedluk and Lertsatitthanakorn (2014) crushed the HDPE char into powder and produced briquettes and used 1 kg for water heating (from room temperature to boiling temperature) within 13 min. Produced char has potential to be used in other environmental applications such as adsorption of heavy metals from municipal and industrial wastewaters and toxic gases (Heras et al. 2014). According to Lopez et al. (2009), thermal activation of char at 900 °C for 3 h increased its BET surface area and pore volume upto 55 and 44 %, respectively. Moreover, it decreased the pore size upto 5 %. Steam activation of char produced from waste tires decreased the sulphur content of char, thus making it a more environmental friendly product.

3.3 Gases

Gases are produced as by-products of pyrolysis of different feedstock. By using the mass balance formula, it was estimated that 1 kg of feedstock can produce 13–26.9 % gases by weight from the pyrolysis process (Lopez et al. 2011a; Williams 2006). The yield of produced gases is highly affected by the process temperature. Lopez et al. (2009) reported that an increase in temperature will increase the yield of gases and decrease the yield of liquid oil. The major gases are H_2, CO and CO_2. In addition, PVC feedstock produces HCl gas that is hazardous gas and causes apparatus corrosion (Chen et al. 2014a, b; Lopez et al. 2012). The BET surface area of catalysts also affects the yield of gas production. Higher BET areas provide more contact for feedstock with the catalyst, resulting in higher gas yield (Syamsiro et al. 2014).

4 Conclusion and Future Perspective

The tremendous increase in plastic consumption worldwide for a wide range of products is causing serious waste disposal and environmental issues. This increases the importance of plastic recycling and treatment technologies to deal with such problematic waste in an environmental friendly manner. Pyrolysis is a promising technology used to convert waste plastic into liquid oil and other valuable by-products such as char and gases under controlled conditions and is considered to be a relatively more environmental friendly technology than uncontrolled incineration. The pyrolysis products yield depend on a number of process parameters such as temperature, heating rate, moisture contents, retention time, type of plastic and particle size. A yield of up to 80 % of liquid oil by weight can be achieved from plastic waste. The produced liquid oil has similar characteristics to the conventional diesel including, density (0.8 kg/m^3), viscosity (up to 2.96 mm^2/s), cloud point (−18 °C), flash point (30.5 °C) and energy content (41.58 MJ/kg), and can be used as an energy source. Char produced from the pyrolysis can be activated at standard conditions to be used in wastewater treatment, heavy metals removal, and smoke and odor removal. The produced gases from pyrolysis are H_2, CO and CO_2 and can be used as energy carriers. High temperature and retention time are the main limitations of pyrolysis of plastic waste, which need to be optimized to make the process more economically and environmentally favorable.

A number of catalysts have been used in pyrolysis process resulting in improved liquid oil yield and quality, however exploration and utilization of cheaper catalysts such as natural zeolites requires more intense research. Moreover, catalysts modification require further attention to improve their performance to optimize the overall pyrolysis process. Since this book chapter focused on the case study of KSA, it is important to mention that intensive research and development work is still required to understand the full scope and potential of pyrolysis of plastics, in terms of technical, economic and environmental issues using life cycle assessment (LCA) tools in relation to local social and environmental conditions, and for adapting this technology as a WTE technology in KSA (Shahzad et al. 2015; Rathore et al. 2016; Demirbas et al. 2016b; Lai et al. 2016; Nizami and Ismail 2013).

References

Abbas-Abadi MS, Haghighi MN, Yeganeh H, Macdonald AG (2014) Evaluation of pyrolysis parameters on polypropylene degradation products. J Anal Appl Pyrol 109:272–277

Achilias DS, Roupakias C, Megalokonomos P, Lappas AA, Antonakou EV (2007) Chemical recycling of plastic waste made from polyethylene (LDPE and HDPE) and polypropylene (PP). J Hazard Mater 149:536–542

Aguado J, Serrano DP, Escola JM, Garagorri E (2002) Catalytic conversion of low density polyethylene using a continuous screw kiln reactor. Catal Today 75:257–262

Aguado J, Serrano DP, Escola JM (2008) Fuels from waste plastics by thermal and catalytic processes: a review. Ind Eng Chem Res 47:7982–7992

Ahmad L (2015) "Plastic Recycling" [online]. Available at http://www.lotfi.net/recycle/plastic.html 14 June 2015

Al-Salem SM, Lettieri P, Baeyens J (2010) The valorization of plastic solid waste (PSW) by primary to quaternary routes: from re-use to energy and chemicals. Prog Energy Combust Sci 36:103–129

Antoniou N, Zabaniotou A (2013) Features of an efficient and environmentally attractive used tyres pyrolysis with energy and material recovery. Renew Sustain Energy Rev 20:539–558

Arena U, Mastellone ML (2006) Fluidized bed pyrolysis of plastic wastes. In: Scheirs J, Kaminsky W (eds) Feed stock recycling and pyrolysis of waste plastics: converting waste plastics into diesel and other fuels. Wiley, West Sussex, pp 435–447

Ashworth DC, Elliott P, Toledano MB (2014) Waste incineration and adverse birth and neonatal outcomes: a systematic review. Environ Int 69:120–132

Baghurst DR, Mingos DMP (1992) Superheating effects associated with microwave dielectric heating. J Chem Soc Chem Commun 674–677

Beneroso D, Bermúdez JM, Arenillas A, Menéndez JA (2015) Comparing the composition of the synthesis-gas obtained from the pyrolysis of different organic residues for a potential use in the synthesis of bioplastics. J Anal Appl Pyrol 111:55–63

Buekens AG, Huang H (1998) Catalytic plastics cracking for recovery of gasoline-range hydrocarbons from municipal plastic wastes. Resour Conserv Recycl 23:163–181

Chen C, Jin Y, Chi Y (2014a) Effects of moisture content and CaO on municipal solid waste pyrolysis in a fixed bed reactor. J Anal Appl Pyrol 110:108–112

Chen D, Yin L, Wang H, He P (2014b) Pyrolysis technologies for municipal solid waste: a review. Waste Manag 34:2466–2486

Ciobanu G, Ignat D, Carja G, Ratoi S, Luca C (2008) Zinc—modified forms of zeolites by wet impregnation method. Chem Bull "POLITEHNICA" Univ (Timişoara) 53(67): 1–2

Dai X, Yin X, Wu C, Zhang W, Chen Y (2001a) Pyrolysis of solid biomass in a circulating fluidized bed reactor. Acta Energiae Solaris Sinca 22:124–130

Dai X, Yin X, Wu C, Zhang W, Chen Y (2001b) Pyrolysis of waste tires in a circulating fluidized-bed reactor. Energy 26:385–399

Decree R (2010) Royal Decree establishing King Abdullah city for atomic and renewable, No. A 35. Available from http://www.energy.gov.sa/_pdf/KACARE%20Royal%20Decree%20english.pdf. 17 April 2010

Demirbas A (2004) Pyrolysis of municipal plastic wastes for recovery of gasoline range hydrocarbons. J Anal Appl Pyrol 72:97–102

Demirbas A (2010) Fuels from biomass, biorefineries. Springer, London, pp 33–73

Demirbas A, Rehan M, Al-Sasi BO, Nizami AS (2016a) Evaluation of natural gas hydrates as a future methane source. Pet Sci Technol doi:10.1080/10916466.2016.1185442

Demirbas A, Bafail A, Nizami AS (2016b) Heavy oil upgrading: Unlocking the future fuel supply. Petroleum Science and Technology 34(4):303–308

Eqani SAMAS, Khalid R, Bostan N, Saqib Z, Mohmand J, Rehan M, Ali N, Katsoyiannis IA, Shen H (2016) Human lead (Pb) exposure via dust from different land use settings of Pakistan: A case study from two urban mountainous cities. Chemosphere 155:259–265. http://dx.doi.org/10.1016/j.chemosphere.2016.04.036

Evans A, Strezov V, Evans T (2009) Assessment of sustainability indicators for renewable energy technologies. Renew Sustain Energy Rev 13:1082–1088

FAO: Food and Agriculture Organization (2010) Energy supply and demand: trends and prospects. Report available online: ftp://ftp.fao.org/docrep/fao/010/i0139e/i0139e03.pdf

Fonts I, Azuara M, Gea G, Murillo MB (2009) Study of the pyrolysis liquids obtained from different sewage sludge. J Anal Appl Pyrol 85(1–2):184–191

Gao F (2010) Pyrolysis of waste plastics into fuels. PhD Thesis. University of Canterbury Available from http://ir.canterbury.ac.nz/bitstream/10092/4303/1/Thesis_fulltext.pdf

Gardy J, Hassanpour A, Lai X, Rehan M (2014) The influence of blending process on the quality of rapeseed oil-used cooking oil biodiesels. Inter Sci J (J Environ Sci) 3:233–240

George AS (1994) Plasma arc heating—an overview. Ind Heating 61:57–60

Hamid SH, Amin MB, Maadhah AG (2000) Handbook of polymer degradation, 2nd edn. Marcel and Decker, New York, p 335

Heras F, Jimenez-Cordero D, Gilarranz MA, Alonso-Morales N, Rodriguez JJ (2014) Activation of waste tire char by cyclic liquid-phase oxidation. Fuel Process Technol 127:157–162

Hernández MR, Gómez A, García AN, Agulló J, Marcilla A (2007) Effect of the temperature in the nature and extension of the primary and secondary reactions in the thermal and HZSM-5 catalytic pyrolysis of HDPE. Appl Catal A 317:183–194

Huang H, Tang L (2007) Treatment of organic waste using thermal plasma pyrolysis technology. Energy Convers Manag 48:1331–1337

IEA: International Energy Agency (2014) World energy outlook 2014 Factsheet, how will global energy markets evolve to 2040? Available from http://www.worldenergyoutlook.org/media/weowebsite/2014/141112_WEO_FactSheets.pdf

Isioma N, Muhammad Y, Sylvester O'D, Innocent D, Linus O (2013) Cold flow properties and kinematic viscosity of biodiesel. Univers J Chem 1(4): 135–141

Jamradloedluk J, Lertsatitthanakorn C (2014) Characterization and utilization of char derived from fast pyrolysis of plastic wastes. Procedia Eng 69:1437–1442

Ji L, Hervier A, Sablier M (2006) Study on the pyrolysis of polyethylene in the presence of iron and copper chlorides. Chemosphere 65:1120–1130

Jung SH, Cho MH, Kang BS, Kim JS (2010) Pyrolysis of a fraction of waste polyethylene for the recovery of BTX aromatics using a fluidized bed reactor. Fuel Process Technol 91:277–284

KACARE: King Abdullah City of Atomic and Renewable Energy (2012) Renewable energy-waste to energy. A pillar of the sustainable energy kingdom. In: Presentation in first international environment conference. King Fahd Civic Centre, Yanbu Al Sinaiyah, KSA 20–21 Nov 2012. http://www.energy.gov.sa/_pdf/KACARE%20Royal%20Decree%20english.pdf

Kaminsky W (2006) Feedstock Recycling and Pyrolysis of Waste Plastics, Wiley, Hoboken, New Jersey, pp 435–447

Khan MS, Kaneesamkandi Z (2013) Biodegradable waste to biogas: renewable energy option for the Kingdom of Saudi Arabia. Int J Innov Appl Stud 4(1):101–113

Kim SS, Agblevor FA, Lim J (2009) Fast pyrolysis of chicken litter and turkey litter in a fluidized bed reactor. J Ind Eng Chem 15:247–252

Kim JS, Cho MH, Jung SH (2010) Pyrolysis of mixed plastic wastes for the recovery of benzene, toluene and xylene (BTX) aromatics in a fluidized bed and chlorine removal by applying various additives. Energy Fuels 24:1389–1395

Kumar PA, Mishra DK, Mishra BG, Singh RK (2010) Effect of sulphuric acid treatment on the physicochemical characteristics of kaolin clay. Colloids Surf A: Physicochem Eng Asp 1–3 (20):98–104363

Kumar S, Singh RK (2011) Recovery of hydrocarbon liquid from waste high density polyethylene by thermal pyrolysis. Braz J Chem Eng 28(4):659–667

Lai L, Kumar S, Chintala R, Owens VN, Schumacher J, Nizami AS, Lee SS, Rafique R (2016) Modeling the impacts of temperature and precipitation changes on soil CO2 fluxes from a Switchgrass stand recently converted from cropland. J Environ Sci 43:15–25

Lee KH (2009) Thermal and catalytic degradation of pyrolytic oil from pyrolysis of municipal plastic wastes. J Anal Appl Pyrol 85:372–379

Lee KH (2012) Effect of zeolite type on catalytic upgrading of pyrolysis wax oil. J Anal Appl Pyrol 94:209–214

Lee KH, Shin DH (2007) Characteristics of liquid product from the pyrolysis of waste plastic mixture at low and high temperatures: influence of lapse time of reaction. Waste Manag 27:168–176

Lerici LC, Renzini MS, Pierella LB (2015) Chemical Catalyzed Recycling of Polymers: Catalytic Conversion of PE, PP and PS into Fuels and Chemicals over H-Y. Procedia Mater Sci 8:297–303

Li AM, Li XD, Li SQ, Ren Y, Chi Y, Yan JH, Cen KF (1999) Pyrolysis of solid waste in a rotary kiln: influence of final pyrolysis temperature on the pyrolysis products. J Anal Appl Pyrol 50:149–162

Li SQ, Yan JH, Li RD, Chi Y, Cen KF (2002) Axial transport and residence time of MSW in rotary kilns Part I. Experimental. Powder Technol 126:217–227

Li SQ, Ma LB, Wan W, Yan Q (2005) A mathematical model of heat transfer in a rotary kiln thermo-reactor. Chem Eng Sci 28:1480–1489

Lopez G, Olazar M, Artetxe M, Amutio M, Elordi G, Bilbao J (2009) Steam activation of pyrolytic tyre char at different temperatures. J Anal Appl Pyrol 85:539–543

López A, de Marco I, Caballero BM, Laresgoiti MF, Adrados A (2010) Pyrolysis of municipal plastic wastes: influence of raw material composition. Waste Manag 30:620–627

Lopez A, Marco ID, Caballero BM, Adrados A, Laresgoiti MF (2011a) Deactivation and regeneration of ZSM-5 zeolite in catalytic pyrolysis of plastic waste. Waste Manag 31:1852–1858b

Lopez A, Marco ID, Caballero BM, Laresgoiti MF, Adrados A, Aranzabal A (2011b) Catalytic pyrolysis of plastic wastes with two different types of catalytic: ZSM-5 zeolite and Red Mud. Appl Catal B: Environ 104:211–219c

López A, De Marco I, Caballero BM, Laresgoiti MF, Adrados A (2011c) Influence of time and temperature on pyrolysis of plastic wastes in a semi-batch reactor. Chemical Engineering Journal 173(1):62–71

Lopez A, Marco ID, Caballero BM, Laresgoiti MF, Adrados A (2012) Catalytic stepwise pyrolysis of packaging plastic waste. J Anal Appl Pyrol 96:54–62d

Luo S, Xiao B, Hu Z, Liu S (2010) Effect of particle size on pyrolysis of single-component municipal solid waste in fixed bed reactor. Int J Hydrogen Energy 35:93–97

Mani M, Nagarajan G, Sampath S (2011) Characterization and effect of using waste plastic oil and diesel fuel blends in compression ignition engine. Energy 36(1):212–219

Marcilla A, Beltrán MI, Navarro R (2009) Evolution of products during the degradation of polyethylene in a batch reactor. J Anal Appl Pyrol 86:14–21

Marculescu C, Antonini G, Badea A, Apostol T (2007) Pilot installation for the thermo-chemical characterization of solid wastes. Waste Manag 27:367–374

Mastral FJ, Esperanza E, Berrueco C, Juste M, Ceamanos J (2003) Fluidized bed thermal degradation products of HDPE in an inert atmosphere and in air–nitrogen mixtures. J Anal Appl Pyrol 70:1–17

Mastral FJ, Esperanza E, Garcia P, Juste M (2002) Pyrolysis of high-density polyethylene in a fluidised bed reactor. Influence of the temperature and residence time. J Anal Appl Pyrol 63:1–15

MEP: Ministry of Economy and Planning (2010) The nine development plan 2010–2014. Ministry of Economy and Planning Documents, Riyadh, KSA

Miskolczi N, Angyal A, Bartha L, Valkai I (2009) Fuel by pyrolysis of waste plastics from agricultural and packaging sectors in a pilot scale reactor. Fuel Process Technol 90:1032–1040

Nizami AS, Ouda OKM, Rehan M, El-Maghraby AMO, Gardy J, Hassanpour A et al (2015a) The potential of Saudi Arabian natural zeolites in energy recovery technologies. Energy 2015, 1–10. http://dx.doi.org/10.1016/j.energy.2015.07.030

Nizami AS, Rehan M, Ismail IMI, Almeelbi T, Ouda OKM (2015b) Waste biorefinery in Makkah: A solution to convert waste produced during Hajj and Umrah Seasons into wealth. Conference: 15th Scientific Symposium for Hajj, Umrah and Madinah visit. Held in May 2015 in Madinah, Saudi Arabia. doi: 10.13140/RG.2.1.4303.6560

Nizami AS, Rehan M, Ouda OKM, Shahzad K, Sadef Y, Iqbal T, Ismail IMI (2015c) An argument for developing waste-to-energy technologies in Saudi Arabia. Chem Eng Trans 45:337–342. http://dx.doi.org/10.3303/CET1545057

Nizami AS, Ismail IMI (2013) Life-cycle Assessment of Biomethane from Lignocellulosic Biomass. Chapter in 'Life cycle assessment of renewable energy sources'. Green Energy and Technology book series. 2013, DOI: 10.1007/978-1-4471-5364-1_4. Publisher: Springer-Verlag London Ltd

Nizami AS, Shahzad K, Rehan M, Ouda OKM, Khan MZ, Ismail IMI, Almeelbi T, Basahi JM, Demirbas A (2016) Developing Waste Biorefinery in Makkah: A Way Forward to Convert Urban Waste into Renewable Energy. Appl Energy http://dx.doi.org/10.1016/j.apenergy.2016.04.116

Oasmaa A, Czernik S (1999) Fuel oil quality of biomass pyrolysis oils state of the art for the end users. Energy Fuels 13:914–921

Ohmukai Y, Hasegawa I, Mae K (2008) Pyrolysis of the mixture of biomass and plastics in countercurrent flow reactor Part I: experimental analysis and modeling of kinetics. Fuel 87:3105–3111

Onwudili JA, Insura N, Williams PT (2009) Composition of products from the pyrolysis of polyethylene and polystyrene in a closed batch reactor: Effects of temperature and residence time. J Anal Appl Pyrol 86:293–303

Ouda OKM, Cekirge HM, Raza SA (2013) An assessment of the potential contribution from waste-to-energy facilities to electricity demand in Saudi Arabia. Energy Convers Manag 75:402–406

Ouda OKM, Raza SA, Al-Waked R, Al-Asad JF, Nizami A-S (2015) Waste-to-energy potential in the Western Province of Saudi Arabia. J King Saud Univ Eng Sci. doi:http://dx.doi.org/10.1016/j.jksues.2015.02.002

Ouda OKM, Raza SA, Nizami AS, Rehan M, Al-Waked R, Korres NE (2016) Waste to energy potential: A case study of Saudi Arabia. Renewable Sustainable Energy Rev 61:328–340. http://dx.doi.org/10.1016/j.rser.2016.04.005

Panda AK, Singh RK (2013) Experimental optimization of process for the thermo-catalytic degradation of waste polypropylene to liquid fuel. Adv Energy Eng (AEE) 1(3):74–84

Panda AK, Singh RK, Mishra DK (2010) Thermolysis of waste plastics to liquid fuel. A suitable method for plastic waste management and manufacture of value added products—A world prospective. Renew Sustain Energy Rev 14:233–248

Patni N, Shah P, Agarwal S, Singhal P (2013) Alternate Strategies for Conversion of Waste Plastic to Fuels. ISRN Renewable Energy. http://dx.doi.org/10.1155/2013/902053

Perugini F, Mastellone ML, Arena U (2005) A life cycle assessment of mechanical and feedstock recycling options for management of plastic packaging wastes. Environ Prog 24(2):137–154

Rahmanian N. Ali SHB, Homayoonfard M, Ali NJ, Rehan M, Sadef Y, Nizami AS (2015) Analysis of Physiochemical Parameters to Evaluate the Drinking Water Quality in the State of Perak, Malaysia. J Chem, vol. 2015, Article ID 716125:1–10. http://dx.doi.org/10.1155/2015/716125

Rathore D, Nizami AS, Singh A, Pant D (2016) Key issues in estimating energy and greenhouse gas savings of biofuels: challenges and perspectives. Biofuel Res J 10:380–393. DOI: 10.18331/BRJ2016.3.2.3

Rehan M, Nizami AS, Shahzad K, Ouda OKM, Ismail IMI, Almeelbi T, Iqbal T, Demirbas A (2016) Pyrolytic liquid fuel: a source of renewable energy in Makkah. Accepted in Energy Sources, Part A: Recovery, Utilization, and Environmental Effects. doi:10.1080/15567036.2016.1153753

Sánchez ME, Menéndez JA, Domínguez A, Pis JJ, Martínez O, Calvo LF, Bernad PL (2009) Effect of pyrolysis temperature on the composition of the oils obtained from sewage sludge. Biomass Bioenerg 33:933–940

Sadaf Y, Nizami AS, Batool SA, Chaudhary MN, Ouda OKM, Asam ZZ, Habib K, Rehan M, Demibras A (2015) Waste-to-energy and recycling value for developing integrated solid waste management plan in Lahore. Energy Sources, Part B: Economics, Planning, and Policy. doi:10.1080/1556249.2015.105295

Sarker M, Kabir A, Rashid MM, Molla M, Mohammad ASMD (2011) Waste polyethylene terephthalate (PETE-1) conversion into liquid fuel. J Fundam Renew Energy Appl 1:5

Serrano DP, Aguado J, Escola JM, Rodriguez JM (2002) European trends in the feedstock recycling of plastic wastes, Stud. Surf Sci Catal 142-B: 77–84

Serrano DP (2007) The current status of plastics recycling in Europe. In: Proceedings of ISFR, Jeju, Korea September 16–20

Shah SH, Khan ZM, Raja IA, Mahmood Q, Bhatti ZA, Khan J, Farooq A, Rashid N, Wu D (2010) Low temperature conversion of plastic waste into light hydrocarbons. J Hazard Mater 179:15–20

Shahzad K, Rehan M, Ismail IMI, Sagir M, Tahir MS, Bertok B, Nizami AS (2015) Comparative life cycle analysis of different lighting devices, Chem Eng Trans, 45:631–636. doi: 10.3303/CET1545106

Sharma BK, Moser BR, Vermillion KE, Doll KM, Rajagopalan N (2014) Production, characterization and fuel properties of alternative diesel fuel from pyrolysis of waste plastic grocery bags. Fuel Process Technol 122:79–90

Siddiqui MN, Redhwi HH (2009) Pyrolysis of mixed plastics for the recovery of useful products. Fuel Process Technol 90:545–552

Sriningsih W, Saerodji MG, Trisunaryanti W, Triyono Armunanto R, Falah II (2014) Fuel production from LDPE plastic waste over natural zeolite supported Ni, Ni-Mo, Co and Co-Mo metals. Procedia Environ Sci 20:215–224

Syamsiro M, Hu W, Komoto S, Cheng S, Noviasri P, Prawisudha P, Yoshikawa K (2013) Co-production of liquid and gaseous fuels from polyethylene and polystyrene in a continuous sequential pyrolysis and catalytic reforming system. Energy Environ Res 3(2). ISSN 1927-0569

Syamsiro M, Saptoadi H, Norsujianto T, Noviasri Cheng S, Alimuddin Z, Yoshikawa K (2014) Fuel oil production from municipal plastic wastes in sequential pyrolysis and catalytic reforming reactors. Energy Procedia 47:180–188

Tahir MS, Shahzad K, Shahid Z, Sagir M, Rehan M, Nizami AS (2015) Producing methane enriched biogas using solvent absorption method. Chem Eng Trans, 45:1309–1314. doi:10.3303/CET1545219

Thorat PV, Warulkar S, Sathone H (2013) Thermofurl—"Pyrolysis of waste plastic to produce liquid hydrocarbons". Adv Polym Sci Technol Int J. ISSN 2277-7164

UNWWD (2014) The United Nations world water development report. Water and energy facts and figures. Available from http://unesdoc.unesco.org/images/0022/002269/226961E.pdf

US-EIA (2014) US energy information administration. Country analysis brief, Saudi Arabia Available from: http://www.eia.gov/countries/cab.cfm?fips=sa

Walendziewski J (2002) Engine fuels derived from waste plastics by thermal treatment. Fuel 81:473–481

Wang JL, Wang LL (2011) Catalytic pyrolysis of municipal plastic waste to fuel with nickel-loaded silica-alumina catalysts. Energy Sources Part A 33:1940–1948

Wang LH, Zhang Y, Song LN (2006) Experimental research on pyrolysis process of waste rubber. Liaoning Gongcheng Jishu Daxue Xuebao (Ziran Kexue Ban)/J.Liaoning Tech Univ (Natural Science Edition) 25: 336–338

Wang WL, Chang JM, Cai LP, Shi SQ (2014) Quality improvement of pyrolysis oil from waste rubber by adding sawdust. Waste Manag 34:2603–2610

Wang ZM, Huang SS (2008) Novel 300 kW arc plasma inverter system based on hierarchical controlled building block structure. China Weld (English Edition) 7:12–16

WEC (2013) World Energy Council for sustainable energy: world energy resources survey. Available from https://www.worldenergy.org/wp-content/uploads/2013/09/Complete_WER_2013_Survey.pdf

WHO: World Health Organization (2014) Available Online http://www.who.int/gho/urban_health/situation_trends/urban_population_growth_text/en/

Williams PT (2005) Waste treatment and disposal. Wiley, Chichester, UK. https://nebm.ist.utl.pt/repositorio/download/2507

Williams PT (2006) Yield and composition of gases and oils/waxes from the feedstock recycling of waste plastic. In: Scheirs J, Kaminsky W (eds) Feeds tock recycling and pyrolys is of waste plastics: converting waste plastics into diesel and other fuels. Wiley, West Sussex, pp 285–309

Williams PT, Williams EA (1999) Interaction of plastics in mixed-plastics pyrolysis. Energy Fuels 13:188–196

Wongkhorsub C, Chindaprasert N (2013) A comparison of the use of pyrolysis oils in diesel engine. Energy Power Eng 5:350–355

Xiong S, Zhuo J, Zhang B, Yao Q (2013) Effect of moisture content on the characterization of products from the pyrolysis of sewage sludge. J Anal Appl Pyrol 104:632–639

Yang X, Sun L, Xiang J, Hu S, Su S (2013) Pyrolysis and dehalogenation of plastics from waste electrical and electronic equipment (WEEE): a review. Waste Manag 33:462–473

Yin C (2002) Microwave-assisted pyrolysis of biomass for liquid biofuels production. Bioresour Technol 120:273–284

Yoshioka T, Grause G, Eger G, Kaminsky W, Okuwaki A (2004) Pyrolysis of polyethylene terephthalate in a fluidised bed plant. Polym Degrad Stab 86:499–504

Zhao L, Chen DZ, Wang ZH, Ma XB, Zhou G (2011) Pyrolysis of waste plastics and whole combustible components separated from municipal solid wastes: comparison of products and emissions. In: Proceedings of the thirteen international waste management and landfill symposium, Sardinia, 3–6 Oct 2011, pp 117–118

Zhou G, Li J, Yu Y, Li X, Wang Y, Wang W, Komarneni S (2014) Optimizing the distribution of aromatic products from catalytic fast pyrolysis of cellulose by ZSM-5 modification with boron and co-feeding of low-density polyethylene. Appl Catal A 487:45–53

Turning Food Waste into Biofuel

Wen-Jing Deng

Abstract Food waste (FW) causes economic and environmental problems worldwide. Currently, most food waste is landfilled or incinerated for possible energy recovery. However, these methods have serious adverse effects on the environment. FW is nutritionally rich and offers a unique microbial feedstock for the production of numerous valuable bioproducts. The aim of this review is to investigate the technologies used to convert FW to forms of renewable energy such as biodiesel, ethanol, hydrogen and methane. Life-cycle assessment is performed to examine and compare the environmental effects of various methods of FW conversion.

Keywords Food waste · Biodiesel · Ethanol · Hydrogen · Methane

1 Introduction

Food waste (FW) is organic waste discharged from various sources, such as food-processing plants, domestic and commercial kitchens, cafeterias and restaurants. It comprises the 'end products of various food-processing industries that have not been recycled or used for other purposes', which are regarded as 'the non-product flows of raw materials whose economic value is less than the cost of collection and recovery for reuse; therefore discarded as waste' (Ezejiofor et al. 2014). According to the Food and Agricultural Organisation (FAO), between 1.3 and 1.6 billion tonnes of food, such as fresh vegetables, fruit, and meat, bakery and dairy products, are lost along the food-supply chain each year. This accounts for one third of the food produced globally for human consumption. Every year, food

W.-J. Deng (✉)
Department of Science and Environmental Studies,
The Education University of Hong Kong, Tai Po, NT, Hong Kong
e-mail: wdeng@ied.edu.hk

O.P. Karthikeyan et al. (eds.), *Recycling of Solid Waste for Biofuels and Bio-chemicals*, Environmental Footprints and Eco-design of Products and Processes, DOI 10.1007/978-981-10-0150-5_13

worth billions of dollars is wasted by the developed economies of the world, whereas the greatest loss occurs during food production in low-income countries. However, the problem does not end with food wastage. More than 95 % of FW ends up at landfill sites, where it is converted into methane, carbon dioxide and other greenhouse gases (GHGs) by anaerobic digestion (AD). The negative consequences for the environment are severe. FW generation is expected to increase over the next 25 years due to economic development and population growth, mainly in Asian countries. In addition, when FW is buried at landfill sites, its high energy value and the opportunity to obtain a carbon-rich feedstock through regeneration are lost. Kiran et al. (2014) summarised typical foods wasted in Asia-Pacific countries and around the world, as shown in Table 1.

In a report by the FAO, food produced but not eaten was estimated to add an annual 3.3 billion tonnes of GHGs to the planet's atmosphere. In addition to its adverse environmental effects, the wastage of food costs producers (excluding fish and seafood producers) $750 billion per year. With global population growth and industrialisation, the demand for energy around the world is rapidly increasing. World energy consumption is expected to increase by 50 % to 180,000 GWh/yr by 2020 (Fernando et al. 2006), primarily due to the rising demand in rapidly growing Asian countries such as China and India (Khanal 2008).

Fossil fuels, such as petroleum, natural gas and coal, are non-renewable, limited energy sources, and as such cannot meet the continuously increasing demand for energy. According to a report released by the Intergovernmental Panel on Climate Change (IPCC 2014), fossil-fuel combustion already accounts for 57 % of the emissions responsible for global warming. Therefore, it is increasingly important to search for new sustainable and renewable sources that both provide sufficient energy and are minimally harmful to humans and the environment. FW is rich in nutrients, mainly carbohydrate polymers (starch, cellulose and hemicellulose), proteins, lipids, organic acids, smaller inorganic parts and other compounds. Sugar and protein make up 35.5–69 and 3.9–21.9 % of FW, respectively. The effects of global climate change, the rapid economic development of emerging markets and the gradual but inexorable depletion of fossil-fuel resources have directed researchers' attention to FW as a sustainable resource providing renewable carbon for the industrial production of chemicals, polymers and other materials, energy and fuel. FW has been used as a unique microbial feedstock for the production of numerous valuable bioproducts, such as methane, hydrogen, ethanol, biopolymers and bioplastics. FW can also be used to produce high-value materials such as organic acids, biodegradable plastics and enzymes. However, the market demand for these chemicals is much smaller than that for biofuel. Therefore, we focus on techniques developed to produce biofuels such as biodiesel, ethanol, hydrogen and methane from FW. We review the existing literature in this area and compare numerous studies of the conversion of FW to various types of fuel. In addition, life-cycle assessment is conducted to examine and compare the environmental effects of several conversion methods.

Table 1 Typical wasted foods in several Asia-Pacific countries and around the globe by Kiran et al. (2014)

Waste (KT)	World	Asia	South-eastern Asia	Australia	Cambodia	China	Indonesia	Japan	Malaysia	New Zealand	North Korea	Philippines	South Korea	Thailand	Vietnam
Cereal	95,245	52,374	12,599	1380	506.1	18,990	4.6	413.4	183.4	28.6	253.0	215.7	628.4	1999	2706
Rice	26,738	22,668	10,792	0.4	506.0	6046	3.3	139.4	50.2	NR	NR	162.7	458.2	1997	2478
Sugar	459.9	188.9	151.7	93.6	NR	0.4	NR	20.8	NR	NR	NR	NR	NR	151.7	NR
Pulses	2735	1134	241.6	36.0	0.9	142.3	38.0	7.1	NR	1.2	10.3	NR	2.0	7.0	8.6
Oil crops	18,424	13,590	2515	3.9	3.8	9017	2238	69.6	1.4	0.1	15.2	NR	12.7	159.4	30.5
Vegetable oil	616.1	269.3	116.9	NR	NR	133.4	NR	13.0	116.9	NR	NR	NR	NR	NR	NR
Vegetables	81,441	59,949	2710	54.1	46.9	39,286	755.0	1224	64.8	73.2	414.2	242.5	1555	339.5	777.2
Beans	1049	447.3	218.1	1.1	0.9	49.1	37.2	6.5	NR	0.2	10.3	2.2	1.6	3.7	5.2
Onions	5891	3877	186.0	14.6	NR	139.5	2107	99.9	68.1	NR	NR	3.5	6.9	5.5	22.7
Peas	412.7	145.1	2.1	7.2	NR	39.9	NR	0.4	NR	1.1	NR	0.3	0.1	0.1	NR
Tomatoes	12,874	7415	104.2	NR	NR	3181	85.3	100.7	1.6	9.5	8.3	9.9	57.6	7.3	NR
Potatoes	62,229	12,912	466.1	23.6	NR	7501	250.0	177.0	NR	10.9	156.0	34.4	95.3	9.0	83.3
Fruits	53,796	28,328	4529	30.9	30.5	8323	2706	749.0	89.1	43.4	153.5	1183	276.6	786.4	531.0
Apples	5742	4116	13.2	5.9	NR	3192	3.1	84.6	NR	22.4	72.8	3.8	49.0	1.2	5.1
Banana	13,532	8544	1896	5.4	7.8	949.3	637.4	213.0	56.1	7.6	NR	901.3	NR	153.7	137
Coconut	3038	2488	2159	NR	NR	20.5	2066	NR	1.3	NR	NR	7.8	NR	69.1	0.9
Pineapple	1829	579	431.9	NR	2.2	97.7	NR	15.4	NR	0.3	NR	109.9	2.8	189.5	50
Coffee	105.0	33.3	28.3	NR	NR	0.03	20.9	NR	0.6	NR	NR	6.4	NR	NR	NR
Milk	16,560	10,887	183.3	NR	1.5	1447	45	NR	3.8	164.8	4.9	NR	42.4	25.2	9.5
Cream	33.9	0.1	NR	NR	NR	0.1	NR	NR	NR	NR	NR	NR	NR	NR	NR
Butter	84.0	1.7	NR	NR	NR	NR	NR	NR	NR	NR	NR	NR	NR	23.1	NR
Animal fats	174.1	1.8	NR	NR	NR	0.1	NR	NR	NR	NR	NR	NR	NR	NR	NR
Meat	1184	183.2	NR	NR	NR	NR	NR	107.2	NR	NR	NR	NR	107.2	23.1	NR
Offal	63.0	19.6	NR	8.7	NR	NR	NR	NR	NR	NR	NR	NR	NR	NR	NR
Poultry meat	97.5	61.2	NR	NR	NR	NR	34.5	NR	NR	NR	NR	NR	NR	23.1	NR
Total FW (MT)	1300	278	≥ 79.3	≥ 6.34	2.50	82.80	≥ 30.90	16.40	3.36	≥ 1.25	5.19	≥ 12.00	4.91	8.6	11.55

FW Food waste; *KT* Kilotons; *MT* Million tonnes; *NR* Not reported

2 Biodiesel Production

The possibility of using biodiesel as a fuel was first recognised during global efforts to identify and develop alternative energy resources in response to the worldwide depletion of non-renewable energy sources (Berchmans and Hirata 2008). The demand for biodiesel is also increasing as a result of the adverse environmental effects of emissions from conventional fossil-fuel combustion and a decline in domestic oil production (Mondala et al. 2009). Biodiesel, which has been shown to have great potential as an alternative fuel, can be produced from FW by either direct transesterification, using alkaline or acid catalysts, or the transesterification of microbial oils by various oleaginous microorganisms. Commercially, biodiesel is produced by transesterification in the presence of alcohol and a catalyst. This process involves the conversion of triglycerides (oil) to methyl esters (biodiesel) and a by-product (glycerol) (Low et al. 2011). Biodiesel is biodegradable and non-toxic, releases little sulphur when burned, has a carbon monoxide-free and aromatic-free emission profile, and may even benefit the environment by providing a means of recycling spent oils and fats (Siles et al. 2010). Waste cooking oil (WCO) comprises various types of residue from kitchens, restaurants, food factories and even human and animal waste; it is not only detrimental to human health but causes environmental pollution. The production of biodiesel from FW as a partial substitute for petroleum diesel is one means of addressing the twin problems of environmental pollution and energy shortage. Chen et al. (2008) examined the use of immobilised *Candida lipase* to produce biodiesel from WCO in a three-step fixed-bed reactor. The main product was biodiesel, consisting of fatty acid methyl esters. The microbial oils produced by many yeast strains can be used as a substitute for plant oils, due to their similar fatty-acid compositions, or as raw materials for biodiesel production. Papanikolaou et al. (2011) cultivated five strains of *Aspergillus sp.* and one strain of *Penicillium expansum* fungi on waste cooking olive oil to produce lipid-rich biomass. The amount of lipid accumulation was found to depend on the culture used; one strain of *Aspergillus sp.* accumulated up to 64.0 % (w/w) of lipids in dry fungal mass, with a productivity of 0.74 g/g. Pleissner et al. (2013) recovered glucose, free amino nitrogen and phosphate from FW by fungal hydrolysis using *Aspergillus awamori* and *Aspergillus oryzae* as raw materials for biodiesel production. The production yield was 31.9 % for glucose, 0.28 % for free amino nitrogen and 0.38 % for phosphate. Ultimately, the fermentation process yielded 10–20 g of biomass, rich in carbohydrates, lipids, proteins and saturated and polyunsaturated fatty acids. FW hydrolysate was thus found to offer a suitable culture medium and nutrient source for fungal and yeast cultivation. The conventional process flow of biodiesel production is shown in Fig. 1 (Daud et al. 2015). Other studies on biodiesel production from FW arelisted in Table 2.

The other important material used to produce biodiesel is palm oil. More than 90 % of Southeast Asia's biodiesel is produced by transesterification, with palm oil as the raw material. This simple and rapid chemical reaction yields a high rate of

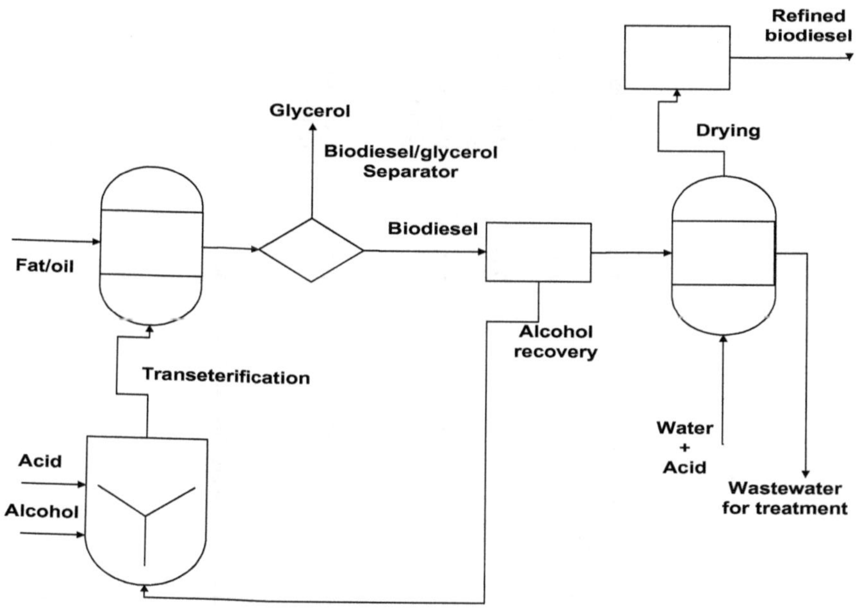

Fig. 1 Conventional process flow of biodiesel production (Daud et al. 2015)

conversion of oil (triglycerides) to biodiesel (methyl esters) (Marchaetti et al. 2007). However, the untreated biodiesel produced by transesterification contains several impurities. A traditional purification method is wet washing: water or a weak acid is used to remove some of the excess contaminants and leftover production chemicals from the biodiesel. Although this method has several advantages, adding water to the process has many disadvantages, such as increased cost and production time, the generation of a highly polluting effluent (wastewater) that must be treated prior to environmental discharge and a significant loss of biodiesel in the wastewater phase (Berrios and Skelton 2008; Canakci et al. 2001). It has been observed that biodiesel production exceeding 350,000 L/day can yield more than 70,000 L of biodiesel-contaminated wastewater per day (Jaruwat et al. 2010). It is thus necessary to develop a more efficient technique for wastewater treatment in biodiesel plants.

3 Ethanol Production

The demand for ethanol has recently increased around the world, due to its wide range of potential industrial applications. Ethanol is one of the most important renewable fuels. Its use helps to reduce the pollution generated by the worldwide utilisation of fossil fuels, and incurs smaller costs than other fuel types. Ethanol is mainly used as a chemical feedstock to produce ethylene, which is a key raw

Table 2 Biodiesel production from FW

Waste	Microorganism	Vessel type	Pre-treatment	Conditions	Duration (h)	Y (g cell/g FW)	Y (g/g FW)	Reference
Waste, cooking olive oil	A. niger NRRL363	SmF-250 mL flasks	Filtration	28 °C, pH 6, 200 rpm	5	1.2	0.6	Papanikolaou et al. (2011)
Waste, cooking olive oil	A. niger NRRL363	SmF-250 mL flasks	Filtration	28 °C, pH 6, 200 rpm	8	1.15	0.74	Papanikolaou et al. (2011)
FW	Schizochytrium mangrovei	SmF-2 L bioreactor	Fungal hydrolysis by A. oryzae and A. awamori, autolysis	25 °C, pH 6.5, 400 rpm	4	NR	NR	Pleissner et al. (2013)
FW	Chlorella pyrenoidosa	SmF-2 L bioreactor	Fungal hydrolysis by A. oryzae and A. awamori, autolysis	28 °C, pH 6.5, 400 rpm	4	NR	NR	Pleissner et al. (2013)

FW Food waste; *Y* Yield; *P* Productivity; *SmF* Submerged fermentation; *l* Specific growth rate; *A* Aspergillus; *NR* Not reported material in the production of polyethylene and other plastics, with a market demand of more than 140 million tonnes per year. Therefore, methods of producing bioethanol from cheap feedstock have gained much attention from researchers (Lundgren and Hjertberg 2010). Traditionally, bioethanol has been produced from starch-rich crops such as potato, rice, and sugar cane (Thomsen et al. 2003). Commercial enzymes, particularly *Saccharomyces cerevisiae*, can easily be used to convert starch to glucose and ferment the resulting glucose to ethanol. Another important resource for bioethanol production is cellulose, but the hydrolysis of cellulose is a more complex process. If FW contain large quantities of cellulosic feedstock, hydrolysis is much more difficult. Many studies have been performed to identify low-cost and high-efficiency methods of producing ethanol from abundant and cheap forms of waste such as lignocellulosic waste, municipal waste and FW (Jensen et al. 2011). Table 3 displays the findings obtained on the processes and yield associated with producing ethanol from different types of FW.

Materials

Various types of FW, such as starchy products from bakeries and food-processing plants, cheese whey from dairy-processing industries, mandarin waste and banana peel, can be used as raw materials for ethanol production. Over the last two decades, numerous studies have been conducted to examine the use of various forms of FW to produce ethanol. For example, Kumar et al. (1998) used cheese whey as a raw material, and Walker et al. (2012) used restaurant waste.

Enzymes

Three commercial enzyme solutions, fungal α-amylase, cellulose and glucoamylase, are widely used in the production of ethanol. Their optimum temperature ranges from 50 to 60 °C for fungal α-amylase, from 55 to 60 °C for cellulose and from 55 to 60 °C for glucoamylase. Their optimum pH ranges from 4.0 to 6.5 for fungal α-amylase, from 4.8 to 5.2 for cellulose and from 4.0 to 4.5 for glucoamylase. A mixture of the three enzymes is most effective for substrates with a large molecular weight.

Pre-treatment

The results of many studies indicate that pre-treatment may not be necessary to convert FW to ethanol prior to enzymatic hydrolysis. Several researchers have found that instead of pre-treatment, autoclaving FW before fermentation is often necessary to improve product yield and purity, albeit at the cost of energy and water consumption. It should be noted that thermal treatment may partially degrade sugars and other nutritional components, thereby reducing the amount of useful sugars and amino acids produced (Sakai et al. 2006). Fresh and wet FW has been reported to be more effective than re-wetted dried FW (Kim et al. 2005), mainly due to the smaller specific surface area of the dried substrate, which reduces the efficiency of the reaction between enzymes and substrate. Therefore, it is preferable to use FW without a drying pre-treatment as long as microbial contamination is manageable. In the absence of thermal sterilisation, acid-tolerant ethanol producing microorganisms

Table 3 Ethanol production from FW (Kiran et al. 2014)

Waste	Method	Vessel type	Pre-treatment	Microorganism	Duration (h)	Y (g RS/100 g FW)	Y (g/g FW)	Reference
Bakery waste	Simultaneous	14 L fermenter	None	*S. cerevisiae*	14	54	0.25	Kumar et al. (1998)
FW	Repeated batch simultaneous	1 L fermenter with 0.8 L working vol	None	*S. cerevisiae* ATCC26602	264	12.3	0.06	Ma et al. (2007)
Mandarin waste, banana peel	Simultaneous	500 mL flask	Drying, steam explosion	*S. cerevisiae* Anr, *Pachysolen tannophilus*	24	25.2	0.11	Sharma et al. (2007)
FW	Separate	500 mL flask100 mL working vol	None	*S. cerevisiae* KA4	16	23.4	0.12	Kim et al. (2008)
FW	Simultaneous	Flask with 100 g FW	None	*S. cerevisiae*	48	11.25	0.08	Ma et al. (2007)
FW	Separate continuous	Tower shaped reactor, 0.45 L working vol	LAB spraying	*S. cerevisiae* strain KF-7	15	11.7	0.03	Tang et al. (2008)
FW	Simultaneous	Flask with 100 g FW	None	*S. cerevisiae*	67.6	34.8	0.23	Wang et al. (2008)
FW	Continuous simultaneous	Fermenter with 4.3 kg FW	LAB spraying	*S. cerevisiae* KF7	25	36.4	0.09	Koike et al. (2009)
FW	Simultaneous	1 L fermenter with 0.8 L working vol	None	*S. cerevisiae* KRM-1	48	8.9	0.06	Ma et al. (2009)

(continued)

Table 3 (continued)

Waste	Method	Vessel type	Pre-treatment	Microorganism	Duration (h)	Y (g RS/100 g FW)	Y (g/g FW)	Reference
FW	Repeated batch simultaneous	250 mL flask 150 mL working vol	None	*Zymomonas mobilis* GZNS1	14	15.4	0.07	Ma et al. (2008)
FW	Simultaneous	250 mL flask 200 mL working vol	None	*S. cerevisiae*	48	60	0.36	Hong et al. (2011)
FW	Separate	5 L fermenter with working volume of 3 L	None	*S. cerevisiae*	24	27	0.16	Kim et al. (2011)
FW	Synchronous saccharification	Fermenter with 200 g FW	None	*Saccharomyces italicus* kJ	352	12.5	NR	Li et al. (2011)

NR Not reported; *FW* Food waste; *LAB* Lactic acid bacteria; *RS* Reducing sugar; *Y* Yield; *P* Productivity such as *Zymomonas mobilis* have been used to ferment FW, because acidic conditions prevent microbial contamination and putrefaction (Koike et al. 2009; Ye et al. 2008).

Enzyme Digestion

As yeast cells cannot ferment starch or cellulose directly into bioethanol, the efficiency of the conversion of FW to ethanol depends on the extent of carbohydrate saccharification (Tubb 1986). A mixture of α-amylase, β-amylase, and glucoamylase of various origins is best suited for use with substrates with a large molecular weight. Pullulanase has recently also been added to the list of saccharifying enzymes (Tomasik and Horton 2012). Small fermentable sugars (e.g. maltose, amylose, glucose, maltose syrup and fructose) can be produced during saccharification, and forms of cellulase and xylanase, such as endoglucanase, exoglucanase, β-glucosidase and β-xylosidase, can also be used to improve the hydrolysis of cereals to convert starches to glucose.

Procedure

After enzyme digestion, samples are inoculated with *Saccharomyces cerevisiae* (yeast) to convert the glucose produced in the previous step to ethanol. Fed-batch and simultaneous saccharification and fermentation (SSF) methods have been developed to achieve high glucose concentrations and thus a high ethanol yield from low-cost FW. The fed-batch culture method is commonly used to produce highly concentrated reducing sugars that can easily be fermented to ethanol. Compared with the batch-culture method, the fed-batch configuration has been found to significantly improve the performance of saccharification and subsequently ethanol fermentation. In one study, for instance, the glucose bioconversion yield was shown to reach 92 % of its theoretical value (Yan et al. 2012). SSF is another promising means of converting cellulose into ethanol. During SSF, a cellulose-hydrolysing enzyme complex (cellulase) is combined with an ethanol-producing organism to simultaneously hydrolyse cellulose to glucose and convert glucose to ethanol in the same reactor. It is vital to identify the optimal fermentation conditions, because enzymes and fermenting microorganisms may have different optimum pH values and temperatures. Ma et al. (2009) investigated the SSF of kitchen garbage by the acid-tolerant *Zymomonas mobilis* without sterilisation. The ethanol yield was 10.08 g/L h; 15.4 g sugar was used per 100 g of garbage; and 0.49 g ethanol per g sugar was obtained within 14 h.

Methods of ethanol production have not differed significantly between studies listed in Table 3 (Kiran et al. 2014). Walker et al. (2012) used restaurant waste such as corn, potatoes and pasta to produce ethanol. The FW was converted to ethanol in two steps: first, the two-part enzymatic digestion of starch using α-amylase and glucoamylase, followed by the fermentation of the resulting sugars to ethanol using yeast. Low ethanol concentrations (0.8 %) were achieved due to the low initial composition of starch in the FW. Ethanol concentration increased with increasing enzyme dosage. The process flow is shown in Fig. 2.

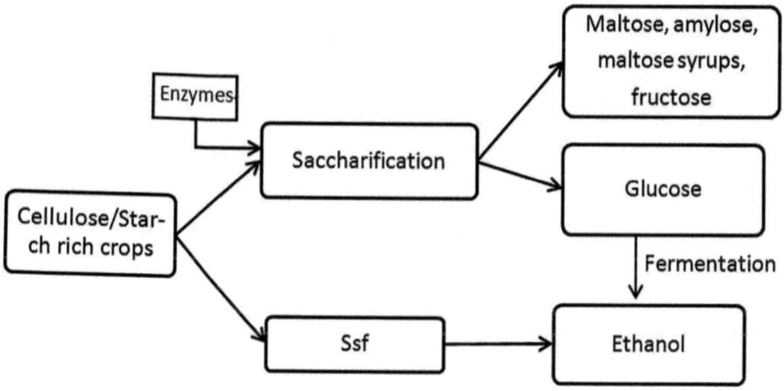

Fig. 2 The process flow of ethanol production from FW

4 Hydrogen Production

Hydrogen is a promising alternative to fossil fuels, as it is clean, renewable and has a high energy yield. Anaerobic fermentation is an environmentally friendly and energy-saving method of biohydrogen production. The anaerobic acidification of FW produces various organic acids, H_2, CO_2 and other intermediates. As the reactions involved in hydrogen production are rapid and do not require solar radiation, they can be used to treat large quantities of organic waste. Hydrogen is used as a compressed gas, with an energy yield (142.35 kJ/g) approximately 2.1 times greater than that of methane and 2.8 times greater than that of gasoline. Carbohydrate-rich FW is suitable for hydrogen production. In a review of earlier studies on the production of hydrogen from FW (Table 4), Kiran et al. (2014) reported that hydrogen yields ranged from 0.9 to 8.35 mol/mol hexose (Patel et al. 2012). Although hydrogen can be produced thermos-chemically and radio-lytically, the processes involved are not very economical. Hydrogen can be produced by water electrolysis only in areas in which cheap electricity is readily available. Therefore, the use of bacteria to produce hydrogen directly from renewable biomass and water has considerable potential to increase the use of hydrogen as a fuel. Hydrogen production by microorganisms can be divided into two main categories: one involves photosynthetic bacteria cultured under anaerobic light conditions, and the other involves chemotrophic bacteria. Chemotropic bacteria can generate hydrogen without photoenergy (Gray and Gest 1965). Shimizu et al. (2008) investigated hydrogen fermentation from FW in the absence of inocula. Hydrogen production occurred at an initial FW pH of 7.0–9.0 and an initial temperature of 22–35 °C (maximum production of 40 ml-H_2/g-TS at an initial pH of 9.0 and a temperature of 35 °C). In the next step, the pH was controlled by the addition of a NaOH solution. The optimal pH for hydrogen production was 5.0–6.0, due to the acceleration of butyrate fermentation in this range. Many factors may affect hydrogen production, such as materials, pre-treatment type and process configuration. The process flow of hydrogen production from FW is shown in Fig. 3.

Table 4 Hydrogen production from FW

Waste	Vessel type	Pretreatment	Inoculum	Duration (d)	OLR (kg VS/m^3 d)	OLR (kg COD/m^3)	Y (mL/g VS)	Reference
FW	Leaching bed reactor with 3.8 L working vol	None	HSSS	7	NR	NR	160	Han and Shin (2004)
FW with sludge	415 mL bottle with 200 mL working vol	None	HSSS	3	NA	NA	67	Kim et al. (2005)
FW	Bioreactor with 3 L working vol	None	Anaerobic SS	5	8	NR	125	Shin et al. (2004)
FW	Bioreactor with 3 L working vol	None	Anaerobic SS	60	3	NR	NR	Shin et al. (2005)
FW	CSTR with 10 L working vol	None	SS	150	38.4	64.4	283	Youn et al. (2005)
FW	1 L bioreactor with 500 mL working vol	None	Anaerobic SS	2	NA	NA	57	Chu et al. (2008)
FW	7.5 L bioreactor with 3 L working vol	Heat pre-treatment (90 °C, 20 min)	SS	3	NA	NA	153.5	Pan et al. (2008)
FW	ASBR with 4.5 L working vol	None	HSSS	NR	NR	NR	80.9	Kim et al. (2008)
FW	Bioreactor with 1 L working vol	None	SS	2	NA	NA	NR	Kim et al. (2011)
FW	Rotating drum with 200 L working vol	None	30	22.65	NR	NR	65	Li and Fang (2007)

FW Food waste; *Y* Yield; *P* Productivity; *SS* Seed sludge; *HSSS* Heat shocked seed sludge; *d* Day; *min* Minute; *NR* Not reported; *NA* Not applicable

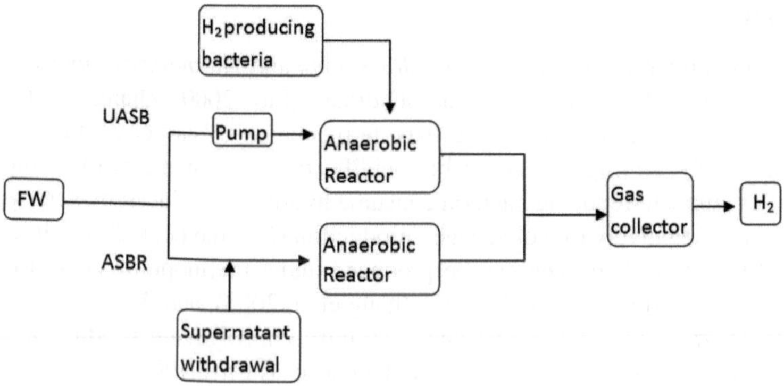

Fig. 3 The process flow of hydrogen production from FW

Materials

It has been reported that carbohydrate-based waste yields 20 times more hydrogen than that produced from fat-based and protein-based waste (Show et al. 2012). This difference may be due to the consumption of hydrogen to form ammonium using nitrogen generated from protein biodegradation. *Clostridium sporogenes* (*Clostridium sp.*) yields more hydrogen (1.61–2.36 mol/mol glucose) than other anaerobic bacteria, such as *Enterobacter* and *Bacillus* (Hawkes et al. 2002). Therefore, carbohydrate-based waste may be the ideal raw material for hydrogen production.

Pre-treatments

Many researchers have examined the use of mixed cultures to produce hydrogen from FW. However, hydrogen generated by *Clostridium* and *Enterobacter* is often used by hydrogenotrophic bacteria (Li and Fang 2007). The conditions of hydrogen-producing bacteria are an important determinant of hydrogen yield. FW itself may also be a source of hydrogen-producing microflora. Kim et al. (2008) tested several pre-treatments to select the optimal FW microflora for use in hydrogen production. Lactic-acid bacteria were found to be the most abundant species in untreated FW, whereas hydrogen-producing bacteria were dominant in pre-treated FW. Heat treatment is an effective means of suppressing lactate production and increasing hydrogen/butyrate production. Heat treatment is the most commonly used method of screening hydrogen-producing bacteria, despite its potential to increase cost. Researchers have investigated heat treatments ranging from 75 °C (Chang et al. 2002) to 121 °C (Wang et al. 2003) and from 15 min (Lay et al. 1999) to 2 h (Fan and Chen 2004). To date, optimal temperature and duration of heat treatments have not been investigated in detail. However, heat treatment is likely to increase costs in large-scale operations. Luo et al. (2010) investigated several methods of inoculum-based pre-treatment, and concluded that pre-treatment is not particularly important, as it has only short-term effects on hydrogen production.

Procedure

Fermentative microorganisms such as *Clostridium* and *Thermoanaerobacterium* are able to produce hydrogen from carbohydrates (Lay 2000; Zhang et al. 2003). Hydrogen can be generated by the dark fermentation of carbohydrate-rich waste (Khanal 2008). During the anaerobic acidification of organic wastes, methanogenesis or sulphate-reducing bacteria consume hydrogen produced by acidogenesis, contributing negatively to biohydrogen production (Mizuno et al. 2000). It has been reported that more hydrogen can be produced under thermophilic conditions than mesophilic conditions (Yu et al. 2002; Zhang et al. 2003), and that microorganisms such as *Thermoanaerobacterium* and *Thermosaccharolyticum* produce almost as much hydrogen through thermophilic acidogenesis as *Clostridium butyricum* (Ueno et al. 2001). In addition, thermophilic conditions have been reported to inhibit methanogenesis (Ueno et al. 1996). The whole metabolic pathway is affected by several factors, such as pH, temperature, carbon source, nutrients and retention time (Wang and Wan 2009).

Effective pH control is crucial to improve the germination of *Clostridium sp.*, and to initiate and operate a hydrogen-producing bioprocess (Han and Shin 2004). When FW is degraded during fermentation, pH decreases due to the production of volatile fatty acids (VFAs). Maintaining appropriate pH levels is crucial to ensure stable fermentation (Hawkes et al. 2002). In recent studies of the effects of pH on dark hydrogen fermentation, optimal pH values have been found to depend on experimental factors such as initial pH values, operational pH values, substrates, reactors, temperature and seed sludge (Wang and Wan 2009).

The effects of temperature on hydrogen production have been examined in many studies. Most researchers have found that hydrogen yield increases with temperature. Yu et al. (2002) found that the hydrogen yield from winery wastewater at 55 °C was 38 % higher than that at 20 °C. Morimoto et al. (2004) reported that the hydrogen yield from glucose at 60 °C (218 ml H_2/g hexose) was 60 % higher than that at 50 °C (136 ml H_2/g hexose). Valdez-Vazquez et al. (2005) reported that hydrogen-production yield at 55 °C was 437 ml H_2/g hexose, substantially higher than the yield at 37 °C (201 ml H_2/g hexose). Temperature has also been shown to affect the metabolic pathway, due to a change in the composition of by-products: mainly acetate and butyrate, but also propionate and ethanol. However, the results of studies investigating the effects of temperature on hydrogen production are not consistent.

Various fermentation systems have been developed to produce H_2 from FW, such as batch fermentation, semi-continuous fermentation, continuous fermentation and single- or multiple-stage fermentation (Hallenbeck and Ghosh 2009). High H_2 production rates have been reported in anaerobic sequencing batch reactors and up-flow anaerobic sludge blanket (UASB) reactors, due to their high biomass concentrations. During these processes, solid retention time (SRT) determines substrate-uptake efficiency, microbial size and composition, and metabolic pathways. A long SRT favours the growth of H_2 consumers, while a short SRT may reduce substrate uptake efficiency, active biomass retention and subsequently the efficiency of the overall process. If the optimal SRT can be achieved at a low hydraulic retention time (HRT), the productivity and technical feasibility of the hydrogen-production process are enhanced (Kim et al. 2008). Various optimal HRT values have been reported. Kim et al. (2008) investigated the effects of an SRT of 24–160 h and an HRT of 24–42 h on hydrogen production from FW. The maximum hydrogen yield, 80.9 mL/g volatile solid (VS), equivalent to 1.12 mol/mol hexose, was obtained at an SRT of 126 h and an HRT of 33 h.

As the conversion of FW into hydrogen has been shown to be highly dependent on operational conditions, effective control of operational conditions is essential to achieve efficient hydrogen production. A two-stage process may be possible: harvesting hydrogen at the acidification stage of anaerobic treatment and using the remaining acidification products, such as acetate and butyrate, to produce methane.

5 Methane Production

The biomethanation of FW has received attention for its potential to reduce the harmful effects of waste generation on the environment and human health. Biomethanation could play an important role in waste management, as it incurs low costs, produces little residual waste and is a renewable energy source (Morita and Sasaki 2012). In addition to producing biogas, biomethanation yields a nutrient-rich digestate as a by-product, which can be used as a fertiliser or soil conditioner. Optimal methane production is only achieved when microorganisms interact (Chartrain et al. 1993). Imbalances between microbial groups may not only reduce methane production but lead to process failure (Lee et al. 1999), due to the accumulation of intermediate compounds that inhibit methanogens. Several parameters affect methane yield and process stability, such as moisture, VS content, nutrient content, particle size and biodegradability. Omar et al. (2008) investigated the anaerobic treatment of cattle manure in a laboratory-scale bioreactor, and observed an improvement in biogas yield of up to 0.207 m^3/kg VS added, with an average methane content of 65 %, as a result of adding palm oil mill effluent as an inoculum. Lee et al. (1999) used a 5 L continuously fed digester to convert FW to methane, resulting in 70 % VS conversion with a methane yield of 440 mL/g VS. Gunaseelan (2004) reported methane production from 54 types of FW and vegetable waste ranging from 180–732 ml/g VS, depending on the origin of the waste.

AD

Process configuration is a very important determinant of the efficiency of methane production. The processes reported in the literature fall into two main categories: single-stage AD and two-stage AD. Single-stage AD is widely used for municipal solid waste (MSW) treatment. All of the reactions take place simultaneously in a single reactor, and the system experiences less frequent technical failures and incurs smaller investment costs than its two-stage counterpart. In contrast, two-stage AD is often used to produce hydrogen and methane in separate reactors (Chu et al. 2008).

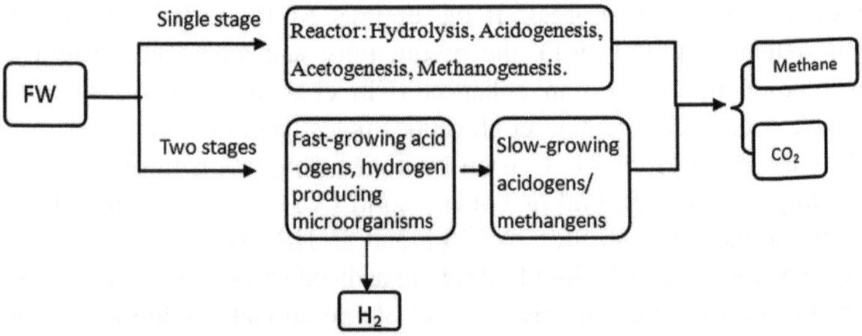

Fig. 4 The process flow of methane production from FW

During the first stage, fast-growing acidogens and hydrogen-producing microorganisms are enriched to produce hydrogen and VFAs. During the second stage, slow-growing acetogens and methanogens are built up and used to convert the VFAs to methane and carbon dioxide. Anderson et al. (1994) compared traditional single-phase AD with two-phase AD, and found the latter to perform better in terms of acid production and methane production. Speece (1996) investigated the AD of ethanol, and found that the maximum load and removal capacity of chemical oxygen demand (COD) during two-phase AD were twice those during single-phase AD (Speece 1996). Lee and Chung (2010) also proved that the two-stage process of hydrogen/methane fermentation has significantly greater energy-recovery potential than methane-only fermentation. Park et al. (2008) compared the use of single-phase and two-phase thermophilic systems (SPS and TPS, respectively) to produce methane from artificial kitchen waste. In both the SPS and TPS, the highest methane recovery ratio of 90 %, measured by dichromate COD (CODcr) analysis, was observed at an organic loading rate of 15 g CODcr/L d. The process flow of methane production from FW is shown in Fig. 4.

Reactor choice

To attain high loading, immobilise microbial consortia and stabilise methanogenesis, packed bed reactors (PBRs) and fixed bed systems have been developed. Kastner et al. (2011) compared two methods of generating biogas from organic waste materials to identify the most appropriate process, reactor system and parameters. Fermentation experiments were carried out in two types of reactors: a continuous stirred tank reactor (CSTR) and a fluidised bed reactor (FBR). Biogas fermentation yielded 670 NL/kg VS with the CSTR and 550 NL/kg VS with the FBR. Biogas productivity was 3.9 NL/L d with the CSTR and 3.4 NL/L d with the FBR. The average methane concentration was approximately 60 % for both reactor systems. Parawira et al. (2005) investigated the mesophilic AD of solid potato waste in two systems: a solid-bed reactor for hydrolysis/acidification connected to a UASB methanogenic reactor, and a solid-bed reactor connected to a methanogenic PBR packed with wheat straw biofilm carriers. Although the second system degraded the organic materials more rapidly than the first system, its methane yield was 390 mL/g VS, and cumulative methane production was the same for the two systems. Moon and Song (2011) investigated the enzymatic solubilisation and methane-production potential of FW by enzymatically hydrolysing FW liquor in a UASB reactor. They identified the optimum conditions for FW hydrolysis in the reactor, and showed that methane production via enzymatic hydrolysis of FW in a UASB reactor may offer a novel method of obtaining high-value biogas by AD.

Increasing the proportion of methane in biogas is an important means of increasing energy generation and reducing the amount of CO_2 released. Biogas is typically composed of 60 % methane and 40 % CO_2. The energy value of biogas is determined by its methane concentration. The GHG heating factor of methane is 21 times higher than that of CO_2. Combustion of biogas converts methane into CO_2 and reduces GHG emission by more than 20 fold. When methane is extracted from waste and used to produce heat and/or electricity, the waste does not degrade in an open environment, thereby reducing direct methane atmospheric emissions. Moreover, due to its high energy yield, the use of biogas is likely to supplant that of fossil fuels, which are the main contributor to GHG emissions. Studies on methane production from FW are listed in Table 5.

6 Life-cycle Assessment of Conversion Methods

Kalogo et al. (2007) conducted life-cycle assessment to compare the environmental effects of waste-derived ethanol with those of gasoline, corn ethanol and crop-cellulosic ethanol. The authors also compared these effects with the environmental consequences of landfilling solid waste. An E85 waste-ethanol blend (blended with 15 % gasoline) were found to reduce GHG emissions by 65 and 58 % compared with gasoline and corn ethanol, respectively. Converting organic waste to ethanol was shown to save 397–1830 MJ fossil energy per metric ton of waste, whereas landfilling consumes 177–577 MJ fossil energy per metric ton of MSW. Ebner et al. (2015) analysed the life-cycle GHG emissions associated with a novel method of converting food-processing waste into ethanol (EtOH) and two co-products: compost and animal feed. The life-cycle GHG emissions associated with the ethanol-production process were 1458 gCO_2e/L EtOH. In a study conducted in California, Chester and Martin (2009) assessed the production potential of ethanol from MSW. They measured the cost, energy and GHG emissions associated with the conversion of MSW to ethanol using the dilute-acid process developed by the US National Renewable Energy Laboratory (NREL) (Aden et al. 2002). Schmitt et al. (2012) also used the NREL's model as the basis for a new model of lignocellulosic biorefinery conversion. They assessed the technical and environmental factors associated with converting three lignocellulosic resources into ethanol.

Djomoa and Blumbergab (2011) used life-cycle assessment to quantify and compare the energetic and environmental performance of hydrogen produced from wheat straw (WS–H_2), sweet sorghum stalk (SSS–H_2) and potato steam peel

Table 5 Methane production from FW

Waste (KT)	Inoculum	Pretreatment	Process type	Vessel type	Duration	OLR (kg VS/m^3 d)	Biogas yield (mL/g VS)	CH_4 yield (mL/g VS)	% CH4	Efficiency (VS%)	Reference
Fruit and vegetable waste	Cow manure	None	Two stage	Bioreactor with 0.5 L working vol	29	1–9	NR	530	70	95.1	Lee et al. (1999)
FW	Anaerobic SS	Freeze drying of waste	Two stage	UASB with 8 L working vol	120	1.04	NR	277–482	NR	90	Heo et al. (2004)
FW	Anaerobic SS	None	Two stage	Continuous pilot scale 5 tons/d capacity	90	7.9	NR	440	70	70	Gunaseelan et al. (2004)
Fruit and vegetable waste	Anaerobic SS	None	Single stage	Serum bottles with 135 mL vol	100	NA	NR	180–732	NR	NR	Cho et al. (1995)
FW and activated sludge	Anaerobic SS	None	Single stage	Semi continuous reactor with 3.5 L working vol	250	2.43	NR	321	64.4	55.8	Parawira et al. (2005)
Potato waste	Anaerobic SS	None	Two stage	Packed bed with 1 L working vol	38	NR	NR	390	82	NR	Zhang et al. (2007)
FW	Anaerobic SS	None	Two stage	Bioreactor with 12 L working vol	60	8	NR	NR	68.8	86.4	Youn et al. (2005)
FW	Bacteria isolated from landfill soil and cow manure	None	Single stage	3 Stage semi continuous with 8 L working vol	30	NR	NR	NR	67.4	NR	Kim et al. (2008)
FW	Anaerobic SS	None	Single stage	Batch	28	NA	600	440	73	81	Forster-Carneiro et al. (2008)
FW	SS	None	Two stage	CSTR with 10 L working vol	150	6.6	NR	464	80	88	Kim et al. (2011)

FW Food waste; *SS* Seed sludge; *SsF* Simultaneous saccharification fermentation; *NR* Not reported; *NA* Not applicable

($SPP–H_2$). When the co-product was not considered, the GHG emissions were 5.60 kg CO_2eq kg^{-1} H_2 for $WS–H_2$, 5.32 kg CO_2eq kg^{-1} H_2 for $SSS–H_2$, and 5.18 kg CO_2eq kg^{-1} H_2 for $SPP–H_2$. These biohydrogen pathways reduced GHG emissions by 52–56 % compared with hydrogen production from diesel, and by 54–57 % compared with the production of hydrogen by steam-methane reforming. The energy ratios (ERs) were also comparable: 1.08 for $WS–H_2$, 1.14 for $SSS–H_2$ and 1.17 for $SPP–H_2$. Therefore, a shift from $SPP–H_2$ to $WS–H_2$ is not expected to affect the ERs and GHG emissions of these biohydrogen pathways.

Patterson et al. (2013) compared the environmental burdens associated with the production of biomethane vehicle fuel and a biohydrogen/biomethane blend from FW and wheat feed, based on data from two laboratory experiments. The two-stage system provided a high hydrogen yield (84.2 l H_2 kg^{-1} VS added) for FW treated by batch processes, but had a lower overall energy output than the single-stage system. A reduction in the environmental burden relative to that of diesel was achieved, partly due to the diversion of waste from landfill. The semi-continuously fed two-stage system gave a low hydrogen yield (7.5 l H_2 kg^{-1} VS added) for the wheat feed, but had a higher overall energy output. Both processes reduced the fossil-fuel burden, and further improvements to process efficiency will help to achieve a lower CO_2 burden than that of diesel.

7 Conclusions

Increasing attention has been paid to the management of FW due to the economic value and positive environmental effects of obtaining fuel from FW. We have discussed various means of bioconverting FW to ethanol, hydrogen, methane and biodiesel. Although the conversion of FW to fuel has been shown in many papers to be technically feasible, the related technology is still not mature. There is an urgent need to identify low-cost and environmentally beneficial approaches to the bioconversion of FW. The efficiency and economy of fuel production from FW could also be improved by conducting optimisation studies and other research on the integration of various value-added product manufacturing processes.

Acknowledgements This work was supported by the Hong Kong Early Career Start/General Research Fund (ECS/GRF 845212).

References

Aden A, Ruth M, Ibsen K, Jechura J, Neeves K, Sheehan J, Wallace R (2002) Lignocellulosic biomass to ethanol process design and economics utilizing co-current dilute acid prehydrolysis and enzymatic hydrolysis for corn stover. National Renewable Energy Laboratory

Anderson GK, Kasapgil B, Ince O (1994) Microbiological study of two stage anaerobic digestion start-up. Water Res 28:2383–2392

Berchmans HJ, Hirata S (2008) Biodiesel production from crude Jatropha curcas L. seed oil with a high content of free fatty acids. Bioresour Technol 99(6):1716–1721
Berrios M, Skelton RL (2008) Comparison of purification methods for biodiesel. Chem Eng J 144:459–465
Canakci M, Gerpen JV (2001) Biodiesel production from oils and fats with high free fatty acids. J Agric Saf Health 44(6):1429–1436
Chang WS, Hong S, Park J (2002) Effect of zeolite media for the treatment of textile wastewater in a biological aerated filter. Process Biochem 37:693–698
Chartrain M, Katz L, Marcin C, Thien M, Smith S, Fisher F, Goklen K, Salmon P, Brix T, Price K, Greasham R (1993) Purification and characterization of a novel bioconverting lipase from Pseudomonas aeruginosa MB 5001. Enzyme Microb Technol 15:575–580
Chester M, Martin E (2009) Cellulosic ethanol from municipal solid waste: a case study of the economic, energy, and greenhouse gas impacts in California. Environ Sci Technol 43 (14):5183–5189
Chen YM, Xiao B, Chang J, Wang XW (2008) Synthesis of biodiesel from waste cooking oil using immobilized lipase in fixed bed reactor. Energy Conver Manage 50(3):668–673
Chu CF, Li YY, Xu KQ, Ebie Y, Inamori Y, Kong HN (2008) A pH- and temperature-phased two-stage process for hydrogen and methane production from food waste. Int J Hydrogen Energy 33(18):4739–4746
Daud NM, Abdullah SRS, Hasan HA (2015) Production of biodiesel and its wastewater treatment technologies: A review. Process Saf Environ Prot 94:487–508
Djomoa SN, Blumbergab D (2011) Comparative life cycle assessment of three biohydrogen pathways. Bioresour Technol 102(3):2684–2694
Ebner JH, Rodrigo A, Labatut RA, Rankin MJ, Pronto JL, Gooch CA (2015) Lifecycle greenhouse gas analysis of an anaerobic codigestion facility. Environ Sci Technol 49:199–208
Ezejiofor TIN, Enebaku UE, Ogueke C (2014) Waste to wealth- value recovery from agrofood processing wastes using biotechnology: a review. Br Biotechnol J 4(4):418–481
Fan KS, Chen YY (2004) H2 production through anaerobic mixed culture: effect of batch S0/X0 and shock loading in CSTR. Chemosphere 57:1059–1068
Fernando S, Hall C, Saroj J (2006) NOx Reduction from Biodiesel Fuels. Energ Fuel 20:376–382
Gray CT, Gest H (1965) Biological formation of molecular hydrogen. Science 148:186–192
Gunaseelan VN (2004) Biochemical methane potential of fruits and vegetable solid waste feedstocks. Biomass Bioenergy 26(4):389–399
Hallenbeck PC, Ghosh D (2009) Advances in fermentative biohydrogen production: the way forward? Trends Biotechnol 27(5):287–97
Han SK, Shin HS (2004) Biohydrogen production by anaerobic fermentation of food waste. Int J Hydrogen Energy 29:569–577
Hawkes FR, Dinsdale R, Hawkes DL, Hussy I (2002) Sustainable fermentative hydrogen production: challenges for process optimization. Int J Hydrogen Energ 27(11/12):1339–1347
IPCC (2014). Climate change 2014: Mitigation of climate change
Jaruwat P, Kongjao S, Hunsom M (2010) Management of biodiesel wastewater by the combined processes of chemical recovery and electrochemical treatment. Energ Convers Manage 51 (3):531–537
Jensen JW, Felby C, Jørgensen H (2011) Cellulase hydrolysis of unsorted MSW. Appl Biochem Biotechnol 165:1799–1811
Kalogo Y, Habibi S, MacLean H, Joshi S (2007) Environmental implications of municipal solid waste-derived ethanol. Environ Sci Technol 41(1):35–41
Khanal SK (2008) Anaerobic biotechnology for bioenergy production, 1st edn. Wiley-Blackwell, Oxford, pp 189–190
Kim KC, Kim SJ, Kim MJ, Kim SJ (2005) Saccharification of food wastes using cellulolytic and amylolytic enzymes from *Trichodermaharzianum* FJ1 and its kinetics. Biotechnol Bioprocess Eng 10(1):52–59
Kim DH, Kim SH, Shin HS (2008) Hydrogen fermentation of food waste without inoculum addition. Enzyme Microbial Technol 45(3):181–187

Kim JH, Lee JC, Pak D (2011) Feasibility of producing ethanol from food waste. Waste Manag (Oxford) 31(9–10):2121–2125
Kiran EU, Trzcinski AP, Ng WJ, Liu Y (2014) Bioconversion of food waste to energy: a review. Fuel 134:389–399
Koike Y, An MZ, Tang YQ, Syo T, Osaka N (2009) Production of fuel ethanol and methane from garbage by high efficiency two-stage fermentation process. J Biosci Bioeng 108(6):508–512
Kumar JV, Shahbazi A, Mathew R (1998) Bioconversion of solid food wastes to ethanol. Analyst 123(3):497–502
Lay JJ (2000) Modeling and optimization of anaerobic digested sludge converting starch to hydrogen. Biotechnol Bioeng 68(3):269–278
Lay JJ, Lee YJ, Noike T (1999) Feasibility of biological hydrogen production from organic fraction of municipal solid waste. Wat Res 33(11):2579–2586
Lee JP, Lee JS, Park SC (1999) Two-phase methanization of food wastes in pilot scale. Appl Biochem Biotechnol 77–79:585–93
Lee YW, Chung JW (2010) Bioproduction of hydrogen from food waste by pilot-scale combined hydrogen/methane fermentation. Int J Hydrogen Energ 35:11746–11755
Li CL, Fang HHP (2007) Fermentative hydrogen production from wastewater and solid wastes by mixed cultures. Crit Rev Environ Sci Technol 37(3):1–39
Li H, Yang L, Kim YJ, Kim SJ (2011) Continuous ethanol production by the synchronous saccharification and fermentation using food wastes. Korean J Chem Eng 28(4):1085–1089
Low SC, Gan GK, Cheong KT (2011) Separation of methyl ester from water in a wet neutralization process. J Sustain Energy Environ 2:15–19
Lundgren A, Hjertberg T (2010) Ethylene from renewable resources. Surf Renew Resour 109–126
Luo G, Xie L, Zou Z, Wang W, Zhou Q (2010) Evaluation of pretreatment methods on mixed inoculum for both batch and continuous thermophilic biohydrogen production from cassava stillage. Bioresour Technol 101(3):959–964
Ma KD, Wakisaka M, Kiuchi T (2007) Repeated-batch ethanol fermentation of kitchen refuse by acid tolerant flocculating yeast under the non-sterilized condition. J Food Eng 8(4):275–279
Ma H, Wang Q, Zhang WY, Xu WL, Zou DX (2008) Optimization of the medium and process parameters for ethanol production from kitchen garbage by Zymomonasmobilis. Int J Green Energy 5(6):480–490
Ma K, Wakisaka M, Sakai K, Shirai Y (2009) Flocculation characteristics of an isolated mutant flocculent *Saccharomyces cerevisiae* strain and its application for fuel ethanol production from kitchen refuse. Bioresour Technol 100(7):2289–2292
Marchaetti JM, Miguel VU, Errazu AF (2007) Possible methods for biodiesel production. Renew Sustainable Energy Rev 11:1300–1311
Mondala A, Liang K, Toghiani H, Hernandez R, French T (2009) Biodiesel production by in situ transesterification of municipal primary and secondary sludges. Bioresour Technol 100 (3):1203–1210
Moon HC, Song IS (2011) Enzymatic hydrolysis of foodwaste and methane production using UASB bioreactor. Int J Green Energy 8(3):361–371
Morimoto M, Atsuko M, Atif A AY, Ngan MA, Fakhru'l-Razi A, Iyuke SE, Bakir AM (2004) Biological hydrogen production of hydrogen from glucose by natural anaerobic microflora. Int J Hydrogen Energ 29(7):709–713
Morita M, Sasaki K (2012) Factor influencing the degradation of garbage in methanogenic bioreactors and impacts on biogas formation. Appl microbiol biotechnol 94(3):575–582
Mizuno O, Dinsdale R, Hawkes FR, Hawkes DL, Noike T (2000) Enhancement of hydrogen production from glucose by nitrogen gas sparing. Bioresour Technol 73:59–65
Omar R, Harun RM, Mohd Ghazi TI, Wan Azlina WAKG, Idris A, Yunus R (2008) Anaerobic treatment of cattle manure for biogas production. In: Proceedings Philadelphia, annual meeting of American institute of chemical engineers, Philadelphia, USA, pp 1–10
Papanikolaou S, Dimou A, Dakas S, Diamantopoulou P, Philippoussis A, Galiotou-Panayotou M, Aggelis G (2011) Biotechnological conversion of waste cooking olive oil into lipid-rich biomass using Aspergillus and Penicillium strains. J Appl Microbiol 110:1138–1150

Parawira W, Mutro M, Read JS, Mattiasson B (2005) Profile of hydrolases and biogas production during two-stage mesophilic anaerobic digestion of solid potato waste. Process Biochem 40:2945–2952

Park Y, Hong, F, Cheon J, Hidaka T, Tsuno H (2008) Comparison of thermophilic anaerobic digestion characteristics between single-phase and two-phase systems for kitchen garbage treatment. J Biosci Bioeng 105(1):48–54

Patel SKS, Kumar P, Kalia VC (2012) Enhancing biological hydrogen production through complementary microbial metabolisms. Int J Hydrogen Energy 37(14):10590–10603

Patterson T, Esteves S, Dinsdale R, Guwy A, Maddy J (2013) Life cycle assessment of biohydrogen and biomethane production and utilisation as a vehicle fuel. Bioresour Technol 131:235–245

Pleissner D, Lam WC, Sun Z, Lin CSK (2013) Food waste as nutrient source in heterotrophic microalgae cultivation. Bioresour Technol 137:139–146

Sakai K, Ezaki Y (2006) Open L-lactic acid fermentation of food refuse using thermophilic Bacillus coagulans and fluorescence in situ hybridization analysis of microflora. J Biosci Bioeng 101(6):457–463

Schmitt E, Bura R, Gustafson R, Cooper J, Vajzovic A (2012) Converting lignocellulosic solid waste into ethanol for the State of Washington: an investigation of treatment technologies and environmental impacts. Bioresour Technol 104:400–409

Shimizu S, Fujisawa A, Mizuno O, Kameda T, Yoshioka T (2008) Fermentative hydrogen production from food waste without inocula. In: 5th International Workshop on Water Dynamics. AIP Conf Proc 987:171–174

Show KY, Lee DJ, Tay JH, Lin CY, Chang JS (2012) Biohydrogen production: current perspectives and the way forward. Int J Hydrogen Energy 37:15616–15631

Siles JA, Martín MA, Chica AF, Martín A (2010) Anaerobic co-digestion of glycerol and wastewater derived from biodiesel manufacturing. Biores Technol 101(16):6315–6321

Speece RE (1996) Anaerobic biotechnology for industrial wastewaters. Archae Press, Nashville

Tang YQ, Koike Y, Liu K, An MZ, Morimura S, Wu XL, Kida K (2008) Ethanol production from kitchen waste using the flocculating yeast Saccharomyces cerevisiae strain KF-7. Biomass Bioenergy 32(11):1037–1045

Thomsen AB, Medina C, Ahring BK (2003) Biotechnology in ethanol production. New Emerg Bioenergy Technol 2:40–44

Tomasik P, Horton D (2012) Enzymatic conversions of starch. Adv Carbohydr Chem Biochem 68:59–436

Tubb RS (1986) Amylolytic yeasts for commercial applications. Trends Biotechnol 4(4):98–104

Ueno Y, Otsuka S, Morimoto M (1996) Hydrogen production from industrial wastewater by anaerobic microflora in chemostat culture. J Ferment Bioeng 82(2):194–197

Ueno Y, Haruta S, Ishii M, Igarashi Y (2001) Characterization of a microorganism isolated from the effluent of hydrogen fermention by microflora. J Biosci Bioeng 92(4):397–400

Valdez-Vazquez I, Rios-Leal E, Esparza-Garcia F, Cecchi F, Poggi-Varaldo HM (2005) Semi-continuous solid substrate anaerobic reactors for H2 production from organic waste: mesophilic versus thermophilic regime. Int J Hydrogen Energ 30:1383–1391

Walker K, Vadlani P, Madl R, Hohn KL (2012) Ethanol Fermentation from Food Processing Waste. Environ Progress Sustain Energ 32(4):1280–1283

Wang J, Wan W (2009) Factors influencing fermentative hydrogen production: a review. Int J Hydrogen Energy 34(2):799–811

Yan S, Yao JM, Yao LM, Zhi ZJ, Chen XS, Wu JY (2012) Fed batch enzymatic saccharification of food waste improves the sugar concentration in the hydrolysates and eventually the ethanol fermentation by *Saccharomyces cerevisiae* H058. Brazil Arch Biol Technol 55(2):183–192

Ye ZL, Zheng Y, Li YH, Cai WM (2008) Use of starter culture of *Lactobacillus plantarum* BP04 in the preservation of dining-hall food waste. World J Microbiol Biotechnol 24(10):2249–2256

Yu H, Zhu Z, Hu W, Zhang H (2002) Hydrogen production from rice winery wastewater in an up flow anaerobic reactor by using mixed anaerobic cultures. Int J Hydrogen Energy 27:1359–1365

Zhang T, Liu H, Fang HHP (2003) Biohydrogen production from starch in wastewater under thermophilic condition. J Environ Manage 69(2):149–156

Zhong WZ, Zhang ZZ, Luo YJ, Sun SS, Qiao W, Xiao, M (2011) Effect of biological pretreatments in enhancing corn straw biogas production. Bioresour Technol 102:11177–11182

Solidification and Stabilization of Tannery Sludge

K. Sri Bala Kameswari, Pendem Rohit Babu, B. Lekshmi and Chitra Kalyanaraman

Abstract The tanning industry occupies a unique place in the industrial map of India. Nearly 2000 tanneries are in operation in India, with a total processing capacity of 700,000 tonnes of hides and skins per annum. During treatment of tannery wastewater, primary (chemical) and secondary (biological) sludge are generated. Safe disposal of sludge generated during treatment of tannery wastewater is a major concern from an environmental point of view. At present, the sludge is being disposed off in secured landfill facilities. Each tannery is spending about Rs. 750 to Rs. 1000 per tonne for disposal of sludge into secured landfill facilities, which includes transportation of sludge from the tannery to the secured landfill facility, loading and unloading. Solidification and Stabilization (S/S) is the Best Available Treatment Technology (BATT). Hence, in the present study, in order to utilize the sludge generated during treatment of tannery wastewater, S/S studies were carried out for encapsulation of chromium-bearing sludge. The solidification process was carried out using binding materials such as cement and lime in various combinations. Various performance tests were carried out on the S/S blocks to understand the leaching behavior of chromium by conducting leaching tests viz., Toxicity Characteristic Leaching Procedure (TCLP), EP ToX and ANS Leaching Tests and compressive strength of S/S blocks were determined.

Keywords Chromium · Leaching tests · Sludge · Solidification and stabilization

K. Sri Bala Kameswari (✉) · B. Lekshmi · C. Kalyanaraman
Environmental Technology Division, CSIR-Central Leather Research Institute,
Adyar, Chennai 600 020, India
e-mail: sribala_k@yahoo.com

P.R. Babu
Shanmugha Arts, Science, Technology and Research Academy, Sastra University,
Thanjavur 613 402, India

O.P. Karthikeyan et al. (eds.), *Recycling of Solid Waste for Biofuels and Bio-chemicals*, Environmental Footprints and Eco-design of Products and Processes, DOI 10.1007/978-981-10-0150-5_14

1 Introduction

Leather industry has gained high socio-economic relevance in India. The leather sector has contributed significantly to economic growth by providing job opportunities. The leather industry in India is spread out in unorganized sectors. Small scale, cottage and artisan sectors account for over 90 % of the total production. The main reason for development and growth of the leather industry in the country is its large animal population. Primarily, tanning operations are concentrated in the regions where availability of good quality water and raw materials are abundant. In India, tanning clusters are located in the states of Tamil Nadu, Andhra Pradesh, Punjab, West Bengal and Uttar Pradesh and a few isolated tanneries are in Bihar, Madhya Pradesh and Kerala. The location of various tanning clusters in India is presented in Fig. 1.

The tannery operation consists of converting the raw skin, a highly putrescible material, into leather, a stable material, which can be used in the manufacture of a wide range of products. The whole process involves a sequence of complex chemical reactions and mechanical processes (UNEP 1991). Amongst these, tanning is the fundamental stage which gives leather its stability and essential characters. Raw hides/skins after various steps of pre and post tanning processes are converted into a final product with specific properties: stability, appearance, water resistance, temperature resistance, elasticity and permeability for perspiration.

In the tanning process, collagen fibres are stabilized by the tanning agents, such that the hide is no longer susceptible to putrefaction or rotting. Collagen fibres are stabilized by the cross-linking action of the tanning agents. After tanning, the hides or skins are not subject to putrefaction, their dimensional stability, resistance to mechanical action and heat resistance increase (Andres 1995; HMIP 1995). The most commonly used tanning agent is basic chromium sulphate ($Cr(OH)SO_4$). A high proportion (80–90 %) of leather produced today is tanned using chromium(III) salts (Black et al. 2013).

For processing one tonne of raw material about 2–5 kg of chromium is used in the main chrome-tanning process and about 1–2 kg of chromium is used in the post-tanning process. Chromium uptake is about 60–80 %, with the remainder being discharged along with wastewater (Ludvik and Buljan 2000). The tanning process is the main source of chromium in the effluent, but effluent from the post-tanning processes would also contain chromium when re-tanning is resorted to. Smaller amounts of chromium arise as a result of leaching of chromium during the wet process stages, which follow chromium tanning or re-tanning. Tannery effluents are treated to meet standards specified for discharges to surface waters. Physico-chemical treatment of tannery wastewater includes oxidation, precipitation, sedimentation, flotation, equalizing flows and neutralization. It is mainly performed to remove organic matter, sulphide from beam house effluents and chromium(III) from tanning and post-tanning operations. Chromium(III) salts in the environment

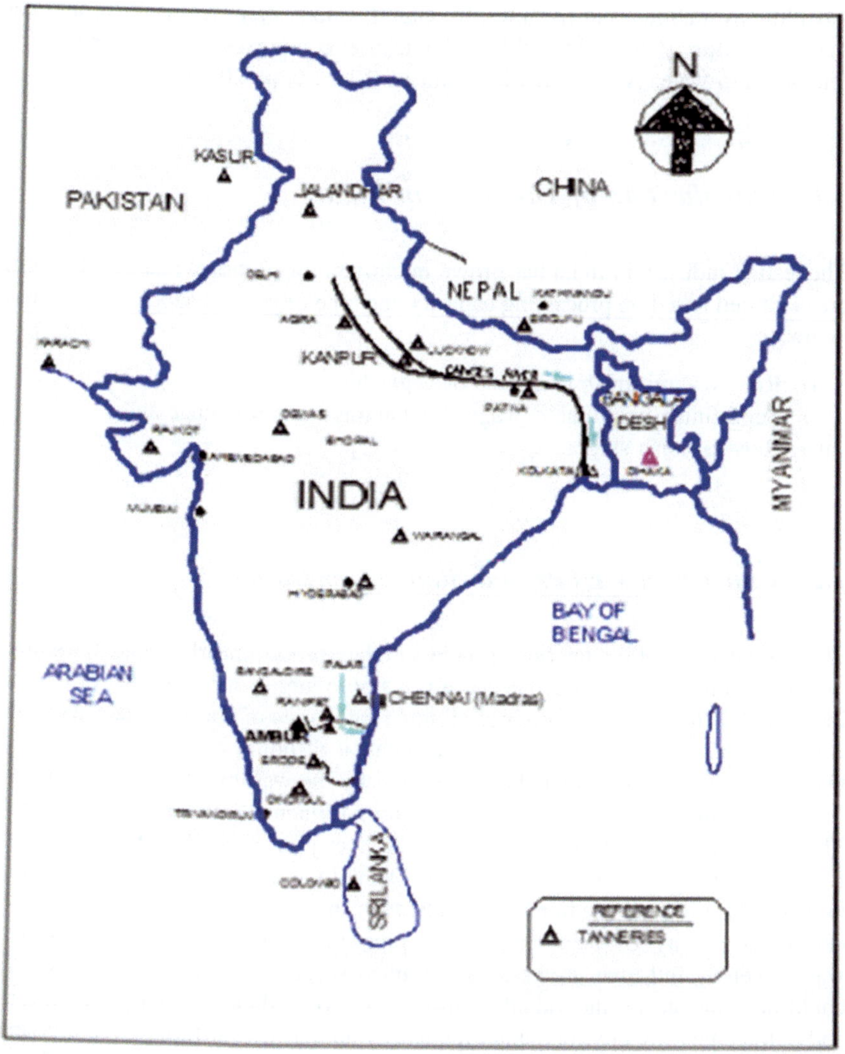

Fig. 1 Location of tanneries in India

have low availability and mobility. In natural waters, they are adsorbed onto particulate matter or form insoluble polynucleate complexes. Most chromium(III) released into water is deposited in sediment (Black et al. 2013).

1.1 Classification of Tanneries in India

The leather industry in India has grown in clusters for historical reasons. Tanneries are classified based on processing capacity and type of process adopted, as depicted below:

(i) Raw to semi-finish (wet-blue or vegetable tanning) stage
(ii) Semi-finish (wet-blue or vegetable tanning) to finish stage and
(iii) Raw to finish stage.

1.2 Concept of CETPs and Sludge Generation

There are nearly 2000 tanneries in India and they process nearly 700,000 tonnes of raw hides/skins per year. Leather has always been largely a by-product of the meat industry. Leather tanneries produce all three categories of waste: wastewater, solid waste and air emissions. However, management of liquid and solid waste are the two important environmental challenges. Mostly the wastewater is treated either in individual effluent treatment plants or Common Effluent Treatment Plants (CETPs). The concept of CETP was introduced in the year 1984 by the Ministry of Environment and Forest (MoEF), New Delhi to treat wastewater from a large number of small and medium-scale industries in clusters. This concept was conceived as a way of achieving end-of-pipe treatment of combined industrial wastewater by full-time professionally trained specialists at lower unit costs than could be achieved by individual industries and to facilitate discharge, monitoring and enforcement by environmental regulatory authorities. In India there are nearly 200 individual Effluent Treatment Plants (ETPs) and 17 Common Effluent Treatment Plants (CETPs) exclusively operated for treatment of tannery wastewater. The process flow diagram for treatment of tannery wastewater is presented in Fig. 2.

During treatment of tannery wastewater, both chemical and secondary biological sludge are generated. The quantity and quality of sludge varies, depending on the tanning process adopted. More sludge is generated from 'raw to finish' operations as compared to 'semi-finish to finish' process. The chemical sludge contains chromium in the form Cr(III) and its presence in the sludge is mainly due to usage of basic chromium sulphate during tanning process. As per Hazardous Waste (Management and Handling) Rules 2008, the sludge from wastewater treatment

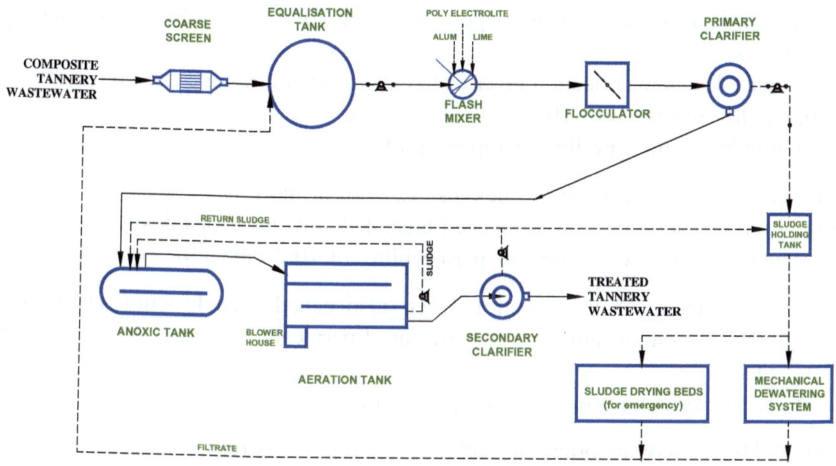

Fig. 2 A typical process flow diagram for treatment of tannery wastewater

plants and chromium bearing waste (Trivalent chromium above 5000 mg/kg) are considered as hazardous.

"Hazardous Waste" means any waste which by reason of its physical, chemical, reactive, toxic, flammable, explosive or corrosive characteristics causes danger or is likely to cause danger to human health or the environment, whether alone or when in contact with other wastes or substances, and may cause, or significantly contribute to, an increase in mortality or an increase in serious irreversible, or incapacitating reversible, illness; potential hazard to human health or the environment when improperly treated, stored, transported or disposed off or otherwise managed.

Landfilling is the ultimate disposal process for Municipal Solid Waste (MSW) management. Even though landfilling is the least desired option in the hierarchy of solid waste management, it is an unavoidable component in MSW Management. As per Municipal Solid wastes (Management and Handling) Rules, 2000, site selection for proposed landfill site, facilities required at landfill site, specification for landfilling, water quality monitoring, ambient air quality monitoring, plantation at landfill site, closure of landfill site,/post closure are very important aspects in the design and implementation of sanitary landfills. In case of sanitary landfills for disposal of MSW, single layer liner system is recommended. The liner shall be a composite barrier having 1.5 mm HDPE or equivalent having a permeability of less than 1×10^{-7} cm/s (CPCB Document—PROBES/124/2008-2009). In India, the design concepts for sanitary landfill is different for hazardous wastes disposal landfills. The present study deals with solidification and stabilization of chromium bearing sludge generated during treatment of tannery wastewater and falls under the category of hazardous wastes.

In India, landfills with double liner system have been recommended by the Central Pollution Control Board (CPCB), New Delhi, for disposal of hazardous wastes. The liner systems are specified by CPCB, New Delhi (CPCB

Document-Hazardous Waste Management Series: HAZWAMS/2010-2011) and are as follows:

Primary leachate collection layer of thickness 30 cm or more and co-efficient of permeability in excess of 10^{-2} cm/s (10^{-4} m/s).

A primary composite liner comprising of

(i) HDPE geo-membrane of thickness 1.5 mm or more
(ii) A compacted clay (or compacted amended soil) layer of thickness 45 cm or more having a coefficient of permeability of 10^{-7} cm/s (10^{-9} m/s) or less.

A secondary leachate collection layer (also called leak detection layer) of a thickness of 30 cm or more and co-efficient of permeability in excess of 10^{-3} cm/s (10^{-5} m/s).

A secondary composite liner comprising of

(i) A HDPE geomembrane of thickness 1.5 mm or more
(ii) A compacted clay layer of thickness 45 cm or more having a co-efficient of permeability of 10^{-7} cm/s (10^{-9} m/s) or less.

In addition to a double liner system other components viz., leachate monitoring system, gas collection system, leachate treatment and post-closure measures are essential.

During the treatment of tannery wastewater, primary sludge is generated from primary clarifiers and secondary sludge is generated from biological treatment units. The disposal of sludge from wastewater treatment presents highly complex problems due to increasingly stringent environmental regulations and industrial growth that have markedly increased the disposal requirements. At present, after dewatering of the sludge, it is disposed off in decentralized secured landfill facilities (SLF), designed and implemented in Tamil Nadu and is the patented technology of CSIR-CLRI, Chennai. These SLF's have been implemented as an integral part of CETP's operating for treatment of tannery wastewater. Landfilling is the least desired option in the hierarchy of solid waste management and landfills are considered as temporary storage facilities only. Generally, landfills are designed for a certain time period e.g., 10 years was considered for the tannery sector, and once the SLF is filled then another site has to be identified. With the growing demand and scarcity of land, it is very difficult to identify sites for implementation of secure landfill facilities for disposal of sludge generated from tanneries.

Disadvantages of landfills are: (1) large land requirement, with time it will be more and more difficult to find suitable sites; and (2) secondary pollution. Experience has demonstrated that percolate and gas of refuse harm the surrounding environment if they are not disposed properly. The reuse of wastewater sludge in construction materials would not only alleviate disposal problems but also provide economic, ecological and energy-saving advantages. Landfilling scenario faces the highest cost. The investment costs for gasification is very high. Therefore, the total cost is always higher for incineration and landfilling.

Immobilization of organic or inorganic contaminates of tannery waste is essential prior to their disposal or for utilization of sludge. Otherwise, physical and chemical changes, due to weathering or other effects, can start the release of contaminants from the waste beyond an unacceptable limit. This may result in contamination of the soil, surface and groundwater. In case of inorganic waste, as in the present case, there are chances of oxidation of Cr(III) (present in the sludge) into Cr(VI) due to prevailing environmental conditions. Solidification and Stabilization (S/S) would provide dual benefits, viz., (i) immobilization of toxic elements and (ii) utilization of solidified mass as construction materials.

1.3 Solidification and Stabilization (S/S) Process

Solidification refers to a category of waste treatment that is being used increasingly to treat a wide variety of wastes, both solid and semi-solid. Solidification refers to a process in which materials are added to the waste to produce a solid. Generally, the solidification process is designed and used to accomplish one or more of the following objectives.

- Reduce contaminant mobility
- Improve the handling and physical characteristics of the waste.

In the present study, the dewatered sludge was collected from two different CETPs and was characterized for chromium content. Then solidification and stabilization studies were carried out for encapsulation of chromium in the cement matrix. Further the process was evaluated by conducting various leaching studies and the compressive strength of S/S mortar blocks was also evaluated.

2 Overview on the Solidification and Stabilization Process

The S/S technology was initially been developed as a treatment concept for hazardous waste prior to landfilling. At present, it is also applied as a remediation technology for contaminated soils, especially in the USA and the UK, and for a wide variety of contaminants such as organics and heavy metals (Pensaert et al. 2008). S/S remedies are designed to reduce the flux of contamination that leaches from a contaminant source to be within acceptable parameters set forth in a site-specific remediation goal. To determine site- and contaminant-specific potential effectiveness of S/S technology, performance specifications should be used to ensure effectiveness of the remedy. Comprehensive treatability studies are essential for providing site-specific information to evaluate the S/S technology. Based on treatability studies, design parameters and scale up for full-scale implementation can be formulated (The Interstate Technology and Regulatory Council Solidification/Stabilization Team 2011).

High-Calcium Fly Ashes and their potential influence on ettringite formation in cementitious systems was investigated and reported that, the amount of ettringite formed decreased with increasing C_3A content and increased with increase in anhydrite content in the ash (Tishmack et al. 1999). For different types of wastes such as fine-grained soils, coarse-grained soils, metal-bearing sludges, nitrate salts from processing operations, chloride salts from off-gas treatment systems and incinerator, fly and bottom ash have been evaluated for application of the S/S process. The S/S process was assessed using grout/portland cement stabilization, sulfur polymer encapsulation (SPE), polymer encapsulation (PE) and phoenix ash technology (PAT) considering factor such as waste form and engineering criteria (US EPA 1996). For metal-bearing sludges, the S/S process is an environmental friendly disposal option (Leong and Laortanakul 2002). The effect of chromium on Ordinary Portland cement (OPC) hydration was evaluated by continuous observation of early hydration and it was reported that addition of chromium does not affect the mechanism of the hydration process, but it does affect the kinetic and dynamics of the cement hydration process (Barbir et al. 2012).

The Stabilization/Solidification (S/S) process, as a pre-landfill waste treatment technology, has been investigated for arsenic-containing solid wastes using Portland cement, fly ash and polymeric materials and reported that S/S can convert hazardous industrial wastes into a stable form of waste safe for disposal (Singh and Pant 2006). Solidification/Stabilisation (S/S) to treat synthetic drill cuttings as a pre-treatment to landfilling or for potential re-use as construction products was explored by Al-Ansary and Al-Tabbaa (2007). An industrial byproduct, anhydrite $CaSO_4$ was used as a binder for S/S of heavy metal-containing sludges (Andres et al. 1998). Alkali-activated pulverized fuel ash was used as a cementitious binder for S/S of electroplating sludge (Asavapisit and Chotklang 2004). A study was carried out to determine the strength, leachability and microstructure characteristics of cement-based solidified plating sludge using OPC and pulverized fuel ash as solidification binders (Asavapisit et al. 2005). Batchelor (2006) reviewed the S/S process with cement as a binder. Containment of chromium and zinc, present in ferrochromium flue dust, was solidified and stabilised using cement (Cohen and Petrie 1997).

In a diffusion-controlled scenario, the mechanism of chromium release from cement-based solidified wastes containing hexavalent chromium was investigated and it was reported that the water cement ratio is the controlling factor for the leaching behaviour of chromium in cement specimens (Bobirica et al. 2010). A chromium-rich toxic waste generated by the tanning process of leather was studied in terms of its composition, stability and potential use as a ceramic pigment for the glass and ceramic industries by Abreu and Toffoli (2009). Chromium-containing pigments were synthesised from chromium galvanic sludges and the characteristics of synthesized pigments were similar to the colour developed by the pigments obtained, starting from pure Cr_2O_3 (Andreola et al. 2008). Utilisation of tannery sludge as raw material for clay products has been investigated by Basegio et al. (2002).

3 Materials and Methods

3.1 Characterization of Sludge Samples

Sludge samples were collected from two different CETPs. The samples were transported to the laboratory in plastic bags. All samples, kept in the laboratory, were air-dried at room temperature. Samples were then stored at room temperature in plastic bags until required. The dry mass of the sludge samples were determined by oven drying them for 24 h at 105 °C as per standard methods (20th edition APHA 1998). The Total Solids (TS), Volatile Solids (VS) content in the sludge samples were measured as per standard methods (20th edition APHA 1998). After removing moisture content, chromium concentrations were measured as per standard methods (20th edition, method 3500-Cr. B, APHA 1998). In order to confirm the presence of Chromium(VI), colour development using diphenyl carbazide was followed. The pH of the sludge samples were measured as per ASTM method—D 4972 (ASTM 2007). The bulk density of the sludge was measured by using ASTM—Method D2937. Loss on ignition (LOI) tests were conducted by following ASTM method–D7348 (ASTM 2007).

3.2 Cone Penetration Test

Cone Penetration Tests (CPT) were conducted on solidified mortar blocks as preliminary screening tests in order to screen the mix proportions of cement: lime: sludge. The test procedure given by the U.S. Army (1972) was followed and the details of mix proportions are given in Table 1.

The CPT is a quick screening test for evaluation of compressive strength on solidified waste. This test involves forcing the cone penetro-meter into the solidified waste after 48 h of curing and then measuring the penetration resistance. Based on penetration resistance of the cone on solidified waste, binder–sludge ratios are arrived at.

Table 1 Details of mix proportion

Sludge code	Mix proportion (C:L:S)	Cement (in grams)	Lime (in grams)	Sludge (grams)	Water/binder plus sludge ratio
S1	1:1:2	100	100	200	0.55
	1:1:4	100	100	400	0.58
	1:1:6*	100	100	600	0.60
S2	1:1:2	100	100	200	0.55
	1:1:4	100	100	400	0.58
	1:1:6*	100	100	600	0.60

Note *: Was not considered further because of poor strength and higher cone penetration values
C Cement; *L* Lime; *S* Sludge

3.3 Solidification of Sludge Samples

Based on preliminary results obtained from cone index test, the sludge samples were solidified using cement (C) and lime (L) in various combinations in conjunction with sludge (S) with water/binder plus sludge ratio (v/w) of 0.55–0.60.

Two sizes of moulds were selected i.e., (i) 30 mm × 50 mm and (ii) 70.6 × 70.6 mm. The first one is not a standard size for conducting Unconfined Compressive Strength (UCS) of mortar cubes. This size was specifically selected for conducting leaching studies for long-term analysis purpose. The standard size of moulds i.e., 70.6 × 70.6 mm, was used for conducting UCS test. The mortar cubes in the moulds containing the sludge–binder mixture were cured at room temperature. The mortar cubes were removed from the moulds and allowed to cure under the same conditions of temperature and humidity. For each mix proportion, six number of mortar blocks were moulded and S/S blocks are shown in Fig. 3.

3.4 Compressive Strength Test

The S/S specimens require a minimum strength to bear their self-weight and the overburden pressure when they are disposed of in secured landfills. Unconfined compressive strength test was performed for mortar blocks of size 70.6 mm × 70.6 mm after 28 days of curing. The mortar blocks were loaded and the

Fig. 3 Curing of S/S mortar blocks

characteristic compressive strength was calculated based on the load at which the specimen started cracking.

3.5 *Leaching Studies*

Leaching studies are essential in order to assess the performance of the S/S process for effectiveness in encapsulation of the contaminant i.e., chromium in the present study. After curing for a period of 7 days, the solidified moulds were subjected to leaching studies using protocols like Toxicity Characteristic Leaching Procedure (TCLP), Extraction Procedure Toxicity (EP ToX) Test and American Nuclear Society (ANS) leach test in different leaching media. Toxicity Characteristic Leaching Procedure Test (TCLP), Extraction Procedure Toxicity (EP ToX) Test and American Nuclear Society (ANS) leach tests were carried out on the S/S mortar blocks. Details of the leaching protocols are presented below.

3.5.1 Toxicity Characteristic Leaching Procedure (TCLP) Test

The Toxicity Characteristic Leaching Procedure (TCLP) protocol simulates extreme exposure conditions for the solidified/stabilized material, by exposing the sample to a highly agile environment in which the contaminant becomes vulnerable to leaching. During the TCLP test, S/S mortar blocks were crushed and the size was reduced to less than 9.5 mm or surface area per gram of material equal to or greater than 3.1 cm^2. Size reduction is necessary to increase the surface area of the particles. Extraction media were selected based on the pH of the waste or S/S mortar blocks. The leaching medium was prepared as per the procedure given in U.S. EPA (SW-846; method 1311, US EPA 1992). During the TCLP test, solid to liquid, i.e., leaching medium, ratio of 1:20 was maintained and the TCLP studies were carried out in Zero Headspace Extractor (ZHE) vessel fitted in a Millipore Rotatory Agitator. The TCLP is designed to determine the mobility of both organic and inorganic analytes present in liquid, solid and multi-phase wastes. After the test, the leaching medium was filtered through 45 μm Whatman filter paper and the leachate was analyzed for pH, hardness, chromium content, conductivity and total dissolved solids (TDS). TCLP studies were carried out without and with S/S of sludge samples. Chromium content was analyzed as per Standard Methods (method 3500-Cr B, APHA 1998).

3.5.2 EP ToX Test and ANS Leach Test

The procedure for the Extraction Procedure Toxicity (EP Tox) was very similar to the TCLP, except that the leaching medium was distilled water and the duration of extraction was 24 h instead of 18 h in case of the TCLP test. The TCLP and EP

Table 2 Details of leaching tests

Name of the test	Particle size	Leaching medium	Solid:Liquid ratio	Duration of each extraction (h)	No. of extractions	Remark
TCLP	<9.5 mm	Acetic acid	1:20	18	1	Not for volatiles
Extraction Procedure Toxicity (EP ToX)	<9.5 mm	Water	1:20	24	1	Not for volatiles
American Nuclear Science Leach test (ANS Leach test)	Monolith (0.2 < l: d < 0.5)	Deionised water (or) simulated ground water	Volume:surface = 10 cm	2.7 h, 1, 2, 3, 4, 5, 14, 28, 43, 90 days	12	Can obtain effective diffusion coefficient and leachability index

ToX tests were single extraction tests, whereas the ANS leach test involved 12 extractions stretching over a period of 90 days. The duration of extractions varied widely from 2 h in the beginning of the test to about 90 days towards the end. The leaching medium was distilled water and the monolith moulds were subjected to the ANS leach test with a liquid to solid ratio of 10:1. The details of the leaching tests are summarized and presented in Table 2.

4 Results and Discussion

4.1 *Sludge Characteristics*

Dewatered sludge samples were collected from two different CETPs and were analyzed for pH, total solids, volatile solids and chromium content as per procedures given in Standard Methods, 20th Edition (APHA 1998). The density of sludge samples is also assessed. Characteristics of sludge samples are given in Table 3.

18.9 and 29.2 mg/g of chromium was present in the sludge samples, respectively, which is more than the permissible level of 5000 mg/kg as per hazardous waste management and handling rules (2008). The bulk density of the sludge was 0.8 and 0.9 g/cc and loss on ignition (LOI) was 51 and 40 % respectively, indicating that more organic matter is present in the sludge samples.

4.2 *TCLP Studies on Sludge Samples*

Sludge characteristics showed that the chromium content exceeded the permissible level of 5000 mg/kg as per hazardous waste management and handling rules (2008). TCLP studies were carried out on sludge samples as per US EPA protocol in order to determine leachability of chromium under the experimental condition. The leachate was analyzed for chromium concentration and results are presented in Table 3. Chromium concentrations in the leachate exceeded the regulatory level of 5 mg/L as per Characteristic Leaching Procedure Test Protocol. In order to arrest the mobility of chromium, pre-treatment was necessary. Hence, in order to arrest the mobility of chromium, sludge was solidified and stabilized using cement and lime as binding materials; a physical binding process.

4.3 *Studies on S/S Mortar Blocks*

The details of mix proportions of mortar blocks are presented in Table 1. After curing for 7 days, the mortar blocks with different mix proportions were ground

Table 3 Characteristics of sludge samples and chromium concentration in the leachate

Parameter	Sludge (S1)	Sludge (S2)
pH	8.6	8.9
Total solids (mg/g)	434.7	890.4
Volatile solids (mg/g)	221.5	354.9
Total chromium (mg/g)	18.9	29.2
Density (g/cc)	0.80	0.90
Loss on ignition (LOI, %)	51	40
Chromium concentration in the leachate (mg/g)	1.92	1.98

into powder with a particle size of less than 9.5 mm and were subjected to leaching studies as per TCLP and EP ToX and some mix proportion mortar blocks were subjected to long term leaching as per ANS leach test without crushing.

4.3.1 Leaching Studies After S/S of Sludge

Leachability studies were carried out using TCLP and EP ToX protocols separately for various mix proportions of mortar blocks. After leaching studies, leachate samples were collected and were characterized for pH, hardness, chromium, Oxidation Reduction Potential (ORP), Total Dissolved Solids (TDS) and conductivity. Leachate characteristics after TCLP and EP ToX studies are presented in Table 4.

TCLP and EP ToX leaching studies showed that chromium did not leach when exposed to acetic acid or water as a leaching medium. The increased values of hardness, conductivity and TDS in the leachate samples were due to solubility of

Table 4 Leachate characteristics after TCLP and EP ToX studies

Sludge	Mix proportion (C:L:S)	pH	Hardness (mg/L as $CaCO_3$)	Conc. of Cr (mg/L)	Conductivity (mS/cm)	TDS (g/L)
Leachate characteristics after TCLP studies						
S1	1:1:2	11.0	4220	BDL	7.89	4.23
	1:1:4	11.8	3680	BDL	7.26	3.87
S2	1:1:2	11.3	5160	BDL	12.05	6.63
	1:1:4	11.9	5590	BDL	11.07	6.08
Leachate characteristics after EP ToX studies						
S1	1:1:2	11.3	5160	BDL	12.05	6.63
	1:1:4	11.9	5590	BDL	11.07	6.08
S2	1:1:2	11.2	1710	BDL	9.11	4.94
	1:1:4	11.2	2110	BDL	9.25	5.02

Note: *BDL* Below Detectable Limit

calcium present in binding materials (viz., cement and lime) in the leaching medium. Also, by comparing the chromium concentration in leachate samples after TCLP studies, without and with S/S of sludge samples, it was evident that chromium present in sludge samples was immobilized arresting leaching of chromium after S/S of sludge. However, TCLP and EP ToX protocols determine single time point leaching behavior and may not give a clear idea about the performance of the S/S process. It is very difficult to interpret the data with real time scenarios with regard to the underlying release-controlling mechanism (e.g., equilibrium or mass transfer) or rate of leaching. In real time, the controlling factors such as buffering capacity, groundwater movement and contact of monolithic nature of S/S materials with the movement of groundwater will affect the release of contaminants into groundwater (The Interstate Technology and Regulatory Council Solidification/Stabilization Team 2011). Immobilization of chromium-laden electroplating sludge was assessed for compressive strength and leachability and reported that chromium is well retained within the solid matrix of cement (Sophia and Swaminathan 2005).

4.3.2 Long-Term Leaching Studies

Long term leaching studies were carried out on S/S samples by adopting the ANS test protocol. The ANS leach test involved 12 extractions stretching over a period of 90 days. Duration of extractions varied widely, from 2 h in the beginning of the test to about 90 days towards the end. The leaching medium was distilled water and the monolith moulds were subjected to ANS leach tests with a liquid to solid ratio of 10:1. After conducting leaching studies, leachants were analyzed. ANS leach tests were carried out on mortar blocks prepared with C:L:S (1:1:2). At different intervals of leaching period, leachate samples were collected and were analyzed for parameters like pH, hardness, chromium content, Total Dissolved Solids (TDS) and conductivity. The results are presented in Table 5.

No chromium leaching was detected for each extraction period. The cumulative characteristics after 90 days of leaching for the parameters like hardness, conductivity and TDS were 2460 mg/L, 19.45 mS/cm and 10.08 g/L respectively indicated that the TDS values had increased due to solubility of calcium present in cement and lime.

4.4 Compressive Strength

For the two different sludge samples, two combinations of mix proportions were tested. Compressive strength was assessed after 28 days of curing and the results are tabulated in Table 6.

Table 5 Characteristics of leachate after ANS protocol

Extraction number	Duration of each extraction	Parameter				
		pH	Hardness (mg/L as $CaCO_3$)	Conc. of Cr (mg/L)	Conductivity (mS/cm)	TDS (g/L)
1	2 h	8.9	500	BDL	1.324	0.655
2	7 h	9.3	490	BDL	1.354	0.714
3	1 day	9.1	510	BDL	1.312	0.723
4	2 day	9.7	80	BDL	1.687	0.842
5	3 day	9.7	60	BDL	1.865	0.934
6	4 day	9.7	170	BDL	2.140	1.077
7	5 day	10.5	100	BDL	2.100	1.100
8	14 day	10.5	120	BDL	1.687	0.842
9	19 day	10.5	110	BDL	1.865	0.934
10	28 day	10.4	120	BDL	1.354	0.714
11	43 day	10.5	100	BDL	1.312	0.723
12	90 day	10.5	100	BDL	1.450	0.820
Cumulative Characteristics after 90 days of leaching period		9.9*	2460		19.45	10.08

Note *: pH—average value

As per US EPA guidelines (SW-846), solid wastes for disposal in landfills must have a minimum compressive strength of 4 kg/cm^2. All mortar blocks exceeded the 4 kg/cm^2 limit. As the sludge content increased, a decrease in the compressive strength was observed. Comprehensive treatability studies are essential for providing site-specific information to evaluate the S/S technology. Based on the treatability studies, design parameters and scale up for full-scale implementation can be formulated (The Interstate Technology and Regulatory Council Solidification/Stabilization Team 2011). Limitations in the present study were mainly due to the presence of organic matter i.e., if the organic content present in the sludge increases, a decrease in compressive strength was observed. Hence, it is necessary to reduce or remove organic matter prior to S/S for encapsulation of metal-bearing sludges.

Table 6 Compressive strength of S/S mortar blocks

Sludge code	Proportion (C:L:S)	Characteristic compressive strength (kg/cm^2)
S1	1:1:2	12.05
	1:1:4	8.42
S2	1:1:2	14.02
	1:1:4	9.50

5 Summary and Conclusions

Based on preliminary characterization of sludge samples, trivalent chromium concentrations in the sludge were found to exceed 5000 mg/kg. From TCLP studies on sludge samples, the chromium concentration in the leachate was observed to exceed the permissible limit of 5 mg/L. To arrest leaching of chromium from sludge samples, sludge has to be encapsulated. Solidification and Stabilization (S/S) of sludge was done using cement, lime as binding materials. Hence during S/S, chromium was encapsulated into a solid matrix, which arrested the mobility of chromium. Long-term leaching studies i.e., with the ANS test, verified encapsulation of chromium. Hence S/S of chromium-bearing sludges is an ideal treatment technology. In the present study, minimum compressive strength of 8.42 kg/cm^2 was obtained after 28 days of curing. Whereas as per USEPA, a minimum compressive strength of 4 kg/cm^2 after 28 days of curing. The S/S mortar blocks can be utilized as construction materials especially for filling of low lying areas or as filling material in general. Hence they can be utilized as an alternative to conventional construction materials.

Acknowledgments The authors wish to thank Council of Scientific and Industrial Research (CSIR), India for financial support of the study under the ZERIS—XII Five Year Plan Network project. The authors would like also to thank the Director, Central Leather Research Institute (CLRI) India for permitting to publish this work.

References

Abreu MA, Toffoli SM (2009) Characterization of a chromium-rich tannery waste and its potential use in ceramics. Ceram Int 35(2009):2225–2234

Al-Ansary MS, Al-Tabbaa A (2007) Stabilisation/solidification of synthetic petroleum drill cuttings. J Hazard Mater 141(2007):410–421

Andreola F, Barbieri L, Bondioli F, Cannio M, Ferrari AM, Lancellotti L (2008) Synthesis of chromium containing pigments from chromium galvanic sludges. J Hazard Mater 156:466–471 (2007)

Andres et al (1998) Experimental Study of the waste binder anhydrite in the solidification/Stabilization process of heavy metal sludges. J Hazard Mater 57:155–168

Andres, *Umwelthandbuchfür die ledererzeugendenBetriebe, draft version*, 1995

APHA (1998) Standard Methods for the Examination of Water and Wastewater, In: LS Clesceri, AE Greenberg, AD Eaton (eds) American Public Health Association, Washington DC

Asavapisit S, Chotklang D (2004) Solidification of electroplating sludge using alkali-activated pulverized fuel ash as cementitious binder. Cem Concr Res 34(2004):349–353

Asavapisit S, Naksrichumb S, Harnwajanawong N (2005) Strength, leachability and microstructure characteristics of cement-based solidified plating sludge. Cem Concr Res 35(2005):1042–1049

ASTM D2937-10, Standard test method for density of soil in place by the drive-cylinder method

ASTM International, D 4972-01 Standard Test Methods for pH of Soils. 2007

ASTM International, D 7348-13 Standard test methods for Loss on Ignition (LOI) of solid combustion residues, 2007

Barbir D, Dabic P, Krolo P (2012) Stabilization of chromium salt in ordinary portland cement. Sadhana 37(6):731–737 (December 2012, Indian Academy of Sciences)

Basegio T, Berutti F, Andre´a Bernardes A, Bergmann CP (2002) Environmental and technical aspects of the utilisation of tannery sludge as a raw material for clay products. J Eur Ceram Soc 22(2002):2251–2259

Batchelor B (2006) Overview of waste stabilization with cement. Waste Manag 26(2006):689–698

Black M, Canova M, Rydin S, Scalet BM, Roudier S, Sancho LD (2013) Best Available Techniques (BAT) reference document for the tanning of hides and skins

Bobirica C, Bobirica L, Stanescu R, Constantinescu I (2010) Leaching Behavior Of Cement-Based Solidified Wastes Containing Hexavalent Chromium. UPB Sci Bull Ser B 72(1). ISSN 1454-2331

Cohen B, Petrie JG (1997) Containment ofchromium and zinc in ferrochromium flue dust by cement based solidification. Can Metall Q 36(4):251–260

CPCB Document-Programme Objective Series: PROBES/124/2008-2009, Guidelines and Check-list for evaluation of MSW Landfills proposals with Information on existing landfills, Central Pollution Control Board, New Delhi, August, 2008

CPCB Document-PROBES/124/2008-2009, Protocol for performance evaluation and monitoring of the common hazardous waste treatment storage and disposal facilities including common hazardous waste incinerators, Central Pollution Control Board, New Delhi, May 24, 2010

HMIP (1995) Chief Inspectors' Guidance to Inspectors—Processing of Animal Hides and Skins, Process Guidance, Her Majesty's Inspectorate of Pollution, UK

Leong ST, Laortanakul P (2002) An environmentally optimized solidification/stabilization for disposal of heavy metal sludge wastes. Thammast J Sci Techol 7(3) (Sept–Dec 2002)

Ludvik J, Buljan J (2000) Chrome management in the tanyard. United Nations Industrial Development Organization

Method. 1311, U.S. Environmental Protection Agency, Toxicity Characteristic Leaching Procedure. Washington, DC. 1992

Pensaert S, De Groeve S, Staveley C, Menge P, De Puydt S (2008) Immobilisation, Stabilisation, Solidification: a New Approach for the Treatment of ontaminated Soils. Case studies: London Olympics & Total Ertvelde, 15de Innovatieforum Geotechniek–8 oktober 2008

Singh TS, Pant KK (2006) Solidification/stabilization of arsenic containing solid wastes using portland cement, fly ash and polymeric materials. J Hazard Mater 131(1–3):29–36

Sophia AC, Swaminathan K (2005) Assessment of the mechanical stability and chemical leachability of immobilized electroplating waste. Chemosphere 58:75–82

The Interstate Technology and Regulatory Council Solidification/Stabilization Team (2011) Technical/Regulatory Guidance- Development of Performance Specifications for Solidification/Stabilization

Tishmack JK, Olek J, Diamond S (1999) Characterization of high-calcium fly ashes and their potential influence on ettringite formation in cementitious systems. Cem Concr Aggregates CCAGDP 21(1):82–92

U.S. Army, Office, Chief of Engineers (1972) Laboratory Soils Testing, Engineer Manual 1110-2-1906. U.S. Army Crops of Engineers, Washington, DC

UNEP (1991) Tanneries and the environment: a technical guide to reducing the environmental impact of Tannery Operations. United Nations Environment Programme, Industry and Environment, Programme Activity Centre (UNIDO, IE/PAC), Paris

US EPA (402-R-96-014, June 1996), Stabilization/Solidification Processes for mixed waste, Center For Remediation Technology and Tools Radiation Protection Division, Office of Radiation and Indoor Air, U.S. Environmental Protection Agency, 401 M Street, S.W., Washington, DC 20460

Odour Pollution from Waste Recovery Facilities

Nastaein Qamaruz-Zaman and Nurashikin Yaacof

Abstract The accumulation and treatment of waste in a centralised facility has the potential to generate and become a source of odour nuisance in a community. Positive engagement with the public and developing their understanding of odour pollution can help resolve conflicts between the facility siting and operation and vicinity to nearby residences. This chapter unravels the potential of odour generation and common odorous compounds from different waste to energy recovery facilities including anaerobic digestion, incineration and refuse-derived fuel (RDF) plants. The olfactometer and its principals of operation will be described to justify the applicability of the equipment in odour impact assessments. Finally, the chapter also present technologies of odour control and life cycle approach in determination of the suitable control technologies.

Keywords Odour · Biofilter · Olfactometer · Wet scrubber · FIDOL

1 Introduction

Waste recovery facilities are important treatment plants for the purpose of continuous energy production from biomass and wastes. As the main sources of feed materials are organic, one of the challenges faced by these facilities is odour risk from handling and decomposition of the organic materials.

There are diverse publications on waste recovery facilities including refuse derived fuel (RDF), anaerobic digester and incineration plants, mostly focusing on the design of the systems itself and process effectiveness, rather than odour

N. Qamaruz-Zaman (✉) · N. Yaacof
School of Civil Engineering, Universiti Sains Malaysia, 14300 Nibong Tebal, Penang, Malaysia
e-mail: cenastaein@usm.my

N. Qamaruz-Zaman
Solid Waste Management Cluster, Universiti Sains Malaysia, Engineering Campus, 14300 Nibong Tebal, Penang, Malaysia

O.P. Karthikeyan et al. (eds.), *Recycling of Solid Waste for Biofuels and Bio-chemicals*, Environmental Footprints and Eco-design of Products and Processes, DOI 10.1007/978-981-10-0150-5_15

pollution at these facilities. This chapter aims to address this gap in knowledge by exposing odour issue at the waste recovery facilities covering topics on odour nuisance, assessment and control methods to contain odour pollution.

2 Background of Odour Pollution

2.1 *Definition of Odour*

Odour is a major environmental nuisance in the environment mainly cause by livestock and industries. It is widely recognised as offensive air pollution due to human activity usually in populated areas. Odours that result directly or indirectly from human activities causing an adverse effect are often classified as contaminants and are subject to regulation (Nicell 2009). By definition, odour is the aggregate of a mixture of gases on the sense of smell. Its strength is determined by using odour detection threshold (ODT). The ODT is the number of dilutions with odour-free air required for an odour to be just detected by 50 % of the odour panel or until the least definitely perceptible odour is achieved.

2.2 *Odour Effect on Environment and Public Health*

Several studies have concluded that offensive odour influences human health depending on the odour character and the level of odour intensity. A study by Stellacci et al. (2010) identified a greater number of odour-associated symptoms such as nausea, headache, lack of appetite and, more rarely, other acute and even chronic health effects occurring in living area near sewage treatment plants or waste sites (Stellacci et al. 2010). Several self-reported physical symptoms were associated with odour annoyance. These included unusual shortness of breath, eye irritation, hoarseness/dry throat, toothache, unusual tiredness, fever/shivering, joint—and muscular pain (Aatamila et al. 2011). Prolonged exposure to odour can cause undesired reactions ranging from emotional stresses such as states of anxiety, unease, headache or depression to physical symptoms such as eye irritation, respiratory problems, nausea or vomiting (Sironi et al. 2010).

2.3 *Odour from a Human Perspective*

Measurements of odour concentrations alone are insufficient to assess human perception of odour hence odour samples need to be evaluated by using the sensory perception to strengthen the results. There are four factors in evaluating odour

Table 1 Human perspective on odour (DEHP 2015)

Human perspective major	Details
Odour detection threshold (ODT)	ODT referring to the minimum concentration of odorant stimulus necessary for detection in some specified percentage of the test population. The odour concentration of a sample can be characterised by the number of dilutions to reach this detection threshold. Odour concentration is the most common attribute used to quantify odours
Intensity	Refers to the perceived strength or magnitude of the odour sensed. Intensity increases linearly with increasing odour concentration. Odour intensity is usually assessed by using seven levels of the intensity from scale 0 = no odour to 7 = extremely strong odour
Hedonic tone	Judgement of the relative pleasantness or unpleasantness of an odour. It is assessed on a nine point hedonic tone scale from very pleasant, +4 to offensive, −4
Odour quality	Simply a qualitative description of what the odour smells like

samples which are ODT, intensity, hedonic tone and odour quality. Table 1 summarises the details of the four major human perspectives.

FIDOL are the factors that are usually used in understanding the odour annoyance. When an individual exposed to odour perceives this as unwanted, it is argued that the following factors are the main determinants (Belgiorno et al. 2013):

- Frequency of the exposure
- Intensity of the odour
- Duration of exposure to the odour
- Offensiveness of the odour
- Location (Tolerance and expectation of the exposed subjects)

The individuals affect towards odour can be influenced by these factors and can be used as a basis for odour investigations and impact assessments (Nicell 2009).

The first factor is frequency. Frequency can determine the pleasantness of the odour among the residents. It is also a suitable predictor of annoyance and both frequency and intensity have been reported to affect annoyance of waste odours (Aatamila et al. 2011). The second factors is intensity. Odour intensity is the relative perceived psychological strength of an odour that is above its detection threshold and is independent of the knowledge of the odour concentration (Belgiorno et al. 2013). Intensity is measured by using a seven scale: 0-no odour, 1-very faint odour, 2-faint odour, 3-distinct odour, 4-strong odour, 5-very strong odour and 6-extremely strong odour and its relationship with the odour concentration is usually being calculated by using linear regression.

Next is the duration. Duration refers to the elapsed time over which an odour is experienced and is related to the type and location of the source as well as the local meteorology. Odours can be experienced intermittently for short periods or endured for periods of continuous and lasting duration (Nicell 2009). Normally, the longer the duration, the higher the impact of the odour to the human olfactory sense because it can cause an individual to suddenly change their activities. Sometimes, however, longer duration of exposure to the odour did not change human activities according to Quabach et al. (2014), the perception of odour, the time interval of 1 h is not representative because the odour perception depends on the duration of one breath (Quabach et al. 2014).

Besides that, odour duration will have a stronger effect on human activities if the offensiveness is high as odours are highly variable in their offensiveness. Offensiveness is the fourth factor in FIDOL, sometimes referred to as "hedonic tone". Offensiveness is the subjective rating of the pleasantness or unpleasantness of an odour. While it is generally recognized that the odour threshold is a very useful indicator of odour strength and that with increasing odour concentration there is a greater impact, the threshold fails to account for the hedonic character of the odour (Nicell 2009). As a rule of thumb, hedonic tone is likely to decrease with concentration or intensity. The higher the intensity of the odour the more likely it is that the odour is experienced as unpleasant.

The last factor is location. The location here is more focused on the type of area near to the odour sources affecting surrounding communities' lifestyle, works or visits, the type of activity they are engaged in, and the sensitivity of the receiving environment. In general, the degree of impact of an odour is directly related to the expectations of people who live, occupy, or visit within a region. For example, odours associated with industrial operations are more likely to be tolerated in industrialized zones, where there is an expectation that such activities will take place. Or, if a person who is exposed to an odour associates it with a natural occurrence, such as mudflats, swamps, or seaweed, or with rural agricultural activities, they often do not consider the odour to be offensive or objectionable (Nicell 2009).

3 Odour Sources and Its Compounds

3.1 Odour Sources

When measuring odours, it is not sufficient to measure odour concentration in isolation, but it is necessary to account for the air flow associated with the monitored odour source, as, in most cases, these parameters are related to each other (Belgiorno et al. 2013). There are three types of odour sources which are point sources, area sources and volume sources. Each source has its own sampling method as the sources are different in area, volume and flow rate. A point source is a source where odour is emitted from small or single points such as stack and scrubber, area sources are sources of wide and open areas and usually originate from extended solids or liquid areas such as ponds and volume source is the source with the odour being emitted from a close area such as a building.

3.2 Odorous Compounds

Odorous compounds are the compounds that have smell as a characteristic. Some countries use the compounds as surrogates in determining odour concentration, for example Japan. In the Japanese odour regulation, there are 22 substances of odourous compounds for determining the odour concentration. Table 2 summarises the 22 substances and concentrations in the Japanese odour regulation.

Table 3 summarises the odour characteristics of some of the odorous compounds and their corresponding odour detection threshold. In a study by Huang et al. (2015), concentrations of reduced sulfur compounds (RSCs), [hydrogen sulfide (H_2S), methyl mercaptan (CH_3SH), dimethyl sulfide (DMS), and dimethyl disulfide (DMDS)], nitrogenous compounds (NCs) [ammonia (NH_3) and trimethylamine (TMA)], and carbonyl compounds [acetaldehyde and butyraldehyde] were determined by instrumental analysis in the industrial area in Busan, Korea (Huang et al. 2015). While a study by Nimmermark at an animal farm showed that the content of volatile fatty acids (VFA) has a higher influence in aerated pig slurry (Nimmermark 2011).

Table 2 Odorous compounds in Japan regulation

Substance	Limit concentration (ppm)
Ammonia	1–5
Methyl mercaptan	0.002–0.01
Hydrogen sulfide	0.02–0.2
Dimethyl sulfide	0.01–0.2
Dimethyl disulfide	0.009–0.1
Trimethylamine	0.005–0.07
Acetaldehyde	0.05–0.5
Propionaldehyde	0.05–0.5
Butyl aldehyde	0.009–0.08
Isobutyl aldehyde	0.02–0.2
Valeraldehyde	0.009–0.05
Isovaleraldehyde	0.003–0.01
Isobutyl alcohol	0.9–20
Ethyl acetate	3–20
Methyl isobutyl ketone	1–6
Toluene	10–60
Styrene	0.4–2
Xylene	1–5
Propionic acid	0.03–0.2
Butyric acid	0.001–0.006
Valeric acid	0.0009–0.004
Isovaleric acid	0.001–0.01

Table 3 Detection threshold and odour description of some odorous compounds (WEF 1995)

Compound name	Odour detection threshold (ppm v/v)	Odour description
Acetaldehyde	0.067	Pungent, fruity
Allyl mercaptan	0.0001	Disagreeable, garlic
Ammonia	17	Pungent, irritating
Benzyl mercaptan	0.0002	Unpleasant, strong
n-Butyl amine	0.080	Sour, ammonia
Chlorine	0.080	Pungent, suffocating
Di-isopropyl amine	0.13	Fishy
Dimethyl sulphide	0.001	Decayed cabbage
Diphenyl sulphide	0.0001	Unpleasant
Ethyl amine	0.27	Ammonia-like
Ethyl mercaptan	0.0003	Decayed cabbage
Hydrogen sulphide	0.0005	Rotten eggs
Methyl mercaptan	0.0005	Rotten cabbage
Phenyl mercaptan	0.0003	Putrid, garlic
Propyl mercaptan	0.0005	Unpleasant
Pyridine	0.66	Pungent, irritating
Skatole	0.001	Faecal, nauseating
Sulphur dioxide	2.7	Pungent, irritating

Gas chromatography–mass spectrometry (GC–MS) is widely applicable to characterise chemical compounds present in volatile organic compounds (VOCs) and volatile sulphur compounds (VSCs) such as those used in the study by Orzi et al. (2010) and Tsai et al. (2009). However, the relatively low concentrations (typically at ppbv and pptv levels) of the chemical compounds makes it difficult to identify the odorants present without first using a separation technique. Common absorbent in gas chromatography such as Tenax G-C are capable of trapping a wide range of organic compounds at ambient temperatures. However, the application of Tenax G-C is limited and does not work for polar molecules. An alternative absorbent such as Poropak Q is more suitable for trapping polar or low boiling point gases.

For hydrogen sulphide gas, the odorant is a common trait of smell from sewage treatment works associated with anaerobic conditions. Detectors having gold film or a lead acetate tape are typically used to measure H_2S at the site. Gold-film detectors work by utilising resistance changes caused by adsorbed H_2S molecules. Lead acetate, on the other hand, recognizes changes in optical density of the lead acetate due to the exposure to H_2S.

4 Odour Impact Assessment Method and the Equipment

4.1 Introduction

Various methods have been developed in assessing odour concentrations. Since odour management depends on measurements and some are based on human sensitivity, the right method needs to be chosen according to the suitability for the situation. Methods can be categorised into two types; the sensory method is based on the human olfactory system and the analytical method which uses instruments to measure odour.

Of all methods available, olfactometry is the most widely accepted method. Europe, New Zealand, Australia and US are examples of countries where olfactometry is the chosen method in odour assessment. Like regulations and standards, use of olfactometry is differently regulated in each country. EN13725:2003: *Air Quality-Determination of Odour Concentration by Dynamic Olfactometry* is the standard of olfactometry set for the user in Europe. This standard was set by the Comitee Europeen de Normalisation (CEN). In the US, the ASTM International standard ASTM E679-04: *Standard Practice for Determination of Odour and Taste Threshold by a Forced-Choice Ascending Concentration Series Method of Limits* and ASTM E544-99 (2004): *Standard Practice for Referencing Suprathreshold Odour intensity* are the standards to be followed. Other countries like Germany and France also have their own olfactometry standard. Table 4 summarises the list of olfactometry methods in each reviewed country.

The standard unit for odour concentrations is known as odour unit. The odour unit is defined by the EN 13725 standard as 1 OU_E a European Odour Unit per cubic metre, which is a sensory measurement that relies on a physiological response when odour is detected by the nose when exposed to a sample at a particular concentration.

The DEFRA (UK) Odour Guide for Local Authorities (DEFRA 2010) gave several odour concentration thresholds as a guideline to indicate the level of odour exposure to the receptors (Table 5).

Table 4 List of Olfactometry standard

Countries	Standard
Europe	EN13725:2003
United States	ASTM D-1391 (1978), ASTM E679-91 (1991)
Germany	VDI 3881 (1980), VDI 3882 (I) (1992)
France	AFNOR-X-43-101 (1986)
Netherland	NVN 2820 (1996)

Table 5 Indication on the level of odour exposure according to measured odour concentrations

Odour concentration, OU m^{-3}	Exposure level
1	Point of detection
5	Faint odour
10	Distinct odour

A concentration of 1 OU m^{-3} indicates a very weak odour and is usually difficult to be detected by the majority of the population that is people with normal odour sensitivity. Odours of 5–60 OU m^{-3} usually are background odours like the smell of traffic, grass mowings, vegetation etc.

4.2 VDI 3940

The VDI 3940 standard series (VDI 3940 1993, 2006a, b) describe the methodology for assessing the odour of ambient air around selected sources in a grid and in a plume. There are a few countries that adopted this method for assessing odour nuisance such as Germany, Austria, France and the Netherlands (Sowka 2010). There are two types of measurement in VDI 3940.

The first is the grid measurement following the VDI 3940 Part 1. According to the standard part one of a VDI 3940 series, a grid measurement is a technique in which a team of assessors registers odours in following measuring grids within the area under study for a fixed period of time (e.g. 6 months). The results obtained are used for assessing the impact of odours on air quality (Sowka 2010). It is recommended that the area under study be a circle, with the source of odour emissions in its center. The research area is covered uniformly with spaced grid points. Measuring squares are created by connecting 4 points. The grid step is very important for a proper evaluation of odour impact. It is recommended that the step set up initially be equal to 250 m. Depending on the conditions and needs, higher (up to 500 m) or lower (e.g. 125, 100 and 50 m) steps are also acceptable.

In the case of high point sources, odour impact is determined at a distance of 250 m from the edge of the plant and a grid size is 250 m × 250 m is recommended. For low point and fugitive sources, located less than 250 m from the nearest buildings, it is possible to reduce the size of squares to cover all possible places where the odour distribution is not uniform. In such situations, the sizes of the squares are 50 m × 50 m. The size of the squares will increase with distance. According to the VDI 3940 Part 1, standard measurement should be carried out on different days and should be planned ahead so that each of the measuring points of the grid is independent of each other. To obtain representativeness of test results, before starting the measurements, it is necessary to select regular and non-random days of measurements. Ultimately, 52 or 104 measurements should be performed in 6 month or in 12 month periods, respectively.

Second is plume measurement using the VDI 3940 Part 2. According to the VDI 3940 Part 2, a plume measurement is carried out in order to obtain information on the impact of odour within a plume range. The study will take place under specified weather condition. Compared to the grid measurement, the percentage occurrence of odour in a single measurement is a variable in the calculation of odour impact. Normally, a plume measurement is related to a specific industrial facility. It is necessary to gather information on the manufacturing process and as well the product type and also the operating hours of the main emission sources. Odour from the emitter permeates the air and this process is directly dependent on wind speed and direction and atmospheric stability. Hence, it is necessary to measure some meteorological parameters in a given area during the sampling period. According to German standards, these parameters are as follows: wind speed and direction, air stability and ambient temperature. It is also important to register episodic events like rain, fog or snow during the measurements.

According to VDI 3940, Part 2, before choosing the measurement points it is necessary to determine the extent of an odour plume. Then, the approximate wind direction is determined. Assessors move away from the plant downwind to the end of a plume. Reaching the plant upwind, they can assess the distance from an odour source at which it is perceptible. The plume boundary is reached when the percentage odour time reaches a predetermined percentage (10 %). A minimum of three intersection lines is required according to the VDI standard for plume measurements. At least five measurement points for an intersection line crosses and is plotted by five assessors. The distances between intersections of lines and measurement points depend on the anticipated plume size, which can be affected by the height of emission sources, by the odour of air flux, current weather conditions and topography. For consecutive measurements of intersection line, the positions of an assessor should be changed so that no single assessor takes the same position during plume measurements.

4.3 *Olfactometer*

An olfactometer is a sensorial technique that use the human nose as a sensor, such as dynamic olfactometry and usually follow the standard and regulation set by EN13725 (2003). Dynamic olfactometry involves continuous dilution of a known flow of the sample with a known flow of odourless air. The static method on the other hand requires dilution of an odorous air sample in a known odourless air volume to be done first before being presented to the odour panel, usually in batches (Littaru 2007).

The principle of the olfactometer is to characterize the odours by referring directly to their effects on a panel of qualified examiners. In addition, as an olfactometer use the human olfactory system and several factors may influence odour perception. The most important one is the variability of human olfaction between different subjects. This problem is minimized by using a panel composed by several examiners, selected with precise criteria in order to choose people with a standardized olfaction, and by averaging the single examiners' responses (Capelli et al. 2008). The input layer of each olfactory bulb contains about 1000 spherical structures called glomeruli. Within each glomerulus, less than 100 second-order olfactory neurons (Mitral cells and Tufted cells) receive input from ~25,000 olfactory receptor cells.

The odour panellists are selected based on a reference substance (*n*-butanol) and only those meeting the average n-butanol odour threshold ranging between 20–80 ppb are considered as fit odour panellists. The n-butanol odour allows a tuning of the olfaction of the panellists, as well as a finer definition of the Odour Unit. It is assumed that 1 Odour Unit OU m^{-3} corresponds to a concentration of 20 ppb of the reference substance (*n*-butanol).

4.4 Method of Odour Sampling

The assessment of odour samples using olfactometers involves first the collection of the odorous gas samples from the source using proper equipment usually into a sample bag or sample canisters which is then connected to the measuring instrument for analysis.

In addition to odour concentration reported as OU m^{-3}, the odour emission rate is also another important parameter considering that in most cases odour concentration and airflow are linked. Odour emission rate is the product of airflow and odour concentration and its values are used as input data for designing odour control systems, or as inputs for odour dispersion modelling.

For point sources such as stacks, sampling is done by inserting a sampling probe into the stack and withdrawing the sample from the air flow. The flow rate is the product of velocity and cross-sectional area of the stack which can be determined using anemometers or pitot tubes at several positions across the area of the stack (DEFRA 2010).

For area sources such as wastewater treatment ponds, landfills etc. there is no clear indication of airflow over the surface. Hence, "hood" methods as shown in

Fig. 1 Odour sampling at an effluent pond using flux hood and vacuum chamber

Fig. 1 are usually used. In the "hood" methods, a hood or wind tunnel is placed over the surface of the odour emission. Then odourless air (or nitrogen gas) is blown through the apparatus at a flowrate between 5 and 24 L/min (Capelli et al. 2013). The emission rate is then obtained considering the airflow through the hood and the odour concentration from the sample collected from the hood.

4.5 *Dispersion Modelling*

As people begin to pay attention to odour problems, continuous odour monitoring technologies are necessary to understand and identify strategies to reduce the impact of odour emissions on local communities. Researchers working in the environmental field, especially majoring in air pollution are interested in measuring concentration of air pollutants continuously at different time scales (Zou et al. 2010). A great deal of work is necessary for determining pollutant concentration. As manual monitoring and observing pollutant concentration are costly to install and maintain, researchers try to develop alternatives. Therefore, air dispersion modelling was created to make the monitoring and observing work easier and cheaper. A dispersion model is a set of mathematical equations used for determining the fate of pollutants emitted in the atmosphere. It also calculates the concentration level at any point in space depending on the meteorological

Table 6 The functions of odour dispersion modelluing

Purpose	References
(1) Assessing the compliance of concentrations from emissions of planned facilities with air quality guidelines or standards	• Ozkurt et al. (2013) • Lee et al. (2012)
(2) Determining appropriate stack height	• Lateb et al. (2011) • Canepa (2004) • Pregger and Friedrich (2009)
(3) Assessing the contribution of individual plants to the overall concentrations	• Ainslie and Jackson (2009)
(4) Designing ambient air monitoring network	• Elkamel et al. (2008) • Lozano et al. (2010)
(5) Forecasting pollution episodes	• Saide et al. (2011) • Finardi et al. (2008) • Kim Oanh et al. (2005)
(6) Estimating the influence of geophysical factors on dispersion	• Tartakovsky et al. (2013) • Levy et al. (2002) • Indumati et al. (2009) • Song et al. (2009)

information at that point (Foundation Course of Air Quality Management in Asia 2008) Air dispersion modelling helps to determine the following issues highlighted in Table 6.

4.6 Electronic Nose

The first 'model electronic' nose, based on a small chemo-electronic sensor array, was developed in a study by Persaud and Dodd (1982). Although this system comprised only three chemically active sensors, it provided the generic architecture used in most systems today. This architecture, modelled on the limited understanding of the biological olfactory system at that time, and in particular the work of Deutsch, relied upon an array of partially specific sensors and a primitive method of data fusion, able to provide discrimination between a number of simple odorants (Pearce 1997).

Electronic noses were initially introduced as instruments able to mimic the human olfactory system. They include indeed similar corresponding components: the array of chemical sensors, the data processing unit and the pattern recognition engine respectively, for the olfactory receptor cells, the olfactory bulb and the brain. But, of course, the sensors of the electronic nose can respond to both odorous and odourless volatile compounds. It can be applied to any source which releases volatiles, whether they smell or not, provided that this occurs within the sensitivity range of the sensors, which is rather narrow compared to the human nose. So, the initial enthusiasm for e-nose as an instrument to assess the odour is no counterpart for a more realistic and scientific approach.

The instrument continues to be used as "electronic nose" to detect odour emissions, but, rather to try to compare it to the human nose, it is used more often as a complementary tool with respect to chemical or sensitive methods (Romain et al. 2008). In many cases, it would be advisable to analyze air continuously, in order to detect the odours from an industrial source and to determine exactly when such odours are perceived. Since it is not always possible to correlate the chemical composition of a gaseous mixture with its odour properties, analytical measurements are not suitable for odour determination. A method for continuous odour monitoring could be based on the use of electronic noses.

4.7 Surveying and Odour Diaries

Odour needs the right assessment method and steps. There are on site measurement, dispersion modelling and survey method to be chosen. Those steps can be run simultaneously or independently dependant on the situation of the odour problem. Planning is important in odour assessment for optimal results. The appropriate tools need to be considered because each tool has its own advantages/disadvantages.

Prediction tools such as dispersion modelling are viewed as less effective by some compared to real-time tools such as direct sensory assessment and retrospective techniques such as complaints. The assessment usually starts with complaints, followed by investigation and choosing the assessment tools. The UK, New Zealand and the US all start their assessment by conducting a complaint evaluation.

In the UK direct measurement of the percentage of people annoyed are at the start of assessment. Complaints are usually not handled following a standard method and done through local authorities, while direct measurement is usually done using The Standardised Telephone Questionnaire (STQ). Beside complaint analysis and telephone survey, in China questionnaire surveys are used for odour investigations (MoEGJ 2015).

Another method is using odour diaries. Odour diaries are based on the FIDOL concept. Odour diaries are suitable for recording odour episodes in a short time. They are also suitable for the evaluation by a group of people tasked with constant monitoring. For example, in New Zealand, they run odour diaries if several complaints are registered for the same places. Those individuals will receive a set of tables that need to be filled. Typically odour complaint data will not be more accurate in low population areas, when there are more than one odour source and when complaint records cannot be validated with wind data. Therefore population size and number of odour sources play an important role in the first step of odour impact assessment.

5 Waste Recovery Facilities (WRF)

5.1 Introduction to WRF

Nowadays one of the most important challenges for chemical engineering is to convert wastes into raw materials for other processes. This is the case of municipal solid waste (MSW), continuously increasing with more than 260 million of metric tons generated last year in Europe. The main objective of these facilities is to divide the received MSW stream into its different fractions in order to subject them to specific treatments. Organic fraction of municipal solid waste (OFMSW) and recyclable materials such as paper, plastic containers, cans, etc. are removed; the rest of the waste stream, known as reject fraction, is usually disposed in landfills (Montejo et al. 2011). Waste recovery facilities (WRF) are specialized plants that receive, separate and prepare recyclable materials for marketing to end-user manufacturers. The WRF receives and sorts through both municipal and commercial waste to recover recyclable materials, including:

- Wood/greenwaste processed for compost and woodchips
- Metal—ferrous/metallic items
- Plastic—many grades

- Glass—all colours
- Paper—newspaper, junk mail, phone books, magazines, scrap paper, paperboard and cardboard

Materials that cannot be recycled are taken to the landfill.

An example is the waste recovery facility at Placer County, California, USA where approximately 50 % of the waste is diverted from going to the landfill, helping Placer County comply with a state mandated recycling rate (WMA 2015). Urban solid waste management is a critical issue in most countries, requiring an integrated effective system approach. In this framework, one of the options advocated by waste management planners and government regulations is the recovery of material and energy from municipal solid waste (MSW) through production of refuse-derived fuel (RDF) (Caputo and Pelagagge 2002).

Not all odorous substance are found in one single source as shown in Table 7. For example, odour compounds detected at an anaerobic digestion plant and plastic waste recycling facility varied. There is no universal odorous compounds for all

Table 7 Odorous compounds at waste recovery facilities

Facility	Odourants	Concentration	Unit
OFMSW anaerobic digestion plant (Orzi et al. 2010)	Aliphatic hydrocarbons (total)	0.48–30.2	ppbv
	Aromatic hydrocarbons (total)	19.3–55.1	ppbv
	Butanone	12–556	ppbv
	Alcohols (total)	38.5–762	ppbv
	2-butanol	17–208	ppbv
	Ethanol	4–365	ppbv
	Propanol	36–121	ppbv
	Ethers (total)	1.72–78.8	ppbv
	Esters (total)	12.7–355.3	ppbv
	Methyl acetate	55–158	ppbv
	Terpenes (total)	414–4750	ppbv
	α—Pinene	26–80	ppbv
	Limonene	178–4389	ppbv
	ρ—Cymene	53–134	ppbv
	Nitrogen compounds (total)	3.42–7.11	ppbv
	Sulphur compounds (total)	8.1–79.6	ppbv
	Halogenated compounds (total)	0–0.08	ppbv
	Carboxylic acids	0–164.7	ppbv
	Total VOC	1178–6005	ppbv
Plastic waste recycling facility (Tsai et al. 2009)	Total hydrocarbons	2623–9873	μgm^{-3}
	Total ketones	3772–15,626	μgm^{-3}
	Total esters	1344–2440	μgm^{-3}
	Total aldehydes	46–259	μgm^{-3}
	Acetonitrile	64–67	μgm^{-3}
	Acrylonitrile	11–49	μgm^{-3}

waste recovery facility, suggesting that investigators need to have an open mind in the assessment. Otherwise, some important data on the compounds actually causing the odour issue at the facility under investigation might be missed.

5.2 *Refuse-Derived Fuel (RDF)*

Energy-intensive industries like cement, paper, chemical production or power generation are some of the proponents for refuse-derived fuel (RDF), either as co-combustion in modified existing plants or as mono-combustion in specifically designed boilers. Installing material recovery facilities (MRFs) in a solid waste management system could be a feasible alternative to achieve sustainable development goals in urban areas if current household and curb side recycling cannot prove successful in the long run.

In the processing of RDF, biodrying is sometimes utilized to dry the waste in order to increase energy content (EC) (Tambone et al. 2011). For example, a minimum net heating value (NHV) of 15,000 kJ kg^{-1} w.w. must be met for MSW-derived SRF under Italian law.

In addition, biodried material, because of low potential microbial activity, guarantees low risk of potential waste self-heating and self-ignition, and a strong reduction of potential odour emissions, allowing easy storage. Odour is related to the presence of degradable OF in waste that under optimal condition, i.e. moisture, pH, nutrients, and anoxic or partially anoxic environments, can ferment to produce odours.

In a laboratory study of six samples of fresh and bio-dried MSW, it was observed that biodrying halved potential odour impacts. This was due to the low microbial that occurred in the biodried material because of the low moisture content. The very good correlation found for RDRI versus OU ($r = 0.93$, $P < 0.005$, $n = 6$) confirmed this finding (Tambone et al. 2011).

5.3 *Anaerobic Digestion*

The process of decomposing organic fraction of municipal solid waste—OFMSW, energy crops, and agro-industrial wastes using the activity of specific microorganisms in the absence of oxygen, is called anaerobic digestion. The end product includes the production of biogas (mainly methane and carbon dioxide) that can be used for electricity generation).

In a biogas or anaerobic digestion plant, the main group of odourants emitted are volatile organic compounds (VOC) which contributed significantly to odour issue at MSWs biogas plants. Ketones, alkanes, alkenes, alcohols, acids, terpenes (pinene, cymene and limonene) and organic sulphur compounds are among the VOCs commonly reported. This is in agreement with a study of a full-scale anaerobic digestion plant located in Northern Italy treating around 30,000 Mg of waste per year. The VOCs present in the air of the ingestate consisted mostly of terpenes (61 %), alcohols (18 %), and esters (9 %). Fresh wastes (ND) were associated with the presence of the terpenes, limonene and beta-pinene, both widespread in fruits, vegetables, and pine species. Digested wastes contained a high presence of p-cymene ranging between 53–134 ppbv (Orzi et al. 2010).

The Ecoparc-2 in Barcelona, Spain, receives waste material including MSW, as well as green waste from parks and gardens of a number of municipalities of the metropolitan area of Barcelona to process in its 240,000 tons/year anaerobic reactor. Odour emission of the industrial complex exceeded the German criteria for industrial areas (15 % odour hours), while those found in the most immediate urban zone were below the criteria for residential areas (10 % odour hours). An odour exposure study was conducted using field panel observations around the Ecoparc-2, according to the German standard EN3940.

Use of anaerobic sequencing batch reactors (ASBR) has been shown in the study by Ndegwa et al. (2008) to contain odour issues known to be associated with anaerobic digestion when treating dilute animal waste. The ASBR operates in four sequences: feed, react, settle, and decant in a cyclic mode. Essentially ASBR separates the SRT and HRT in the same reaction chamber to allow solids to have longer retention times while the easily degradable liquid spends a much shorter time in the reactor. A higher removal of volatile fatty acid is thus achieved whereby the levels of odorous compounds and potential of odour generation from the resulting effluents are subsequently reduced. VFA contents of the effluents decreased to below 230 mg/L; the level of VFA below which odour problems are not expected, and certainly less than the 520 mg/L threshold of offensive odours. The treatment of dilute swine slurries in ASBRs for the purpose of odour control is most effective at lower operating temperature and cycle-frequency between one to three times.

5.4 *Incineration*

Incineration can reduce waste volume up to 95 %, although is not a very popular process. Nevertheless, the more stringent requirements on air pollution can be controlled using the existing technology correctly. Incineration of MSW is the most implemented treatment option in populous countries such as China. Japan is the country with the highest number of waste incineration plants, about 1900 facilities, 190 of them with power generation; it is followed by countries of the European Union, mainly France, and the United States. According to previous works on integrated waste management, MSW has a high calorific value, allowing the incineration with greater energy recovery, although, in accordance with current strategies, incineration should be done once recyclable materials have been recovered, i.e. to the reject fraction from MBT plants also named refuse derived fuel (RDF).

The main advantages of RDF incineration over using MSW as fuel are better efficiency of energy recovery and better quality of flue gases by significant reduction of heavy metals in the fly ashes. Both advantages are closely related to the composition of incinerator feedstock, however, the latest data of MSW characterizations in Spain were published in 1999 and for RDF have never been available (Montejo et al. 2011). The rising prices of raw materials and the depletion of landfill space have resulted in an increasing concern for material recovery and reuse. On the other hand, thermal treatment by using incineration technology has been proven as an attractive method of waste disposal for many years due to the primary advantages of hygiene control, volume reduction, and energy recovery. Previous experience in solid waste management has been that solid waste pre-sorting prior to incineration is solely a function of material recycling (Chang and Chang 1998).

In China, incineration is a significant component of an integrated waste management program for large cities especially in the eastern and coastal provinces with dense population and lack of adequate sites for landfill. Since the first MSW incinerator was established in Shenzhen City of the Guangdong Province in 1988, more and more 101 MSW incinerators were established by the end of 2010. The combined normal rated treatment capacity of 101 MSW incinerators was 85 kg/day in 2010, reflecting a growth of 5.7 times compared with the year of 2003.

Huang et al. (2015) explored public acceptance of waste incineration projects in China and noted the public's concern over odour exposure from incineration plants. Two incineration projects in the Jiangsu Province valued at RMB 270.50 million and RMB 500 million, anonymously known as Project A and Project B, respectively, were compared. Before Project A began operation, nearby residents raised concern over odour release from the incinerator. This worry was brushed aside by the government and an enforced approach was adopted in Project A which ultimately turned into aggressive confrontations between the public and the government. In Project B, however, the government in charge successfully reconciled the conflicts by re-settling the households affected by the incinerator odour. It was concluded that in events of complaints, the government's flexibility in and readiness for problem-resolving could increase public's acceptance. Finally, for Projects A and B, the nearby households were compensated with resettlement, in fact, the influence of the odour was underestimated as odour spread far beyond the compensated regions.

6 Technologies of Odour Control

There are many types of odour control technologies in the market. Those technologies were chosen according to suitability in an industry. Odour control technologies that are usually been used are such as biofilters, activated carbon, wet scrubber, thermal oxidation and oxidization chemical.

One of the common methods in odour control is using biofilters. From the name of the technologies, it concludes that biofilters use living organisms for controlling pollutants by capturing and degrading these. There are two types of the filters used: organic, such as bark and wood chips, and synthetic, such as granulated carbon or plastics (Bindra et al. 2015). As stated by Luo and Lindsey (2006), biofilters are an efficient and practical technology for gas cleaning and can reduce odours to acceptable levels, but sometimes, they are unreliable because of operation under sub-optimal conditions (Luo and Lindsey 2006).

Biofilters need to be kept humid by using a system consisting of a filter bed of organic/inorganic material, but not necessarily subject to continuous water spraying like wet scrubbers. Instead of media like bark or plastic, wet scrubbers use absorbance of pollutants into a liquid. Acid, water and $NaClO_2$ solution are examples of liquids used in wet scrubbers (Chien et al. 2015; Park et al. 2015). Wet scrubbers have a higher removal efficiency compare to packed towers in removing acid and basic ammonia gases, and efficiency is needed to meet emission regulations, especially for high-volumetric flow rates and low-concentration soluble gas pollutants (Chien et al. 2015).

Adsorption is operational in most natural physical, biological and chemical system and widely used in industries. The most common adsorbent materials are alumina silica, metal hydroxides and activated carbon (Malik et al. 2002). The adsorbent adsorbs pollutants which is typically used remediation of wastewater (Gurses et al. 2006). Adsorption occurs in three steps. In the beginning, the adsorbate disperses from the major body of the stream to the external surface of the adsorbent particle. Then, the adsorbate migrates from the relatively small area of the external surface to the pores within each adsorbent particle. The majority of adsorption usually occurs in these pores because there is the majority of available surface area. In the final step, the contaminant molecule adheres to the surface in the pores.

Other technologies in removing odour are by using thermal or chemical oxidation. Both processes will react with contaminants and oxidising them to harmless compounds. The only difference of the two technologies is the agent used in the process. Thermal oxidation uses high temperature as the agent while chemical oxidation use chemical substances. Thermal oxidation has a disadvantage in removing sulfur containing compounds. It needs to be used together with biofiltration which produces a large amount of wastes that require further treatment resulting in high operating cost (Antonopoulou et al. 2014).

In order to choose the right type of odour control technology, the following factors need to be considered:

(i) odour removal efficiency,
(ii) composition and concentration of air flow,
(iii) capital and O&M cost,
(iv) reliability,
(v) energy and water usage, and
(vi) space requirement.

Table 8 summarises various odour control technologies and methods that are in place at waste recovery facilities, while advantageous and disadvantageous of the technologies are given in Table 9.

Table 8 Evaluation of odour control technologies/treatment methods (Bindra et al. 2015)

System	Odour removal	Capital cost	O&M cost	Reliability	Space requirement	Energy usage	Chemical	Water usage
Biofilters	>90 %	Mod	Low	Good	High	High	Low	Mod
Activated carbon	99 %	Mod	Low	Excellent	Low	Low	High	None
Packed-bed wet scrubbers	>90 %	Mod	High	Good	Mod	Mod	Mod	Mod
Fine mist wet scrubbers	>90 %	High	High	Mod	Mod	Mod	High	Mod
Thermal oxidizers	>95 %	Very high	High	Mod	Low	Very high	High	None
Oxidization chemicals	99 %	None	Very high	Mod	None	None	Very high	None
Masking agents	N/A	None–low	Mod	Mod	None	None	High	None

Mod Moderate

Table 9 Major advantages and disadvantages of odour control technologies (Bindra et al. 2015)

Technology	Advantages	Disadvantages
Biofilters	• No chemicals required • Odour compounds are destroyed not just transferred to different media • Simple technology	• High concentration air streams can cause acidification of media • High energy usage • Large land area required for system • Water usage, potential for leachate
Activated carbon	• No chemicals or pumps • Simple system and O&M • Small space requirement • Good for higher molecular weight gases	• Only cost effective for low concentration air streams • Solid waste produced (used carbon) • Not effective for ammonia
Packed-bed wet scrubbers	• Effective for high concentration H_2S air • Good for large air volumes • Proven and tested technology	• Potential use of hazardous chemicals • Intermittent cleaning required • Sophisticated instrumentation required
Fine mist wet scrubbers	• Higher odour absorption efficiency than packed tower • Bigger range of odour compounds treated than packed tower • Ability to treat higher air flows • Effective for high concentration H_2S air	• Potential use of hazardous chemicals • Water softening required • Sophisticated instrumentation required • Continuous waste discharge
Thermal oxidizers	• Excellent for VOC's • Good for concentrated air streams • Odour compounds are destroyed • Small size	• Use of fossil fuels (oil or gas) • Can result in formation of SO_2 and NO_x emissions • High capital and O&M costs
Oxidizing chemicals	• Simple O&M • No capital cost • Can treat variable air streams	• Chemical usage can be very high and expensive • Oxidizing agents will oxidize non-odour compounds as well, thus increases costs • Oxidizing agents may kill aerobic microbes and alter composting process
Masking agents	• Good for variable odour incidents • No capital cost	• Not useful for VOC's • Odour is just masked, not treated/destroyed • Masking agents can be perceived as nuisance odour

Odour still remains a major concern for organics processing facilities and is one of the most problematic issues that such facilities face in Canada and around the world. This study was intended to look at what technologies can be used to effectively treat odours from an efficiency, cost and reliability bases and what impact these treatment technologies have on the environment. In terms of odour treatment efficiency all three systems are similar and provide good odour treatment for odour compounds of concern. Effectively implemented either of the 3 systems will be effective odour control options.

Bindra et al. (2015) conducted a Life Cycle Analysis (LCA) comparing packed-bed wet scrubber systems, organic (wood-chip media) biofilters and inorganic (synthetic media) biofilter systems based on the aspect of efficiency, cost and reliability together with the impact of these treatment technologies on the environment. The study concluded that a biofilter system is more advantageous than the packed-bed system when considering the effect of preliminary cost estimates, O&M, flexibility of the system, hazardous chemical production and ease of operation. In terms of the biofilter media, inorganic media is preferred over wood chip media as the latter needs less frequent material replacement. The packed-bed system presents the lowest environmental impact based on the LCA findings. With biofilter systems a sizable portion of their environmental impact is due to the system being energy intensive due to the reliance on the fan suction system energy.

7 Conclusions

Odour is an issue at several waste recovery facilities such as biogas, incineration and refuse-derived fuel plants. Various techniques such as biofilters, wet scrubbers, activated carbon and even odour masking has been shown to be effective at treating odour at the waste recovery facilities. The selection of the appropriate odour control technology should also consider potential environmental effects from the use of such systems at the treatment plants. Examples include accidental release of poisonous gases e.g. when sodium hypochlorite is used in wet scrubbers catches fire, the need for disposal of spent biofilter media and activated carbon, among others.

Acknowledgement The assistance provided by the Fundamental Research Grant Scheme reference no. 2013/0191/F0723 is greatly acknowledged.

References

Aatamila M, Verkasalo PK, Korhonen MJ, Suominen AL, Hirvonen M-R, Viluksela MK, Nevalainen A (2011) Odour annoyance and physical symptoms among residents living near waste treatment centres. Environ Res 111:164–170

Ainslie B, Jackson PL (2009) The use of an atmospheric dispersion model to determine influence regions in the Prince George, B.C. airshed from the burning of open wood waste piles. J Environ Manage 90:2393–2401

Antonopoulou M, Evgenidou E, Lambropoulou D, Konstantinou I (2014) A review on advanced oxidation processes for the removal of taste and odor compounds from aqueous media. Water Res 53:215–234

Belgiorno V, Naddeo V, Zarra T (2013) Odour Impact Assesment Handbook. John Wiley & Sons, Ltd., West Sussex, United Kingdom

Bindra N, Dubey B, Dutta A (2015) Technological and life cycle assessment of organics processing odour control technologies. Sci Total Environ 527–528:401–412

Canepa E (2004) An overview about the study of downwash effects on dispersion of airborne pollutants. Environ Model Softw 19:1077–1087

Capelli L, Sironi S, Del Rosso R, Cèntola P, Grande M (2008) A comparative and critical evaluation of odour assessment methods on a landfill site. Atmos Environ 42(30):7050–7058

Capelli L, Sironi S, Del Rosso R (2013) Odor sampling: techniques and strategies for the estimation of odor emission rates from different source types. Sensors 13(1):938–955

Caputo AC, Pelagagge PM (2002) RDF production plants: II economics and profitability. Appl Therm Eng 22:439–448

Chang YH, Chang NB (1998) Optimization analysis for the development of short-term solid waste management strategies using presorting process prior to incinerators. Resour Conserv Recycl 24(1):7–32

Chien C-L, Tsai C-J, Sheu S-R, Cheng Y-H, Stari AM (2015) High-efficiency parallel-plate wet scrubber (PPWS) for soluble gas removal. Sep Purif Technol 142:189–195

DEFRA (2010) Odour guidance for local authorities. Department for Environment, Food and Rural Affairs, United Kingdom

DEHP (2015) Department of environment and heritage protection, guideline: odour impact assessment from developments. Queensland Government. https://www.ehp.qld.gov.au/licences-permits/business-industry/pdf/guide-odour-impact-assess-developments.pdf. Accessed 31 May 2015 (Last updated 24 March 2015)

Elkamel A, Fatehifar E, Tehari M, Al-Rashidi MS, Lohi A (2008) A heuristic optimization approach for air quality monitoring network design with the simultaneous consideration of multiple pollutants. J Environ Manage 88:507–516

Finardi S, De Maria R, D'Allura A, Cascone C, Calori G, Lollobrigida F (2008) A deterministic air quality forecasting system for Torino urban area, Italy. Environ Model Softw 23:344–355

Gurses C, Dogar M, Yalçın M, Açıkyıldız R, Bayrak S, Karaca (2006) The adsorption kinetics of the cationic dye, methyleneblue, onto clay. J Hazard Mater 131(1–3):217–228

Huang Y, Ning Y, Zhang T, Fei Y (2015) Public acceptance of waste incineration power plants in China: comparative case studies. Habitat Int 47:11–19

Indumati S, Oza RB, Mayya YS, Puranik VD, Kushwaha HS (2009) Dispersion of pollutants over land–water–land interface: study using CALPUFF model. Atmos Environ 43:473–478

Kim Oanh NT, Chutimona P, Ekbordin W, Supat W (2005) Meteorological pattern classification and application for forecasting air pollution episode potential in a mountain-valley area. Atmos Environ 39:1211–1225

Lateb M, Masson C, Stathopoulos T, Bédard C (2011) Effect of stack height and exhaust velocity on pollutant dispersion in the wake of a building. Atmos Environ 45:5150–5163

Lee G, You SI, Ritchie SG, Saphores J-D, Jayakrishnan R, Ogunseitan O (2012) Assessing air quality and health benefits of the clean truck program in the Alameda corridor, CA. Transp Res Part A 46:1177–1193

Levy JI, Spenglera JD, Hlinkab D, Sullivan D, Moon D (2002) Using CALPUFF to evaluate the impacts of power plant emissions in Illinois: model sensitivity and implications. Atmos Environ 36:1063–1075

Littaru P (2007) Environmental odours assessment from waste treatment plants: dynamic olfactometry in combination with sensorial analysers "electronic noses". Waste Manag 27(2): 302–309

Lozano A, Usero J, Vanderlinden E, Raez J, Contreras J, Navarrete B, Bakouri HE (2010) Optimization of the design of air quality monitoring networks and its application to NO_2 and O_3 in Jaen. Spain, Microchem J 96:406–411

Luo J, Lindsey S (2006) The use of pine bark and natural zeolite as biofilter media to remove animal rendering process odours. Bioresour Technol 97:1461–1469

Malik DJ, Strelko V Jr, Streat M, Puziy AM (2002) Characterization of novel modified active carbons and marine algal biomass for the selective adsorption of lead. Water Res 36:1527–1538

MoEGJ (Ministry of the Environment Government of Japan) (2015) The offensive odour control law in Japan, office of odor, noise and vibration environmental management Bureau. https://www.env.go.jp/en/laws/air/offensive_odor/all.pdf. Accessed 31 May 2015

Montejo C, Costa C, Ramos P, Marquez MDC (2011) Analysis and comparison of municipal solid waste and reject fraction as fuels for incineration plants. Appl Therm Eng 31:2135–2140

Ndegwa PM, Hamilton DW, Lalman JA, Cumba HJ (2008) Effects of cycle-frequency and temperature on the performance of anaerobic sequencing batch reactors (ASBRs) treating swine waste. Bioresour Technol 99(6):1972–1980

Nicell JA (2009) Assessment and regulation of odour impacts. Atmos Environ 43:196–206

Orzi V, Cadena E, D'Imporzano G, Artola A, Davoli E, Crivelli M, Adani F (2010) Potential odour emission measurement in organic fraction of municipal solid waste during anaerobic digestion: relationship with process and biological stability parameters. Bioresour Technol 101 (19):7330–7337

Ozkurt N, Sari D, Akalin N, Hilmioglu B (2013) Evaluation of the impact of SO2 and NO2 emissions on the ambient air-quality in the Çan-Bayramiç region of northwest Turkey during 2007–2008. Sci Total Environ 456–457:254–266

Park H-W, Choi S, Park D-H (2015) Simultaneous treatment of NO and SO_2 with aqueous $NaClO_2$ solution in a wet scrubber combined with a plasma electrostatic precipitator. J Hazard Mater 285:117–126

Pearce TC (1997) Computational parallel between the biological olfactory pathway and its analogue "The electronic nose": Part II. Sensor-based machine olfaction. Biosystems 41:69–90

Persaud K, Dodd GH (1982) Analysis of discrimination mechanisms in the mammalian olfactory system using a model nose. Nature 299:352–355

Pregger T, Friedrich R (2009) Effective pollutant emission heights for atmospheric transport modelling based on real-world information. Environ Pollut 157:552–560

Quabach ES, Piringer M, Petz E, Schauberger G (2014) Comparability of separation distances between odour sources and residential areas determined by various national odour impact criteria. Atmos Environ 95:20–28

Romain A-C, Delva J, Nicolas J (2008) Complementary approaches to measure environmental odours emitted by landfill areas. Sens Actuators B 131:18–23

Saide PE, Carmichael GR, Spak SN, Gallardo L, Osses AE, Mena-Carrasco MA, Pagowski M (2011) Forecasting urban PM10 and PM2.5 pollution episodes in very stable nocturnal conditions and complex terrain using WRF-Chem CO tracer model. Atmos Environ 45:2769–2780

Sironi Selena, Capelli Laura, Centola Paolo, Del Rosso Renato, Pierucci Sauro (2010) Odour impact assessment by means of dynamic olfactometry, dispersion modelling and social participation. Atmos Environ 44:354–360

Song S-K, Shon Z-H, Kim K-H (2009) Photochemical oxidation and dispersion of gaseous sulfur compounds from natural and anthropogenic sources around a coastal location. Atmos Environ 43:3015–3023

Sowka I (2010) Assessment of air quality in terms of odour according to selected European guidelines: grid and plume measurement. Environ Prot Eng 36:134–141

Stellacci P, Liberti L, Notarnicola M, Haas CN (2010) Hygenic sustainability of site location of wastewater treatment plants. A case study. I. Estimating odour emission impact. Desalination 253:51–56

Tambone F, Scaglia B, Scotti S, Adani F (2011) Effects of biodrying process on municipal solid waste properties. Bioresour Technol 102(16):7443–7450

Tartakovsky D, Broday DM, Stern E (2013) Evaluation of AERMOD and CALPUFF for predicting ambient concentrations of total suspended particulate matter (TSP) emissions from a quarry in complex terrain. Environ Pollut 179:138–145

Tsai C-J, Chen M-L, Chang K-F, Chang F-K, Mao I-F (2009) The pollution characteristics of odor, volatile organochlorinated compounds and polyclic hydrocarbons emitted from plastic waste recycling plants. Chemosphere 74(2009):1104–1110

VDI 3940 (1993) Determination of odorants in ambient air by field inspections. Verein Deutscher Ingenieure, Berlin, Beuth Verlag

VDI 3940 Part 1 (2006a) Measurement of odour impact by field inspection—measurement of the impact frequency of recognizable odours—grid measurement. Verein Deutscher Ingenieure, Berlin, Beuth Verlag

VDI 3940 Part 2 (2006b) Measurement of odour impact by field inspection—measurement of the impact frequency of recognizable odours—plume measurement. Verein Deutscher Ingenieure, Berlin, Beuth Verlag

WMA (Waste Management Authority) (2015) Western placer. http://www.wpwma.com/facilities.html. Accessed 31 May 2015 (Last updated 20 May 2015)

Zou B, Zhan FB, Wilson JG, Zeng Y (2010) Performance of AERMOD at different time scale. Simul Model Pract Theor 18:612–623